Ergebnisse der Mathematik und ihrer Grenzgebiete

3. Folge · Band 31
A Series of Modern Surveys in Mathematics

Springer

Berlin
Heidelberg
New York
Barcelona
Budapest
Hong Kong
London
Milan
Paris
Santa Clara
Singapore
Tokyo

Valerij V. Kozlov

Symmetries, Topology and Resonances in Hamiltonian Mechanics

Springer

Author

Valerij V. Kozlov
Department of Mathematics
Moscow State University
119899 Moscow
Russia

Translators

S.V. Bolotin
Department of Mathematics
Moscow State University
119899 Moscow
Russia
e-mail: bolotin@nw.math.msu.su

D. Treshchev
Department of Mathematics
Moscow State University
119899 Moscow
Russia
e-mail: dtresch@nw.math.msu.su

Yuri Fedorov
B-1113 Lenin Hills
Moscow State University
117234 Moscow
Russia
e-mail: fedorov@nw.math.msu.su

With 32 Figures

Mathematics Subject Classification (1991):
58F05, 58F07, 70Hxx

ISBN 3-540-57039-X Springer Verlag Berlin Heidelberg New York

Library of Congress Cataloging-in-Publication Data.
Kozlov, V.V. (Valeriĭ Viktorovich) Symmetries, topology, and resonances in Hamiltonian mechanics / Valery V. Kozlov. p. cm. Includes index.
ISBN 3-540-57039-X – ISBN 0-387-57039-X
1. Hamiltonian systems. 2. Symmetry (Physics) 3. Topology. 4. Mathematical physics. I. Title.
QC20.7.H35K69 1995
531′.11′0151474–dc20

Printed in Germany

Typeset with TEX: Data conversion by Lewis & Leins GmbH, Berlin
SPIN: 10015730 41/3020 - 5 4 3 2 1 0 - Printed on acid-free paper

Mephistopheles
Ich wünschte nicht, Euch irrezuführen.
Was die Wissenschaft betrifft,
Es ist so schwer, den falschen Weg zu meiden,
Es liegt in ihr so viel verborgenes Gift,
Und von der Arzenei ists kaum zu unterscheiden.
Am besten ists auch hier, wenn Ihr nur Einen hört
Und auf des Meisters Worte schwört.
Im ganzen: haltet Euch an Worte !
Dann geht Ihr durch die sichre Pforte
Zum Tempel der Gewißheit ein.

Schüler
Doch ein Begriff muß bei dem Worte sein.

Faust I. Goethe. Goethe Werke B. 3.
Insel Verlag, Frankfurt am Main 1977

Preface

The problem of exact integration of equations of dynamics has been one of the most popular fields of research since Newton's famous "Philosophiae naturalis principia mathematica". The principal idea in the investigation is the general notion of symmetry. When solving the problem of a mass point motion in a central field, Newton already used the concept of symmetry: by factorizing orbits of the rotation group he reduced this problem to a one-dimensional motion in a potential field. Later, Lagrange and Jacobi noticed that the classical integrals of the n-body problem are related to invariance of the equations of motion with respect to Galileo's transformation group. This fundamental remark was generalized by Emmy Noether: to each transformation group conserving the Hamiltonian action there corresponds an integral of the motion equations. The converse is also correct: a phase flow of a Hamiltonian system with a known additional integral as a Hamiltonian takes solutions of the original equations of motion into solutions of the same equations. This gives the idea of the proof of the well-known Liouville theorem on the complete integrability of Hamilton's equations: the phase flows of involutive integrals commute pairwise and generate an Abelian symmetry group of maximal possible dimension on their joint level manifolds.

At first, the question of exact integration was treated only in an analytical way: to find explicit expressions for first integrals and solutions. However, after Poincaré's work it became clear that the integrability phenomenon is closely connected with the global behavior of phase trajectories. When studying a dynamical system "in the whole" it is essential to know its topology. Quite recently it was discovered that the complicated topological structure of the configuration space is not consistent with integrability of the motion equations of the corresponding dynamical system. On the other hand, as shown by Poincaré, obstructions to integrability of Hamiltonian systems are resonance phenomena connected with the destruction of invariant resonant tori under perturbations. An analytical aspect of this fact is the well-known problem of small divisors in celestial mechanics. The other known obstacles to integrability – splitting of asymptotic surfaces and branching of solutions in the complex time plane – are also closely connected with resonances.

This book makes the first attempt to systematize the results on integrability of Hamiltonian systems obtained during the past 10–15 years as well as to give a modern interpretation of classical findings in this field.

The contents of the book are as follows. In the Introduction we give a historical survey of the integrability problem in dynamics. The basic notions of Hamiltonian mechanics are explained in Chapter I. Chapter II is devoted to the methods of exact integration of Hamiltonian systems: here we discuss various concepts of integrability of such systems. In Chapter III we indicate rough obstructions to integrability expressed in terms of topological invariants of the configuration space. In Chapters IV-VIII we discuss resonance phenomena in connection with the integrability problem. The methods given there enable one to prove rigorously the nonintegrability of many urgent problems of dynamics. Special attention is given to the phenomenon of stochastization of Hamiltonian systems under a small perturbation of the Hamiltonian function.

Our explanations use various mathematical techniques. However, those notions that go beyond the standard university course can be found in the book itself. Therefore, reader's persistence and patience is all that is needed. The book is intended primarily for young mathematicians and physicists who have the possibility of advancing this fascinating field, which has many important problems still unsolved.

Moscow, Summer 1995 Valerij V. Kozlov

Table of Contents

Introduction

1. In 1834 Hamilton represented the differential equations of classical mechanics, i.e., the Lagrange equations

$$\frac{d}{dt}\frac{\partial L}{\partial \dot{q}} = \frac{\partial L}{\partial q} ,$$

in "canonical" form

$$\dot{q} = \frac{\partial H}{\partial p} , \quad \dot{p} = \frac{\partial H}{\partial q} . \tag{1}$$

Here $p = \partial L/\partial \dot{q}$ is the generalized momentum and the Hamilton function $H = p\dot{q} - L$ is the "total energy" of the mechanical system.

"In part he had been anticipated by the great French mathematicians: for Poisson in 1809 had taken the first step of introducing a function [1]

$$\sum_{r=1}^{n} p_r \dot{q}_r - T ,$$

had expressed it in terms of variables $q_1, \ldots, q_n, p_1, \ldots, p_n$, and actually derived half of Hamilton's equations; while Lagrange in 1810 had obtained a particular set of equations (for the variations of elements of an orbit) in the Hamiltonian form, the disturbing function taking the place of the function H. Moreover the theory of non-linear partial differential equations of the first order had led to systems of ordinary differential equations possessing this form: for, as was shown by Pfaff in 1814-1815 and Cauchy in 1819 (completing earlier work of Lagrange and Monge), the equations of the characteristics of a partial differential equation

$$f(x_1, x_2, \ldots, x_n, p_1, p_2, \ldots, p_n) = 0 , \tag{2}$$

where

$$p_s = \frac{\partial z}{\partial x_s} ,$$

are

$$\frac{dx_1}{\partial f/\partial p_1} = \cdots = \frac{dx_n}{\partial f/\partial p_n} = \frac{dp_1}{-\partial f/\partial x_1} = \cdots = \frac{dp_n}{-\partial f/\partial x_n} .$$

1 T is the kinetic energy of the system.

Hamilton's investigation was extended to the cases when the kinetic potential contains time by Ostrogradski (1848-1850) and Donkin (1854) " (Whittaker [233][2]).

2. The problem of solving Hamiltonian systems (not yet written in the canonical form) had already been discussed by the Bernoulli brothers, Clairaut, D'Alambert and, of course, Lagrange in their works connected with the application of Newton's ideas and principles to various problems of mechanics. Only problems that can be solved by a finite sequence of algebraic operations and "quadratures"—indefinite integrals of known functions—were regarded as solvable (integrable). However, the most urgent problems of dynamics (say, the n body problem) turned out to be "nonintegrable" (more exactly, they had not been integrated). It is said that Alexis Clairaut, who had spent a lot of time trying to solve the equations of the three body problem in connection with the theory of motion of the Moon, finally gave up this task saying "Let somebody else integrate it, who can". Only in the simplest cases, when a system has one degree of freedom ($n = 1$) or it splits into several independent one-dimensional systems, had integration turned out to be possible due to the existence of integrals of total energy type ($H = \text{const}$).

3. Hamilton (1834) and Jacobi (1837) developed a general method for solving equations of dynamics based on special canonical coordinates.

The idea of the Hamilton–Jacobi method goes back to Pfaff's and Cauchy's works on the theory of characteristics. Transformation of independent variables $p, q \to P, Q$ of the form

$$p = \frac{\partial S}{\partial q}, \quad Q = \frac{\partial S}{\partial P}; \qquad S(P, q) : \mathbb{R}^{2n} \to \mathbb{R}$$

takes the canonical equations (1) to the canonical equations

$$\dot{P} = -\frac{\partial K}{\partial Q}, \quad \dot{Q} = \frac{\partial K}{\partial P} \tag{3}$$

with the Hamilton function

$$K(P, Q) = H(p, q)\big|_{P,Q} .$$

If the function K does not depend on Q, then equations (3) can be immediately integrated:

$$P = P_0, \quad Q = Q_0 + t\frac{\partial K}{\partial P}\Big|_{P_0} .$$

Therefore, the problem of solving the canonical equations (1) is reduced to searching for the "generating" (characteristic) function $S(P, q)$ satisfying the nonlinear

2 "It would be very useful to give a detailed critical analysis of the historical development of studying this question. Almost all references regarding the origins of fundamental mathematical concepts in analytical dynamics are wrong" (Wintner [234]).

Hamilton–Jacobi equation

$$H\left(\frac{\partial S}{\partial q}, q\right) = K(P) ,$$

which is a special case of Eq. (2). The most effective way of solving the Hamilton–Jacobi equation is the method of separation of variables. This method is non-invariant in principle, so it requires great analytical skill to choose appropriate variables. In order to emphasize this fact, Jacobi wrote: "The main difficulty in integrating given differential equations is to introduce suitable variables which cannot be found by a general rule. Therefore, we must go in the opposite direction and, after finding some remarkable substitution, look for problems to which it could be successfully applied" ("Vorlesungen über Dynamik"). As an example of such a "remarkable" substitution, Jacobi introduced the elliptic coordinates. By means of these coordinates (and their degenerations) Jacobi and his successor Neumann solved several new problems of dynamics. Among them we note the problem of geodesics on an ellipsoid and the point mass motion on the multidimensional sphere in a force field with a quadratic potential. Later Liouville (1849) and Stäckel (1891) gave a sufficiently general form for Hamiltonians that admit separation of variables.

If a problem has been solved by the Hamilton–Jacobi method, then the functions $P_1(p, q), \ldots, P_n(p, q)$ are first integrals. It is easy to show that they are in involution, that is, their Poisson bracket

$$\{P_i, P_j\} = \sum_{s=1}^{n} \left(\frac{\partial P_i}{\partial q_s}\frac{\partial P_j}{\partial p_s} - \frac{\partial P_j}{\partial q_s}\frac{\partial P_i}{\partial p_s} \right)$$

equals zero identically. This idea was developed by Bour and Liouville in 1855. Using the Hamilton–Jacobi method, it was proved that the Hamilton equations with n degrees of freedom can be integrated provided there are n independent integrals in involution. As a matter of fact, this is an invariant formulation of the Hamilton–Jacobi method. The proof of the Liouville theorem is based on the following argument. Let $H = P_1, P_2, \ldots, P_n$ be a set of independent integrals in involution. If, for example,

$$\det \left\| \frac{\partial P_i}{\partial p_j} \right\| \neq 0 ,$$

then from the equations $P_1(p, q) = c_i$ $(1 \leq i \leq n)$ one can determine (at least locally) the functions $p_j = f_j(q, c)$. Since the integrals P_i are in involution, the 1-form $\sum f_i(q, c)\, dq_i$ is closed, therefore, it is locally a complete differential (in q) of a function $S(q, c)$. Since $H(\partial S/\partial q, q) = c_1$, the function S represents a complete integral of the Hamilton–Jacobi equation.

Within the framework of this approach, in the works of Kovalevskaya, Clebsch, Chaplygin, Steklov, and others, a series of new problems of mechanics was solved, some of them being quite nontrivial. It is worth mentioning that in these classical works the Hamiltonian structure of the motion equations was not used. The integrability condition and the integration itself were based only on the Euler–Jacobi method of the integrating multiplier. Recall that to apply this method to integrate a system of n differential equations, the latter must have an integral invariant and $n-2$ independent integrals. For this reason, the integrability of several problems of dynamics was not noticed. The most striking example is the rotation of a rigid body around a fixed point in a gravitational field of distant centers. In this problem Brun [42] found three integrals which are not sufficient for application of the Euler–Jacobi method. Nevertheless, by virtue of the involution property of the integrals found by Brun, the problem is integrable by the Liouville theorem. The explicit integration of the Brun problem was recently performed by Bogoyavlensky [27].

It is obvious that the Liouville theorem (as well as the Euler–Jacobi theorem) gives only a theoretical possibility of solving differential equations exactly. Their explicit integration is an independent task, often nontrivial. While solving the equations of rotational motion of a rigid body (1758) and the equations of the problem on two fixed attracting centers (1760), Euler for the first time encountered the inversion problem for elliptic integrals. This stimulated his interest in elliptic functions for which he obtained the addition formula during these years. Explicit integration of motion equations in other classical problems has led to general Abelian functions. The fact that at that time this was not self-evident can be seen in Kovalevskaya's letter to Mittag-Leffler (1886): "Not later than this summer he (Picard) looked on with disbelief when I told him that functions of the form

$$y = \frac{\theta(Cx + A, C_1x + A_1)}{\theta_1(Cx + A, C_1x + A_1)}$$

can be very helpful for integrating some differential equations". In works by Kovalevskaya and her successor Kötter the integration technique had reached a high degree of perfection. Later this skill was "lost". In Chaplygin's works the solution was reduced to Abelian integrals only, without explicitly expressing dynamical variables in terms of theta-functions. Nowadays the technique of integration in theta-functions has revived on a new theoretical level (see [4,71]).

4. In almost all solved problems the first integrals turn out to be either rational functions or simply polynomials. Therefore, they can be continued to the complex domain of the phase variables p, q as single-valued holomorphic or meromorphic functions. A single-valued Hamiltonian generates a "complexified" Hamiltonian system. Then the solutions, as functions of complex time (or some auxiliary variable), often turn out to be meromorphic. As examples, one may indicate the Jacobi problem concerning the motion of a point mass on a triaxial ellipsoid, the Kovalevskaya top, the Clebsch case in the problem of a rigid body motion in an ideal fluid. Moreover, Kovalevskaya's and Lyapunov's investigations of the classical heavy top problem showed that the general solution of the motion equations is expressed in terms of single-valued functions of complex time only in the cases when an additional polynomial integral exists. In this connection an interesting problem appeared. It concerns the relationship between the existence of single-valued holomorphic integrals and branching of solutions in the complex time plane. Formulation of this problem goes back to Painlevé.

5. In later works the attention of mathematicians concentrated on qualitative research of Hamiltonian systems solvable by the Hamilton–Jacobi method, and first of all by the method of separation of variables. Especially for integrable systems the "angle–action" variables came into use. They were introduced by Delaunay for studying astronomical perturbations in celestial mechanics. Later they turned out to be very suitable for the old form of quantum mechanics, since, according to the Bohr–Sommerfeld quantization rule, "each "angle–action" variable must be an integer multiple of the Planck constant" (Synge [223]). The quantization conditions were first formulated for systems with separated variables, but eventually it became evident that in the most general situation the joint levels of a complete set of integrals in involution are homeomorphic to multidimensional tori, that in the corresponding angle variables the motion on the tori is described by a quasi-periodic law, and that the action variables are the integrals

$$\frac{1}{2\pi}\oint p\,dq$$

taken along distinct cycles wrapped around the torus in different ways (see Chap. II).

In a neighborhood of each n-dimensional torus of a complete integrable system with n degrees of freedom, it is possible to introduce the canonical angle–action variables $I_1, \ldots, I_n$; $\varphi_1, \ldots, \varphi_n \bmod 2\pi$, the Hamilton function H depending on I only. In these variables the Hamilton equations take the following simple form

$$\begin{aligned} \dot{I}_1 = 0\,, \quad \dot{\varphi}_1 &= \frac{\partial H}{\partial I_1}\,, \\ \ldots \quad \ldots \quad & \ldots \quad \ldots \\ \dot{I}_n = 0\,, \quad \dot{\varphi}_n &= \frac{\partial H}{\partial I_n}\,. \end{aligned} \tag{4}$$

Here $\varphi_1, \ldots, \varphi_n$ are the angle coordinates on the invariant n-dimensional torus that vary uniformly with time. A Hamiltonian system is called non-degenerate if

$$\det\left\|\frac{\partial^2 H}{\partial I_i \partial I_j}\right\| = \frac{\partial(\omega_1, \ldots, \omega_n)}{\partial(I_1, \ldots, I_n)} \neq 0\,.$$

For a non-degenerate system one can pass to new (non-canonical) coordinates $\mathcal{I}_k(I)$, $\varphi_j \bmod 2\pi$ in which Eq. (4) take on the "universal" form

$$\dot{\mathcal{I}}_k = 0\,, \quad \dot{\varphi}_j = \mathcal{I}_j\,; \qquad k, j = 1, \ldots, n\,.$$

Imitating Leo Tolstoy, one may say that all integrable Hamiltonian systems are alike, while each nonintegrable system is nonintegrable in its own way.

6. Lagrange and Jacobi made the fundamental remark that the existence of ten classical integrals in the n gravitating body problem follows from the fact that the equations of motion are invariant with respect to the action of the ten-parameter Galileo group. Later this observation was generalized by Noether (1918): if the action functional

$$\int_{t_1}^{t_2} L(\dot{q}, q)\, dt$$

is invariant with respect to a group action $q \to g^{\alpha}(q)$, then the Lagrange equations

$$\frac{d}{dt}\frac{\partial L}{\partial \dot{q}} = \frac{\partial L}{\partial q}$$

admit the integral $p \cdot v = \text{const}$, where $p = \partial L/\partial \dot{q}$ is the canonical momentum (impulse), and $v(q) = d/dt|_{\alpha=0}(g^{\alpha}(q))$ a vector field generating the symmetry group.

By factoring the orbits of the action of the symmetry group one can lower the order of a system of differential equations. As examples, we mention passing to the barycentric frame of reference and the well-known Jacobi result on "the elimination of the nodes" in the many body problem. Developing these ideas, Sophus Lie proved the nonintegrability by quadratures of a system of n differential equations that admits an $(n-1)$-dimensional solvable symmetry group. The algebraic analog of Lie's theorem is the well-known Galois group of substitutions of roots of a polynomial.

In Hamiltonian mechanics the symmetry groups generated by Hamiltonian systems play a special role: given functions F and H in involution, the phase flow of a system with Hamiltonian F transforms the solutions of the Hamilton equations with Hamiltonian H to solutions of the same equations. Thus, the problem of symmetry groups of Hamilton equations contains, as a special case, the problem of the existence of a first integral. The Noether symmetries are generated by the linear integrals $F = p \cdot v(q)$.

7. On the other hand, the efforts made by Klairaut, Lagrange, Poisson, Laplace, and Gauss in order to obtain approximate solutions of applied problems in celestial mechanics led at last to the creation of perturbation theory. The solutions of the motion equations were sought in the form of a series in degrees of a small parameter (for example, in the Solar system such a parameter is the ratio of the mass of Jupiter to the mass of the Sun). Afterwards, Delaunay, Hilden, and Lindschtedt modified the perturbation theory by using the Hamilton–Jacobi method. Namely, suppose that $H = H_0 + \varepsilon H_1 + \varepsilon^2 H_2 + \cdots$ $(\varepsilon \ll 1)$ and the unperturbed problem with the Hamiltonian H_0 is integrable. Then one seeks the characteristic function S in the form of a series $S_0 + S_1 + \ldots$ that satisfies the equation

$$H_0\left(\frac{\partial S}{\partial q}, q\right) + \varepsilon H_1\left(\frac{\partial S}{\partial q}, q\right) + \cdots = K_0(P) + \varepsilon K_1(P) + \cdots, \tag{5}$$

where the functions K_i are to be determined later. According to the proposition above, the functions S_0 and K_0 can be found from equation (5) for $\varepsilon=0$. For the case $i \geq 1$ the functions S_i and K_i are determined by a recursion procedure; the arbitrariness in their definition can be eliminated by using the condition that so called "secular" terms are absent.

Therefore, the perturbed problem can be regarded as "solved" provided the series of the perturbation theory are defined correctly and converge. Their convergence would imply a series of important consequences (in particular, the perpetual stability of the Solar system). Anticipating, we mention Poincaré's disappointing result: in view of the existence of so called small divisors (small denominators) the series of the perturbation theory are generally divergent. Moreover, the series of the improved perturbation theory developed by Poincaré and Bolind, according to which the solutions are expanded not in degrees of ε, but in degrees of $\sqrt{\varepsilon}$, are divergent as well. Note that if the series of the perturbation theory converge, then the motion equations have a complete set of integrals in involution which can be represented as a converging power series in ε (or in $\sqrt{\varepsilon}$).

Subsequently (in 1916–1927), Whittaker, Cherry, and Birkhoff obtained analogous results for Hamiltonian systems in a neighborhood of an equilibrium position and periodic trajectories. They showed that in the general case there exists a canonical transformation given by a formal power series after which the Hamilton equations can be easily integrated. Hamiltonian systems that admit a convergent Birkhoff transformation are sometimes called integrable according to Birkhoff (Birkhoff integrable). In this case there also exists a complete set of independent commutative integrals of a special form.

8. Once mathematicians had realized the impossibility of solving classical equations of dynamics in a closed form, strict results regarding their nonintegrability began to appear. Apparently, the first was Liouville's theorem (1841) concerning the unsolvability of the equation $\ddot{x} + tx = 0$ by quadratures. More exactly, there does not exist a field that contains all solutions of the Liouville equation and that can be obtained from the field of rational functions of t by a sequence of finite algebraic extensions, adjoining integrals and exponents of integrals (see [107]). In 1887 Bruns proved a theorem on the nonexistence of algebraic integrals independent of the classical ones in the three body problem [43]. "Unfortunately, in his proof was a big gap to cover which was a delicate matter"—wrote Poincaré in his "Analytical Resume". "I was happy to put this wonderful and skilled Brun proof beyond any criticism". Here Poincaré meant his work [204].

The Bruns–Poincaré theorem was generalized by Painlevé (1898) to the case when the integrals are algebraic only in the velocities of three gravitating bodies. Afterwards analogous results were obtained by Husson (1906) and other authors in the dynamics of a rigid body with a fixed point.

One may however agree with Wintner [234] that these elegant negative results have no importance in dynamics, since they do not take into account the peculiarities of the behavior of phase trajectories. As far as first integrals are concerned, locally, in a neighborhood of a non-singular point, a complete set of independent integrals always exists. Whether they are algebraic or transcendent depends explicitly on the choice of independent variables. Therefore, the problem of the existence of integrals makes sense only when it is considered in the whole phase space or in a neighborhood of the invariant set (for example, an equilibrium position or a periodic trajectory).

In spite of the evident insufficiency of algebraic integrals for the purposes of dynamics, results of that kind remained quite popular for a long time. For example, in 1936 Karl Siegel felt it was necessary to prove the Bruns–Poincaré theorem for the restricted three body problem.

9. A productive formulation of the integrability problem for the Hamilton equations and the first nontrivial results in this field were obtained by Poincaré. In the work "On the three body problem and equations of dynamics"(1890) he investigated the complete integrability of the "basic problem of dynamics". Here we mean Hamiltonian systems that appear in perturbation theory: the Hamilton function is expanded in a series in degrees of a small parameter $H = H_0 + \varepsilon H_1 + \varepsilon^2 H_2 + \cdots$, where H_0 is the Hamiltonian of a completely integrable system. Poincaré found necessary conditions for the existence of additional integrals in the form of a power series $\sum F_k \varepsilon^k$, whose coefficients F_k are analytic functions in a neighborhood of the invariant tori of the unperturbed system (for $\varepsilon = 0$).

In particular, it follows from his results that the power series in different versions of the perturbation theory are generally divergent. Poincaré has also indicated qualitative phenomena in the behavior of phase trajectories which prevent the appearance of new integrals. Among these phenomena are the creation of isolated periodic solutions and splitting of asymptotic surfaces. Poincaré applied his general methods to different versions of the n body problem. It turned out that except for the known classical conservation laws the motion equations have no new integrals that are analytic in the masses of planets.

In 1941–1954 Siegel investigated the integrability of a Hamiltonian system near stable equilibrium positions. He proved that in a typical situation the Hamilton equations do not have a complete set of analytical integrals and the Birkhoff transformation is divergent. The idea of Siegel's proof goes back to Poincaré's investigations: it is based on a thorough study of families of non-degenerate long-periodic solutions.

In the 20th century, following Poincaré's work, it became gradually clear that the impossibility of continuing locally existing integrals to integrals defined "in the whole" is connected with the intricate behavior of phase trajectories lying on level surfaces of those integrals which are present, but fail to form a set sufficient for complete integrability. Simply speaking, on an integral surface there must exist trajectories that are everywhere dense in a certain field. Levi-Civita proposed referring to systems that possess m integrals, but not $m+1$ integrals, defined "in the whole" as m-imprimitive. Here integrability problems interact with ergodic theory. An example is Hopf's theorem, proved in 1939 and asserting that a geodesic flow on any compact surface of negative curvature is ergodic. To study geodesics on such surfaces, Birkhoff, Morse, and Hedlund created symbolic dynamics, enabling one to describe the complex behavior of the trajectories in terms of probability. However, as was noted by Poincaré "... trajectories in the three body problem[3] are comparable not with geodesic lines on a surface of negative curvature, but, on the contrary, with those on a convex surface ... unfortunately, this problem is much more complex ...". Here zones of quasi-random behavior of phase trajectories alternate and coexist with fields of trajectories of "regular" type. For a discussion of these questions we refer the reader to Kolmogorov's lecture [119] and Moser's book [183]. A direct application of the idea of intricate behavior of phase trajectories to the three body problem is considered in Alekseev's works [5].

10. In rigid body dynamics the efforts of Bobyliev, Steklov, Chaplygin, Goryachev, and other mathematicians have produced several "particular cases of integrability". We mean exact particular solutions that are obtainable by quadratures. Trajectories of such solutions are, as a rule, closed. At the beginning of the century the following point of view was very popular: given a sufficiently large set of partial solutions, one may obtain a picture of the general case of the rotations of a rigid body by interpolation. This idea was expressed by Klein and Sommerfeld in the book [111]. However, further development of the theory of dynamical systems did not confirm this point of view. As a matter of fact, in the phase space of the problem of rotations of a rigid body around a fixed point there are almost always zones of quasi-random motion. The existence of such zones can in no way be deduced from the existence of a great number of periodic or quasi-periodic trajectories. In this book we do not consider questions connected with the search for exact particular solutions.

11. Nowadays Poincaré's ideas have been developed further and new phenomena in the behavior of Hamiltonian systems have been discovered. Among these are
- intricate topological structure of a configuration space,
- splitting and transverse intersection of asymptotic manifolds,
- branching of solutions in the complex time plane,
- quasi-random oscillations,
- small divisors in higher approximations of the perturbation theory.

Using these concepts, mathematicians have managed to prove rigorously the absence of nontrivial integrals and symmetry groups in several classical problems of dynamics: rotation of a heavy non-symmetric body around a fixed point, motion of a rigid body in an ideal fluid, the problem of four point vortices in the plane, and many other questions.

In each of these problems the nonintegrability result is based on the analysis of the qualitative behavior of phase trajectories. Finally, in our opinion, an independent part of the theory of Hamiltonian systems has taken shape. It has its own specific problems, methods, and results. The aim of this book is to give a systematic exposition of modern ideas and results of this theory.

---

3 as well as many other problems of dynamics

I Hamiltonian Mechanics

There are various approaches to the theory of Hamiltonian systems. They are described in [12, 74, 224, 233]. In this chapter we recall the definition of the basic concepts of Hamiltonian mechanics, and also present several examples of Hamiltonian systems which will be used in the sequel.

1 The Hamilton Equations

1.1 We begin with the axiomatic definition of the Poisson bracket; thisidea probably goes back to Dirac [67]. Let M be an even-dimensional manifold. We denote the set of all infinitely differentiable functions $f : M \to \mathbb{R}$ by $C^\infty(M)$. A symplectic (canonical) structure Σ on M is a bilinear mapping $\{\,,\,\} : C^\infty(M) \to C^\infty(M)$ which satisfies the following conditions:

1) $\{f, g\} = -\{g, f\}$ (antisymmetry),
2) $\{fg, h\} = f\{g, h\} + g\{f, h\}$ (the Leibnitz rule),
3) $\{\{f, g\}, h\} + \{\{g, h\}, f\} + \{\{h, f\}, g\} = 0$ (the Jacobi identity),
4) if a point $m \in M$ is not critical for the function f, then there exists a smooth function g such that $\{f, g\}(m) \neq 0$ (non-degeneration).

A pair (M, Σ) is called a symplectic (canonical) manifold. The function $\{f, g\}$ is called the Poisson bracket of functions f and g. The Poisson bracket turns the linear space $C^\infty(M)$ into an infinite-dimensional Lie algebra over the field $\mathbb{R}$. Its center (the set of elements that commute with all elements of the algebra) consists of constant functions only.

Theorem (Darboux). *In a small neighborhood of an arbitrary point on M there exist local coordinates* $(x_1, \ldots, x_n, y_1, \ldots, y_n)$ $(2n = \dim M)$ *such that*

$$\{f, g\} = \sum_{s=1}^{n} \left(\frac{\partial f}{\partial x_s} \frac{\partial g}{\partial y_s} - \frac{\partial f}{\partial y_s} \frac{\partial g}{\partial x_s} \right) .$$

Coordinates x, y are called symplectic (canonical). The proof of the Darboux theorem can be found in [12,87].

1.2 Let $H : M \to \mathbb{R}$ be a smooth function. We define a Hamiltonian system on (M, Σ) with the Hamilton function H as a differential equation

$$\dot{F} = \{F, H\} , \qquad F \in C^\infty(M) . \tag{1.1}$$

Its solutions are smooth mappings $m : \Delta \to M$ (Δ is an interval in $\mathbb{R}$) such that

$$\frac{dF(m(t))}{dt} = \{F, H\}(m(t)) , \qquad t \in \Delta .$$

In the symplectic coordinates x, y, the equation (1.1) is equivalent to $2n$ canonical *Hamilton equations* :

$$\dot{x}_i = \{x_i, H\} = \frac{\partial H}{\partial y_i} , \quad \dot{y}_i = \{y_i, H\} = -\frac{\partial H}{\partial x_i} , \qquad 1 \leq i \leq n .$$

These equations can be written in a more compact form by introducing the skew-symmetric matrix

$$J = \begin{pmatrix} 0 & I \\ -I & 0 \end{pmatrix} ,$$

where I is the $n \times n$ unit matrix. If we put $(x, y) = z$, then

$$\dot{z} = J \frac{\partial H}{\partial z} . \tag{1.2}$$

The manifold M is called the space of states or the phase space of the Hamiltonian system (1.1), and $(\dim M)/2$ is the number of degrees of freedom. Often it is necessary to consider non-autonomous systems where the Hamiltonian H depends on time explicitly.

For the sake of convenience we denote the vector field (1.2) by $v_H(z)$.

1.3 A diffeomorphism $\phi : M \to M$ is called canonical if it preserves the Poisson bracket: $\{f \circ \phi, g \circ \phi\} = \{f, g\}(\phi(m))$. Canonical diffeomorphisms of the symplectic manifold (M, Σ) form, of course, a group [1]. The phase flow g_H^t of any Hamiltonian system on M is a one-parameter subgroup of canonical diffeomorphisms of M.

In local symplectic coordinates, the condition for the mapping $\phi : x, y \to X, Y$ to be canonical can be represented by any of the following conditions

1) for any cycle (contour) γ

$$\oint_\gamma y\, dx = \oint_\Gamma Y\, dX \ (= \oint_\Gamma Y(x, y)\, dX(x, y)) ,$$

where Γ is the image of γ under the mapping ϕ,

2) $D^T J D = J$, D being the Jacobian matrix of the mapping ϕ.

---

1 "... each time you need to deal with some object Σ endowed with a structure, try to define the group of its automorphisms ... You can hope that in this way you will understand deeply the inner structure of the object Σ". (H. Weyl "Symmetry").

In the new coordinates $(X, Y) = Z$ equations (1.2) have again the Hamiltonian form

$$\dot{Z} = J\frac{\partial K(Z)}{\partial Z},$$

where $K(Z) = H(z)$.

It is useful to study canonical diffeomorphisms by means of generating (characteristic) functions. Let, for example, $\det \|\partial X/\partial x\| \neq 0$. In this case the equation $X = X(x, y)$ can be resolved (at least locally) with respect to x, and one can regard X, y as "independent" variables. Hence, $x = x(X, y)$, $Y = Y(X, y)$. If we put

$$S = \int_{X_0, y_0}^{X, y} x\,dx + Y\,dX$$

(the value of the integral does not depend on the integration path), then

$$x = \frac{\partial S}{\partial y}, \quad Y = \frac{\partial S}{\partial X}.$$

The function $S(X, Y)$ is called the generating function of the canonical mapping ϕ. If, for example, ϕ is the identity mapping, then $S = Xy$.

1.4 Let γ be a closed curve in the extended phase space $M \times \mathbb{R} = \{z, t\}$ of a Hamiltonian system $\dot{z} = JH'(z, t)$. Each point $(z_0, t_0) \in \gamma$ defines a unique regular curve $z = z(t)$, where $z(\cdot)$ is the solution of this system with the initial condition $z(t_0) = z_0$. A set of these curves covers a cylindrical surface Π in $M \times \mathbb{R}$ called the tube of trajectories. According to the Poincaré–Cartan theorem [51], the integral

$$\int_\gamma y\,dx - H\,dt$$

takes the same value for all homologous closed curves γ' on Π (similarly "cycling around" the tube of trajectories Π).

Let two closed curves γ_1 and γ_2 be sections of the surface Π produced by the planes t_1 and t_2 respectively. Then

$$\int_{\gamma_1} y\,dx = \int_{\gamma_2} y\,dx.$$

This result was first obtained by Poincaré [205, III]. In particular, if γ is a closed curve on M, and g_H^t is a phase flow of a Hamiltonian system, then the integral of the 1-form $y\,dx$ along the closed curve $g_H^t(\gamma)$ does not depend on t. From here we can deduce the "basic theorem of Hamiltonian mechanics": the phase flow of a Hamiltonian system is a family of canonical transformations.

1.5 A symplectic structure on M can be defined by means of a symplectic atlas which is a set of maps consistent with each other, and such that the change from one map to another is a smooth canonical mapping. Let, for example, M be the cotangent bundle $T^*\mathcal{N}$ of a smooth manifold $\mathcal{N}$. A symplectic structure on $T^*\mathcal{N}$ is specified by the set of local coordinates x, y, where x are local coordinates on N, and y are components of linear differential forms on $T^*\mathcal{N}$ in the basis dx.

There is another well-known way of defining a symplectic structure and a Hamiltonian system. Here the basic object is a closed non-degenerate 2-form Ω on an even-dimensional manifold M. The form Ω allows us to construct a natural isomorphism of the tangent $T_x\mathcal{N}$ and cotangent $T_x^*\mathcal{N}$ spaces: a vector $\xi \in T_x\mathcal{N}$ corresponds to a 1-form $\phi \in T_x^*\mathcal{N}$ in accordance with the rule $\phi_\xi(\eta) = \Omega(\xi, \eta)$, $\eta \in T_x\mathcal{N}$. Let J be the inverse mapping, H a smooth function on M (possibly depending on time). Since the differential dH is a covector (an element of $T_x^*\mathcal{N}$), we conclude that $J\,dH$ is a smooth vector field. The corresponding differential equation on M

$$\dot{x} = v_H(x) \tag{1.3}$$

is equivalent to (1.1).

Indeed, let F and G be smooth functions on M. The function $\Omega(v_F, v_G)$ is well-defined and, as one can check, it satisfies the properties 1)–4) of the Poisson bracket. Thus, $\Omega(v_F, v_G)$ is the Poisson bracket of functions F, G, and we may denote it by $\{F, G\}$. In view of the identity

$$\{F, H\} = \Omega(v_H, v_F) = dF(J\,dH) = dF(v_H)\ ,$$

we can rewrite equation (1.3) in the form (1.1).

Conversely, let F be a function on M. It follows from properties 1) and 3) of the Poisson bracket that $v_F = \{F, \cdot\}$ is a derivation, i.e., a tangent vector to M. Since the Poisson bracket is non-degenerate, all tangent vectors can be represented in the same form. Let G be one more such function, and $v_G = \{G, \cdot\}$ the corresponding tangent vector. We define a 2-form Ω by the relation

$$\Omega(v_G, v_F) = \{F, G\}\ .$$

This form is obviously bilinear, skew symmetric, and non-degenerate. The Jacobi identity implies that Ω is closed ($d\Omega = 0$). In the symplectic coordinates x, y, the form Ω takes the canonical form

$$\Omega = \sum dy_i \wedge dx_i\ .$$

This is one of the equivalent formulations of the Darboux theorem. In the coordinates x, y, the Eqs. (1.3) have the form (1.2). In addition, we note that J is the matrix of the operator J represented in terms of the symplectic coordinates.

Thus, a symplectic structure on M can be defined by a non-degenerate 2-form. Abusing the notation, we shall call the form Ω the *symplectic structure* as well. In the sequel we shall describe Hamiltonian systems in various ways.

1.6 Suppose that smooth functions H and F commute (i.e., they are "in involution"): $\{F, H\} = 0$. Then F is a first integral of the canonical system with Hamiltonian H and vice versa. The phase flows g_H^t and g_F^t also commute on M.

Since

$$\{\{F,G\},H\} = \{\{F,H\},G\} - \{\{G,H\},F\}\,,$$

integrals of any Hamiltonian system form a subalgebra in the Lie algebra of all smooth functions on M (the Poisson theorem).

1.7 A natural mechanical system is a set $(\mathcal{N}, T, V)$, where $\mathcal{N}$ is a smooth manifold (configuration space), T a Riemannian metric on $\mathcal{N}$ (kinetic energy) and V a smooth function on $\mathcal{N}$ (potential of a force field). Motions of such a system are smoothmappings $q(t) : \mathbb{R} \to \mathcal{N}$ which are extremals of the action functional

$$\int_{t_1}^{t_2} L(\dot{q}(t), q(t))\, dt\,.$$

Here $\dot{q}(t)$ is a tangent vector to $\mathcal{N}$ at the point $q(t)$ and $L = T - V$ the Lagrange function. Evolution of the local coordinates q on $\mathcal{N}$ is described by the Euler–Lagrange equation

$$\frac{d}{dt}\frac{\partial L}{\partial \dot{q}} = \frac{\partial L}{\partial q}\,.$$

Consider the natural mapping $T\mathcal{N} \to T^*\mathcal{N} : (q, \dot{q}) \to (q, p)$ generated by the Riemannian metric, where

$$p = \frac{\partial L}{\partial \dot{q}}\,.$$

Obviously, p is a linear form on $T_q\mathcal{N}$. Since the quadratic form T is positive definite, the linear mapping $\dot{q} \to p$ is an isomorphism of the linear spaces $T_q\mathcal{N}$ and $T^*\mathcal{N}$.

Consider the total energy of the system $H : T^*\mathcal{N} \to \mathbb{R}$ defined by the expression

$$H(p,q) = p\dot{q} - L|_{\dot{q}\to p} = \frac{\partial T}{\partial \dot{q}}\dot{q} - T + V = T + V|_{p,q}\,.$$

Theorem (Poisson–Hamilton). *The functions $q(t)$, $p(t)$ satisfy the canonical equations*

$$\dot{p} = -\frac{\partial H}{\partial q}\,, \quad \dot{q} = \frac{\partial H}{\partial p}\,.$$

Analogous representations are valid for more general Lagrangians. Suppose that L, as a function of the velocities, is convex (the quadratic form $(L''_{\dot{q},\dot{q}}\xi, \xi)$ is positive definite), and increases at infinity faster than any linear function ($L/|\dot{q}| \to \infty$ as $|\dot{q}| \to \infty$). The Legendre transformation

$$(q, \dot{q}) \to (q, p)\,, \quad L \to H\,,$$

$$\dot{p} = -\frac{\partial H}{\partial q}\,, \qquad H = p\dot{q} - L|_{\dot{q}\to p}$$

is defined in the whole. Besides, the Hamiltonian H is also convex in the momenta p and increases at infinity faster than any linear function.

One should not think that the Hamilton equations appear in mechanics only as an equivalent representation of the Lagrange equations. Here is a simple example. Consider a two-dimensional flow of an incompressible fluid. Let a, b be components of the velocity field v of the fluid particles in Cartesian coordinates x, y. From the incompressibility condition $\Delta \cdot v = 0$ it follows that the 1-form $a\,dy - b\,dx$ for all values of t is a differential of a function $\Psi(x, y, t)$. The equations of motion of fluid particles can be represented in the Hamiltonian form

$$\dot{x} = \Psi'_y\,, \quad \dot{y} = -\Psi'_x$$

with the Hamiltonian Ψ. In hydrodynamics the function Ψ is called the stream function: if the flow is stationary, the particles move along the curves Ψ=const.

1.8 Natural systems are "reversible": their Lagrangians are invariant with respect to the change $t \to -t$. Hence, together with a motion $q(t)$ we have the motion $q(-t)$.

The first generalization of reversible mechanical systems consists systems with *gyroscopic forces*. They may have variousstructures. Gyroscopic forces appear as a result of a transformation to a rotating reference frame, and also after lowering the number of degrees of freedom in systems with symmetries (see, for example, [14, Chap. III]), or in the case of motion of a charged particle in a magnetic field. Here is a formal definition.

Let $\mathcal{N}$ be the configuration space of a natural system and $y_1, \ldots, y_n$ be the momenta. The coordinates x, y are canonical, and in these variables the symplectic structure Ω has the standard form

$$\Omega = \sum dy_i \wedge dx_i\,.$$

In addition, we consider one more closed 2-form on $\mathcal{N}$:

$$\Gamma = \sum \Gamma_{ij}(x)\,dx_i \wedge dx_j\,, \qquad d\Gamma = 0\,.$$

In mechanics this form is called a form of gyroscopic forces. The sum $\Omega + \Gamma$ defines a new symplectic structure on the cotangent bundle of $\mathcal{N}$. Given a function H on $T^*\mathcal{N}$, the pair $(\Omega+\Gamma, H)$ defines a Hamiltonian system with the Hamiltonian H; this system will be called a system with gyroscopic forces. It is clear that the existence of such forces does not change the total energy H. We can apply the Darboux theorem for the 2-form $\Omega+\Gamma$ and represent it in canonical form. Namely, since the form Γ is closed, we can write locally $\Gamma = dF$, $F = \sum F_k(x)\,dx_k$. Thus, in the coordinates x, y we have

$$\Omega + \Gamma = \sum dy_i \wedge dx_i + \sum dF_i \wedge dx_i = \sum d(y_i + F_i) \wedge dx_i\,.$$

Therefore, the variables defined by

$$x'_k = x_k\,, \quad y'_k = y_k + F_k(x_1, \ldots, x_n)\,; \qquad 1 \le k \le n$$

are canonical coordinates for the new symplectic structure. In the new variables the Hamilton equations have the canonical form with the Hamiltonian $H(x', y' - F) = H(x, y)$.

Now consider the case when a closed form of gyroscopic forces is exact: $\Gamma = d\Phi$, where Φ is a 1-form on $\mathcal{N}$. Then the equations of motion can be represented as the Lagrange equations with the globally defined Lagrangian

$$L = \frac{1}{2}\langle \dot{x}, \dot{x} \rangle + \langle v(x), \dot{x} \rangle - V(x) .$$

Here the metric $\langle \ , \ \rangle$ defines the kinetic energy of the system, the 1-form $\langle v, \dot{x} \rangle$ is exactly the form Φ and v is a vector field on $\mathcal{N}$. The Hamiltonian has the form $H = H_2 + H_1 + H_0$, where H_s is a homogeneous form of degree s in the momenta, and

$$H_0 = \frac{1}{2}\langle v, v \rangle + V .$$

1.9 Consider a generalization of the variational problem described in §1.6. This generalization goes back to Lagrange. Let $q : [t_1, t_2] \to \mathcal{N}$ be an extremal of the action functional

$$\int_{t_1}^{t_2} L \, dt , \qquad L = T - V$$

in the class of curves with fixed end points which satisfy the equations

$$a_1 \cdot \dot{q} = \cdots = a_m \cdot \dot{q} = 0 . \tag{1.4}$$

Here $a_1, \ldots, a_m$ are smooth covector fields on $\mathcal{N}$ which are linearly independent at any point, and $m < \dim \mathcal{N}$. Following the method of Lagrange multipliers, we introduce new coordinates $\lambda_1, \ldots, \lambda_m$ and the Lagrangian $\mathcal{L} = L - \sum \lambda_i (a_i \cdot \dot{q})$. It can be shown (see, for example, [24]) that extremals of the variational problem can be obtained from the following system of Euler–Lagrange equations

$$\frac{d}{dt}\frac{\partial \mathcal{L}}{\partial \dot{q}} = \frac{\partial \mathcal{L}}{\partial q} , \qquad \frac{d}{dt}\frac{\partial \mathcal{L}}{\partial \dot{\lambda}} = \frac{\partial \mathcal{L}}{\partial \lambda} . \tag{1.5}$$

The second group of equations is equivalent to the constraint equations (1.4). To represent the system (1.5) in the form of the Hamilton equations, we put

$$p = \frac{\partial \mathcal{L}}{\partial \dot{q}} = \frac{\partial T}{\partial \dot{q}} - \sum \lambda_i a_i , \qquad a_j \cdot \dot{q} = 0 , \qquad 1 \leq j \leq m . \tag{1.6}$$

If the form T is positive definite and the covectors a_j are linearly independent, then one can obtain $\dot{q}$ and λ from this system of equations as functions of p and q. Now put

$$H = p\dot{q} - L|_{\dot{q} \to p} = T + V|_{\dot{q} \to p} . \tag{1.7}$$

The variables p and q, as functions of t, satisfy the canonical Hamilton equations with the Hamiltonian H. The proof can be found, for example, in [133, (I)].

Consider the particular case of integrable constraints defined by

$$f_1(q) = \cdots = f_m(q) = 0 . \tag{1.8}$$

In the configuration space $\mathcal{N}^n$, these equations define a submanifold of dimension $n - m$. The extremals of the variational problem corresponding to the constraints (1.8) are motions of the holonomic mechanical system with $n - m$ degrees of freedom. According to §1.6, the equations of motion can be represented as $2(n-m)$ Hamilton equations. However, there is also a different approach. Differentiating Eqs. (1.8) with respect to t, we represent them in the form of (1.4):

$$\frac{\partial f_1}{\partial q}\dot{q} = \cdots = \frac{\partial f_m}{\partial q}\dot{q} = 0 . \tag{1.9}$$

Using (1.6), we introduce canonical variables. Then the equations of motion can be written in the form of $2n$ Hamilton equations which can be regarded as equations in "excessive" variables. In fact, this result was obtained by Suslov [222].

As an example, consider the motion of a point with unit mass on a smooth regular surface $\Sigma = \{f(r) = 0\}$ in the Euclidean space $\mathbb{R}^3 = \{r\}$ under the action of a force with potential $V(r)$. According to (1.5), we put

$$p = \dot{r} + \lambda\frac{\partial f}{\partial r} , \qquad \lambda = \frac{(p, f'_r)}{(f'_r, f'_r)} . \tag{1.10}$$

The motion of the point is described by the Hamilton equations

$$r' = H'_p , \quad \dot{p} = -H'_r , \qquad H = \frac{1}{2}|\dot{r}|^2 + V = \frac{1}{2}(p \times n)^2 + V , \tag{1.11}$$

where n is the unit vector normal to the surface Σ. Therefore, equations (1.10) are defined by this surface itself and do not depend on the form of the equation $f = 0$ defining Σ.

Equations (1.10) possess the energy integral H and the "geometric" integral $F = f(r)$. In the standard symplectic structure $dp \wedge dr$, the Poisson bracket $\{H, F\}$ equals zero. Let $g(\dot{r}, r)$ be a first integral of the "classical" equations of motion

$$\ddot{r} = \frac{\partial V}{\partial r} + \lambda\frac{\partial f}{\partial r} , \qquad f(r) = 0 ,$$

and G be a function of g represented in terms of the canonical variables in accordance with (1.9). Then, obviously, $\{H, F\} = 0$, and it is easy to check that H and F are in involution.

In general, Eqs. (1.4) are not integrable (i.e., they cannot be represented in the form (1.8)). Herz called such constraints non-holonomic. One should not think that in this case the canonicalequations with the Hamiltonian (1.6, 1.7) describe the motion of a holonomic system with Lagrangian L and constraints (1.4). The classical non-holonomic equations

$$\frac{d}{dt}\frac{\partial L}{\partial \dot{q}} - \frac{\partial L}{\partial q} = \sum \mu_i a_i , \qquad a_j \cdot \dot{q} = 0 . \tag{1.12}$$

are different from (1.5). Since the phase flow of the system (1.1) does not necessarily have an absolutely continuous invariant measure, this system generally cannot be reduced to the Hamiltonian form. Equations (1.5) give the basis of the vakonomic mechanics which was developed in [133].

In the extensive monograph of Griffiths [93] the geometry of the Hamiltonian formalism has been presented for the general variational Lagrange problem, and several variational problems have been solved (however, the author is incorrect in saying that he has, therefore, solved some problems of non-holonomic mechanics).

2 Euler–Poincaré Equations on Lie Algebras

2.1 Let $u_1, \dots, u_n$ be independent tangent vector fields on an n-dimensional manifold $\mathcal{N}$. At each point of $\mathcal{N}$ the commutators are linear combinations of the basis vectors $\{u_k\}$:

$$[u_i, u_j] = \sum c_{ij}^k(q) u_k .$$

If f is a smooth function on $\mathcal{N}$, then

$$\dot{f} = \frac{\partial f}{\partial q} \cdot \dot{q} = \sum u_i(f)\omega_i ,$$

where $u_i(f)$ is the derivative of f by the vector field u_i. The variables ω are linear functions of $\dot{q}$. They are called quasivelocities. Now represent the Lagrangian in terms of q and ω: $\mathcal{L}(\omega, q) = L(\dot{q}, q)$.

In the new variables the Lagrange equations take the following form:

$$\frac{d}{dt}\frac{\partial \mathcal{L}}{\partial \omega_k} = \sum_{i,j} c_{ik}^j \frac{\partial \mathcal{L}}{\partial \omega_j}\omega_i + u_k(\mathcal{L}) , \qquad 1 \le k \le n . \tag{2.1}$$

These were first obtained by Poincaré [205]. If the vector fields u_k are equal to some coordinate vector fields $\partial/\partial q_k$, then the Poincaré equations transform to the ordinary Lagrange equations. One should keep in mind that the system (2.1) is not complete: to obtain a complete system we must add the relation between ω and $\dot{q}$.

2.2 Suppose that the Lagrangian $\mathcal{L}$ is a convex function in ω, and $\mathcal{L}$ grows faster at infinity than any linear function. Now make the Legendre transformation:

$$m_k = \partial\mathcal{L}/\partial\omega_k , \qquad \mathcal{H} = m \cdot \omega - \mathcal{L}\big|_{\omega \to m} .$$

Then, as we know,

$$\omega_k = \partial\mathcal{H}/\partial m_k , \qquad u_k(\mathcal{L}) = -u_k(\mathcal{H}) .$$

In the new variables equations (2.1) take the following form:

$$\dot{m}_k = \sum c_{ik}^j m_j \frac{\partial \mathcal{H}}{\partial m_i} - u_k(\mathcal{H}) , \qquad 1 \le k \le n . \tag{2.2}$$

These were first introduced by Chetayev [58].

2.3 Now let $\mathcal{N}$ be a Lie group G and let $u_1, \ldots, u_n$ be independent left-invariant vector fields on G. Then $c_{ij}^k = \text{const}$. Suppose that the Lagrangian L is reduced to the kinetic energy T which is defined by a left-invariant metric $\langle\, , \,\rangle$ on G. Since

$$\dot{q} = \sum u_i(q)\omega_i ,$$

we have

$$L = \frac{1}{2}\langle \dot{q}, \dot{q}\rangle = \frac{1}{2}\langle \sum u_i\omega_i, \sum u_j\omega_j\rangle = \frac{1}{2}\sum I_{ij}\omega_i\omega_j ,$$

where $I_{ij} = \langle u_i, u_j\rangle = \text{const}$ by the fact that the metric is left-invariant. In this case equations (2.1)–(2.2) are as follows:

$$\dot{m}_i = \sum c_{ik}^l m_l \omega_k , \qquad m_s = \sum I_{sp}\omega_p . \tag{2.3}$$

When represented in terms of ω or m, these can be regarded as the equations on the Lie algebra g of the group G or on the dual vector space g^* respectively.

We call Eqs. (2.3) the *Euler–Poincaré equations*. For example, consider the special case when the group G is $SO(3)$ (the group of rotations of a rigid body in three-dimensional Euclidean space). It is well known that the corresponding Lie algebra $g = so\,(3)$ is isomorphic to the algebra of vectors in the three-dimensional oriented Euclidean space endowed with the standard vector product. As basis left-invariant vector fields on G we take the vector fields u_i generated by rotations with unit angular velocity about three orthogonal coordinate axes attached to the rigid body. Then

$$[u_1, u_2] = u_3 , \quad [u_2, u_3] = u_1 , \quad [u_3, u_1] = u_2 .$$

It is easy to see that in this case Eqs. (2.3) coincide with the dynamical Euler equations

$$\frac{d}{dt}\frac{\partial T}{\partial \omega} = \omega \times \frac{\partial T}{\partial \omega} ,$$

or

$$I\dot{\omega} = \omega \times I\omega . \tag{2.4}$$

Here ω and $I = \|I_{ij}\|$ are the angular velocity vector and the inertia tensor respectively. This observation is due to Poincaré [205].

2.4 Equations (2.3) represent a part of the Hamiltonian system describing the geodesic flow of the left-invariant metric I_{ij}. To calculate the Poisson bracket of two functions F and G on the dual space g^*, we consider the Hamiltonian system with Hamiltonian F and calculate the derivative of G along the vector field defined by F. In the variables q, m this Hamiltonian system takes the form of the *Chetayev equations* (2.2)

$$\dot{m}_k = \sum c_{ik}^j m_j \frac{\partial F}{\partial m_i} .$$

Since the function G depends only on m, the other part of the Hamiltonian system can be omitted. Hence

$$\dot{G} = \{G, F\} = \sum c_{ik}^{j} m_j \frac{\partial F}{\partial m_i} \frac{\partial G}{\partial m_k} . \tag{2.5}$$

It follows that the Poisson bracket of functions on g^* is a function on g^* as well. This bracket satisfies properties 1)–3) of the Poisson bracket, but it can be degenerate (since we consider a special class of functions on $T^*G = G \times g^*$). The bracket (2.5) is called the *Lie–Poisson bracket.* It was first introduced by Lie in his theory of transformation groups. If F and G are linear in the components of the "momentum" m, then the bracket $\{F, G\}$ is also linear in m. Therefore the space of linear functions on g^* is a Lie algebra with respect to the Lie–Poisson bracket. This algebra is obviously isomorphic to the algebra g^*.

In the Euler problem on free rotations of a rigid body the Lie–Poisson bracket is given by the equations

$$\{m_1, m_2\} = m_3 , \quad \{m_2, m_3\} = m_1 , \quad \{m_3, m_1\} = m_2 . \tag{2.6}$$

This bracket is degenerate: the function $k^2 = m_1^2 + m_2^2 + m_3^2$ commutes with each function on $(so(3))^*$.

With respect to this bracket the Euler–Poincaré equations are Hamiltonian:

$$\dot{m}_i = \{m_i, H\} , \qquad H = \langle I^{-1} m, m\rangle /2 . \tag{2.7}$$

However, these equations are not "actual" Hamiltonian equations, due to the degeneracy of the bracket $\{\cdot\,,\,\cdot\}$. Let $F_1, \ldots, F_m$ be integrals of Eqs. (2.7), independent at each point of the invariant manifold

$$M_c = \{m \in g^* : F_i = c_i, \quad 1 \le i \le m\} ,$$

and commuting with every function on g^*. Let $\{\cdot\,,\,\cdot\}'$ and H' be the restrictions of the Lie–Poisson bracket $\{\cdot\,,\,\cdot\}$ and the Hamiltonian function H on M_c. The restriction of the system (2.7) on M_c again has the Hamiltonian form:

$$\dot{F} = \{F, H'\}', \qquad F : M_c \to \mathbb{R} .$$

The bracket $\{\cdot\,,\,\cdot\}'$ satisfies properties 1)–3) of the Poisson bracket. If it happens that $\{\cdot\,,\,\cdot\}$ is non-degenerate on M_c, we obtain an ordinary Hamiltonian system on the symplectic manifold $(M_c, \{\cdot\,,\,\cdot\})$ with the Hamiltonian function H'. The theory of the reduction of Euler–Poincaré equations to Hamiltonian equations is given, for example, in [12, Appendix 2].

Now we return to the Euler problem. In a certain orthogonal frame related to the principal inertia axes the quadratic form $T = (I\omega, \omega)/2$ can be written as follows:

$$T = \frac{1}{2}(I_1\omega_1^2 + I_2\omega_2^2 + I_3\omega_3^2) .$$

In this frame the Euler equations (2.4) take the form

$$I_1\dot{\omega}_1 = (I_3 - I_2)\omega_2\omega_3 , \quad I_2\dot{\omega}_2 = (I_1 - I_3)\omega_1\omega_3 , \quad I_3\dot{\omega}_3 = (I_2 - I_1)\omega_2\omega_1 .$$

They have the angular momentum integral $k^2 = (I_1\omega_1)^2 + (I_2\omega_2)^2 + (I_3\omega_3)^2$. Recall that this function commutes with all functions on the dual space $(so(3))^*$. For $c > 0$ the level surface $M_c = \{\omega : k^2 = c^2\}$ of this integral is a two-dimensional sphere. We will now show that the restriction of the Lie–Poisson bracket (2.6) on M_c defines the standard symplectic structure (the 2-form of the oriented area on the sphere M_c). Let $F = f_1m_1 + f_2m_2 + f_3m_3$ be a linear function with constant coefficients. The operator $v_F = \{F, \cdot\}$ is a differentiation. Representing it in the form

$$\sum_{i=1}^{3} a_i \frac{\partial}{\partial m_i} ,$$

we find that the numbers a_i are equal to components of the vector $m \times f$. Similarly, the vector $m \times g$ corresponds to the function $G = \sum g_j m_j$. Since

$$\Omega(v_G, v_F) = \{F, G\} = \left(m, \frac{\partial F}{\partial m} \times \frac{\partial G}{\partial m}\right) = (m, f \times g) , \tag{2.8}$$

it follows that the value of the 2-form Ω on the pair of vectors $m \times f$ and $m \times g$ is equal to the mixed product of the vectors m, f, and g. Suppose that f and g are tangent to the sphere $M_c = \{m^2 = c^2\}$. Then the vectors $\xi = m \times f$ and $\eta = m \times g$ are also tangent to M_c. From (2.8) we obtain

$$\Omega(\eta, \xi) = \left(\frac{n}{|m|}, \xi \times \eta\right) , \tag{2.9}$$

where n is the unit vector normal to M_c. This formula defines the standard area 2-form on M_c up to multiplication by a constant factor.

According to the Darboux theorem, it is possible to represent the Euler equations on M_c in the canonical Hamiltonian form. This can be done explicitly by introducing the special symplectic (canonical) variables $l \bmod 2\pi$, L ($|L| \le c$), by the formulas

$$I_1\omega_1 = \sqrt{c^2 - L^2}\sin l , \quad I_2\omega_2 = \sqrt{c^2 - L^2}\cos l , \quad I_3\omega_3 = L .$$

In these variables the Euler equations have the canonical form

$$\dot{l} = \frac{\partial H'}{\partial L} , \quad \dot{L} = -\frac{\partial H'}{\partial l} , \qquad H' = \frac{1}{2}\left(\frac{\sin^2 l}{I_1} + \frac{\cos^2 l}{I_2}\right)(c^2 - L^2) + \frac{L^2}{2I_3} .$$

The phase portrait of the function H' is reproduced in Fig. 1. Identifying points of the strip $|L| \le c$ with l-coordinates differing by 2π and all points of the lines $L = c$ and $L = -c$, we obtain the sphere M_c with the well known picture of the Poinsot polodia. It can also be shown that written in terms of the variables L, l, the symplectic structure (2.9) is exactly equal to $dL \wedge dl$.

2.5 The Euler–Poincaré equations cannot be transformed to the Hamiltonian form for every Lie algebra g. The obstruction is the nonexistence of an invariant measure. Now we investigate this problem more closely.

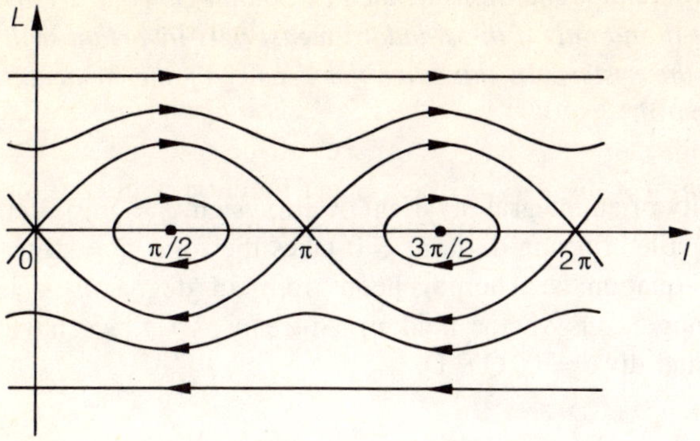

Fig. 1.

Let $f : \mathbb{R}^n = \{z\} \to \mathbb{R}$ be a nonnegative integrable function. The measure $d\mu = f(z)\, d^n z$ is called absolutely continuous, if for any measurable set $D \subset \mathbb{R}^n$ with positive Lebesgue measure, the integral

$$\operatorname{mes}(D) = \int_D f\, d^n z$$

is positive. Let $\dot{z} = v(z)$ be a dynamical system and g^t its phase flow. The measure $d\mu$ is called *an invariant measure* of the dynamical system, if $\operatorname{mes}(g^t(D)) = \operatorname{mes}(D)$ for any measurable set D and all $t \in \mathbb{R}$. If f is a positive function of class C^1, the invariant measure is called an integral invariant.

According to Liouville's theorem, the measure $f\, d^n z$ is an integral invariant if and only if

$$\operatorname{div}(fv) = \sum_{i=1}^{n} \frac{\partial (f v_i)}{\partial z_i} = 0\,.$$

In particular, the phase flow of a Hamiltonian system conserves the standard measure in the phase space $\mathbb{R}^n = \{x, y\}$ (here x, y are canonical coordinates).

Following [141], consider the existence problem for an invariant measure of the Euler–Poincaré Eqs. (2.8) on the algebra $g = \{\omega\}$.

Theorem 1. *The Euler–Poincaré equations have an integral invariant if and only if the group G is unimodular.*

Recall that a group G is unimodular if there exists a bi-invariant measure on G. The criterion for this is the following: $\sum_k c_{ik}^k = 0$ for all i.

To prove that the condition of Theorem 1 is sufficient, we calculate the divergence of the right-hand side of the system (2.3). It is equal to $\sum_k c_{ik}^k \omega_i$. Therefore, by the Liouville theorem, the measure $d^n\omega$ is invariant with respect to the phase flow of the system (2.3). The condition of Theorem 1 is necessary by the following

Proposition 1. *A system of differential equations defined by a homogeneous vector field has an integral invariant if and only if the standard measure is invariant with respect to the phase flow of the system. In this case the density of the invariant (the function f) is a first integral.*

Proof. Let $f > 0$ be the density of an integral invariant of the system $\dot{z} = v(z)$. Let $f = \exp(-w)$. Then the Liouville criterion $\operatorname{div}(fv) = 0$ takes the form $\dot{w} = \operatorname{div} v$. The right-hand side of this equation is a homogeneous form of degree $m - 1$ (m is the degree of the homogeneous vector field v). Since $w \in C^1$, we have $\dot{w} = O(|z|^m)$. Hence $\dot{w} = 0$, and $\operatorname{div} v = 0$. Q.E.D.

If the dimension of g is low, one can give more detailed information on the invariant measure of the system (2.3). If $n = 2$ and g is a non-Abelian group, then Eqs. (2.3) do not have an invariant measure with an integrable (or even smooth) density.

Indeed, there exist some basis vectors e_1, e_2 such that $[e_1, e_2] = e_1$ (see [106]). Therefore, equations (2.3) take the form

$$\dot{m}_1 = m_1\omega_2 , \qquad \dot{m}_2 = -m_1\omega_1 . \tag{2.10}$$

All points of the line $m_1 = I_{11}\omega_1 + I_{12}\omega_2 = 0$ (and only these points) are the equilibrium positions. The phase trajectories are arcs of the ellipses $\sum I_{ij}\omega_i\omega_j =$ const. The phase portrait of the system (2.10) is given in Fig.2. It is clear that each domain $D \in \mathbb{R}^2 = \{\omega_1, \omega_2\}$ approaches the line $m_1 = 0$ as $t \to \infty$ asymptotically. Therefore, this system cannot possess an absolutely continuous invariant measure.

For $n = 3$, the condition of Theorem 1 may not be fulfilled only for solvable algebras. These can be defined by the relations

$$[e_1, e_2] = 0 , \quad [e_1, e_3] = \alpha e_1 + \beta e_2 , \quad [e_2, e_3] = \gamma e_1 + \delta e_2 ,$$

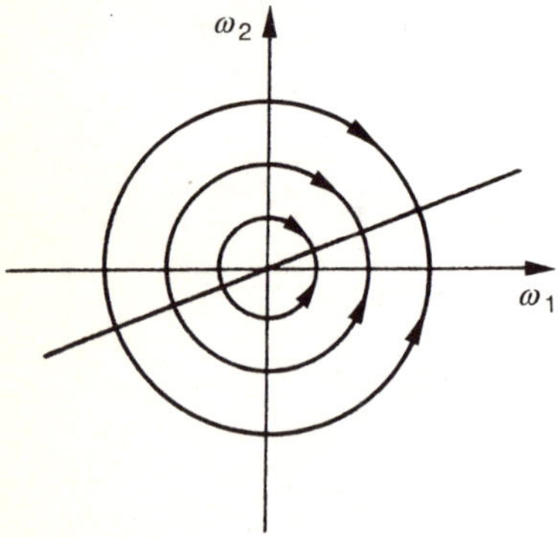

Fig. 2.

where $\{e_1, e_2, e_3\}$ is a basis in g, and the matrix

$$A = \begin{pmatrix} \alpha & \beta \\ \gamma & \delta \end{pmatrix}$$

is non-degenerate (see [106]). The eigenvalues of A can be: (a) real numbers with the same signs, (b) real numbers with different signs, (c) complex numbers with a nonzero real part, (d) purely imaginary numbers.

Proposition 2. *In case* (*d*) *the Euler–Poincaré equations have an integral invariant, in case* (*b*) *there exists no integral invariant, however, the system possesses an invariant measure with density of any finite smoothness, whereas in cases* (*a*) *and* (*c*) *there is no invariant measure with integrable density.*

In case (d) the algebra g satisfies the condition of Theorem 1. The reason for the existence of an invariant measure of finite smoothness in case (b) can be seen easily by considering the equations $\dot{x} = 2x,\ \dot{y} = -y$, which have an invariant measure with density $|x|^s|y|^{2s+1}$ for all $s > 0$. In cases (a) and (c), on each ellipsoid $\sum m_i\omega_i \geq 0$ of the energy integral there exists an asymptotically stable equilibrium position. We emphasize that the conditions given above are determined only by the structure of the algebra g and do not depend on the choice of the left-invariant metric.

A non-unimodular group is known to have an unimodular normal divisor of codimension one [40]. Let $\{e_k\}$ be a basis in g such that the vectors $e_1, \ldots, e_{n-1}$ form a basis in the corresponding ideal of the algebra g and the vector e_n is orthogonal to $e_1, \ldots, e_{n-1}$ with respect to the metric I_{ij}.

Proposition 3. *If all the eigenvalues of the matrix* $A = \|c^s_{nk}\|$ $(k, s \leq n)$ *lie in the left* (*or right*) *complex half-plane, then equations* (2.3) *do not possess an invariant measure with integrable density.*

This statement follows from the fact that the equilibrium position $m_s = 0$ $(s \leq n)$, $m_n = \text{const} \neq 0$ is asymptotically stable (as $t \to +\infty$, or $t \to -\infty$) on the corresponding level surface of the energy integral.

3 The Motion of a Rigid Body

3.1 Suppose that a rigid body rotates around a fixed point in force field with potential V. Let α, β, γ be vectors of a fixed orthonormal frame considered in a moving frame attached to the body. Since these vectors determine uniquely the position (attitude) of the body with respect to the fixed frame, the potential V can be regarded as a function of α, β, γ. Now write down the Poincaré equations, regarding the group $SO(3)$ as the configuration space of the body. Again (as in §2.3), let u_1, u_2, u_3 denote the left-invariant vector fields on $SO(3)$ generated by the permanent rotations of the body about the principal inertia axes with unit angular velocities. We are going to calculate the derivatives $u_i(V)$ of the potential V along u_i. The evolution of α, β, γ with respect to the body is described by the Poisson equations

$$\dot{\alpha} = \alpha \times \omega\,, \quad \dot{\beta} = \beta \times \omega\,, \quad \dot{\gamma} = \gamma \times \omega\,, \tag{3.1}$$

where ω is the angular velocity vector. Let $\omega = \omega' = (1, 0, 0)$. Then

$$\begin{aligned} u_1(V) = \dot{V} &= \left(\frac{\partial V}{\partial \alpha}, \dot{\alpha}\right) + \left(\frac{\partial V}{\partial \beta}, \dot{\beta}\right) + \left(\frac{\partial V}{\partial \gamma}, \dot{\gamma}\right) \\ &= \left(\omega', \frac{\partial V}{\partial \alpha} \times \alpha + \frac{\partial V}{\partial \beta} \times \beta + \frac{\partial V}{\partial \gamma} \times \gamma\right). \end{aligned}$$

Analogous formulas for $u_2(V), u_3(V)$ are omitted. Putting in (2.1) $\mathcal{L} = \mathcal{T} - \mathcal{U}$, we arrive at the equations

$$\frac{d}{dt}\frac{\partial T}{\partial \omega} + \omega \times \frac{\partial T}{\partial \omega} = \alpha \times \frac{\partial V}{\partial \alpha} + \beta \times \frac{\partial V}{\partial \beta} + \gamma \times \frac{\partial V}{\partial \gamma}\,. \tag{3.2}$$

Considering them together with the Poisson equations, we obtain a closed system describing rotations of the rigid body. These equations were derived by Lagrange in his "Analytical mechanics" (1788). If the force field is axisymmetric (say, the potential V depends on γ only), then from (3.1, 3.2) we obtain a closed system of the Euler–Poinsot equations

$$I\dot{\omega} + \omega \times I\omega = \gamma \times \frac{\partial V}{\partial \gamma}\,, \quad \dot{\gamma} = \gamma \times \omega\,. \tag{3.3}$$

3.2 Suppose that a fixed point of the body coincides with its center of mass, and the body rotates in the field of gravitational attraction of distant bodies. The corresponding potential V can be closely approximated by the quadratic form

$$\varepsilon_1(I\alpha, \alpha) + \varepsilon_2(I\beta, \beta) + \varepsilon_3(I\gamma, \gamma)\,, \tag{3.4}$$

where I is the inertia operator of the body and the constants ε_i depend on the mass distribution of the distant gravitating bodies (see, for example, [224]). If the latter are represented by a single attracting center, then in (3.4) we must put $\varepsilon_1 = \varepsilon_2 = 0$. This problem was first considered by Tisserand (1872), who established its integrability by using the Euler–Jacobi method of the integrating multiplier. The general case of the potential (3.4) was investigated by Brun [42]. He found three integrals in involution, which are sufficient for complete integrability of the system (3.1, 3.2). The explicit integration of the Brun problem in terms of θ-functions was performed by Bogoyavlensky [27].

The problem of solving Eqs. (3.1), (3.2) of rigid body motion in various force fields was considered in [29,91,236].

3.3 Suppose that the body rotates in a homogeneous force field. Let ε be the weight of the body, and r the position vector of its center of mass with respect to the moving frame. In this problem $V = \varepsilon(r, \gamma)$, and the Euler–Poinsot equations (3.3) have the form

$$I\dot{\omega} + \omega \times I\omega = \varepsilon\gamma \times r\,, \quad \dot{\gamma} = \gamma \times \omega\,. \tag{3.5}$$

These equations depend on the six parameters I_1, I_2, I_3 , $\varepsilon r_1, \varepsilon r_2, \varepsilon r_3$, where I_i are the principal moments of inertia and r_i the coordinates of the center of mass with respect to the inertia axes.

3.4 Consider the multiplicative group $Sp(1)$ of the quaternions $q = \chi + \xi\vec{i} + \eta\vec{j} + \zeta\vec{k}$ normalized by $\chi^2 + \xi^2 + \eta^2 + \zeta^2 = 1$. To each quaternion q there corresponds a linear mapping T_q of the algebra K of all quaternions into itself defined by the relation

$$T_q(r) = qrq^{-1}\,, \qquad r \in K\,.$$

It is easy to check that T_q maps the set of pure quaternions ($\zeta = 0$) into itself. If we identify this set with the Euclidean space $\mathbb{R}^3$, then T_q will be an orthogonal transformation $\mathbb{R}^3 \to \mathbb{R}^3$. Now consider a rigid body with a fixed point. Its rotation from an original position to an arbitrary one is defined by some orthogonal transformation which, in turn, corresponds to some quaternion $q \in Sp(1)$. Thus, to each quaternion $q \in K$ there corresponds a position of a rigid body with a point fixed. The quaternion $-q$ corresponds to the same position. These observations go back to Gauss. Therefore, in the problem of the rotation of a rigid body the variables $(\chi, \xi, \eta, \zeta) \in \mathbb{R}^4$ can be regarded as the "excessive" coordinates.

Now we use the theory given in §1.8. Note that

$$\begin{aligned} \omega_1 &= 2(\chi\dot{\xi} - \xi\dot{\chi} + \zeta\dot{\eta} - \eta\dot{\zeta})\,, \\ \omega_2 &= 2(\chi\dot{\eta} - \eta\dot{\chi} + \xi\dot{\zeta} - \zeta\dot{\xi})\,, \\ \omega_3 &= 2(\chi\dot{\zeta} - \zeta\dot{\chi} + \eta\dot{\xi} - \xi\dot{\eta}) \end{aligned} \tag{3.6}$$

(see [233]). According to (1.6), the generalized momenta $p_\chi, p_\xi, p_\eta, p_\zeta$ are defined by the expressions

$$\begin{aligned} p_\chi &= \frac{\partial T}{\partial \chi} - \lambda\chi\,, \quad p_\xi = \frac{\partial T}{\partial \xi} - \lambda\xi\,, \\ p_\eta &= \frac{\partial T}{\partial \eta} - \lambda\eta\,, \quad p_\zeta = \frac{\partial T}{\partial \zeta} - \lambda\zeta\,, \\ &\chi\dot{\chi} + \xi\dot{\xi} + \eta\dot{\eta} + \zeta\dot{\zeta} = 0\,. \end{aligned}$$

From here, taking (3.6) into account, we find

$$\begin{aligned} 2I_1\omega_1 f &= p_\xi\chi - p_\chi\xi + p_\eta\zeta - p_\zeta\eta\,, \\ 2I_2\omega_2 f &= p_\eta\chi - p_\chi\eta + p_\zeta\eta - p_\eta\zeta\,, \\ 2I_3\omega_3 f &= p_\zeta\chi - p_\chi\zeta + p_\xi\eta - p_\eta\xi\,, \end{aligned}$$

where $f = \chi^2 + \xi^2 + \eta^2 + \zeta^2$.

Therefore, in the new variables $\chi, p_\chi, \ldots$ the equations of motion have the canonical form with the Hamiltonian

$$H = \frac{1}{8f^2}\left[\frac{(p_\xi\chi - p_\chi\xi + p_\eta\zeta - p_\zeta\eta)^2}{I_1} + \frac{(p_\eta\chi - p_\chi\eta + p_\zeta\eta - p_\eta\zeta)^2}{I_2} + \frac{(p_\zeta\chi - p_\chi\zeta + p_\xi\eta - p_\eta\xi)^2}{I_3}\right] + V(\chi, \xi, \eta, \zeta) .$$

Consider the case when the force field is invariant with respect to the group g of rotations about some fixed axis l. Denote by $\gamma_1, \gamma_2, \gamma_3$ the direction cosines of this axis with respect to the inertia axes. One can show that

$$\gamma_1 = 2(\xi\zeta - \eta\chi) , \quad \gamma_2 = 2(\xi\chi + \eta\zeta) , \quad \gamma_3 = \chi^2 + \zeta^2 - \xi^2 - \eta^2 . \tag{3.7}$$

The criterion for the potential V to be g-invariant is that the projection of the angular momentum to the axis l is constant:

$$M = I_1\omega_1\gamma_1 + I_2\omega_2\gamma_2 + I_3\omega_3\gamma_3 = \text{const} .$$

Represented in the canonical variables $\chi, p_\chi, \ldots$ the momentum M equals $(p_\eta\chi - p_\chi\zeta + p_\eta\xi - p_\xi\eta)/2$. Using this linear integral, one can decrease the number of degrees of freedom by one.

3.5 Consider some theoretical aspects of lowering the order of *Hamiltonian systems with a symmetry*. Let the Hamilton equations

$$\dot{q}_i = \frac{\partial H}{\partial p_i} , \quad \dot{p}_i = \frac{\partial H}{\partial q_i} , \quad 1 \le i \le n$$

possess the linear integral $F = \sum f_i(q)p_i$. The latter naturally corresponds to a one-parameter symmetry group g^s of the configurational space $\mathcal{N}$ – the phase flow of the system

$$\frac{dq_i}{ds} = \frac{\partial F}{\partial p_i} = f_i(q) , \quad 1 \le i \le n . \tag{3.8}$$

The orbits of the group g^s, being the trajectories of this system, can be straightened locally: in a neighborhood of a non-singular point one can choose coordinates $Q_1, \ldots, Q_n$ such that

$$\frac{dQ_i}{ds} = 0 , \quad 1 \le i \le n-1 ; \qquad \frac{dQ_n}{ds} = 1 .$$

The functions $Q_1, \ldots, Q_{n-1}$ are first integrals of equations (3.8), while $Q_n(q)$ satisfies

$$\sum \frac{\partial Q_n}{\partial q_i} f_i = 1 . \tag{3.9}$$

Since $\det \|\partial Q/\partial q\| \neq 0$, there exists a canonical transformation $(p, q) \to (P, Q)$ with the generating function $S(q, P) = \sum P_i Q_i(q)$. In the new coordinates

$$F = \sum f_i p_i = \sum f_i \frac{\partial Q_j}{\partial q_i} P_j = P_n \, .$$

Besides, since F is a first integral of the system with the Hamiltonian H, $H(P, Q)$ does not depend on Q_n. Thus, having fixed the value of the linear integral $F = Q_n$, we decrease the order of the original Hamiltonian system. The variables $Q_1, \dots, Q_{n-1}$ "enumerate" the orbits of the group g.

It is clear that the coordinates $Q_1, \dots, Q_{n-1}$ used for the reduction of the system are not determined uniquely; we can add to them arbitrary first integrals of (3.9). The Hamiltonian of the reduced system generally depends on the choice of the solution Q_n of (3.9). If the value of the integral F equals zero, then the Hamiltonian of the reduced system is determined uniquely on the cotangent bundle of the locally reduced configuration space, whose points are the orbits of the group g. In some cases the reduction for $F = 0$ can be performed not only locally, but "in the whole".

3.6 For the motion of a rigid body in an asymmetric force field Eqs. (3.8) have the following form

$$\begin{aligned} \chi' &= \frac{\partial M}{\partial p_\chi} = -\zeta/2 \, , & \zeta' &= \frac{\partial M}{\partial p_\zeta} = \chi/2 \, , \\ \xi' &= \frac{\partial M}{\partial p_\xi} = -\eta/2 \, , & \eta' &= \frac{\partial M}{\partial p_\eta} = \xi/2 \, . \end{aligned} \tag{3.10}$$

The behavior of the phase trajectories of this system (the orbits of the group g) is rather simple: they are the great circles of the three-dimensional spheres $S_r^3 = \{\chi^2 + \xi^2 + \eta^2 + \zeta^2 = r^2\} \subset \mathbb{R}^4$. The quotient space S_r^3/g (the set of orbits of g on S_r^3) is a two-dimensional sphere S_r^2. It can be regarded as the standard sphere $\{\gamma_1^2 + \gamma_2^2 + \gamma_3^2 = r\} \subset \mathbb{R}^3$. Indeed, the functions $\gamma_1, \gamma_2, \gamma_3$ (defined in (3.7)) form a set of independent integrals of Eqs. (3.10), and the points on the sphere $\gamma_1^2 + \gamma_2^2 + \gamma_3^2 = r$ correspond one-to-one to the great circles of S_r^3. This fibration of a three-dimensional sphere to two-dimensional ones is known in geometry as the Hopf fibration.

We can regard $\gamma_1, \gamma_2, \gamma_3$ as the "excessive" coordinates of the reduced system. According to §3.5, to write down its Hamiltonian we have to solve (3.9), which takes the form

$$-\frac{\partial \varphi}{\partial \chi}\zeta + \frac{\partial \varphi}{\partial \zeta}\chi - \frac{\partial \varphi}{\partial \xi}\eta + \frac{\partial \varphi}{\partial \eta}\xi = 2 \, .$$

One of its solutions is

$$\varphi = \arctan \frac{\zeta}{\chi} + \arctan \frac{\eta}{\xi} \, .$$

Note that any solution of this equation has singularities in $\mathbb{R}^4$.

The canonical variables p_1, p_2, p_3 conjugate to $\gamma_1, \gamma_2, \gamma_3$ can be found from the following system

$$2(p_1\zeta + p_2\chi - p_3\xi) + M\frac{\partial\varphi}{\partial\xi} = p_\xi\,,$$

$$-2(p_1\chi - p_2\zeta + p_3\eta) + M\frac{\partial\varphi}{\partial\eta} = p_\eta\,,$$

$$2(p_1\xi + p_2\eta + p_3\zeta) + M\frac{\partial\varphi}{\partial\zeta} = p_\zeta\,,$$

$$-2(p_1\chi - p_2\xi - p_3\chi) + M\frac{\partial\varphi}{\partial\chi} = p_\chi\,.$$

Using these expressions, we calculate

$$\begin{aligned} I_1\omega_1 &= \frac{1}{2f}(p_\xi\chi - p_\chi\xi + p_\eta\zeta - p_\zeta\eta) \\ &= \frac{1}{2f}\left[2(p_2\gamma_3 - p_3\gamma_2) + M\left(\chi\frac{\partial\varphi}{\partial\xi} - \xi\frac{\partial\varphi}{\partial\chi} + \zeta\frac{\partial\varphi}{\partial\eta} - \eta\frac{\partial\varphi}{\partial\zeta}\right)\right] \\ &= \frac{p_2\gamma_3 - p_3\gamma_2}{f} + \frac{M\gamma_1}{\gamma_1^2 + \gamma_2^2}\,, \\ I_2\omega_2 &= \frac{p_3\gamma_1 - p_1\gamma_3}{f} + \frac{M\gamma_2}{\gamma_1^2 + \gamma_2^2}\,, \\ I_3\omega_3 &= \frac{p_1\gamma_2 - p_2\gamma_1}{f}\,. \end{aligned} \tag{3.11}$$

Here $f^2 = \gamma_1^2 + \gamma_2^2 + \gamma_3^2$. The Hamiltonian of the reduced system equals $H = (I_1\omega_1^2 + I_2\omega_2^2 + I_3\omega_3^2)/2 + V$, where $I_1\omega_1, I_2\omega_2, I_3\omega_3$ must be replaced by their expressions in terms of p_i, γ_i given in (3.11). If $M = 0$, then the Hamiltonian takes the simplest form

$$\begin{aligned} H = \frac{1}{2f}&\left[\frac{(p_2\gamma_3 - p_3\gamma_2)^2}{I_1} + \frac{(p_3\gamma_1 - p_1\gamma_3)^2}{I_2} + \frac{(p_1\gamma_2 - p_2\gamma_1)^2}{I_3}\right] \\ &+ V(\gamma_1, \gamma_2, \gamma_3)\,. \end{aligned} \tag{3.12}$$

In the Euler case (when $V \equiv 0$), as we know, the square of the angular momentum

$$\begin{aligned} (I_1\omega_1)^2 + (I_2\omega_2)^2 + (I_3\omega_3)^2 = \frac{1}{2f}[&(p_2\gamma_3 - p_3\gamma_2)^2 \\ &+ (p_3\gamma_1 - p_1\gamma_3)^2 + (p_1\gamma_2 - p_2\gamma_1)^2] \end{aligned}$$

is constant. It is interesting to note that this function is the Hamiltonian of the canonical equations describing the inertial motion of a point with mass $m = 2$ on the sphere $\gamma_1^2 + \gamma_2^2 + \gamma_3^2 = 1$ and written in terms of the "excessive" coordinates γ_i, p_i $(1 \le i \le 3)$ (see (1.10)).

3.7 Now consider the group $E(3)$ of rigid body motion in three-dimensional Euclidean space and its Lie algebra $e(3)$. It is obvious that $\dim E(3) = 6$. Choose some orthonormal frame attached to the body with the origin at some point O. Let v_1, v_2, v_3 be the left-invariant vector fields on $E(3)$ which correspond to the permanent rotations with unit angular velocity about the three orthogonal axes of this frame. Analogously, to the body motions with the velocity vector of the origin O parallel to these axes there correspond the vector fields u_1, u_2, u_3. The fields v_i, u_j are obviously independent everywhere on $e(3)$. One can show that the structure constants of the algebra $e(3)$ are defined by the relations

$$[v_i, v_j] = \varepsilon_{ijk} v_k\,, \quad [v_i, u_j] = \varepsilon_{ijk} u_k\,, \quad [u_i, u_j] = 0\,, \tag{3.13}$$

where ε_{ijk} equals 1 or -1 if the permutation (i, j, k) is even or odd respectively, and $\varepsilon_{ijk} = 0$ if some of the indices coincide.

Let ω be the angular velocity vector of the body and v the linear velocity vector of the point O. If the Lagrangian $\mathcal{L}$ is a function of ω, v only, then the Poincaré Eqs. (2.1) can be written in the form

$$\frac{d}{dt}\frac{\partial \mathcal{L}}{\partial \omega} = \frac{\partial \mathcal{L}}{\partial \omega} \times \omega + \frac{\partial \mathcal{L}}{\partial v} \times v\,, \qquad \frac{d}{dt}\frac{\partial \mathcal{L}}{\partial v} = \frac{\partial \mathcal{L}}{\partial v} \times \omega\,. \tag{3.14}$$

These lead to the Chetayev equations

$$\dot m = m \times \frac{\partial \mathcal{H}}{\partial m} + p \times \frac{\partial \mathcal{H}}{\partial p}\,, \qquad \dot p = p \times \frac{\partial \mathcal{H}}{\partial m}\,, \tag{3.15}$$

where $m = \partial\mathcal{L}/\partial\omega$, $p = \partial\mathcal{L}/v$, $\mathcal{H}$ are functions of m and p. As known from hydrodynamics [158], the equations describing the motion of a rigid body in an unbounded volume of an ideal fluid, which is motionless at infinity and vortex free, have the same form. In the latter case the function $\mathcal{H}$ is a positive definite quadratic form

$$(Am, m)/2 + (Bm, p) + (Cp, p)/2\,. \tag{3.16}$$

The operators A and C are obviously symmetric. Equations (3.15) with the Hamilton function (3.16) were obtained by Kirchhoff (1870). The *Kirchhoff equations*generally contain 21 parameters. By reducing the operator A (or C) to diagonal form this number can be lowered by three. The vectors m and p are called the impulsive pair (moment) and the impulsive force respectively.

Note that the Euler–Poisson equations (3.3) can be written in the form of (3.15) by putting

$$\mathcal{H} = (I^{-1} m, m)/2 + V(p)\,.$$

This observation is due to Steklov, who noticed that the Tisserand problem is a special case of the Kirchhoff problem.

The Lie–Poisson bracket for the algebra $e(3)$ defined by the relations(3.13) is degenerate: the functions (m, p) and p^2 commute with all functions on $e^*(3)$ and are first integrals of the Kirchhoff equations for all Hamiltonians $\mathcal{H}$. Therefore, we can apply the ideas developed in §2.4. Consider the four-dimensional integral manifolds

$$M_c = \{m, p : (m, p) = c_1, \ (p, p) = c_2\} , \qquad c_2 > 0 ,$$

which are obviously diffeomorphic to the tangent bundle of a two-dimensional sphere. The restriction of the Lie–Poisson bracket to M_c is a non-degenerate Poisson bracket which turns M_c into a symplectic manifold. Thus, the Kirchhoff equations on M_c represent a Hamiltonian system with the Hamiltonian $\mathcal{H}$ restricted to M_c. This fact was simultaneously noticed in [193] and, for $c_1 = 0$, in [132]. For this case the construction given above is especially obvious. Put $m = e \times p$. If $(m, p) = 0$ and $(p, p) > 0$, then the vector e exists and is unique up to addition of any vector parallel to p. Let $K(p, e) = H(e \times p)$. We claim that the functions $e(t)$ and $p(t)$ satisfy the canonical equations

$$\dot{e} = -\frac{\partial K}{\partial p} , \quad \dot{p} = \frac{\partial K}{\partial e} .$$

To prove this we calculate

$$\dot{p} = \frac{\partial K}{\partial e} = \frac{\partial H}{\partial m}\frac{\partial m}{\partial e} = p \times \omega .$$

Since $m = e \times p$, we have

$$\dot{m} = \dot{e} \times p + e \times \dot{p} = -\frac{\partial K}{\partial p} \times p + e \times (p \times \omega) ,$$

$$\frac{\partial K}{\partial p} = \frac{\partial H}{\partial p} + \frac{\partial m}{\partial p} = \frac{\partial H}{\partial p} + \omega \times e .$$

Therefore,

$$\begin{aligned} \dot{m} &= -\frac{\partial H}{\partial p} \times p + e \times (\omega \times p) + e \times (p \times \omega) \\ &= p \times \frac{\partial H}{\partial p} + (e \times p) \times \omega = m \times \frac{\partial H}{\partial m} + p \times \frac{\partial H}{\partial p} , \end{aligned}$$

which was to be proved. This "formal" result completes the calculations in §3.6.

3.8 It is known that a rotating "neutral" ferromagnet becomes magnetized along the rotation axis (Barnett's effect in quantum mechanics). The magnetic moment B of the body is related to the angular velocity ω by $B = \Lambda\omega$, where Λ is a symmetric operator. An analogous phenomenon takes place during rotation of a rigid superconductor (the London effect). If the body rotates in a homogeneous magnetic field with strength H, then a magnetic force with torque $B \times H$ acts on the body. Using the angular momentum theorem we write the equations of motion in the form of the Euler–Poisson equations

$$I\dot{\omega} + \omega \times I\omega = (\Lambda\omega) \times H , \quad \dot{H} + \omega \times H = 0 . \qquad (3.17)$$

Here we ignore the gyromagnetic Einstein–de Haas effect (dual to Barnett's effect) which results in spinning the ferromagnet about its axis when it is magnetized. The complete theory of rotation of a rigid body in a magnetic field is given in [49]. However, if $\Lambda = \lambda E$, , $\lambda = \text{const}$, then equations (3.17) are exact. In this important special case they can be rewritten in the more convenient form:

$$I\dot{\omega} + \omega \times I\omega = \varepsilon(\omega \times \gamma)\,, \qquad \dot{\gamma} + \omega \times \gamma = 0\,, \tag{3.18}$$

where $\gamma = H/|H|$, $\varepsilon = \lambda|H|$.

Since $(\omega \times \gamma, \omega) = 0$, the magnetic forces do not produce work and, therefore, they are gyroscopic. By the usual rule we introduce the 2-form Γ of gyroscopic forces by putting $\Gamma(\omega_1, \omega_2) = \varepsilon(\omega_1 \times \gamma, \omega_2) = -\varepsilon(\gamma, \omega_1 \times \omega_2)$. One can check that the form Γ is exact: $\Gamma = d\varphi$, where the 1-form φ is defined by the relation $\varphi(\omega) = -\varepsilon(\gamma, \omega)$. This implies that equations (3.18) can be represented as the Lagrange equations with a globally defined Lagrangian ([138], compare with remarks in §1.7). Indeed, by introducing the function

$$L = (I\omega, \omega)/2 + \lambda(\omega, \gamma)\,,$$

one can represent Eqs. (3.18) in the form

$$\frac{d}{dt}\frac{\partial L}{\partial \omega} + \omega \times \frac{\partial L}{\partial \omega} = \frac{\partial L}{\partial \gamma} \times \gamma\,, \quad \dot{\gamma} + \omega \times \gamma = 0\,.$$

These equations can be regarded as the Poincaré equations on the group $SO(3)$ with the Lagrangian L. To represent these in Hamiltonian form we put

$$m = \frac{\partial L}{\partial \omega} = I\omega + \lambda\gamma$$

and introduce the Hamilton function

$$H(m, \gamma) = (m, \omega) - L(\omega, \gamma)|_{\omega \to m}\,.$$

Written in terms of the variables m, γ, equations (3.18) take the form of the Kirchhoff equations

$$\dot{m} = m \times \frac{\partial H}{\partial m} + p \times \frac{\partial H}{\partial \gamma}\,, \quad \dot{\gamma} = \gamma \times \frac{\partial H}{\partial m}\,, \tag{3.19}$$

with the Hamilton function

$$\begin{aligned} H &= \frac{1}{2}(I^{-1}m, m) - \lambda(I^{-1}m, \gamma) + \frac{\lambda^2}{2}(I^{-1}\gamma, \gamma) \\ &= \frac{1}{2}(I^{-1}(m - \lambda\gamma), (m - \lambda\gamma))\,. \end{aligned} \tag{3.20}$$

This function is nonnegative, but not positive definite. In hydrodynamics, equations (3.19) with the Hamiltonian (3.20) describe the motion of a massless rigid body with the screw symmetry in a boundless volume of an ideal fluid.

3.9 Consider the motion of a rigid body with a fixed point carrying a symmetric rotor whose axis is fixed in the body. Following Kelvin, this dynamical system is called a gyrostat. It has four degrees of freedom and its configuration space is the direct product $SO(3) \times S^1$. The angular momentum of the rotor with respect to the body is constant; we denote it by λ. The total angular momentum of the system with respect to the fixed point equals $m+\lambda = I\omega+\lambda$. If there are no external forces acting on the system, then the angular velocity vector ω satisfies the generalized Euler equation

$$I\dot{\omega} + \omega \times (I\omega + \lambda) = 0 . \qquad (3.21)$$

This was first obtained by Zhukovsky (1885) for the rotation of a rigid body with cavities filled with an ideal incompressible fluid. Later (1895) this equation was integrated in terms of elliptic functions by Vito Volterra in his paper devoted to the theory of evolution of the Earth poles. Equation (3.21) is the Poincaré equation on the algebra $so(3)$ with the Lagrangian

$$T = (I\omega, \omega)/2 + (\lambda, \omega) .$$

Note that the equations of motion of a rigid body carrying a non-symmetric rotor do not have a simple group structure. The Hamiltonian formalism of this complexified problem is given in paper [104].

3.10 In conclusion, consider also Chaplygin's problem of nonholonomic mechanics concerning a dynamically non-symmetric ball rolling without sliding on a horizontal plane. The center of mass of the ball is supposed to coincide with its geometrical center. The dynamics of this problem is described by the following system in $\mathbb{R}^6 = \{\omega, \gamma\}$:

$$m + \omega \times m = 0 , \quad \dot{\gamma} + \omega \times \gamma = 0 , \qquad m = I\omega + ma^2[\gamma \times (\omega \times \gamma)] . \quad (3.22)$$

Here ω, γ, m are respectively the angular velocity, the radius, and the mass of the ball, I is its inertia operator with respect to its center and γ the vertical unit vector. Equations (3.22) possess four integrals

$$(m, \omega) , \quad (m, \gamma) , \quad (m, m) , \quad (\gamma, \gamma) ,$$

and an integral invariant with the density

$$\mathcal{N} = [(ma^2)^{-1} - (\gamma, (I + ma^2E)^{-1}\gamma]^{-1/2} . \qquad (3.23)$$

This enabled Chaplygin to reduce the integration of equations (3.22) to hyperelliptic quadratures (for details see [56]). Some integrable generalizations of Chaplygin's problem are considered in papers [137,170].

4 Pendulum Oscillations

4.1 The motion of a mathematical pendulum of length l in a gravitational force field with acceleration constant g is described by the differential equation

$$\ddot{x} + \omega^2 \sin x = 0 , \quad \omega^2 = g/l ,$$

where x is the angle of deviation of the pendulum from the vertical. If the pendulum energy $h = \dot{x}^2/2 - \omega^2 \cos x$ differs from ω^2, then $\sin x/2$ and $\cos x/2$ are elliptic functions of time. For $h = \omega^2$

$$\sin x/2 = \tanh \omega(t - t_0)\,, \qquad \tanh = \sinh / \cosh\,. \tag{4.1}$$

This formula defines the doubly-asymptotic motion of the pendulum towards the upper equilibrium position. To such motions there correspond the motions along the separatrices on the phase portrait.

4.2 Let the suspension point of a mathematical pendulum of length l perform vertical oscillations according to the periodic law $\varepsilon\xi(t)$, $\varepsilon = \text{const}$. By x we denote the angle of deviation of the pendulum from the vertical. The kinetic energy equals

$$T = v^2/2 = (l^2\dot{x}^2 + \varepsilon^2\dot{\xi}^2 + 2\varepsilon l\dot{x}\dot{\xi}\sin x)/2\,,$$

while the potential energy of the pendulum is

$$V = -g\big(l\cos x + \varepsilon\xi(t)\big)\,.$$

The Lagrange equation

$$\frac{d}{dt}\frac{\partial L}{\partial \dot{x}} = \frac{\partial L}{\partial x}\,, \qquad L = T - V$$

has the following form

$$\ddot{x} + \omega^2\big(1 + \varepsilon f(t)\big)\sin x = 0\,, \tag{4.2}$$

where $\omega^2 = g/l$ and $f = \ddot{\xi}/g$ are periodic functions of time.

This equation is obviously Hamiltonian: the canonical coordinates are $x \bmod 2\pi$, $p = \dot{x}$, while the Hamilton function is

$$H = p^2/2 - \omega^2(1 + \varepsilon f)\cos x\,. \tag{4.3}$$

The configuration space is a circle $S^1 = \{x \bmod 2\pi\}$, and the phase space is a cylinder $S^1 \times \mathbb{R}$.

For $\varepsilon = 0$ we have an integrable problem with one degree of freedom (the mathematical pendulum with constant length).

4.3 There exist equations similar to (4.2) in many problems of mechanics. For example, consider the plane oscillations of a satellite in an elliptic orbit. The equation of oscillations can be written in the following form ([20])

$$(1 + e\cos\nu)\frac{d^2\delta}{d\nu^2} - 2e\sin\nu\frac{d\delta}{d\nu} + \mu\sin\delta = 4e\sin\nu\,. \tag{4.4}$$

Here e is the orbit eccentricity, and μ is a parameter related to the mass distribution of the satellite. The meaning of the variables δ and ν is explained in Fig. 3. This equation can be represented in the Hamiltonian form ([47])

$$\frac{dp}{d\nu} = -\frac{\partial H}{\partial \delta}\,, \qquad \frac{d\delta}{d\nu} = \frac{\partial H}{\partial p}\,,$$

$$H = \frac{1}{2}\Big[\frac{p}{1 + e\cos\nu} - 2(1 + e\cos\nu)\Big]^2 - (1 + e\cos\nu)\mu\cos\delta\,.$$

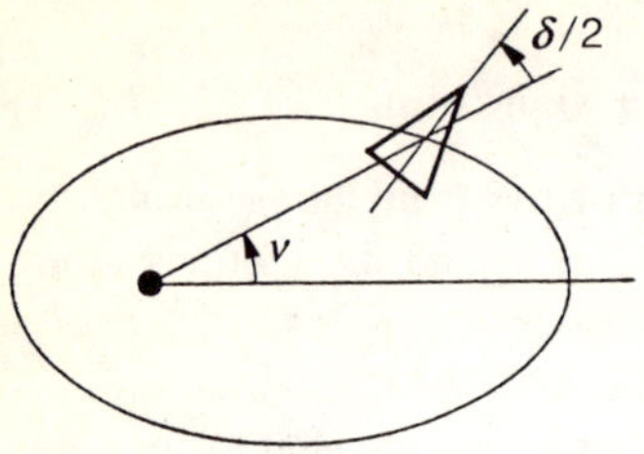

Fig. 3.

For almost circular motion ($e \ll 1$) Eq. (4.4) is close to the oscillation equation of the ordinary pendulum.

4.4 The one-dimensional motion of a charged particle in the field of a wave packet is described by the equation

$$m\ddot{x} = -e\sum_k E_k \sin(\lambda_k x - \omega_k t) . \tag{4.5}$$

Here m, e are the mass and charge of the particle; the sum on the right-hand side of this relation represents a superposition of several plane waves moving with different phase velocities ω_k. This problem has been discussed in the physical literature many times (see, for example, [163, 241]).

In the case of a single wave the change $z = \lambda x - \omega t$ transforms Eq. (4.5) to the equation of ordinary oscillations of a pendulum: $\ddot{z} + \Omega^2 \sin z = 0$, $\Omega^2 = eE\lambda/m$. The change $x \to z$ is equivalent to the transition to the reference frame moving along with the wave.

If there are two waves ($k = 0$ and $k = 1$), then equation (4.5) can be represented in the following form

$$\ddot{z} + \Omega_0^2 \sin z = -\varepsilon\Omega_0^2 \sin\left(\frac{\lambda_1}{\lambda_0} z - \nu t\right) , \tag{4.6}$$

where $z = \lambda_0 x - \omega_0 t$, $\Omega_0^2 = eE_0\lambda_0/m$, $\varepsilon = E_1/E_0$, $\nu = \omega_1 - \lambda_1\omega_0/\lambda_0$. For small values of the dimensionless parameter ε equation (4.6) describes the perturbed motion of the mathematical pendulum.

4.5 Now we write Eqs. (3.5) describing the rotation of a rigid body with a fixed point in the explicit form:

$$\begin{aligned} I_1\dot{\omega}_1 &= (I_2 - I_3)\omega_2\omega_3 - \varepsilon(r_2\gamma_3 - r_3\gamma_2) , \\ I_2\dot{\omega}_2 &= (I_3 - I_1)\omega_3\omega_1 - \varepsilon(r_3\gamma_1 - r_1\gamma_3) , \\ I_1\dot{\omega}_3 &= (I_1 - I_2)\omega_1\omega_2 - \varepsilon(r_1\gamma_2 - r_2\gamma_1) , \end{aligned} \tag{4.7}$$

$$\dot{\gamma}_1 = \omega_3\gamma_2 - \omega_2\gamma_3 , \quad \dot{\gamma}_2 = \omega_1\gamma_3 - \omega_3\gamma_1 , \quad \dot{\gamma}_3 = \omega_2\gamma_1 - \omega_1\gamma_2 . \tag{4.8}$$

Equations (4.7) and (4.8) have
(1) the energy integral $\sum I_i\omega_i^2/2 + \varepsilon \sum r_i\gamma_i$,
(2) the area integral $\sum I_i\omega_i\gamma_i$,
(3) the geometric relation $\sum \gamma_i^2 = 1$.

In the case $I_1 = I_2$, without loss of generality, one may put $r_1 = 0$. In appropriate units of length and mass $I_1 = I_2 = 1$. Let us study the motion of the body with $I_3 = \delta$ and $\varepsilon r_2 = \delta$.

First, to show that such a body does exist, we consider three mutually orthogonal axes x, y, z and put on the x axis two equal masses $\delta/4$ at the points $x = \pm 1$, $y = z = 0$. In addition, two masses $(1/2 - \delta/4)$ are located in a similar way on the z axis, and two masses $\delta(1 \pm 1/g)$ $(g > 1)$ on the y axis at the points $y = \pm 1/2$. It is easy to verify that all the conditions indicated above are valid.

Just for the sake of simplicity, consider the case $r_3 = 0$. Then equations (4.7) turn into

$$\dot{\omega}_1 = (1-\delta)\omega_2\omega_3 - \delta\gamma_3\,, \quad \dot{\omega}_2 = (\delta - 1)\omega_1\omega_3\,, \quad \dot{\omega}_3 = \gamma_1\,. \tag{4.9}$$

Let δ tend to zero. Then Eqs. (4.9) transform to

$$\dot{\omega}_1 = \omega_2\omega_3\,, \quad \dot{\omega}_2 = -\omega_1\omega_3\,, \quad \dot{\omega}_3 = \gamma_1\,. \tag{4.10}$$

These equations together with (4.8) form a closed system defining the restricted problem of the rotation of a heavy rigid body with a fixed point.

The meaning of this restricted problem is the following. As $\delta \to 0$, the rigid body degenerates into a line segment rotating around the fixed point according to the spherical pendulum law. The well-known picture of spherical pendulum motion gives us a clear idea of the nutation and the precession phenomena in rigid body dynamics. One might expect that for $\delta = 0$ the problem of pure rotations becomes meaningless. However, this is not true, since as $\delta \to 0$, the moment of inertia and the moment of the gravitational force relative to the dynamical symmetry axis tend to zero simultaneously. As a result, we obtain a nontrivial equation describing the pure rotation of the body (see below). Note that the transition to the restricted problem in rigid body dynamics is completely analogous to that to the restricted three body problem in celestial mechanics.

As a limit case, from integrals (1–3) of the original problem we find the following integrals of the system (4.8, 4.10)

$$\omega_1^2 + \omega_2^2 = 2h\,, \quad \omega_1\gamma_1 + \omega_2\gamma_2 = c\,, \quad \gamma_1^2 + \gamma_2^2 + \gamma_3^2 = 1\,. \tag{4.11}$$

For $2h > c^2$, relations (4.11) define a three-dimensional integral manifold $M_{h,c}$ in the 6-dimensional phase space of the system (4.8, 4.10). Put

$$\omega_1 = \sqrt{2h}\sin\xi\,, \quad \omega_2 = \sqrt{2h}\cos\xi\,.$$

The variables ξ, $\dot{\xi} = \omega_3$, and γ_3 are coordinates on $M_{h,c}$. It is easy to show that the coordinate ξ satisfies the equation

$$\ddot{\xi} = \frac{c}{\sqrt{2h}} \sin\xi - \sqrt{1 - \frac{c^2}{2h}} \sin(\sqrt{2h}\, t) \cos\xi \,. \tag{4.12}$$

Now represent it in Hamiltonian form:

$$\begin{gathered} \dot{\xi} = H'_\eta \,, \qquad \dot{\eta} = -H'_\xi \,, \\ H = \frac{\eta^2}{2} + \frac{c}{\sqrt{2h}} \cos\xi + \sqrt{1 - \frac{c^2}{2h}} \sin(\sqrt{2h}\, t) \sin\xi \,. \end{gathered} \tag{4.13}$$

We put $(1 - c^2/2h)^{1/2} = \nu$ and regard ν as a small parameter. Note that Eqs. (4.13) make sense also for the case $\nu = 0$, when the degeneration of the manifold $M_{h,c}$ takes place. The Hamiltonian of system (4.13) is given by the expression

$$\begin{gathered} H = H_0 + \nu H_1 + o(\nu) \,, \\ H_0 = \eta^2/2 + \cos\xi \,, \qquad H_1 = \sin\xi \sin(\sqrt{2h}\, t) \,. \end{gathered} \tag{4.14}$$

For $\nu = 0$, we come back to the integrable problem—the mathematical pendulum.

4.6 Consider also the Kirchhoff problem on the motion of a rigid body in an ideal fluid under the condition that the "mixed" terms in the Hamiltonian (3.16) are absent: $B = 0$. We can put

$$2L = (I\omega, \omega) + (\mathcal{J}^{-1} v, v) \,,$$

where $I, \mathcal{J}$ are positive definite symmetric operators, and introduce the variables $\omega, p = \mathcal{J}^{-1} v$. Then (3.14) take the following form

$$I\dot{\omega} = I\omega \times \omega + p \times \mathcal{J} p \,, \quad \dot{p} = p \times \omega \,. \tag{4.15}$$

These can be interpreted as the Euler–Poinsot equations governing rigid body motion in a force field with potential $(\mathcal{J} p, p)/2$. This observation is equivalent to Steklov's analogy (see §2.7).

Suppose that in some orthonormal basis the matrices I and $\mathcal{J}$ have diagonal form: $I = \mathrm{diag}(I_1, I_2, I_3)$, $\mathcal{J} = \mathrm{diag}(\mathcal{J}_1, \mathcal{J}_2, \mathcal{J}_3)$. Following §4.5, we turn to a restricted formulation of the problem. Fix the parameters $I_1 = I_2, \mathcal{J}_3$ and replace $I_3, \mathcal{J}_1, \mathcal{J}_2$ by $\delta I_3, \delta\mathcal{J}_1, \delta\mathcal{J}_2$:

$$\omega_1^2 + \omega_2^2 + \alpha p_3^2 = h \,, \quad \omega_1 p_1 + \omega_2 p_2 = c \,, \quad p_1^2 + p_2^2 + p_3^2 = p^2 \,.$$

Here $\alpha = \mathcal{J}_3/I_1, h, c, p = \text{const}$. Using the equations of motion, it is easy to derive the equations describing the evolution of the doubled "pure rotation" angle $\varphi = 2\arctan(p_1/p_2)$:

$$\ddot{\varphi} + \Lambda(p^2 - u^2)\sin\varphi = \left(\frac{cu}{p^2 - u^2}\right)^{\cdot} \,, \qquad \Lambda = \frac{\mathcal{J}_1 - \mathcal{J}_2}{I_3} \,, \tag{4.16}$$

$$\dot{u}^2 = (h - \alpha u^2)(p^2 - u^2) - c^2 \,. \tag{4.17}$$

In the general case the variable u is an elliptic function of time, and equations (4.16) have a tedious form. Put $hp^2 = c^2 + \nu^2$ and suppose that ν is small. Then the polynomial on the right-hand side of (4.17) has two roots, which are close to zero, and it takes positive values between them. In this case

$$u(t) = \nu u_0(t) + o(\nu)\ , \qquad u_0 = -\cos\sqrt{h - \alpha p^2}(t - t_0)\ , \quad t_0 = \text{const}\ .$$

For small ν, equation (4.16) can be represented in the form

$$\ddot{\varphi} + \Lambda p^2 \sin\varphi = \frac{c\nu}{p^2\sqrt{h - \alpha p^2}} \sin[\sqrt{h - \alpha p^2}(t - t_0)] + o(\nu)\ . \tag{4.18}$$

This equation describes pendulum oscillations under the action of a small periodic external force.

5 Some Problems of Celestial Mechanics

5.1 Consider the motion of a mass point in a gravitational field of n fixed centers. To write the Hamiltonian of this system with two degrees of freedom, it is convenient to use complex numbers. Let $z_1, \dots, z_n$ be different points in the complex plane $\mathbb{C}$. Then the Hamilton function of this problem takes the form

$$H(p, z) = |p|^2/2 + V(z)\ , \qquad p \in \mathbb{C}\ , \quad z \in \mathbb{C} \setminus \{z_1, \dots, z_n\}\ , \tag{5.1}$$

where

$$V(z) = -\sum_{k=1}^{n} \mu_k |z - z_k|^{-1}\ , \qquad \mu_k > 0$$

is the gravitational potential.

For $n = 1$ and $n = 2$ we have the integrable problem studied by Kepler and Euler. In the Kepler problem the additional integral is momentum, while the Euler problem can be integrated by separation of variables. Besides, the Kepler problem is integrable in multidimensional Euclidean space [183]. The motion in Minkowski space, which is interesting in relativistic mechanics, is considered in [152]. The complete integrability of the multidimensional analog of the two fixed center problem seems not to be mentioned in the literature.

5.2 Suppose that the Sun (S) and Jupiter (J) move along circular orbits around their common mass center. We choose the units of length, time, and mass in such a way that the angular velocity of the bodies and the sum of their masses equal one. Then, obviously, the distance between S and J equals 1 as well.

The equations of motion of an asteroid A in the moving frame have the form

$$\ddot{x} - 2\dot{y} = -\frac{\partial V}{\partial x}\ , \quad \ddot{y} + 2\dot{x} = -\frac{\partial V}{\partial y}\ , \tag{5.2}$$

where $-V = (x^2 + y^2) + (1 - \mu)/p_1 + \mu/p_2$, μ is the mass of the asteroid, and p_1, p_2 are distances from A to S and J. Equations (5.2) possess the Jacobi integral

$$H = (\dot{x}^2 + \dot{y}^2)/2 + V(x, y) .$$

These equations can be represented in canonical form with the total energy of the asteroid A as the Hamiltonian. The system (5.2) is known to have five equilibrium positions L_1–L_5 which are called libration points.The equilibria L_1–L_3 lying on the axis S–J were discovered by Euler. They are always unstable. The other two equilibrium positions (discovered by Lagrange) together with the points S and J are the vertices of equilateral triangles. These equilibria are stable in the linear approximation provided the condition $\mu(1 - \mu) < 1/27$ is fulfilled. The problem of their stability in the sense of Lyapunov's definition turned out to be more complicated. Using Kolmogorov's theorem on conservation of quasi-periodic motions, some authors established that the triangular libration points are stable for all μ (satisfying the stability condition in the linear approximation) except two values $\mu_1 = 0.0242938$ and $\mu_2 = 0.0113560\ldots$. For these values the frequencies of the linear oscillations are connected by the resonance relations $1 : 2$ and $1 : 3$. The stability for these exceptional values (in Lyapunov's sense) was proved by Markeev [169].

We can also consider the two-dimensional circular restricted n body problem: $n - 1$ bodies perform a uniform circular motion in one and the same plane around their common mass center, while the last small-mass body moves in this plane in their gravitational field. The complete description of libration points in the restricted n body problem for $n > 3$ is still absent.

The equations of motion in the restricted n body problem turn out to be a Hamiltonian system with gyroscopic forces(in the sense of the definition in §1.7) such that the form of the gyroscopic forces coincides with the customary area 2-form $dx \wedge dy$.

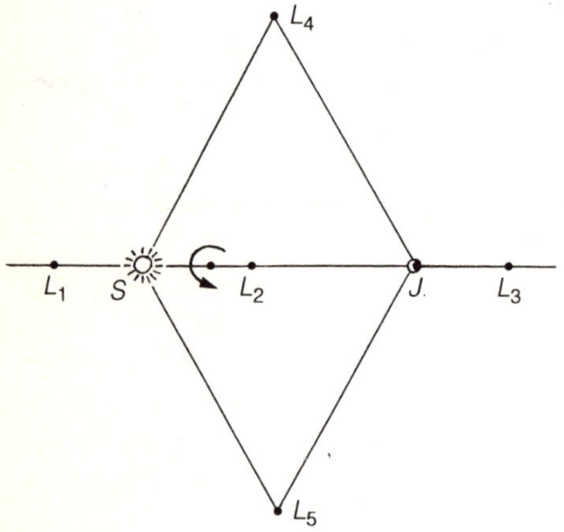

Fig. 4.

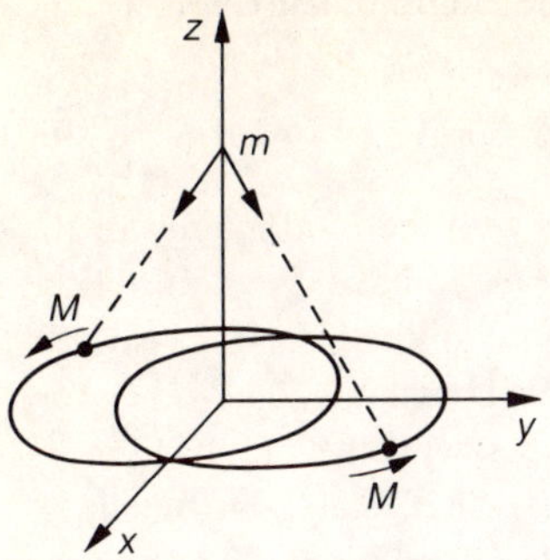

Fig. 5.

5.3 Consider also a version of the restricted three body problem, in which two points move in the x, y plane along elliptic orbits symmetric with respect to the z axis, while the third massless point remains on this axis (see Fig. 5). The motion of the latter point is described by the equation

$$\ddot{z} = -z[z^2 + r^2(t)]^{-3/2} , \tag{5.3}$$

where the distance r from the massive point to the z axis can be found from the relation

$$r = \frac{p}{1 + e\cos\varphi} , \qquad p, e = \text{const} ,$$

and the angle φ is determined by the well-known Kepler formula

$$\tan\frac{\varphi}{2} = \sqrt{\frac{1+e}{1-e}}\tan\frac{u}{2} , \qquad u - e\sin u = nt , \quad n = p^{-3/2} .$$

For $e = 0$ Eq. (5.3) becomes autonomous and, therefore, can be easily integrated by quadratures.

6 Systems of Interacting Particles

6.1 Consider the one-dimensional motion of n particles with unit masses whose pairwise interaction is defined by a potential f. Let $x_1, \ldots, x_n$ be the coordinates of the particles. The equations of motion are Hamiltonian with the Hamilton function

$$H = \frac{1}{2}\sum_{i=1}^{n} y_i^2 + \sum_{i<j} f(x_i - x_j) . \tag{6.1}$$

Usually, the potential f is supposed to be an even function. We may also consider one-dimensional lattices (chains) with particles interacting only with adjacent ones.

The dynamics of a "closed" lattice is governed by the Hamiltonian

$$H = \frac{1}{2}\sum_{i=1}^{n} y_i^2 + \sum_{i=1}^{n} f(x_i - x_{i+1}) , \qquad x_{n+1} = x_n , \tag{6.2}$$

while for an "open" lattice

$$H = \frac{1}{2}\sum_{i=1}^{n} y_i^2 + \sum_{i=1}^{n-1} f(x_i - x_{i+1}) . \tag{6.3}$$

In these cases the potential f is not necessarily even.

Apart from the energy integral, systems with the Hamiltonians (6.1–6.3) possess the momentum integral $P = \sum y_i$. For this reason, they are completely integrable for $n \leq 2$. For $n = 3$ the systems defined by (6.1) and (6.2) coincide.

Below we indicate the most popular examples of systems of interacting particles.

6.2 The dynamics of a chain of particles consecutively connected by elastic strings with elasticity coefficient κ is described by the Hamilton equations

$$\dot{x}_i = \frac{\partial H}{\partial y_i} , \quad \dot{y}_i = -\frac{\partial H}{\partial x_i} ; \qquad i = 1, \dots, n \tag{6.4}$$

with the Hamilton function (6.3), where the potential is given by the well-known Hook's law

$$f(z) = \frac{\kappa}{2} z^2 .$$

In this case Eqs. (6.4) are linear and therefore can be easily integrated. The normal oscillation modes are mutually independent and there is no energy exchange between them.

Fermi, Pasta, and Ulam [77] performed a numerical investigation of the energy exchange between the modes in nonlinear chains with the potentials

$$\frac{\kappa}{2} z^2 + \frac{\kappa\alpha}{3} z^3 , \quad \frac{\kappa}{2} z^2 + \frac{\kappa\alpha'}{4} z^4 , \qquad \alpha, \alpha' = \text{const} . \tag{6.5}$$

In this system one might expect equipartition of energy, which is natural in statistical mechanics. However, this was not confirmed by the investigation: it turned out that only a small part of the energy was exchanged.

6.3 Now let n points be placed on the unit circle and connected consecutively by elastic strings with elasticity coefficient κ. Let z be the angular distance between adjacent points. Then

$$f(z) = 2\kappa \sin^2 \frac{z}{2} = \kappa(1 - \cos z) .$$

Since the potential is defined up to addition of an arbitrary constant, we can put $f = -\kappa \cos$. Let x_i be the angular coordinates of the points. The system is described by equations (6.4) with the Hamiltonian (6.2) for a closed lattice and the potential $f = -\kappa \cos$.

6.4 Substituting the latter expression in (6.1), we arrive at the Grosse–Nevier system which is well-known in theoretical physics. The corresponding stationary Schrödinger equation was solved for some values of the energy.

6.5 Now put in (6.1)

$$f(z) = a/|z|^\alpha \; ; \qquad a, \alpha = \text{const} .$$

The case $\alpha = 1$, $a < 0$ corresponds to a gravitational interaction.

Let $I = \sum x_i^2$ be the moment of inertia of the system with respect to the point $x = 0$. Then

$$\ddot{I} = 2(\sum x_i \dot{x}_i)^\cdot = 2\sum x_i \ddot{x}_i + 2\dot{x}_i^2 = 4T - 2\sum x_i \frac{\partial V}{\partial x_i} . \tag{6.6}$$

Since V is a homogeneous form of degree $-\alpha$, by Euler's formula

$$\sum x_i \frac{\partial V}{\partial x_i} = -\alpha V .$$

Putting $h = T + V$, from (6.6) we obtain Lagrange's formula

$$\ddot{I} = 4h + 2(\alpha - 2)V .$$

For $\alpha = 2$ this yields the moment of inertia of the system

$$I(t) = 2ht^2 + \beta + \gamma \; ; \qquad \beta, \gamma = \text{const} . \tag{6.7}$$

Thus, the case $h < 0$ corresponds to collision of particles, whereas for $h > 0$ they go to infinity. These observations were made by Jacobi. The expression (6.7) along with the energy and the momentum integrals enables one to solve the equations of motion for the case of three particles.

The complete integrability of the system with the potential a/z^2 for all n was established by Calogero. Afterwards, Moser [185] found other integrable cases for $f = a/\sin^2 z$ and $f = a/\sinh^2 x$. Applying Moser's techniques, Calogero generalized these results by proving the integrability of the lattice system with the potential $\wp(z)$ [50]. The potentials a/z^2, $a/\sin^2 z$, and $a/\sinh^2 x$ are known to be degenerate cases of Weierstrass's $\wp$-function.

6.6 In 1967 Toda considered the lattice with the potential

$$f(z) = \frac{a}{b} e^{-bz} + az , \qquad ab > 0 \tag{6.8}$$

(see [228]). For $a, b > 0$ this function corresponds to strong repulsion and weak attraction (see Fig.6), which is relevant to the nature of interactions between atoms. For small z, from (6.8) we obtain the expansion

$$f(z) = \frac{a}{b} + \frac{ab}{2} z^2 - \frac{ab^2}{6} z^3 + \cdots .$$

Thus, for small z *the Toda lattice* can be viewed as a nonlinear lattice with the elasticity coefficient $\kappa = ab$ and the nonlinear parameter $\alpha = -b/2$ in (6.5).

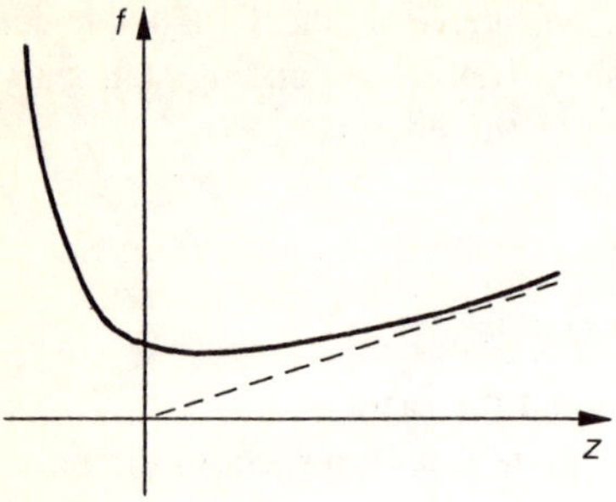

Fig. 6.

For a closed lattice the term az in the potential (6.8) is irrelevant, since

$$\sum_{i=1}^{n}(x_i - x_{i+1}) = 0 , \qquad x_1 = x_{i+1} .$$

For this reason, the Toda lattice is also called the system with exponential interaction.

Bogoyavlensky proposed considering *generalized Toda lattices*.Their dynamics is described by the Hamiltonian

$$H = \frac{1}{2}\sum_{i=1}^{n} y_i^2 + \sum_{k,l=1}^{n+1} b_{k,l} \exp[(a_k, x) + (a_l, x)] . \tag{6.9}$$

Here $a_1, \ldots, a_{n+1}$ are vectors in $\mathbb{R}^n$, and the real coefficients $b_{k,l}$ satisfy the following conditions:

a) for any $y \in \mathbb{R}^n$, $\max_k(a_k, y) > 0$,

b) for all k, the coefficients $b_{k,l}$ are positive.

It is easy to show that after elimination of the motion of the mass center the Hamiltonian of the Toda lattice is reduced to the form (6.9). Hamiltonians of some problems of cosmology have the same form [26].

7 Non-holonomic Systems

As was mentioned in §1.8, in general the equations of motion of non-holonomic systems cannot be reduced to Hamilton equations. However, in several special cases this is possible. Here are some examples.

7.1 Consider a non-holonomic Chaplygin system with two (non-holonomic) degrees of freedom. Here coordinates $q_1, q_2, \ldots, q_n$ can be chosen in such a way that the constraint Eqs. (1.4) take the form

$$\dot{q}_j = a_j\dot{q}_1 + b_j\dot{q}_2 \,, \qquad j = 3, \ldots, n \,, \tag{7.1}$$

and the coefficients a_j, b_j, as well as the Lagrangian $L = T - V$, do not depend explicitly on $q_3, \ldots, q_n$. Chaplygin systems can be defined on a fibrated manifold in invariant geometrical terms.

Equations (1.11) for the coordinates q_1, q_2 are separable, and they can be reduced to the following second order system

$$\begin{gathered} \frac{d}{dt}\frac{\partial L^*}{\partial \dot{q}_1} - \frac{\partial L^*}{\partial q_1} = \dot{q}_2 S \,, \qquad \frac{d}{dt}\frac{\partial L^*}{\partial \dot{q}_2} - \frac{\partial L^*}{\partial q_2} = -\dot{q}_1 S \,, \\ S = \sum_{j=3}^{n} \frac{\partial L}{\partial \dot{q}_j}\left(\frac{\partial a_j}{\partial q_2} - \frac{\partial b_j}{\partial q_1}\right) . \end{gathered} \tag{7.2}$$

Here L^* is a function of $q_1, q_2, \dot{q}_1, \dot{q}_2$ obtained from L by the substitution (7.1); the arguments of S also must be expressed in terms of q_1, q_2.

Chaplygin [55] proved that if the system (7.2) possesses an integral invariant with density $\mathcal{N}(q_1, q_2)$, then the solutions $q_1(t)$, $q_2(t)$ are extremals of the following variational problem

$$\delta \int_{t_1}^{t_2} L^* \, dt = 0 \,, \qquad \int_{t_1}^{t_2} \mathcal{N} \, dt = \text{const} \,. \tag{7.3}$$

If $\mathcal{N} > 0$, then after the change of time

$$d\tau = \mathcal{N} \, dt \tag{7.4}$$

Eqs. (7.2) are reduced to canonical Hamilton equations. To obtain the corresponding Hamiltonian, we must make the change in the expression for L^*, then perform the customary Legendre transformation.

The conditions for the existence of integral invariants of smooth dynamical systems were studied in [140].

7.2 The problem of rolling a dynamically non-symmetric ball on a horizontal plane, which was mentioned in §3.10, is a Chaplygin system with three degrees of freedom. By fixing the value of the area integral and eliminating rotations of the ball about a vertical axis through the contact point, we reduce the number of degrees of freedom to two. One can show that if the value of the area integral equals zero, then the reduced system has the form (7.2). By using the integral invariant (3.23) and a time change this system can be transformed to Hamilton equations.

7.3 Following Suslov [222, Chap.53], consider the motion of a rigid body around a fixed point with the non-holonomic constraint $(a, \omega) = 0$, where the vector a is fixed in the frame attached to the body. Let the body rotate in a homogeneous force field with the potential $V = (b, \gamma)$, b =const. The motion equations (1.11) can be represented as the Euler–Poincaré equations on the Lie algebra $so(3)$:

$$I\dot{\omega} + \omega \times I\omega = \gamma \times b\ , \quad \dot{\gamma} + \omega \times \gamma = 0\ , \qquad (a, \omega) = 0\ . \tag{7.5}$$

It is easy to show that if the vector a is an eigenvector for the inertia operator, then the phase flow of the system (7.5) conserves the standard measure in $\mathbb{R}^6 = \{\omega, \gamma\}$. As noticed in [44], if this condition is not fulfilled, then for $b = 0$ this system does not possess even an absolutely continuous (with respect to the Lebesgue measure in $\mathbb{R}^6 = \{\omega, \gamma\}$) invariant measure. Therefore, we suppose that the vector a is generally directed along one of the principal inertia axes; without loss of generality we put $a = (0, 0, 1)$.

Equations (7.5) were solved in [109] for the case when a and b are orthogonal. Now consider the case $b = \varepsilon a$, $\varepsilon \neq 0$. Then, taking into account the constraint $\omega_3 = 0$, we represent the first two dynamical Eqs. (7.5) in the form

$$I_1\dot{\omega}_1 = \varepsilon\gamma_2\ , \quad I_2\dot{\omega}_2 = -\varepsilon\gamma_1\ .$$

It follows that

$$I_1\ddot{\omega}_1 = \varepsilon\dot{\gamma}_2\ , \quad I_2\ddot{\omega}_2 = -\varepsilon\dot{\gamma}_1\ .$$

Then, by virtue of the Poisson equations $\dot{\gamma}_1 = -\omega_2\gamma_3$, $\dot{\gamma}_2 = \omega_1\gamma_3$, we obtain

$$I_1\ddot{\omega}_1 = \varepsilon\gamma_3\omega_1\ , \quad I_2\ddot{\omega}_2 = -\varepsilon\gamma_3\omega_2\ . \tag{7.6}$$

Using the energy integral

$$(I_1\omega_1^2 + I_2\omega_2^2)/2 + \varepsilon\gamma_3 = h\ ,$$

we can express γ_3 in terms of ω_1, ω_2. Then the equations (7.6) can be rewritten in the form of the Lagrange equations

$$I_1^2\ddot{\omega}_i = \frac{\partial V}{\partial \omega_i} \quad (i = 1, 2)\ , \qquad V = \frac{1}{2}\left(h - \frac{I_1\omega_1^2 + I_2\omega_2^2}{2}\right)^2 .$$

These equations have the energy integral

$$T + V = \text{const}\ , \qquad T = \frac{1}{2}(I_1\omega_1^2 + I_2\omega_2^2)\ ,$$

whose value equals $\varepsilon^2/2$ for actual motions.

The change $I_i\omega_i = m_i$ reduces the problem to that of the motion of a point mass in a potential force field:

$$\ddot{m}_i = \frac{\partial V}{\partial m_i} \quad (i = 1, 2)\ , \qquad V = \frac{1}{2}\left(h - \frac{I_1^{-1}m_1^2 + I_2^{-1}m_2^2}{2}\right)^2 . \tag{7.7}$$

For $I_1 = I_2$ we have a point mass motion in a central force field. In this case the equations are solved in terms of elliptic functions of time.

We note that, in contrast to the theory of Chaplygin's multiplier, the reduction of Eqs. (7.7) to the Lagrange (or Hamilton) equations does not use a change of time.

However, the coordinates in these equations are the components of the angular velocity or angular momentum of the rigid body.

8 Some Problems of Mathematical Physics

8.1 As is known in hydrodynamics [158], the motion of n point (cylinder) vortices in a plane (space) is governed by the following system of $2n$ differential equations

$$\Gamma_s \dot{x}_s = -\frac{\partial H}{\partial y_s}, \quad \Gamma_s \dot{y}_s = -\frac{\partial H}{\partial x_s}, \qquad 1 \leq s \leq n,$$
$$H = \frac{1}{2\pi} \sum_{s \neq k} \Gamma_s \Gamma_k \ln[(x_s - x_k)^2 + (y_s - y_k)^2]^{1/2}, \tag{8.1}$$

where (x_s, y_s) are the Cartesian coordinates of a vortex, and Γ_s is its intensity. Suppose that all the Γ_s's differ from zero. Equations (8.1) are Hamiltonian with respect to the symplectic structure in $\mathbb{R}^{2n} = \{x, y\}$ given by the Poisson bracket

$$\{f, g\} = \sum_s \frac{1}{\Gamma_s} \left(\frac{\partial f}{\partial y_s} \frac{\partial g}{\partial x_s} - \frac{\partial f}{\partial x_s} \frac{\partial g}{\partial y_s} \right).$$

Apart from the Hamilton function H, they have three independent integrals

$$P_x = \sum \Gamma_s x_s, \quad P_y = \sum \Gamma_s y_s, \qquad M = \sum \Gamma_s (x_s^2 + y_s^2)/2.$$

It is evident that

$$\{P_x, P_y\} = -\sum \Gamma_k = \text{const}, \quad \{P_x, M\} = -P_y, \quad \{P_y, M\} = P_x.$$

The functions P_x, P_y are in involution provided that the sum of the intensities does not vanish.

Equations (8.1) can be reduced to the customary canonical equations by putting $\xi_s = \sqrt{\pm\Gamma_s} x_s$, $\eta_s = \sqrt{\pm\Gamma_s} y_s$, $(s = 1, \ldots, n)$. Here the signs $+$ and $-$ correspond to $\Gamma > 0$ and $\Gamma < 0$ respectively. By K we denote the function H represented in terms of ξ. In the new variables Eqs. (8.1) have the form

$$\dot{\xi}_s = \mp \frac{\partial K}{\partial \eta_s}, \quad \dot{\eta}_s = \pm \frac{\partial K}{\partial \xi_s}, \qquad 1 \leq s \leq n.$$

It turns out that the Hamiltonian system (8.1) can be represented as a gradient dynamical system. Let $(\cdot, \cdot)$ be a Riemannian metric on the variety M, and Φ a function on M. Differential equations $\dot{x} = v(x)$ on M are called gradient (evolutionary) equations if

$$(v, \cdot) = d\Phi(\cdot). \tag{8.2}$$

Gradient systems were investigated by Lyapunov in stability theory. Their structural stability was studied by Smale as well as Thom and his successors in catastrophe theory.

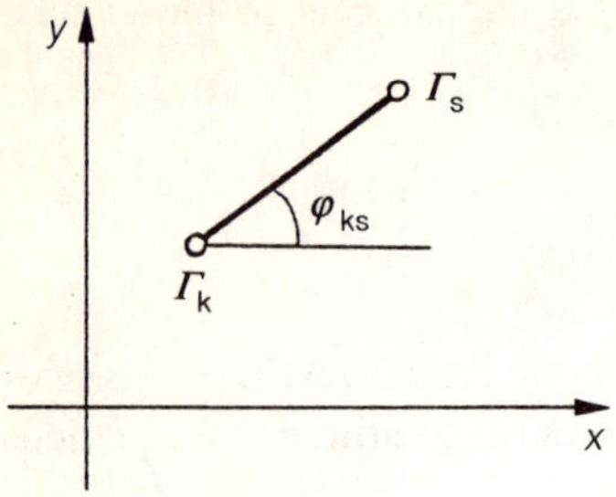

Fig. 7.

Now put

$$\Phi = \frac{1}{2\pi} \sum_{s \neq k} \Gamma_s \Gamma_k \varphi_{sk}\,, \qquad \varphi_{sk} = \arctan \frac{y_s - y_k}{x_s - x_k}\,.$$

Then equations (8.1) can be rewritten in the form of (8.2):

$$\Gamma_s \dot{x}_s = \frac{\partial \Phi}{\partial x_s}\,, \quad \Gamma_s \dot{y}_s = \frac{\partial \Phi}{\partial y_s}\,, \qquad 1 \leq s \leq n\,. \tag{8.3}$$

The Riemannian metric in $\mathbb{R}^{2n} = \{x, y\}$ is defined by thc sum

$$\sum \Gamma_s (dx_s^2 + dy_s^2)\,.$$

It follows from (8.2) that

$$\dot{\Phi} = \left| \frac{\partial \Phi}{\partial x} \right|_*^2 ,$$

where $|\cdot|_*$ is the length of a covector in the dual space. Therefore, if Φ is a single-valued function, then $\Phi(x(t))$ tends either to $+\infty$ or to some constant c for $t \to +\infty$ (when M is compact, c is a critical value of Φ). Thus, for $t \to \pm\infty$ the solution $x(t)$ either tends to infinity or asymptotically approaches the set of critical points of Φ.

In the considered case the function Φ is not single-valued, and therefore the solutions of (8.3) are not asymptotic. However, as t increases, the continuous branch of Φ either goes to infinity or tends monotonically to some constant.

8.2 The two-dimensional (plane) flows of a homogeneous ideal fluid in a potential force field are described by the well-known Euler equations

$$u'_t + u'_x u + u'_y v + f'_x = 0\,, \quad v'_t + v'_x u + v'_y v + f'_y = 0\,. \tag{8.4}$$

Here u, v denote components of the velocity of a fluid particle with coordinates x, y at time t; $f = p/\rho + V$, p is the pressure, ρ the density, V the energy density of the potential field. Equations (8.4) must be considered together with the equation of continuity

$$u'_x + v'_y = 0 \,. \tag{8.5}$$

We seek solutions of the system (8.4), (8.5) in the following form

$$\begin{aligned} u = \Psi'_y \,, \qquad & v = -\Psi'_x + \varepsilon \cos \lambda t \,; \\ \varepsilon, \lambda = \text{const} \,, \quad & f = \xi \sin \lambda t + \eta \cos(\lambda t) + \zeta \,. \end{aligned} \tag{8.6}$$

Here Ψ, ξ, η, ζ are still unknown functions of x, y. The motion of fluid particles is described by the Hamilton equations

$$\dot{x} = H'_y \,, \quad \dot{y} = -H'_x \,; \qquad H = \Psi - \varepsilon x \cos(\lambda t) \,. \tag{8.7}$$

If $\varepsilon = 0$, then the system (8.7) can be easily integrated by quadratures.

The solutions (8.6) turn (8.5) into an identity. Substituting (8.6) into (8.4) and matching the coefficients at $\cos(\lambda t)$, $\sin(\lambda t)$ yields six equations for the determination of four functions Ψ, ξ, η, ζ:

$$\xi'_x = -\lambda + \xi'_y = 0 \,, \tag{8.8}$$

$$\Psi''_{yy} + \eta'_x = -\Psi''_{xy} + \eta'_y = 0 \,, \tag{8.9}$$

$$\Psi''_{xy}\Psi'_y - \Psi''_{yy}\Psi'_x + \zeta'_x = -\Psi''_{xx}\Psi'_y + \Psi''_{xy}\Psi'_x + \zeta'_y = 0 \,. \tag{8.10}$$

It follows from (8.8) that $\xi = \lambda y + \text{const}$. The sufficient condition for the system (8.9), (8.10) to be solvable is the relation

$$\triangle \Psi = \text{const} \,, \tag{8.11}$$

where $\triangle$ is the Laplace operator. To show this we notice that

$$\eta''_{xy} = -\Psi'''_{yyy} \,, \quad \eta''_{yx} = \Psi'''_{xxy} \,.$$

Therefore $\Psi'''_{xxy} + \Psi'''_{yyy} = (\triangle \Phi)'_y = 0$ provided the condition (8.11) is fulfilled. The solvability condition for the system (8.10) can be checked in a similar way.

In particular, if Ψ is a harmonic function, then Euler's equations (8.4) admit the solutions (8.6). For $\varepsilon = 0$ we have a potential flow. There are many examples of stationary flows with harmonic stream functions (see [158]). In the simplest example

$$\Psi = \frac{\Gamma}{2\pi} \ln(x^2 + y^2)^{1/2} \,.$$

This stream function defines vortex with intensity Γ. The vortex pair (with intensities $\Gamma_1 = -\Gamma_2 = \Gamma$) is known to move uniformly in the direction orthogonal to the segment joining the vortex points. The distance between them is constant. Now place these vortices at the points $(0, a)$ and $(0, -a)$ of a moving frame x, y. Then the stream function is defined by

$$\Psi = \frac{\Gamma}{2\pi}\left(\frac{y}{2a} + \ln\left(\frac{x^2 + (y-a)^2}{x^2 + (y+a)^2}\right)^{1/2}\right)$$

(see [158, §155]). In this case the stream lines are presented in Fig. 8.

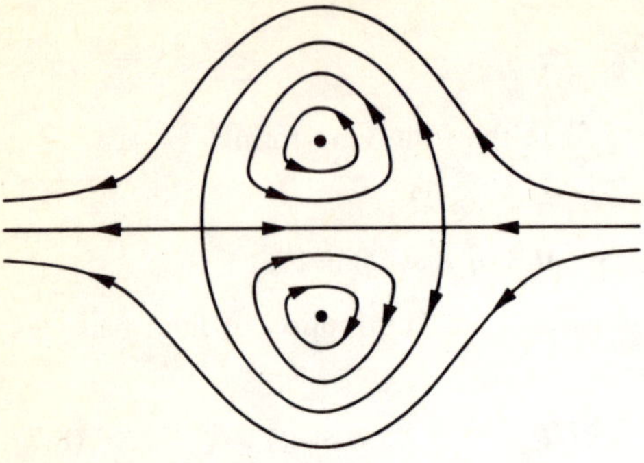

Fig. 8.

8.3 In the paper of Contopoulos [60] dealing with models of galaxies some Hamiltonian systems in a neighborhood of an equilibrium position were considered. These systems were assumed to admit resonance relations between frequencies. The simplest system that has such a property and describes the star motion in a galaxy with cylindrical symmetry is defined by the Hamiltonian

$$H = \frac{1}{2}\left(p_1^2 + p_2^2 + q_1^2 + q_2^2 + 2q_1^2 q_2 - \frac{2}{3}q_2^3 \right) . \tag{8.12}$$

This system was numerically investigated in detail by Hénon and Heiles [99]. In this problem the frequencies of small oscillations are equal. Gustavson's paper [94] duscusses the results of these investigations in connection with the construction of formal integrals of Hamiltonian systems.

The system with the Hamiltonian (8.12) can be obtained in another way by considering the dynamics of a closed three particle lattice with the potential

$$f(z) = \frac{z^2}{2} + \frac{\alpha z^3}{3} \tag{8.13}$$

(compare with (6.5)). To show this, we consider the Hamiltonian (6.3) with the potential (8.13) and perform the canonical change of variables

$$x = A\xi , \qquad y = A\eta ,$$

where A is the orthogonal matrix

$$A = \begin{pmatrix} 1/\sqrt{6} & 1/\sqrt{2} & 1/\sqrt{3} \\ -\sqrt{2/3} & 0 & 1/\sqrt{3} \\ 1/\sqrt{6} & -1/\sqrt{2} & 1/\sqrt{3} \end{pmatrix} .$$

In the new variables ξ, η the Hamiltonian turns into

$$H = \frac{1}{2}(\eta_1^2 + \eta_2^2 + \eta_3^2 + 3\xi_1^2 + 3\xi_2^2) + \frac{3\alpha}{\sqrt{2}}\left(\eta_1^2\eta_2 - \frac{1}{3}\xi_3^2 \right) .$$

By putting $\eta_3 = 0$ (we suppose that the center of mass of the lattice does not move) and making the similarity transformation

$$\xi_1 = \frac{\sqrt{2}}{\alpha} q_1 , \quad \xi_2 = \frac{\sqrt{2}}{\alpha} q_2 , \quad \eta_1 = p_1 , \quad \eta_2 = p_2 ,$$
$$t \to \frac{t}{\sqrt{3}} , \quad H \to \frac{6}{\alpha^2} ,$$

we arrive at the Hamiltonian (8.12).

8.4 The homogeneous double-component classical model of the Yang–Mills equation is connected with the Hamiltonian system defined bythe Hamiltonian

$$H = (p_1^2 + p_2^2)/2 + q_1^2 q_2^2 \tag{8.14}$$

(see [74, 240]). Putting formally $h = 0$, $I_1 = -I_2 = 1/\sqrt{2}$ in (7.7) and making a rotation in the plane (m_1, m_2) by the angle $\pi/4$, we obtain the system with the Hamiltonian (8.14).

9 The Problem of Identification of Hamiltonian Systems

If the differential equations written in some local coordinates on the smooth variety M^{2n} do not have the canonical (Hamiltonian) form, this does not necessarily imply that they are not Hamiltonian, since the local coordinates may not be symplectic. Let us give some examples of dynamical systems whose Hamiltonian properties are not obvious a priori.

9.1 As the first example we consider a linear system with constant coefficients

$$\dot{x} = Ax , \qquad x \in \mathbb{R}^n . \tag{9.1}$$

Suppose that it possesses a first integral $f = (Bx, x)/2$, where B is a non-degenerate symmetric operator, and $\det A \neq 0$. It turns out that
(1) the pair (R^n, Ω), where $\Omega(x', x'') = (BA^{-1}x', x'')$, is a symplectic manifold,
(2) the vector field Ax is Hamiltonian with the Hamilton function f.

Indeed, since f is an integral of equations (9.1), we have $\dot{f} = (Bx, Ax) = (x, BAx) = 0$. Therefore, BA is a skew-symmetric operator. This implies that BA^{-1} is a skew-symmetric operator as well. Since A and B are non-degenerate, the same holds for the external 2-form Ω. Like any 2-form with constant coefficients, this 2-form is closed. It remains to note that

$$\Omega(Ax, \cdot) = (BA^{-1}(Ax), \cdot) = (Bx, \cdot) = df .$$

9.2 Now consider the geometrical Poisson equation of rigid body dynamics

$$\dot{e} = e \times \omega , \tag{9.2}$$

where e and ω are vectors in three-dimensional oriented Euclidean space, and ω is a known function of time. Equation (9.2) possesses the integral $(e, e) = c \geq 0$. Since in rigid body dynamics e is a unit vector, we put $c = 1$. We endow the sphere $S^2 = \{e : (e, e) = 1\}$ with a symplectic structure by putting $\Omega(x', x'') = (e, x' \times x'')$, where x', x'' are vectors tangent to S^2 at the point e. The form Ω, being an oriented area of S^2, is closed and non-degenerate, but it is not exact:

$$\int_{S^2} \Omega = 4\pi \, .$$

The vector field $v = e \times \omega(t)$ is a nonstationary tangent vector field on S^2.

Let us show that Eq. (9.2) is a Hamilton equation on the symplectic manifold (S^2, Ω) with the Hamilton function $H = -(e, \omega(t))$. Indeed, $\Omega(v, \cdot) = (e, (e \times \omega) \times (\cdot)) = (\cdot, e \times (e \times \omega)) = (\cdot, e(\omega, e) - \omega) = -(\cdot, \omega) = dH$.

It is interesting to note that if $e = \xi(t)$ is a solution of Eq. (9.2), then the function $f = (\xi(t), e)$ is its first integral. If ω is a p-periodic function of time, then the mapping for the period of the linear system (9.2) conserves an oriented area of S^2 and, consequently, has at least two distinct fixed points. In this case there exists an integral p-periodic in t.

9.3 Following Birkhoff, we consider also the "generalized Pfaff problem" on stationary curves of the functional

$$P(x(\cdot)) = \int_{-t_1}^{t_2} \left(\sum_{i=1}^{n} u_i \dot{x}_i + B \right) dt \, .$$

Here u_i, B are smooth functions of variables $x_1, \ldots, x_n$, and t (see [23]). It is easy to show that the variational equation $\delta P = 0$ $(\delta x_i(t_1) = \delta x_i(t_2))$ defines the x's as functions of t that satisfy the system of equations

$$\frac{\partial u}{\partial t} + (\operatorname{curl} u)\dot{x} = \frac{\partial B}{\partial x}, \qquad \operatorname{curl} u = \left\| \frac{\partial u_i}{\partial x_j} - \frac{\partial u_j}{\partial x_i} \right\| . \tag{9.3}$$

Indeed, these are the Euler–Lagrange equations

$$\frac{d}{dt}\frac{\partial L}{\partial \dot{x}} = \frac{\partial L}{\partial x}$$

with the Lagrangian $L = u\dot{x} + B$ linear in the velocities.

If $n = 3$, then the action of the operator $\operatorname{curl} u$ on the vector ξ is equivalent to the vector multiplication $\nu \times \xi$ and ν is exactly the curl (rotor) of the vector field u. This explains why we denote the skew-symmetric matrix in (9.3) by $\operatorname{curl} u$ in the multidimensional case.

The matrix $\operatorname{curl} u$ is supposed to be non-degenerate for all values of x and t. Hence, n is even and Eqs. (9.3) define uniquely a nonstationary vector field depending on the variables $x_1, \ldots, x_n$.

If the functions u_j do not depend explicitly on t, then equations (9.3) are obviously Hamiltonian with the Hamilton function B. Here the phase space is

$\mathbb{R}^n = \{x\}$, while the symplectic structure is defined as follows

$$\Omega = d\sum_j u_j\,dx_j = \sum_{i,j} \frac{\partial u_j}{\partial x_i}\,dx_i \wedge dx_j = \sum_{i\leq j} \left(\frac{\partial u_j}{\partial x_i} - \frac{\partial u_i}{\partial x_j}\right) dx_i \wedge dx_j\,.$$

In the general case, when the field u is nonstationary, the form Ω can nevertheless be represented in a stationary form. To show this we consider the system of differential equations

$$\dot{x} = v(x, t)\,,$$

which is defined by (9.3). Let $x(t, z)$ be the solution of this system with the original value $x(0, z) = z$. We regard the mapping $z \to x = x(t, z)$ as a non-autonomous change of variables. Now put

$$u(x, t)\,dx + B(x, t)\,dt = u_*(z, t)\,dz + B_*(z, t)\,dt\,.$$

Since the Euler–Lagrange equations are covariant, in the new variables Eq. (9.3) takes the same form

$$\frac{\partial u_*}{\partial t} + (\operatorname{curl} u_*)\,\dot{z} = \frac{\partial B_*}{\partial z}\,. \tag{9.4}$$

Since $\dot{z} = 0$, we have $\partial u_*/\partial t = \partial B_*/\partial z$. It follows that $\partial \Omega_*/\partial t = 0$, where $\Omega_* = d_z(u_*(z, t)\,dz)$. Thus, in the new variables z the 2-form Ω_* is stationary. According to Poincaré's lemma, there exist locally the covector field $u'(z)$ and the function $S(z, t)$ such that

$$u_*(z, t)\,dz = u'(z)\,dz + d_z S(z, t)\,.$$

As a consequence, Eq. (9.4) takes the following form

$$(\operatorname{curl} u')\,\dot{z} = \frac{\partial B'}{\partial z}\,, \qquad B' = B + \frac{\partial S}{\partial t}\,.$$

This is a Hamiltonian system with the function B as Hamiltonian.

The meaning of the extra term $\partial S/\partial t$ is seen from the following observation: after addition of the differential $dS = \partial S/\partial t\,dt + \partial S/\partial x\,dx$ the variation of the functional $P(x(\cdot))$ does not change.

9.4 In order to represent a dynamical system in Hamiltonian form one has to find two objects: a Hamilton function and an appropriate symplectic structure. It turns out that in an even-dimensional phase space in a small neighborhood of each non-singular point any dynamical system is Hamiltonian. This follows from the fact that the equations of the system written in appropriate local coordinates can be reduced to the form

$$\dot{x}_1 = 1\,, \quad \dot{x}_2 = \cdots = \dot{x}_{2n} = 0\,. \tag{9.5}$$

This fact is known as the *theorem on straightening of trajectories*. The system (9.5) is obviously Hamiltonian: the symplectic structure is defined by the Poisson bracket

$$\{f, g\} = \sum_{i=1}^{n} \left(\frac{\partial f}{\partial x_i} \frac{\partial g}{\partial x_{i+n}} - \frac{\partial f}{\partial x_{i+n}} \frac{\partial g}{\partial x_i} \right),$$

and the Hamilton function $H = x_{n+1}$. This remark was made by Birkhoff [23, Chap. 2].

Thus, the representation of differential equations in Hamiltonian form makes sense either in the neighborhood of an equilibrium position or in a suffuciently large domain of the phase space, where the trajectories enjoy the recursion property (in a neighborhood of a periodic trajectory, for example). Unfortunately, so far this problem has not been studied at all.

II Integration of Hamiltonian Systems

Differential equations and, in particular, the Hamilton equations can be divided into integrable and non-integrable. "However, trying to give an exact definition of integrability, we discover that many different definitions are possible and each of them is interesting from a certain theoretical point of view" (Birkhoff [23]). In this chapter we survey various approaches to the problem of integrability of Hamiltonian systems.

1 Integrals. Classes of Integrals of Hamiltonian Systems

1.1 Recall that a non-constant function $f : M^n \to \mathbb{R}$ is called a *first integral* (or simply integral) of a dynamical system $\dot{x} = v(x)$, $x \in M^n$ if $f(x(t)) \equiv \text{const}$ for all trajectories $x(t)$. If f is differentiable, then this definition can be written as the condition

$$\dot{f} = \frac{\partial f}{\partial x} v \equiv 0 .$$

According to the theorem on straightening of trajectories, in a small neighborhood of any point $x_0 \in M^n$ which is not an equilibrium position, there always exist coordinates $x_1, \ldots, x_n$ in which the differential equations take the simplest form:

$$\dot{x}_1 = 1 , \qquad \dot{x}_2 = \cdots = \dot{x}_n = 0 .$$

Hence, the coordinates $x_2, \ldots, x_n$ form a "complete" set of independent integrals: any integral is a function of $x_2, \ldots, x_n$. Until Poincaré's work mathematicians regarded the problem of solving differential equations as a question of finding explicit expressions for the first integrals only. However, this approach is purely analytical, and its solution is in no way connected with the behavior of phase trajectories. In some cases, one can indicate simple explicit expressions for the local integrals, although in the whole the dynamical system possesses no first integrals at all.

Here is a simple example: let M be an n-dimensional torus with the angular coordinates $(\varphi_1, \ldots, \varphi_n) \bmod 2\pi$. Consider the dynamical system defined by the equations

$$\dot{\varphi}_k = \omega_k = \text{const} , \qquad 1 \leq k \leq n , \tag{1.1}$$

where the frequencies $\omega_1, \ldots, \omega_n$ are incommensurable, i.e., if $\sum k_i \omega_i = 0$ for integer k_i, then all the k_i's must equal zero. Note that almost all points $(\omega_1, \ldots, \omega_n)$

in the space $\mathbb{R}^n$ enjoy this property. Let $f : \mathbb{T}^n \to \mathbb{R}$ be a smooth integral of the system (1.1). Consider its Fourier expansion

$$f = \sum f_k e^{i(k,\varphi)}, \qquad k \in \mathbb{Z}^n$$

and differentiate it with respect to the system (1.1):

$$\dot{f} = \sum f_k i(w, k) e^{i(k,\varphi)} = 0 .$$

Since $(\omega, k) \neq 0$ for all $k \neq 0$, $f \equiv f_0 = \text{const}$. It can be shown that the system (1.1) possesses no non-constant, continuous (or even measurable) integrals. The reason is that every phase trajectory fills the torus $\mathbb{T}^n$ everywhere densely.

On the other hand, for each vector $\omega \neq 0$ there exist $n-1$ linearly independent vectors $a_1, \ldots, a_{n-1} \in \mathbb{R}^n$ orthogonal to ω: $(a_1, \omega) = \cdots = (a_{n-1}, \omega) = 0$. Then the functions

$$f_1 = (a_1, \varphi), \ \ldots, \ f_{n-1} = (a_{n-1}, \varphi)$$

form a "complete" set of independent integrals. However, none of them are single-valued functions on $\mathbb{T}^n$.

1.2 Let $F : M^{2n} \to \mathbb{R}$ be a first integral of the Hamiltonian system $\dot{z} = v_H(z)$. It turns out that if $dF(z_0) \neq 0$, then in a neighborhood of the point $z_0 \in M$ one can choose canonical coordinates $x_1, \ldots, x_n, y_1, \ldots, y_n$ such that $F(x, y) = y_1$[1]. This assertion is a Hamiltonian version of the theorem on straightening of trajectories (the proof can be found, for example, in [224]).

In the coordinates x, y, the function H does not depend on x_1. Therefore, if we put $F = y_1 = \text{const}$, then the system

$$\dot{x}_k = \frac{\partial H}{\partial y_k}, \quad \dot{y}_k = -\frac{\partial H}{\partial x_k} \qquad (k \geq 2)$$

is a Hamiltonian system with $n-1$ degrees of freedom. Thus, the existence of one integral enables us to lower the dimension of the phase space by two (and not by one as in the general case): one decrease is due to fixing the value of F, the second as a result of eliminating the conjugate cyclic variable x_1. However, to use the integral F for lowering the order effectively we must first find an explicit solution of the Hamiltonian system $\dot{z} = v_F(z)$.

This observation can be generalized as follows: if a Hamiltonian system has s independent first integrals, whose pairwise Poisson brackets vanish, then it can be locally reduced to a system with $n-s$ degrees of freedom. Lowering the order of Hamiltonian systems with a non-involutive set of integrals is discussed in the book [14].

In applications one usually deals with analytic Hamiltonian systems: their phase space M^{2n} is endowed with the structure of an analytic manifold, the Poisson

1 In particular, locally (but not in a neighborhood of an equilibrium position) the Hamilton function can always be reduced to the form $H = y_1$.

bracket of any two analytic functions is analytic on M^{2n}, and the Hamiltonian is also an analytic function. In this situation it is natural to consider the existence of integrals which are analytic on M^{2n}. If analytic functions are independent at a point, then the same holds almost everywhere. We denote the class of functions analytic on M^{2n} by $C^\omega(M)$ (or simply C^ω).

However, one should keep in mind that an analytic Hamiltonian system might have integrals of class C^r, but not of class C^{r+1} (the case $r = 0$ is not excluded: we refer to a continuous function as an integral, if it is not constant locally and takes constant values on any trajectory). To show this we consider a non-autonomous Hamiltonian system with one degree of freedom and the Hamilton function

$$H = \alpha y + f(x, t) ,$$

where α is a real parameter and f is an analytic 2π-periodic function of x and t. Since H is periodic in x and t, it is natural to regard the direct product $\mathbb{R} \times \mathbb{T}^2 = \{y; x, t \bmod 2\pi\}$ as the extended phase space.

Now write the Hamilton equations in the explicit form

$$\dot{x} = \alpha , \quad \dot{y} = -\frac{\partial f}{\partial x} = -F(x, t) . \tag{1.2}$$

These equations can obviously be solved by quadratures:

$$x = \alpha t + x_0 , \quad y = y_0 - \int_0^t F(\alpha\tau + x_0, \tau)\, d\tau .$$

We look for a first integral of the system (1.2) in the form $y + g(x, t)$, where $g : \mathbb{T}^2 \to \mathbb{R}$ is a function which satisfies the equation

$$\frac{\partial g}{\partial t} + \alpha \frac{\partial g}{\partial x} = F(x, t) . \tag{1.3}$$

Let

$$F = \sum F_{mn} \exp i(mx + nt) , \quad g = \sum g_{mn} \exp i(mx + nt) .$$

Then

$$g_{mn} = \frac{F_{mn}}{i(m\alpha + n)} .$$

If the function f is analytic, then $|F_{mn}| \le c e^{-\rho(|m|+|n|)}$ for some positive constants c, ρ. On the other hand, as is known from the theory of Diophantine approximations, for almost all α the following estimates (in the sense of the Lebesgue measure) on $\mathbb{R}$ are valid

$$|m\alpha + n| \ge \frac{k}{(|m| + |n|)^\gamma} , \qquad k, \gamma = \text{const} \ge 0 .$$

Hence, the coefficients g_{mn} also diminish exponentially fast as $|m| + |n|$ grows. Therefore, for almost all α the trigonometric series

$$\sum \frac{F_{mn}}{i(m\alpha + n)} e^{i(mx+nt)}$$

converges to an analytic solution of Eq. (1.3). In particular, the Hamilton equations (1.2) admit a single-valued analytic integral.

However, if α can be approximated closely enough by rational numbers, then Eq. (1.3) may have periodic solutions only of finite smoothness, if at all.

To generalize this argument we may indicate an analytic function $f : \mathbb{T}^2 \to \mathbb{R}$ and sets $M_\omega, M_\infty, \dots, M_k, \dots, M_0, M_\emptyset$ everywhere dense in $\mathbb{R}$ such that for $\alpha \in M_\omega$ the Hamilton Eqs. (1.2) have a single-valued analytic integral, for $\alpha \in M_\infty$ there exists an infinitely differentiable integral, but no integral of the class $C^\omega, \dots,$ for $\alpha \in M_k$ there exists an integral of the class C^k, but not of C^{k+1}, for $\alpha \in M_0$ Eqs. (1.2) possess only a locally non-constant continuous invariant function, and for $\alpha \in M_\emptyset$ there do not exist even continuous integrals.

This was conjectured in [135], where it was also proved that the sets M_ω and $M_\emptyset$ are dense everywhere in $\mathbb{R}$. The latter property follows from the existence of a phase trajectory which is dense in the extended phase space. In its complete form this conjecture was proved by Moshchevitin [187], who also indicated the explicit form of the function f and described the structure of the sets $M_\omega, M_\infty, \dots, M_k, \dots, M_0, M_\emptyset$ having power of continuum.

As has been already mentioned in Chap. I, in problems of dynamics the phase space M^{2n} usually coincides with the cotangent bundle of the configuration space $\mathcal{N}^n$, whereas a Hamilton function is usually quadratic in the canonical momenta.

It was noted long ago that the known integrals of equations of dynamics are either polynomials in momenta or functions of such polynomials. For example, the Noether integrals are linear in the momenta, while integrals of Hamiltonian systems solved by separation of variables are quadratic in momenta. Here are several important special cases.

a) First consider an inertial motion. In this case the Hamilton function H coincides with the kinetic energy T. Any analytic integral F can be represented as a series in homogeneous forms of the momenta: $F = \sum F_k$, F_k being a homogeneous form of degree k. From the form of the Hamilton equations

$$\dot{x} = \frac{\partial T}{\partial y}, \quad \dot{y} = -\frac{\partial T}{\partial x}$$

it follows that the total derivative of F_k is a homogeneous function of the momenta $y_1, \dots, y_n$ of degree k. Consequently, each homogeneous form in the expansion $\sum F_k$ is a first integral. Of course, not all of them are independent.

b) Suppose that a Hamiltonian system with the Hamilton function $H = T(x, y) + \varepsilon V(x)$ has an integral as a power series in the parameter ε:

$$F = \sum F_k(x, y)\varepsilon^k . \tag{1.4}$$

This situation is typical for the equations of dynamics. After the change of variables

$$x \to x\,, \quad y \to \sqrt{\varepsilon}y\,, \quad t \to t/\sqrt{\varepsilon} \tag{1.5}$$

the Hamilton equations with the Hamiltonian $T+\varepsilon V$ transform to the equations with the Hamiltonian $T+V$, while the integral (1.4) takes the form

$$\sum F_k(x, \sqrt{\varepsilon}y)\varepsilon^k = \sum \Phi_m(x, y)(\sqrt{\varepsilon})^m\,,$$

where Φ_m are polynomials in the momenta y. Since the new Hamiltonian system does not depend on ε, these polynomials are its integrals.

The converse assertion is also correct: if the system with the Hamiltonian $T+V$ has a polynomial integral, then the system with the Hamiltonian $T+\varepsilon V$ has an integral in the form of the power series (1.4). To prove this, we use the change which is the inverse of (1.5). As a result, in the Hamilton equations the parameter ε appears. After this change the polynomial integral (up to multiplication by an inessential factor) becomes $F+\sqrt{\varepsilon}\Phi$, where F and Φ are functions analytic in ε. It is clear that F and Φ are integrals of the equations with the Hamiltonian $T+\varepsilon V$, where one of the functions $F|_{\varepsilon=0}$, $\Phi|_{\varepsilon=0}$ coincides with the major homogeneous form of the original polynomial integral.

Thus, when studying equations of dynamics it is natural to consider integrals which are polynomials in momenta with smooth and single-valued coefficients on the configuration space. We shall call such integrals polynomial.

Whittaker and Birkhoff studied existence of polynomial integrals of the first and second degree [23,233]. Note that for integrals of small fixed degree this problem can be solved by elementary means. However, if the degree of the integral is not fixed in advance, then this problem becomes more complicated.

Birkhoff considered also the existence of *conditional* polynomial integrals [23, Chap. II]. These are polynomials in momenta which are integrals only for some fixed values of the total energy.

2 Invariant Relations

2.1 Let $\dot{x} = v(x)$, $x \in M^n$ be a dynamical system and $f_1, \dots, f_m$ smooth functions on M^n. Consider the set $I_c \subset M$ defined by the equations

$$f_1 = c_1\,, \quad \dots\,, \quad f_m = c_m\,. \tag{2.1}$$

If I_c is invariant with respect to the phase flow g_v^t, then these equations are called *invariant relations* and thefunctions $f_1, \dots, f_m$ *particular integrals*. An equivalentdefinition is the following: $\dot{f}_k = 0$ on I_c for all $k \leq m$. This definition is also acceptable in the non-autonomous case.

The theory of invariant relations in Hamiltonian systems turns out to be closely connected with the hydrodynamics of an ideal fluid [136,148].

2.2 Let $\mathcal{N}^n$ be the configuration space, and $H(x, y, t) : T^*N \times \mathbb{R}_t \to \mathbb{R}$ a Hamiltonian. Suppose that the Hamilton equations with n degrees of freedom

$$\dot{x} = \frac{\partial H}{\partial y}, \quad \dot{y} = -\frac{\partial H}{\partial x} \tag{2.2}$$

have an invariant manifold $y = u(x, t)$, where u is a smooth covector field on $\mathcal{N}^n$ (or its subset) which may depend on time. We connect with u the $n \times n$ skew-symmetric matrix

$$\operatorname{curl} u = \frac{\partial u}{\partial x} - \left(\frac{\partial u}{\partial x}\right)^T .$$

By analogy with the case $n = 3$, we denote the action of $\operatorname{curl} u$ on the vector w by $\operatorname{curl} u \times w$.

Now introduce the smooth vector field of velocities on $\mathcal{N}^n$ by putting

$$v(x, t) = \dot{x} = \left.\frac{\partial H}{\partial y}\right|_{y=u} .$$

It turns out that the fields u and v satisfy the equations

$$\frac{\partial u}{\partial t} + \operatorname{curl} u \times v = -\frac{\partial B}{\partial x}, \qquad B(x, t) = H|_{y=u} , \tag{2.3}$$

$$\frac{\partial}{\partial t}\operatorname{curl} u + \operatorname{curl}(\operatorname{curl} u \times v) = 0 . \tag{2.4}$$

Indeed, the fact that the manifold $y = u(x, t)$ is an invariant manifold of (2.2) is equivalent to the differential equation

$$\dot{u} = -\left.\frac{\partial H}{\partial x}\right|_{y=u} .$$

This can be represented in the following form

$$\frac{\partial u}{\partial t} + \frac{\partial u}{\partial x}v = -\frac{\partial B}{\partial x} + \left.\frac{\partial H}{\partial y}\right|_{y=u}, \qquad \frac{\partial u}{\partial x} = -\frac{\partial B}{\partial x} + \left(\frac{\partial u}{\partial x}\right)^T v .$$

From here we obtain (2.3). Equation (2.4) arises from the action of the operator curl on the left- and right-hand sides of (2.3).

Conversely, let $u(x, t)$ be a field on M. We put

$$v(x, t) = \left.\frac{\partial H}{\partial y}\right|_{y=u} .$$

If the fields u and v satisfy the "Lamb equations" (2.3), then the manifold $I = \{y = u(x, t)\}$ is an invariant manifold of the Hamiltonian system (2.2). The trajectories lying on I are determined by solving the following equations on $\mathcal{N}$

$$\dot{x} = v(x, t) . \tag{2.5}$$

Equations (2.3) for Hamiltonian systems seem to appear first in the calculus of variations as the condition for fields of extremals to be consistent (see [24] and [80, Chap. X]). A generalization of the Lamb equations to non-Hamiltonian systems is considered in the book [17].

2.3 For the system (2.5), the following "Thompson theorem" is valid: the integral

$$\int_\gamma u\,dx \tag{2.6}$$

taken along any cycle $\gamma \subset \mathcal{N}$ moving with the phase flow conserves its value. This fact follows from Poincaré's theorem on integral invariants of Hamiltonian systems (§1): by the formula $y = u(x, t)$ the cycle γ is lifted to a cycle Γ in the phase space $T^*\mathcal{N}$, and then we use the property that the integral

$$\int y\,dx$$

taken along the moving cycle Γ conserves.

We refer to a covector field u as potential, if $\operatorname{curl} u = 0$; locally $u = \partial\varphi/\partial x$, where φ is a function of x and t. We note the following "Lagrange theorem": a covector field $u(x, t)$ which is potential at $t = 0$ remains potential for any $t > 0$. In this case, the integral (2.6) vanishes for any cycle γ contractible in $\mathcal{N}$ to a point. Lagrange's theorem is a simple corollary of this remark and the Thompson theorem. If a field u is potential, then the invariant n-dimensional manifold $I = \{y = u\}$ is called a Lagrangian manifold.

Substituting $u = \partial\varphi/\partial x$ into (2.3), we obtain the "Lagrange–Cauchy integral"

$$\frac{\partial\varphi}{\partial t} + H\left(x, \frac{\partial\varphi}{\partial x}, t\right) = f\,, \tag{2.7}$$

where f is a function of time. After the gauge transformation

$$\varphi \to \varphi - \int f(t)\,dt\,,$$

f can be replaced by zero. In Hamiltonian mechanics Eq. (2.7) (for $f = 0$) is called *the Hamilton–Jacobi equation.*

2.4 If $\operatorname{curl} u \times w = 0$, then the vector $w \neq 0$ is called a vortex vector. For odd n, such vectors always exist. We shall refer to the field u as non-singular, if the rank of the matrix $\operatorname{curl} u$ equals $n - 1$. For $n = 3$ the field u is non-singular if and only if $\operatorname{curl} u \neq 0$.

For a non-singular field u, the corresponding vortex vectors define at each time a smooth field of directions on $\mathcal{N}$. Its integral trajectories are called vortex lines. It turns out that the phase flow of the system (2.5) takes vortex lines to vortex lines. This is a corollary of the Thompson theorem given in §2.3 and it generalizes the well-known Helmholtz theorem in dynamics of an ideal fluid. The latter asserts that the vortex lines are "frozen": if the particles of fluid form a vortex line at some instant, they form such a line during the whole motion.

The vortex fields $w(x, t)$ are defined up to multiplication by functions of x and t. There are remarkable fields among them (determined up to multiplication by a constant factor). Their characteristic property is described by

Theorem 1. *For odd n in the non-singular case there exists a vortex vector field satisfying the equation*

$$\frac{\partial w}{\partial t} + [v, w] = 0\,, \tag{2.8}$$

where $[\cdot, \cdot]$ *is a commutator of vector fields on* $\mathcal{N}$.

Equation (2.8) is analogous to the Euler equations describing the evolution of the angular momentum of a rigid body.

Now recall the definition of a commutator of vector fields w and v. To each of these fields there corresponds a linear differential operator

$$L_w = \sum w_i \frac{\partial}{\partial x_i}\,, \quad L_v = \sum v_i \frac{\partial}{\partial x_i}\,.$$

The operator

$$L_v L_w - L_w L_v$$

is also a linear differential operator corresponding to the vector field $[v, w]$. The explicit formula for the commutator is

$$L_{[v,w]} = \sum_j \left(\sum_i v_i \frac{\partial w_j}{\partial x_i} - w_i \frac{\partial v_j}{\partial x_i} \right) \frac{\partial}{\partial x_j}\,.$$

Proof of Theorem 1. Since $\operatorname{curl} u \times w = 0$, we have

$$\frac{\partial(\operatorname{curl} u)}{\partial t} \times w + \operatorname{curl} u \times \frac{\partial w}{\partial t} = 0\,.$$

Then, taking into account (2.4), we find

$$\operatorname{curl} u \times \frac{\partial w}{\partial t} - [\operatorname{curl}(\operatorname{curl} u \times v)] \times w = 0\,. \tag{2.9}$$

Besides, $\operatorname{curl} u \times w = 0$ yields

$$\operatorname{curl}(\operatorname{curl} u \times v) \times w - \operatorname{curl}(\operatorname{curl} u \times w) \times v = \operatorname{curl} u \times [w, v]\,.$$

Therefore, (2.9) can be represented in the form

$$\operatorname{curl} u \times \left(\frac{\partial w}{\partial t} + [v, w] \right) = 0\,.$$

Since the field u is non-singular,

$$\frac{\partial w}{\partial t} + [v, w] = \alpha w\,,$$

where α is a function of x and t. Put $w = \rho w_0$, where ρ satisfies the equation

$$\frac{\partial \rho}{\partial t} + \frac{\partial \rho}{\partial x} v = \alpha \rho . \tag{2.10}$$

Then w_0 is the desired vortex field. It remains to prove the solvability of (2.10). By putting $\rho = \exp \xi$ we arrive at the equation

$$\frac{d\xi}{dt} = \frac{\partial \xi}{\partial t} B + \frac{\partial \xi}{\partial x} v = \alpha ,$$

which is easily integrated by the method of characteristics. The theorem is proved.

Let $n = 3$ and $\mathcal{N}$ have the structure of Euclidean space. Then the field u can be identified with a vector field on $\mathcal{N}$, hence $\operatorname{curl} u$ is obviously one of the vortex fields. According to Theorem 1, in this case there exists a vector field defined by Eq. (2.8) such that $\operatorname{curl} u = \alpha w$.

Theorem 2. *The function α satisfies the continuity equation*

$$\frac{\partial \alpha}{\partial t} + \operatorname{div}(\alpha v) = 0 ,$$

where $\operatorname{div} a$ is the trace of the matrix $\partial a / \partial x$.

Corollary. *Equations* (2.5) *possess the integral invariant*

$$\int_D \alpha \, d^3 x .$$

The proof of Theorem 2 is based on application of the well-known formula of vector algebra

$$\operatorname{curl}(a \times b) = [b, a] + a \operatorname{div} b - b \operatorname{div} a .$$

Using this formula and the obvious relation $\operatorname{div} \operatorname{curl} u = 0$, we can represent Eq. (2.4) in the following form

$$\frac{\partial}{\partial t} \operatorname{curl} u + [v, \operatorname{curl} u] + (\operatorname{curl} u) \operatorname{div} v = 0 . \tag{2.11}$$

Put $\operatorname{curl} u = \alpha w$. Then from (2.11) we deduce

$$\alpha \left(\frac{\partial w}{\partial t} + [v, w] \right) + \left(\frac{\partial \alpha}{\partial t} + \frac{\partial \alpha}{\partial x} v + \alpha \operatorname{div} v \right) w = 0 .$$

If the field w satisfies (2.8), then

$$\frac{\partial \alpha}{\partial t} + \frac{\partial \alpha}{\partial x} v + \alpha \operatorname{div} v = \frac{\partial \alpha}{\partial t} + \operatorname{div}(\alpha v) = 0 .$$

Q.E.D.

All the arguments of this subsection follow from the vortex Eq. (2.4) only. These remain valid if we replace (2.4) by the more general equation

$$\frac{\partial A}{\partial t} + \operatorname{curl}(Av) = 0\,, \tag{2.12}$$

where $A = \|A_{ij}\|$ is a skew-symmetric matrix satisfying the condition

$$\frac{\partial A_{ij}}{\partial x_k} + \frac{\partial A_{jk}}{\partial x_i} + \frac{\partial A_{ki}}{\partial x_j} = 0\,. \tag{2.13}$$

The vortex vectors are the eigenvectors of the matrix A corresponding to zero eigenvalues.

Equations (2.12) describe, in particular, the evolution of the vector of magnetic strength in a medium with infinite conductivity. In this situation the vortex lines stand for the strength lines of the magnetic field.

2.5 Consider the stationary case: the field u and the Hamilton function H do not depend explicitly on time. Here we have the "Bernoulli theorem": the function B is constant along the stream lines (the integral trajectories of the field $v(x)$) as well as on the vortex lines. Indeed, since u is stationary, Eq. (2.3) takes the following form

$$\operatorname{curl} u \times v = -\frac{\partial B}{\partial x}\,.$$

If w is a vortex field, then

$$\frac{\partial B}{\partial x} w = -(\operatorname{curl} u \times v) w = (\operatorname{curl} u \times w) v = 0\,.$$

Analogously,

$$\dot{B} = \frac{\partial B}{\partial x} v = -(\operatorname{curl} u \times v) v = 0\,,$$

since the matrix $\operatorname{curl} u$ is skew symmetric.

Let us show that for a non-critical point $x_0 \in \mathcal{N}$ of the function B the vectors $v(x_0), w(x_0) \neq 0$ are linearly independent. Indeed, if these vectors are dependent, then v is a vortex vector. Then for any vector a

$$\frac{\partial B}{\partial x} a = -(\operatorname{curl} u \times v) a = 0\,.$$

Therefore, $dB(x_0) = 0$, which was to be proved.

If $B \neq \text{const}$, then it is natural to consider the distribution of the planes $\Pi(x)$ spanned by the independent vectors $v(x), w(x)$ and tangent to $\mathcal{N}$. Since v and w commute (Theorem 1), by the Frobenius theorem (see [87]), the distribution $\Pi(x)$ is integrable. This means that through each point $x \in \mathcal{N}$ there passes a unique integral surface Σ_x of this distribution tangent to $v(x)$ and $w(x)$. If the fields u, v are complete on Σ, then, as known from topology, Σ is diffeomorphic to one of the following surfaces: $\mathbb{R}^2$ (plane), $\mathbb{R} \times \mathbb{T}$ (cylinder), $\mathbb{T}^2$ (torus) (see, for example, [74]). Besides, represented in some global coordinates on Σ, the stream and the vortex lines turn out to be straight lines. In the general case the surfaces Σ may be immersed in $\mathcal{N}$ in a rather intricate way. The case when Σ is closed in $\mathcal{N}$ is especially interesting. This is always so for $n = 3$: the integral surfaces Σ_x coincide with connected components of the level surfaces of the non-constant function B. To use this phenomenon effectively one should consider a nontrivial question on the existence of an appropriate covector field $u(x)$ on the whole space $\mathcal{N}$.

The Bernoulli theorem can be generalized to the case when the field u is singular: the rank of the matrix $\operatorname{curl} u$ (or, more generally, the rank of the matrix A in Eq. (2.12)) drops more than by one. At each point $x \in \mathcal{N}$ the vortex fields span a linear subspace $W_x \subset T_x\mathcal{N}$. If the rank of the matrix $\operatorname{curl} u$ (or A) is constant, then the dimension of W_x does not depend on x. Thus, a distribution of tangent spaces is defined. According to the Frobenius theorem and by (2.13), this is an integrable distribution. Therefore, the configuration space $\mathcal{N}$ is foliated by smooth regular integral manifolds of the distribution W which have dimension $n - \operatorname{rank} A$. It is natural to call these the vortex manifolds. The phase flow of the system (2.5) transforms them into themselves, while in the stationary case the function B is constant on each of them.

2.6 As an example, consider the Euler problem on free rotation of a rigid body around a fixed point. Here the configuration space $\mathcal{N}$ is the group $SO(3)$. The angular momentum of the body with respect to the space is fixed. As a result, on the group $SO(3)$ a stationary three-dimensional flow arises. One can check that this is a vortex flow. The function B in this problem is constant on $SO(3)$ only for the degenerate case, i.e., when the inertia tensor is spherical. Thus, in the typical situation $\operatorname{curl} u \times v \neq 0$. The stream lines and vortex lines lie on the Bernoulli surfaces $I_c = \{x : B(x) = c\}$. For non-critical values of c the surfaces I_c are diffeomorphic to two-dimensional tori. Note that there are only three critical values: they coincide with the energy of rotation of the body about the principal inertia axes (with the value of the angular momentum fixed).

The field of velocity vectors v and the vortex field w defined on the group $SO(3)$ enjoy several remarkable properties. First, the phase flow of the dynamical system

$$\dot{x} = v(x) \,, \qquad x \in SO(3)$$

preserves the bi-invariant measure on $SO(3)$. The latter is invariant with respect to the left and right actions of this group. In local coordinates – the Euler angles θ, φ, ψ – it has the following form (see [188, Chap.1]):

$$d\mu = \sin\theta \, d\theta \, d\varphi \, d\psi \,.$$

If we put $\operatorname{curl} u = \alpha w$, then α, as a function of the Euler angles, becomes exactly $\sin\theta$ (compare with §2.4).

It turns out that the vortex field w, which commutes with the field v, can be described by the relations

$$\operatorname{curl} u \times w = 0 \,, \qquad \xi(w) = \text{const} \,,$$

where ξ is the 1-form $u\,dx$. The field w has a simple mechanical interpretation: the dynamical system $\dot{x} = w(x)$ generates rotations of the rigid body with angular velocity constant in the space and directed along the angular momentum vector.

All these statements constitute the "vortex" theory of the Euler top. They can be proved by using some local coordinates on the group $SO(3)$ (say, the Euler angles).

The above example is easily generalized to the case of motion on an arbitrary Lie group $\mathfrak{G}$ with a left-invariant Lagrangian. Here the momentum integrals are the Noether integrals (there are $l = \dim \mathfrak{G}$ such integrals) corresponding to left-invariant symmetry fields.

2.7 For Hamiltonian systems with three degrees of freedom there is a "vortex" analog of the Hamilton–Jacobi method. Let $H(x, y)$ be the Hamilton function. The partial differential Eq. (2.3) written for the autonomous case is

$$\operatorname{curl} u \times \left(\left. \frac{\partial H}{\partial y} \right|_{y=u} \right) = -\frac{\partial}{\partial x} \left(H|_{y=u} \right) . \tag{2.14}$$

For example, in the reversible case when

$$H = \frac{1}{2} \sum g^{ij}(x) y_i y_j - V(x) ,$$

the system (2.14) has the following explicit form

$$\sum_{j,k} g^{jk} \left(\frac{\partial u_i}{\partial x_j} - \frac{\partial u_j}{\partial x_i} \right) u_k = -\frac{1}{2} \frac{\partial}{\partial x_i} \left(\sum_{j,k} g^{jk} u_j u_k \right) + \frac{\partial V}{\partial x_i} ,$$

$$1 \le i, j, k \le 3 .$$

Theorem 3. *Suppose that there exists a three-parameter family of solutions* $u(x, \alpha)$ *of the system* (2.14) *with the following properties*

1) $\det \|\partial u / \partial \alpha\| \neq 0$,

2) $d_x H(x, u(x, \alpha)) \neq 0$.

Then the Hamiltonian equations with Hamiltonian H are integrable by quadratures.

Suppose also that there exists a "complete" potential solution of the system (2.14):

$$u(x, \alpha) = \frac{\partial}{\partial x} S(x, \alpha) .$$

Then from (2.14) we obtain the "Cauchy–Lagrange integral"

$$H \left(x, \frac{\partial}{\partial x} S(x, \alpha) \right) = h(\alpha) .$$

This is the stationary Hamilton–Jacobi equation. The function $S(x, \alpha)$ is its complete integral, since

$$\det \left\| \frac{\partial u}{\partial \alpha} \right\| = \det \left\| \frac{\partial^2 S}{\partial x \partial \alpha} \right\| \neq 0 .$$

However, condition 2) of the theorem is not fulfilled for the potential case.

Proof of Theorem 3. According to condition 1), from the equations

$$y_i = u_i(x_1, x_2, x_3, \alpha_1, \alpha_2, \alpha_3) , \qquad 1 \leq i \leq 3$$

one can find (at least locally) α_k as functions of x, y: $\alpha_k = F_k(x, y)$. It follows from §2.2 that these functions are integrals of the Hamiltonian system under consideration. According to condition 2), the functions F_1, F_2, F_3, H are independent. Thus, it only remains to use the well-known Jacobi theorem on integrability of a system of n differential equations possessing an invariant measure and $n - 2$ independent integrals ([105, lecture 12]).

2.8 Equation (2.3) coincides with the Birkhoff Eqs. (9.3) in Chap. I. Therefore, it describes extremals of the variational problem

$$\delta \int_{t_1}^{t_2} u\,dx - B\,dt = 0 , \qquad \delta x(t_1) = \delta x(t_2) = 0 .$$

This variational principle can be deduced from the classical Hamilton principle of stationary action.

If the field u is stationary, then Eqs. (2.5) are Hamiltonian and, therefore, the theory developed above is applicable.

3 Symmetry Groups

3.1 Consider a dynamical system given by the differential equation

$$\dot{x} = v(x) . \tag{3.1}$$

A vector field u commuting with the field v ($[u, v] = 0$) is called *a symmetry field* of the system (3.1). Let us show that the phase flow of the system

$$\frac{dx}{d\tau} = u(x) , \tag{3.2}$$

being the one-parameter group of transformations g_u^τ, takes solutions of the system (3.1) to solutions of the same system. For this purpose we use the theorem on straightening of trajectories: Eqs. (3.2) can be reduced locally to the form

$$\frac{dx_1}{d\tau} = \frac{dx_2}{d\tau} = \cdots = \frac{dx_{n-1}}{d\tau} = 0 , \quad \frac{dx_n}{d\tau} = 1 . \tag{3.3}$$

Since $[u, v] = 0$, the components $v_1, \dots, v_n$ of the vector field v do not depend on x_n. Therefore, after the transformations

$$x_1 \to x_1 \,, \quad \dots \,, \quad x_{n-1} \to x_{n-1} \,, \quad x_n \to x_n + \tau$$

solutions of (3.1) do transform to the same solutions.

The existence of a symmetry group essentially simplifies the study of a dynamical system. For example, written in terms of the same special variables $x_1, \dots, x_n$ the subsystem

$$\dot{x}_k = v_k(x_1, \dots, x_{n-1}) \,, \quad k \le n-1 \tag{3.4}$$

is closed. If the system (3.4) is integrated, then the last variable x_n can be determined by the simple quadrature

$$x_n = \int_0^t v_n(x_1, \dots, x_{n-1})\, dt + x_n^0 \,.$$

From the geometrical point of view, lowering the order of a system with a symmetry group g^τ means a factorization of the phase space by orbits of this group. However, to lower the order constructively one has to find trajectories of the system (3.2) (orbits of the group g_u^τ).

3.2 Suppose that there exists another symmetry field w such that $[u, w] = \lambda u$, where λ is a function of x. Written in the local coordinates $x_1, \dots, x_n$ introduced in (3.3), the components $w_1, \dots, w_{n-1}$ of the field w also do not depend on x_n. It is easy to see that the phase flow of the system

$$\frac{dx_1}{d\alpha} = w_1(x_1, \dots, x_{n-1}) \,, \quad \dots \,, \quad \frac{dx_{n-1}}{d\alpha} = w_{n-1}(x_1, \dots, x_{n-1})$$

is a symmetry group of (3.4). Using this fact one can lower the order of the original system by two.

These observations give rise to the following important construction proposed by Sophus Lie. Let $u_1, \dots, u_n$ be linearly independent symmetry fields of the system (3.1) such that

$$[u_i, u_j] = \sum_k c_{ij}^k u_k \,, \qquad c_{ij}^k = \text{const} \,.$$

Let A be the linear space spanned by the vector fields

$$\sum_{s=1}^n c_s u_s \,, \qquad c_s \in R \,.$$

This $(n-1)$-dimensional space is a Lie algebra with respect to the product $[\ ,\ ]$.

Recall the definition of a solvable Lie algebra. Let B and C be subalgebras of the algebra A. The set $C \subset B$ is called an ideal of the algebra B if for all $f \in C$ and $g \in B$ the commutator $[f, g]$ belongs to C. A Lie algebra is solvable if there exists a sequence $A = A_0 \supset A_1 \supset \cdots \supset A_k = \{0\}$ of subalgebras of A such that A_{i+1}

is an ideal of codimension 1 in A_i ($i = 0, \dots, k-1$). In particular, commutative algebras are solvable ($[f, g] = 0$ for all $f, g \in A$).

Theorem (Lie). *If the system of differential equations* (3.1) *admits an* $(n-1)$*-dimensional solvable algebra of symmetry fields, then it is integrable by quadratures.*

Integration by quadratures implies obtaining the solutions by using "algebraic" operations (including inversion of functions) and "quadratures"—taking integrals of known functions. Formally, this definition of integrability makes sense locally, because solving an equation on a manifold by quadratures implies that this equation can be integrated in any local coordinates. We assume that any transformation from some local coordinates to other ones is an "algebraic" operation.

The Lie theorem has an important

Corollary. *Let* $u_1, \dots, u_n$ *be linearly independent commuting vector fields. Then each of the systems*

$$\dot{x} = u_k(x) \,, \qquad 1 \le k \le n$$

is integrable by quadratures.

The proof of Lie's theorem breaks down into several steps.

a) We are going to find an explicit solution of the equation $V(F) = 0$, where V is the differential operator $v(x)\,\partial/\partial x$. More precisely, we have to find $n-1$ independent solutions $F_1, \dots, F_{n-1}$ of this equation.

Introduce also $n-1$ linear differential operators

$$U_k = u_k(x)\,\partial/\partial x \,.$$

Since the algebra of the vector fields $\sum c_k u_k$ is solvable, using a linear change of variables, one can introduce $n-1$ new vector fields (denoted also by $u_1, \dots, u_{n-1}$) such that

$$\begin{aligned} [u_1, u_j] &= c^1_{1,j} u_1 \,, \\ [u_2, u_j] &= c^1_{2,j} u_1 + c^2_{2,j} u_2 \,, \\ &\dots \quad \dots \quad \dots \\ [u_{n-1}, u_j] &= c^1_{n-1,j} u_1 + \dots + c^{n-1}_{n-1,j} u_{n-1} \,. \end{aligned} \tag{3.5}$$

Here $c^k_{ij} = \text{const}$.

b) Consider the system of equations

$$V(F) = U_1(F) = \dots = U_{n-2}(F) = 0 \tag{3.6}$$

and show that locally it has a solution F without critical points: $dF \neq 0$. This fact easily follows from the Frobenius theorem and the relations (3.5). Its direct proof is as follows. Since the fields $v, u_1, \dots, u_{n-1}$ are linearly independent, $u_{n-2} \neq 0$. Therefore, by the theorem on straightening of trajectories, the field

u_{n-2} can be reduced locally to the form $(0, \ldots, 0, 1)$. By virtue of (3.5), in the new coordinates $x_1, \ldots, x_n$ the components of $v, u_1, \ldots, u_{n-3}$ do not depend on x_n. Thus, these fields can be considered in the $(n-1)$-dimensional space of the variables $x_1, \ldots, x_{n-1}$, and they again satisfy the relations (3.5). Then we can apply the theorem on straightening of trajectories to the field u_{n-3} and reduce its first $n-1$ components to the form $(0, \ldots, 0, 1)$, and so on. As a result, we find the new coordinates $z_1, \ldots, z_n$ for which the vector fields $v, u_1, \ldots, u_{n-2}$ have the following form

$$\begin{aligned} v &= (0, 1, *, \ldots, *) , \\ u_1 &= (0, 0, 1, *, \ldots, *) , \\ &\ldots \quad \ldots \quad \ldots \\ u_{n-2} &= (0, 0, \ldots, 1) . \end{aligned}$$

In the variables $z_1, \ldots, z_n$, for the desired function we can choose $F = z_1$.

c) Let us show that if F is a solution of the system (3.6), then $U_{n-1}(F)$ is also a solution of this system. Indeed, $[V, U_{n-1}] = U_{n-1}V - VU_{n-1} = 0$. Therefore, $V(U_{n-1}(F)) = U_{n-1}(V(F)) = U_{n-1}(0) = 0$. Then, according to (3.6), $U_{n-1}U_1(f) - U_1U_{n-1}(F) = c^1_{1,\,n-1}U_1(F)$. Thus, $U_1U_{n-1}(F) = 0$. Similarly, one can prove that $U_kU_{n-1}(F) = 0$ for all $k \leq n-1$.

Let G be a solution of (3.6), and $dG \neq 0$. Then $\varphi(G) = U_{n-1}(G) \neq 0$ (otherwise $dG = 0$ by linear independence of the fields $v, u_1, \ldots, u_{n-1}$). We put

$$F = \int_{G_0}^{G} \frac{d\xi}{\varphi(\xi)} .$$

Since F is a function of G, the function F is a solution of (3.6) and

$$U_{n-1}(F) = U_{n-1}(G)\frac{dF}{dG} = \frac{U_{n-1}(G)}{\varphi(G)} = 1 .$$

Thus, the system of equations

$$V(F) = U_1(F) = \cdots = U_{n-2}(F) = 0 , \quad U_{n-1}(F) = 1 \tag{3.7}$$

has a solution (at least locally).

d) To find this solution, we put

$$\begin{aligned} V(F) &= v_1\frac{\partial F}{\partial x_1} + \cdots + v_n\frac{\partial F}{\partial x_n} = 0 , \\ U_1(F) &= u_{1.1}\frac{\partial F}{\partial x_1} + \cdots + u_{1.n}\frac{\partial F}{\partial x_n} = 0 , \\ &\ldots \quad \ldots \quad \ldots \\ U_{n-1}(F) &= u_{n-1.1}\frac{\partial F}{\partial x_1} + \cdots + u_{n-1.n}\frac{\partial F}{\partial x_n} = 1 . \end{aligned} \tag{3.8}$$

Since the fields $v, u_1, \ldots, u_{n-1}$ are linearly independent, from the linear system (3.8) we obtain

$$\partial F/\partial x_1 = \xi_1(x)\,, \ \ldots\,, \ \partial F/\partial x_n = \xi_n(x)\,. \tag{3.9}$$

To find these expressions is a purely algebraic problem. Besides, the 1-form $\xi_1\,dx_1 + \cdots + \xi_n\,dx_n$ is locally exact. As is known from analysis, the function F satisfying (3.9) can be recovered by quadratures.

e) Perform the change of variables

$$y_1 = x_1\,, \ \ldots\,, \ y_{n-1} = x_{n-1}\,, \quad y_n = F_1(x_1, \ldots, x_n)\,,$$

where F_1 is a solution of (3.7). In the new variables, we have

$$\begin{aligned} V &= \tilde{v}_1 \frac{\partial}{\partial y_1} + \cdots + \tilde{v}_n \frac{\partial}{\partial y_n}\,, \\ &\ldots \quad \ldots \quad \ldots \\ V_{n-2} &= \tilde{u}_{n-2,1} \frac{\partial}{\partial y_1} + \cdots + \tilde{u}_{n-2,n} \frac{\partial}{\partial y_n}\,. \end{aligned} \tag{3.10}$$

Since $V(y_n) = \cdots = U_{n-2}(y_n) = 0$, we obtain $\tilde{v}_n = \cdots = \tilde{u}_{n-2,n}$ and thus the coordinate y_n can be regarded as a parameter in (3.10). Then, repeating the procedure described in b)–d), we conclude that the system

$$V(F_2) = U_1(F_2) = \cdots = U_{n-3}(F_2) = 0\,, \quad U_{n-2}(F_2) = 1 \tag{3.11}$$

has a solution which can be found by quadratures.

f) The solvability of the system

$$\begin{aligned} V(F_3) = \cdots = U_{n-4}(F_3) = 0\,, &\quad U_{n-3}(F_3) = 1\,, \\ \ldots \quad \ldots \quad \ldots & \\ V(F_{n-1}) = 0, \quad U_1(F_{n-1}) = 1 & \end{aligned} \tag{3.12}$$

can be proved in a similar way.

It follows from (3.7), (3.11), and (3.12) that the functions $F_1, \ldots, F_{n-1}$ are independent first integrals of the original system (3.1). The theorem is proved.

3.3 Suppose that the fields u and v satisfy the relation

$$[v, u] = \mu v + \nu u \tag{3.13}$$

for some constants μ, ν. We are going to show that in this case the order of the system (3.1) can also be lowered.

For this purpose, we use the special coordinates $x_1, \ldots, x_n$ introduced in (3.3). Given a general solution of (3.2), such a reduction can be performed in explicit form. Written in terms of $x_1, \ldots, x_n$ the relation (3.13) turns out to be equivalent to the sequence of equations

$$\frac{\partial v_i}{\partial x_n} = \mu v_i\,, \qquad \frac{\partial v_n}{\partial x_n} = \mu v_n + \nu\,, \tag{3.14}$$

where v_i are components of the field v. From (3.14) we obtain

$$v_i = v_i^0 \exp(\mu x_n) , \quad v_n = v_n^0 \exp(\mu x_n) - \nu/\mu , \tag{3.15}$$

where the functions v_i^0 ($1 \le i \le n$) do not depend on x_n. Putting $dt = [\exp(\mu x_n)]\, ds$, we write the first $n-1$ integrals of the system (3.1) in the form

$$x_1' = v_1^0 , \ \dots , \ x_{n-1}' = v_{n-1}^0 , \tag{3.16}$$

where primes denote differentiation with respect to s. This closed system of equations can be regarded as the result of lowering the order of the original system (3.1).

If the general solution of the system (3.16) is known, then Eqs. (3.1) are integrable by quadratures. To show this we note that since v_n^0 does not depend on x_n, it is sufficient to solve the equation

$$x_n' = -\nu\mu^{-1} \exp(-\mu x_n) + f , \tag{3.17}$$

where f is a known function of s (see (3.15)). After the change $z = \exp(\mu x_k)$, this equation is reduced to $z' = \mu f z - \nu$, which can be easily integrated. Thus, one can express x_i explicitly in terms of s. To determine the functions $x_i(t)$ it is sufficient to invert the integral

$$t = \int e^{\mu x_n}\, ds .$$

In (3.15)–(3.17) we assumed that $\mu \neq 0$. The case $\mu = 0$ is trivial.

Consider more closely the most important case $\nu = 0$. From the last formula in (3.15) it follows that (3.16) can be complemented by the equation

$$x_n' = v_n^0 . \tag{3.18}$$

Since $v_1^0, \dots, v_n^0$ do not depend on x_n, the phase space g_u^τ of the system (3.3) takes solutions of (3.16), (3.18) to the same solutions. Going back to the original time t, we conclude that the same holds for the trajectories (but not the solutions) of the original system (3.1). For this reason the field u can be regarded as a generalized symmetry field of this system.

In accordance with the last remark, the Lie theorem in §3.2 admits the following generalization. Let the linearly independent vector fields $v, u_1^0, \dots, u_n^0$ generate a solvable n-dimensional Lie algebra $[v, u_k] = \lambda_k v$ ($\lambda = \text{const}$). Then the system (3.1) can be solved by quadratures. The reader may prove this generalized theorem by using the method discussed in §3.2.

3.4 According to the theorem on straightening of trajectories, in a small neighborhood of a regular point of the vector field v the system (3.1) has an n-dimensional Abelian symmetry group. Thus, the problem of the existence of a smooth (or analytic) symmetry field makes sense only in the vicinity of an equilibrium position or in the whole phase space.

Here are two examples of dynamical systems which admit nontrivial analytic symmetry fields, but do not have non-constant continuous integrals.

a) Consider a quasi-periodic motion on the n-dimensional torus $\mathbb{T}^n = \{(x_1, \ldots, x_n) \bmod 2\pi\}$ defined by the system $\dot{x}_i = \omega_i$ $(1 \le i \le n)$ with constant frequencies independent over the ring of integer numbers (cf. (1.1)). This system is ergodic on $\mathbb{T}^n$ and thus does not admit even measurable (let alone continuous) integrals. However, each constant vector field on $\mathbb{T}^n$ is a symmetry field of this system.

b) Let $v(x) = Ax$, and all eigenvalues of the constant operator A lie in the left (or right) half-plane. Since the equilibrium position $x = 0$ is asymptotically stable for $t \to +\infty$ (or for $t \to -\infty$), the corresponding system (3.1) does not have non-constant continuous integrals. Indeed, let $f(x)$ be a first integral, x_0 any point in $\mathbb{R}^n$, and $x(t)$ the solution of (3.1) with the initial value $x(0) = x_0$. Since $x(t) \to 0$ as $t \to +\infty$ $(t \to -\infty)$, by continuity of f, $f(x(t)) \to 0$ as $t \to +\infty$ $(t \to -\infty)$. Thus, $f(x_0) = f(0)$ for all x_0, because $f(x(t))$ is constant as a function of t. On the other hand, the field $u(x) = x$ is a symmetry field of the linear system $\dot{x} = Ax$; it generates the group of dilatations $x \to e^\tau x$, $\tau \in \mathbb{R}$.

A more interesting example is the Hamiltonian system considered in §1.3 which generically has integrals of finite smoothness only. However, the non-trivial transformation group

$$y \to y + \alpha\,, \quad x \to x\,, \quad t \to t$$

turns out to be a symmetry group of this system. This group is generated by the vector field $(1, 0, 0)$ (in the coordinates y, x, t).

Speaking about a *nontrivial* symmetry group, we suppose that its field u is linearly independent of the field v. Note that if $u = \lambda(x)v$ and $[u, v] = 0$, then λ is a first integral of the system (3.1).

One should keep in mind that an analytic system of differential equations may have symmetry vector fields of finite smoothness. As an example, consider the following dynamical system on the two-dimensional torus $\mathbb{T}^2 = \{(x_1, x_2) \bmod 2\pi\}$

$$\dot{x}_1 = \omega_1 R\,, \quad \dot{x}_2 = \omega_2 R\,, \tag{3.19}$$

where $\omega_i = \text{const}$. The function R is positive and analytic on $\mathbb{T}^2$. This system has the invariant measure

$$\operatorname{mes}(D) = \iint_{\mathcal{D}} \frac{dx_1\,dx_2}{R}\,,$$

and therefore is locally Hamiltonian. Kolmogorov [116] showed that for almost all ratios ω_1/ω_2 the system is reducible to the form

$$\dot{x}_1 = \lambda_1\,, \quad \dot{x}_2 = \lambda_2\,; \qquad \lambda_i = 4\pi^2\omega_i/\operatorname{mes}(\mathbb{T}^2)\,. \tag{3.20}$$

Therefore, in these cases Eqs. (3.19) admit a nontrivial symmetry field with analytic components. On the other hand, it was shown in [116] that after an appropriate choice of an irrational number ω_1/ω_2 and an analytic function R the system (3.19) can be reduced to the form (3.20) by a k-differentiable (but not $(k+1)$-differentiable) transformation of the torus $\mathbb{T}^2$ into itself.

Suppose that the system (3.19) has a symmetry field with differentiable components X_1 and X_2. These are 2π-periodic in x_1 and x_2. The condition for the vector fields $(\omega_1 R, \omega_2 R)$ and (X_1, X_2) to commute is reduced to the relations

$$\sum \omega_i \frac{\partial X_1}{\partial x_i} = \omega_1 \sum X_i \frac{\partial R}{\partial x_i}, \quad \sum \omega_i \frac{\partial X_2}{\partial x_i} = \omega_2 \sum X_i \frac{\partial R}{\partial x_i}. \tag{3.21}$$

Put $\varphi = \omega_2 X_1 - \omega_1 X_2$. Then

$$\frac{\partial \varphi}{\partial x_1}\omega_1 + \frac{\partial \varphi}{\partial x_2}\omega_2 = 0 .$$

If the ratio ω_1/ω_2 is irrational, then φ = const. Put $X_1 = \omega_1 S$. Then $X_2 = \omega_2 S + \mu$, μ = const. By (3.21), the function S satisfies the equation

$$\sum \omega_i \frac{\partial S}{\partial x_i} = S \sum \omega_i \frac{\partial \mathcal{N}}{\partial x_i} + \mu \frac{\partial \mathcal{N}}{\partial x_2}, \qquad \mathcal{N} = \ln R .$$

We search for its solution in the form of the product KR. To specify the function K we obtain the following equation for the determination of K

$$\frac{\partial K}{\partial x_1}\omega_1 + \frac{\partial K}{\partial x_2}\omega_2 = \mu \frac{\partial F}{\partial x_2}, \qquad F = -\frac{1}{R} .$$

To solve it by Fourier's method, we put

$$K = \sum k_{m_1 m_2} e_{m_1 m_2}, \quad F = \sum f_{m_1 m_2} e_{m_1 m_2};$$
$$e_{m_1 m_2} = \exp[i(m_1 x_1 + m_2 x_2)] .$$

For $|m| = |m_1| + |m_2| \neq 0$, we have

$$k_{m_1 m_2} = \mu m_2 f_{m_1 m_2}/(m_1\omega_1 + m_2\omega_2) .$$

If $\mu = 0$, then K = const, and the symmetry field differs from the original field (3.19) by a constant factor. So, leaving aside this trivial case we shall assume $\mu \neq 0$. For $m_2 \neq 0$ we put

$$|m_2 f_{m_1 m_2}| = e^{-|m|} .$$

Then one can show that depending on the Diophantine properties of the irrational number ω_1/ω_2 the $k_{m_1 m_2}$'s are Fourier coefficients of a function of class C^n, but not C^{n+1} (compare with §3.2).

Thus, after an appropriate choice of an analytical function $F = -1/R$, the axis $\alpha = \omega_1/\omega_2$ breaks up into the sets $M_\omega, M_\infty, \dots, M_n, \dots, M_1, M_\emptyset$ such that for $\alpha \in M_\omega$ the system (3.19) possesses an analytic symmetry field, for $\alpha \in M_\infty$ a symmetry field with infinitely differentiable, but not analytic components, for $\alpha \in M_n$ the system admits a symmetry field of class C^n, but not of class C^{n+1}, …, and, finally, for $\alpha \in M_\emptyset$ it does not admit symmetry fields with differentiable components. All the sets $M_\omega, M_\infty, \dots, M_n, \dots, M_1, M_\emptyset$ have power of continuum and are dense everywhere on the real axis. In addition, the measure of the sets $M_\infty, \dots, M_n, \dots, M_1, M_\emptyset$ equals zero.

One should take into account these remarks when studying the existence of nontrivial symmetry groups of dynamical systems.

3.5 Now we discuss some properties of symmetry groups of Hamiltonian systems. Let F be a first integral of a Hamiltonian system

$$\dot{z} = v_H(z) . \tag{3.22}$$

Then the field $v_F(z)$ is a symmetry field of (3.22). Indeed, let L_H and L_F be differential operators corresponding to the Hamiltonian fields v_H and v_F. Then, by the Jacobi identity for Poisson brackets,

$$[L_H, L_F] = L_{\{F,H\}} . \tag{3.23}$$

Thus, given $\{H, F\} = 0$, we have $[v_H, v_F] = 0$. Note that the same is also true in the more general case $\{H, F\} = \text{const}$.

It is natural to refer to the vector fields generated by integrals F of the Hamiltonian system (3.22) as *Hamiltonian symmetry fields*. Of course, not every symmetry field of a Hamiltonian system is a Hamiltonian symmetry field.

These observations can be generalized as follows. Let f be a closed 1-form in the phase space of a system with the Hamiltonian H. Then, locally $f = dF$, and, therefore, f corresponds to a locally Hamiltonian field v_F with Hamiltonian F. If $\{H, F\} = 0$, then the field v_F is a symmetry field of (3.22). We may call the form f (or the multivalued function F) *a multivalued integral* of the Hamiltonian system with a Hamiltonian H. If the form f is exact, then F is a "globally" single-valued integral.

Here is an example of a multivalued integral. Consider the motion of a charged particle on the flat torus $\mathbb{T}^2 = \{(x, y) \bmod 2\pi\}$ in a constant magnetic field. The equations of motion

$$\ddot{x} - \alpha\dot{y} = 0, \qquad \ddot{y} + \alpha\dot{x} = 0; \qquad \alpha = \text{const}$$

are Hamiltonian. They possess two integrals $\dot{x} - \alpha y$ and $\dot{y} + \alpha x$ linear in the velocities, which are multivalued functions in the phase space $\mathbb{T}^2 \times \mathbb{R}^2$.

3.6 Suppose that there exist n independent functions $F_1 = H, F_2, \dots, F_n$ on the $2n$-dimensional phase space $M^{2n} = \{z\}$ such that

$$\{F_i, F_j\} = \sum_k c_{ij}^k F_k , \qquad c_{ij}^k = \text{const} .$$

Then, clearly, the linear space A spanned by $F_1, F_2, \dots, F_n$ is an n-dimensional Lie algebra. The numbers c_{ij}^k are the structure constants of the algebra A for the basis $F_1, F_2, \dots, F_n$.

Theorem 1 [151]. *Suppose that the following conditions are fulfilled*

1) *at all points of the set* $I_a = \{z : F_1(z) = a_1, \dots, F_n(z) = a_n\}$ *the functions* $F_1, \dots, F_n$ *are independent,*
2) $\sum c_{ij}^k a_k = 0$ *for all* $i, j = 1, .., n$,
3) *the algebra* A *is solvable, and* $\{F_1, F_i\} = c_{1i}^1 F_1$.

Then the solutions of the system $\dot{z} = v_H(z)$ *lying on* I_a *can be found by quadratures.*

The set Π of numbers $(a_1, \dots, a_n)$ that satisfy condition 2) is a linear subspace in $\mathbb{R}^n$ of dimension $\geq \dim A - \dim\{A, A\}$, where $\{A, A\}$ is the commutant of the algebra A, i.e., the set of commutators $\{f, g\}$ for $f, g \in A$. Since A is solvable, $\dim \Pi \geq 1$.

Corollary (Liouville's theorem). *If the functions* $\{F_k\}_1^n$ *are independent and pairwise in involution, then each of the Hamiltonian systems* $\dot{z} = v_{F_k}(z)$ $(1 \leq k \leq n)$ *is integrable by quadratures.*

In this case, obviously, $\Pi = \mathbb{R}^n$. Hamiltonian systems with n degrees of freedom having n independent integrals in involution are called *completely integrable*.

The proof of Theorem 1 relies on the generalized Lie theorem in §3.3. The natural mapping of the Lie algebra of the functions $F_1, \dots, F_n$ onto the Lie algebra of the Hamiltonian vector fields $v_{F_1}, \dots, v_{F_n}$ is an isomorphism because of (3.23) and the fact that the linear combination $\sum \lambda_k F_k$ is identically a constant only for $\lambda_1 = \dots = \lambda_n = 0$ (since $F_1, \dots, F_n$ are functionally independent).

Since $\{F_i, F_j\} = 0$ on the surface I_a (condition 2)), the fields $v_{F_1}, \dots, v_{F_n}$ are tangent to it. It is easily seen that the restrictions of these fields on I_a satisfy the conditions of the generalized Lie theorem. Q.E.D.

It turns out that the Lie theorem, in turn, can be deduced from Theorem 1. To perform this we use Liouville's construction which allows us to include the phase flow of the system (3.1) in the phase flow of a Hamiltonian system of doubled dimension. Let u be a field on an n-dimensional manifold $\mathcal{N}\{x\}$. To this field there corresponds the function $F = y \cdot u(x)$ which is defined on the cotangent fiber bundle $M = T^*\mathcal{N}$ endowed with the natural symplectic structure. The coordinates $y_1, \dots, y_n$ are particular integrals of the Hamiltonian system

$$\dot{z} = v_F(z) , \tag{3.24}$$

since

$$\dot{x} = \frac{\partial F}{\partial y} = u(x) , \quad \dot{y} = -\frac{\partial F}{\partial x} = -\left(\frac{\partial u}{\partial x}\right)^T y .$$

The invariant surface $I = \{y = 0\} \subset M$ is diffeomorphic to $\mathcal{N}$. The diffeomorphism $(x, 0) \to x$ allows us to identify the restriction of the Hamiltonian system (3.24) to I with the original dynamical system on $\mathcal{N}$.

Let $u_1, \dots, u_n$ be fields on $\mathcal{N}$, and $F_1 = y \cdot u_1, \dots, F_n = y \cdot u_n$ the corresponding functions on M. One can check easily that

$$\{F_i, F_j\} = y \cdot [u_i, u_j] .$$

If $[u_i, u_j] = \sum c_{ij}^k u_k$, then $\{F_i, F_j\} = -\sum c_{ij}^k F_k$. The manifold $I = \{x, y : F_1 = \dots = F_n = 0\} = \{y = 0\}$ is invariant for the Hamiltonian system with Hamiltonian F_1. Applying Theorem 1 to the system on I and identifying I with $\mathcal{N}$, we arrive at the generalized Lie theorem.

Thus, Theorem 1 can be viewed as a Hamiltonian version of the Lie theorem.

3.7 As an example consider the motion of three points on a line. Let their interaction be governed by the potential

$$\sum_{i \le j} \frac{a_{ij}}{(x_i - x_j)^2} , \qquad a_{ij} = \text{const} ,$$

where x_i, m_i $(i = 1, 2, 3)$ are the coordinates and masses of the points. Consider three functions

$$F_1 = \sum y_i^2/2m_i + V , \quad F_2 = \sum x_i y_i , \quad F_3 = \sum y_i ,$$

where $y_i = m_i \dot{x}_i$ are the momenta of the points. Clearly, F_1 is the total energy of the system, $2F_2 = \sum (m_i x_i^2)^{\cdot}$, and F_3 is the total momentum. It is easily seen that these functions are independent and

$$\{F_1, F_3\} = 0 , \quad \{F_2, F_3\} = -F_3 , \quad \{F_1, F_2\} = 2F_1 . \tag{3.25}$$

The corresponding Lie algebra A is solvable, since $A \supset B \supset C \supset \{0\}$, where the one-dimensional algebra C is generated by the function F_1, and the two-dimensional algebra B is generated by F_1 and F_3. The subalgebras B and C are codimension-one ideals in A and B respectively.

According to Theorem 1, for zero values of the energy and the momentum integrals the motion of three points can be determined by quadratures. This can be done in an explicit form. Note that in the above example the potential V can be replaced by an arbitrary homogeneous function of degree -2 in $x_i - x_j$.

3.8 A modern systematic approach to applications of the theory of Lie groups to differential equations, with partial differential equations as a special case, is presented in [196]. For methods of exact integration of equations of classical dynamics based on the theory of Lie groups and algebras consult the monograph [196].

4 Complete Integrability

4.1 Let M be a symplectic manifold and $F_1, \dots, F_n$ independent functions on M generating a finite-dimensional subalgebra of the Lie algebra $C^\infty(M)$ (i.e., $\{F_i, F_j\} = \sum c_{ij}^k F_k$, $c_{ij}^k = \text{const}$). At each point $x \in M$ the vectors $\sum \lambda_i V_{F_i}$ span an n-dimensional linear subspace $\Pi(x)$ in $T_x M$. The distribution of the planes

Π is "involutive" (if $X, Y \in \Pi$, then $[X, Y] \in \Pi$). Hence, by the Frobenius theorem, through each point $x \in M$ there passes a maximal integral surface $\mathcal{N}_x$ of the distribution Π. The surfaces $\mathcal{N}_x$ may be immersed in M in a rather intricate way. In particular, they are not necessarily closed. If $n = \dim M/2$, then among the integral manifolds of the distribution Π are closed manifolds $M_a = \{x \in M : F_i(x) = a_i, \sum c_{ij}^k A_k = 0\}$. If $x \in M_a$, then $\mathcal{N}_x$ coincides with one of the connected components of M_a. In particular, when the functions $F_1, ..., F_n$ commute pairwise, M is foliated by the closed manifolds M_a. The structure of the integral manifolds M_a is described by

Theorem 1. *Let smooth functions $F_1, ..., F_n : M \to \mathbb{R}$ be in involution: $\{F_i, F_j\} = 0$ $(1 \leq i, j \leq n)$ and $\dim M = 2n$. If*
1) *they are independent on M_a,*
2) *the Hamiltonian fields v_{F_i} $(1 \leq i \leq n)$ are not constrained on M_a,*

then
1) *each connected component of M_a is diffeomorphic to $\mathbb{R}^k \times \mathbb{T}^{n-k}$ ($\mathbb{T}^1$ is a circumference),*
2) *on $\mathbb{R}^k \times \mathbb{T}^{n-k}$ there exist coordinates $y_1, \ldots, y_k; \varphi_1, \ldots, \varphi_{n-k} \bmod 2\pi$ such that written in terms of these coordinates the Hamilton equation $\dot{x} = v_{F_i}(x)$ takes the following form*

$$\dot{y}_m = c_{mi}\,, \qquad \dot{\varphi}_s = \omega_{si}\,, \qquad c, \omega = \text{const}\,.$$

Here is a sketch of the proof of Theorem 1 (all details can be found, for example, in [12, 75]). Consider n one-parameter groups $g_i^{t_i}$ $(t_i \in \mathbb{R})$ which are the phase flows of n Hamiltonian vector fields $F_1, \ldots, F_n$. Since the functions $F_1, ..., F_n$ are in involution, the fields v_{F_i} are tangent to M_a. Hence, the groups g_i transform the smooth manifolds M_a into themselves, and this defines an action of g_i on M_a. By virtue of condition 2) the actions of $g_i^{t_i}$ are defined for all t_i. Since the fields v_i and v_j commute on M_a, the groups g_i and g_j also commute. Therefore, the action of the n-dimensional Abelian group $\mathbb{R}^n = \{t_1, \ldots, t_n\}$ on M_a is defined by

$$g^{t_1,\ldots,t_n}(x) = g_1^{t_1} \cdots g_n^{t_n}\, x\,.$$

According to item 1), the gradients of the functions $F_1, \ldots, F_n$ are independent at all points of M_a. Hence, on M_a the vector fields $v_1, \ldots, v_n$ are also linearly independent. Since M_a is connected, it follows that the action of the group $\mathbb{R}^n$ is free and transitive. Therefore, M_a is diffeomorphic to the quotient manifold $\mathbb{R}^n/\Gamma$, where Γ is a stationary subgroup of $\mathbb{R}^n$ (it consists of the points $s \in \mathbb{R}^n$ such that $g^s x = x$). Since the fields v_1, v_2 are independent, Γ_k is a discrete subgroup in $\mathbb{R}^n$ isomorphic, as we know, to $\mathbb{Z}^{n-k}$ $(0 \leq k \leq n)$. Thus $M_a \simeq \mathbb{R}^n/\mathbb{Z}^{n-k} = \mathbb{R}^k \times \mathbb{T}^{n-k}$. The "global" coordinates $\varphi \bmod 2\pi$, y are expressed linearly in terms of $t_1, \ldots, t_n$. By putting $t_j = \text{const}$ $(j \neq i)$ we obtain solutions of the Hamiltonian system $\dot{x} = v_i(x)$ in terms of linear functions of the time $t_i = t$.

The Hamiltonian system with the Hamiltonian $H = F_i$ $(1 \leq i \leq n)$ is called *completely integrable*.

4.2 Consider the most interesting case when M_a is compact. Then $k = 0$ and, therefore, $M_a \simeq \mathbb{T}^n$. The straight line (uniform) motion on $\mathbb{T}^n = \{\varphi \bmod 2\pi\}$ according to the law $\varphi_i = \varphi_i^0 + \omega_i t$ $(1 \le i \le n)$ is called quasi-periodic. The numbers $\omega_1, \dots, \omega_n$ are the frequencies of this motion. A torus $\mathbb{T}^n$ is called *non-resonant* if the relation $\sum k_i\omega_i = 0$ with integer $k_1, \dots, k_n$ implies that all k_i equal zero. On non-resonant tori the phase trajectories are dense everywhere. This fact follows from the general Weyl result: let $f : \mathbb{T}^n \to R$ be a Riemann integrable function, and the numbers $\omega_1, \dots, \omega_n$ be rationally independent. Then for any point $\varphi^0 \in \mathbb{T}^n$ there exists the limit

$$\lim_{s\to\infty} \frac{1}{s} \int_0^s f(\omega t + \varphi^0)\, dt\,,$$

which is equal to

$$\frac{1}{(2\pi)^n} \int_{\mathbb{T}^n} f(\varphi)\, d\varphi_1 \dots d\varphi_n\,.$$

In particular, let f be the characteristic function of a Jordan measurable domain $D \subset \mathbb{T}^n$. Applying Weyl's theorem to f, we conclude that the average time that a point of a phase trajectory belongs to the domain D is proportional to the measure of D. This fact describes the property of uniform distribution of trajectories on non-resonant tori. If the torus is *resonant*, then the phase trajectories complete tori of fewer dimensions.

Theorem 2. *Suppose that the conditions of Theorem 1 are satisfied and the integral manifold M_a is compact. Then*

1) *a small neighborhood of M_a in the symplectic manifold M is diffeomorphic to the direct product $D \times \mathbb{T}^n$, where D is a small domain in $\mathbb{R}^n$,*
2) *in $D \times \mathbb{T}^n$ there exist symplectic coordinates $I, \varphi \bmod 2\pi$ $(i \in D, \varphi \in \mathbb{T}^n)$ such that the functions $F_1, \dots, F_n$ depend only on I, and the symplectic structure has the form $dI \wedge d\varphi$.*

In particular, written in terms of $I, \varphi \bmod 2\pi$ the Hamilton function of a completely integrable system, whose invariant manifolds are tori, takes the form $H = H(I)$. Besides,

$$\dot{I} = -\partial H/\partial\varphi = 0\,, \quad \dot{\varphi} = -\partial H/\partial I = \omega(I)\,.$$

Therefore, $I(t) = I_0$, $\omega(I) = \omega(I_0)$. The variables I "enumerating" invariant tori in $D \times \mathbb{T}^n$ and the coordinates φ varying uniformly are called *action* and *angle* variables respectively.

Proof of Theorem 2. The functions $I_i = F_i$ and the angles $\varphi \bmod 2\pi$ existing by Theorem 1 can be regarded as coordinates in a neighborhood of the torus $M_a \simeq \mathbb{T}^n$. Since dF_i are linearly independent, the functions I_i, φ_i $(1 \le i \le n)$ define an isomorphism of a neighborhood of M_a and $D \times \mathbb{T}^n$. Now introduce the non-degenerate matrix of Poisson brackets

$$\begin{pmatrix} \{I_i, I_j\} & \{I_i, \varphi_j\} \\ \{\varphi_i, I_j\} & \{\varphi_i, \varphi_j\} \end{pmatrix} = \begin{pmatrix} 0 & a_{ij} \\ -a_{ij} & b_{ij} \end{pmatrix} .$$

According to Theorem 1, the brackets $\{I_i, \varphi_j\}$ are constant on M_a, therefore $a_{ij} = a_{ij}(I)$. Let us show that the b_{ij}'s also depend only on I. Indeed, by the Jacobi identity

$$\{F_m, \{\varphi_i, \varphi_j\}\} + \{\varphi_i, \{\varphi_j, F_m\}\} + \{\varphi_j, \{F_m, \varphi_i\}\} = 0 .$$

The brackets $\{F_m, b_{ij}\} = \alpha_{ij}^m$ do not depend on φ. On the other hand,

$$\alpha_{ij}^m = \sum_s \frac{\partial b_{ij}}{\partial \varphi_s}\{F_m, \varphi_s\} = \sum_s \frac{\partial b_{ij}}{\partial \varphi_s} a_{ms} .$$

Since $\det \|a_{ms}\| \neq 0$, these expressions enable us to find $\partial b_{ij}/\partial \varphi_s$ as functions of I only. Hence, $b_{ij} = \{\varphi_i, \varphi_j\} = \sum f_{ij}^s(I)\varphi_s + g_{ij}$. Here $f_{ij}^s = 0$, because $d\varphi_i$ are single-valued 1-forms in a neighborhood of M_a.

Now perform a change of action variables $I_s = I_s(\mathcal{J}_1, \ldots, \mathcal{J}_n)$ such that $\{\mathcal{J}_i, \varphi_j\} = \delta_{ij}$. To find this transformation, we have to solve the following system

$$a_{ij}(I) = \{I_i, \varphi_j\} = \sum_s \frac{\partial I_i}{\partial \mathcal{J}_s}\delta_{sj} = \frac{\partial I_i}{\partial \mathcal{J}_j} .$$

This can be resolved under the condition

$$\frac{\partial a_{ij}}{\partial \mathcal{J}_s} = \frac{\partial a_{is}}{\partial \mathcal{J}_j} \iff \sum_k \frac{\partial a_{ij}}{\partial I_k} a_{ks} = \sum_k \frac{\partial a_{is}}{\partial I_k} a_{kj}$$

which follows from the Jacobi identity applied to the functions $I_i, \varphi_j, \varphi_k$.

If the variables φ_i do not commute, then one has to pass to the new angle variables $\psi_i \bmod 2\pi$ by the change $\varphi_i = \psi_i + f_i(\mathcal{J})$. The functions f can be determined from the equations

$$b_{ij} = \frac{\partial f_i}{\partial \mathcal{J}_j} - \frac{\partial f_j}{\partial \mathcal{J}_i} ,$$

which can be resolved locally provided the 2-form $\sum b_{ij} dI_i \wedge dI_j$ is closed. But this form is closed since the original symplectic structure is closed. Thus, the existence of symplectic action–angle variables $\mathcal{J}, \psi \bmod 2\pi$ is proved completely.

Remark. Let p, q be symplectic coordinates in $\mathbb{R}^{2n}$, and $\gamma_1, \ldots, \gamma_n$ basis cycles on M_a depending smoothly on the constants $a = (a_1, \ldots, a_n)$. Since the form $p\,dq - I\,d\varphi$ is closed, the difference

$$\oint_{\gamma_s} p\,dq - \oint_{\gamma_s} I\,d\varphi = \oint_{\gamma_s} p\,dq - 2\pi I_s$$

is constant. Therefore,

$$I_s = \frac{1}{2\pi} \oint_{\gamma_s} p\,dq\,, \qquad 1 \leq s \leq n\,, \tag{4.1}$$

because the action variables themselves are defined by addition of a constant. Formulas (4.1) are most effective in analysing systems with separable variables (see §7).

A Hamiltonian system with a Hamilton function $H(I)$ is called *non-degenerate* (in the domain $D \times \mathbb{T}^n$) if the Jacobian

$$\frac{\partial \omega}{\partial I} = \det \left\| \frac{\partial^2 H}{\partial I^2} \right\|$$

does not vanish in D. In this case almost all (in the sense of the Lebesgue measure) invariant tori are non-resonant, whereas resonant tori are dense everywhere in $D \times \mathbb{T}^n$.

4.3 Following Bohl, a continuous function $t \to g(t)$, $t \in \mathbb{R}$ is called quasi-periodic (conditionally periodic) if $g(t) = f(\omega_1 t, \ldots, \omega_n t)$, where f is some continuous function on the n-dimensional torus $\mathbb{T}^n = \{(\varphi_1, \ldots, \varphi_n) \bmod 2\pi\}$, $\omega_1, \ldots, \omega_n =$ const.

Let $F : M^{2n} \to \mathbb{R}$ be a continuous function on a symplectic manifold, $\dot{x} = v_H(x)$ a completely integrable Hamiltonian system with compact invariant surfaces M_a. Consider the solution of this system with the initial condition $x_0 \in M_a \simeq \mathbb{T}^n$. By Theorem 1, $f(t) = F(x(t, x_0))$ is a quasi-periodic function of time. The numbers $\omega_1, \ldots, \omega_n$ are exactly the frequencies of the quasi-periodic motion on the n-dimensional torus M_a. In particular, if $M^{2n} = \mathbb{R}^{2n}$, then all global canonical coordinates are represented by quasi-periodic functions.

By Weyl's theorem, there exists the time average

$$\lim_{T \to \infty} \frac{1}{T} \int_0^T F(x(t, x_0))\,dt\,, \tag{4.2}$$

which obviously depends on the initial condition x_0. It turns out that if the point x_0 belongs to a non-resonant invariant torus, then the average (4.2) is continuous at this point. Generically, the function (4.2) is not continuous on M^{2n}. The relevant questions are discussed in [131].

4.4 In some problems of Hamiltonian mechanics the number of known integrals exceeds the number of degrees of freedom, but not all integrals commute with each other. Certain integrability conditions for Hamiltonian systems with "excessive" sets of integrals are given in [175, 190].

Theorem 3. *Suppose that a Hamiltonian system on the symplectic manifold M^{2n} possesses $n + k$ integrals $F_1, F_2, \ldots, F_{n+k}$ that are independent on the surface*

$$M_c = \{x \in M^{2n} : F_i(x) = c_i,\ 1 \leq i \leq n + k\}\,,$$

and the rank of the matrix of Poisson brackets

$$\| \{F_i, F_j\} \| \tag{4.3}$$

is constant in a neighborhood of M_c. *Then, if this surface is connected and compact, and the rank of the matrix* (4.3) *does not exceed* $2k$, *then* M_c *is diffeomorphic to an* $(n-k)$*-dimensional torus, and on* M_c *one can choose the angle variables* $\varphi_1, \dots, \varphi_{n-k} \bmod 2\pi$ *such that the Hamilton equations take the following form*

$$\dot{\varphi}_s = \omega_s = \text{const}, \qquad 1 \le s \le n-k .$$

Corollary. *Let a Hamiltonian system with n degrees of freedom have* $2n-2$ *independent integrals. Then connected compact level surfaces of these integrals are two-dimensional tori, and the motion on the tori is quasi-periodic.*

Indeed, since the matrix (4.3) is skew symmetric and the Hamilton function commutes with all integrals, the rank of this matrix does not exceed $2n-4 = 2(n-2) = 2k$. The integrability by quadratures of a Hamiltonian system with n degrees of freedom admitting $2n-2$ independent integrals was established by Jacobi, who used Euler's method of the integrating multiplier [105]. We have already applied the Jacobi theorem in §2.7.

Theorem 3 can be deduced from the Lie–Cartan theorem [14, Chap.3] and the arguments of [190].

5 Examples of Completely Integrable Systems

5.1 The equations of rotation of a heavy rigid body around a fixed point are Hamiltonian on the invariant manifolds

$$\mathcal{I}_c = \{\omega, \gamma : (I\omega, \gamma) = c,\ (\gamma, \gamma) = 1\} .$$

In addition, one extra integral always exists: this is the energy integral. Thus, for complete integrability of these equations on I_c it is sufficient to have one additional integral independent of the three above. As already mentioned, there are 6 parameters in the heavy top problem: three eigenvalues I_1, I_2, I_3 of the inertia operator, and three coordinates r_1, r_2, r_3 of the center of mass relative to the principal axes of the body. The known integrability cases are as follows:

1) Euler's case (1750): $r_1 = r_2 = r_3 = 0$. The additional integral is $(I\omega, I\omega)$.
2) Lagrange's case (1788): $I_1 = I_2,\ r_1, r_2 = 0$. The new integral is $\omega_3 = \text{const}$.
3) Kovalevskaya's case (1889): $I_1 = I_2 = 2I_3,\ r_3 = 0$. The first integral discovered by Kovalevskaya is

$$(\omega_1^2 - \omega_2^2 - \nu\gamma_1)^2 + (2\omega_1\omega_2 - \nu\gamma_2)^2 ,$$

where $\nu = \varepsilon r / I_3,\ r^2 = r_1^2 + r_2^2$.

4) The Goryachev–Chaplygin case (1900): $I_1 = I_2 = 4I_3$, $r_3 = 0$. An additional integral exists under the condition $c = (I\omega, \gamma) = 0$. So, in contrast to cases 1)–3), here an integrable Hamiltonian system exists only on the invariant surface $\mathcal{J}_0$.

Note that all the integrable cases mentioned above form codimension-three manifolds in the 6-dimensional space of the parameters I_i, r_j.

5.2 For the first two cases the equations of motion were closely investigated from various points of view in classical works by Euler, Lagrange, Poisson, and Jacobi. Kovalevskaya's case is nontrivial in many respects. It was found by requiring the solutions of the Euler–Poisson equations to be meromorphic in the complex time plane. The Goryachev–Chaplygin case is considerably simpler: it can be integrated byseparation of variables. To show this we write the Hamiltonian in terms of the special canonical variables L, G, l, g (see §2, Chap. I):

$$H = \frac{G^2 + 3L^2}{8I_3} + \mu\left(\frac{L}{G}\cos l \sin g + \sin l \cos g\right) , \qquad \mu = \varepsilon r .$$

Consider the canonical transformation

$$L = -p_1 - p_2 , \quad G = p_2 - p_1 , \quad q_1 = -l - g , \quad q_2 = g - l .$$

In the new symplectic coordinates

$$H = \frac{p_1^3 - p_2^3}{2I_3(p_1 - p_2)} - \mu\left(\frac{p_1 \sin q_1}{p_1 - p_2} + \frac{p_2 \sin q_2}{p_1 - p_2}\right) .$$

Putting this expression equal to h and multiplying by $p_1 - p_2$, we see that it separates:

$$hp_1 - p_1^3/2I_3 + \mu p_1 \sin q_1 = hp_2 - p_2^3/2I_3 - \mu p_2 \sin q_2 .$$

Put

$$p_1^3/2I_3 - \mu p_1 \sin q_1 - Hp_1 = F , \quad p_2^3/2I_3 + \mu p_2 \sin q_2 - Hp_2 = F . \qquad (5.1)$$

The function F is a first integral of the equations of motion (see §7). In the special canonical variables

$$F = \frac{L(L^2 - G^2)}{8I_3} + \frac{L_2 - G_2}{2G}\mu \sin l \cos g ,$$

while written in terms of the traditional Euler–Poinsot variables ω, γ, it has the form

$$F = -2I_3^2 f , \qquad f = \omega_3(\omega_1^2 + \omega_2^2) + \nu\omega_1\gamma_3 , \qquad \nu = \mu/I_3 .$$

Now we write down a closed system of equations describing the evolution of p_1, p_2:

$$\dot{p}_1 = -\frac{\partial H}{\partial q_1} = -\frac{\mu p_1}{p_1 - p_2}\cos q_1 , \quad \dot{p}_2 = -\frac{\partial H}{\partial q_2} = -\frac{\mu p_2}{p_1 - p_2}\cos q_2 ,$$

or, taking relations (5.1) into account,

$$\dot{p}_1 = \frac{\pm\sqrt{\Phi(p_1)}}{p_1 - p_2}, \quad \dot{p}_2 = \frac{\pm\sqrt{\Phi(p_2)}}{p_1 - p_2}, \tag{5.2}$$

where $\Phi(z) = \mu^2 z^2 - (F + Hz - z^3/2I_3)^2$ is a polynomial of degree 6. The solutions of these equations are expressed in terms of hyperelliptic functions of time. The variables vary in disjoint intervals $[a_1, b_1]$ and $[a_2, b_2]$, where a_i, b_i are adjacent roots of the polynomial Φ, between which it takes positive values.

Now introduce angular variables $\varphi_1, \varphi_2 \bmod 2\pi$ by the formulas

$$\varphi_i = \frac{\pi}{\tau_i} \int_{a_i}^{p_i} \frac{dz}{\pm\sqrt{\Phi(z)}}, \quad \tau_i = \int_{a_i}^{b_i} \frac{dz}{\sqrt{\Phi(z)}}. \tag{5.3}$$

In the new variables Eqs. (5.2) take the following form

$$\dot{\varphi}_i = \frac{\pi}{2\tau_i(p_1(\varphi_1) - p_2(\varphi_2))}, \quad i = 1, 2, \tag{5.4}$$

where $p_i(z)$ are real-valued hyperelliptic functions of period 2π defined by relations (5.3). Since the trajectories of the system (5.4) are straight lines on $\mathbb{T}^2 = \{\varphi \bmod 2\pi\}$, the ratio of frequencies of the corresponding quasi-periodic motions equals the ratio τ_1/τ_2 of the real periods of the hyperelliptic integral

$$\int_{z_0}^{z} \frac{dz}{\sqrt{\Phi(z)}}.$$

This remarkable fact holds also for the Kovalevskaya system. For the details see [131].

5.3 The rigid body motion in an ideal fluid is far richer in integrable cases. Recall that the Kirchhoff equations

$$\dot{m} = m \times \frac{\partial H}{\partial m} + p \times \frac{\partial H}{\partial p}, \quad \dot{p} = p \times \frac{\partial H}{\partial m},$$
$$2H = (Am, m) + (2Bm, p) + (Cp, p)$$

always have three independent integrals $F_1 = H$, $F_2 = (m, p)$, $F_3 = (p, p)$. Thus, the integration of these equations is reduced to looking for a fourth integral independent of F_1, F_2, F_3. Assume that

$$A = \mathrm{diag}(a_1, a_2, a_3), \quad B = \mathrm{diag}(b_1, b_2, b_3), \quad C = \mathrm{diag}(c_1, c_2, c_3).$$

The known integrable cases in this problem are as follows.

1) Kirchhoff's case (1870): $a_1 = a_2$, $b_1 = b_2$, $c_1 = c_2$. The fourth integral is $F_4 = m_3$. The equations are easily integrated in terms of elliptic functions.
2) Clebsch's case (1871): $b_1 = b_2 = b_3 = b$ and

$$a_1^{-1}(c_2 - c_3) + a_2^{-1}(c_3 - c_1) + a_3^{-1}(c_1 - c_2) = 0. \tag{5.5}$$

The extra integral has the form

$$F_4 = m_1^2 + m_2^2 + m_3^2 - \frac{1}{a_1 a_2 a_3}(a_1 p_1^2 + a_2 p_2^2 + a_3 p_3^2) .$$

The parameter b is not essential: it does not appear in the equations.

Since $\{F_1, F_4\} = 0$, we conclude that F_1 is an integral of the Kirchhoff equations with the Hamiltonian $H = F_4$. However, this does not give us a new integrable case, because the coefficients of F_4 also satisfy the condition (5.5). In old papers on hydrodynamics, the case when (5.5) is fulfilled and all numbers a_1, a_2, a_3 are distinct is referred to as the first Clebsch case. If $a_1 = a_2 \neq a_3$, it follows from (5.5) that $c_1 = c_2$. This gives the second Clebsch case (i.e., the particular Kirchhoff case). Finally, for $a_1 = a_2 = a_3$ we have the third integrable Clebsch case (with the function λF_4 , $\lambda = \text{const}$ as the Hamiltonian). Note that the first and third cases are "dual" to each other in the sense that the corresponding phase trajectories may be regarded as different windings of the same invariant tori defined by the integrals H, F_2, F_3, F_4.

3) The Steklov–Lyapunov case (1893):

$$a_1^{-1}(b_2 - b_3) + a_2^{-1}(b_3 - b_1) + a_3^{-1}(b_1 - b_2) = 0 , \tag{5.6}$$

$$c_1 - \frac{(b_2 - b_3)^2}{a_1} = c_2 - \frac{(b_3 - b_1)^2}{a_2} = c_3 - \frac{(b_1 - b_2)^2}{a_3} . \tag{5.7}$$

From here one can conclude that

$$b_j = \mu(a_1 a_2 a_3) a_j^{-1} + \nu , \quad c_1 = \mu^2 a_1 (a_2 - a_3)^2 + \nu' , \ \ldots , \qquad \mu, \nu, \nu' = \text{const} . \tag{5.8}$$

The additional integral is

$$F_4 = \sum_j (m_j^2 - 2\mu(a_j + \nu) m_j p_j) + \mu^2((a_2 - a_3)^2 + \nu'') p_1^2 + \ldots .$$

The parameters ν, ν', ν'' are irrelevant: they are coefficients of the classical Kirchhoff integrals F_1 and F_2.

Under the condition $a_1 \neq a_2 \neq a_3 \neq a_1$, the integral F_4 was found by Steklov. If $a_1 = a_2 \neq a_3$, then (5.6) and (5.7) imply that $b_1 = b_2$ and $c_1 = c_2$. This is Kirchhoff's case. For $a_1 = a_2 = a_3$, the expressions (5.8) give rise to the trivial degenerate case. However, as noticed by Lyapunov, to this case there corresponds the Hamiltonian λF_4 ($\lambda = \text{const}$), and the function F plays the role of an additional integral. Therefore, in the above sense, the Steklov and Lyapunov cases are also "dual" to each other.

4) The particular Chaplygin case (1902):

$$2H = a(m_1^2 + m_2^2 + 2m_3^2) + b(m, p) + a((d + c)p_1^2 + (d - c)p_2^2 + d p_3^2) .$$

The parameters b, d do not contribute to the motion equations. Under the condition $F_2 = (m, p) = 0$ there exists the additional integral

$$F_4 = (m_1^2 - m_2^2 + c p_3^2)^2 + 4 m_1^2 m_2^2 ,$$

whose structure is similar to that of Kovalevskaya's integral.

Note that in the 9-dimensional space of the parameters $a_1, \dots, c_3$ the Kirchhoff, Clebsch, and Steklov–Lyapunov cases form manifolds of one and the same codimension 3.

5.4 The problem of motion of n point vortices in a plane (§8, Chap. I) is completely integrable for $n \le 3$. The case $n = 1$ is trivial; for $n = 2$ independent commuting integrals are, for example, H and M, for $n = 3$ such integrals are H, M, and $P_x^2 + P_y^2$. In the four vortex problem there are as many integrals as degrees of freedom, but not all of them commute.

Consider more closely the case when the sum of the intensities Γ_s is zero. Then the integrals P_x and P_y are in involution. If their values equal zero, the equations of motion of four point vortices turn out to be integrable by Liouville's theorem. The idea of solving this system is based on an appropriate linear canonical transformation which is well known in celestial mechanics in connection with the "elimination" of the motion of the mass center in the n body problem. Namely, without loss of generality we may put $\Gamma_1 = \Gamma_2 = -\Gamma_3 = -\Gamma_4 = -1$. Consider the linear transformation $x, y \to \alpha, \beta$ given by

$$
\begin{aligned}
x_1 &= -\beta_4\,, & y_1 &= -\alpha_3 - \alpha_4 + \beta_2\,,\\
x_2 &= \beta_3 - \beta_4\,, & y_2 &= -\alpha_3 - \beta_1 + \beta_2\,,\\
x_3 &= \alpha_1 + \alpha_2 - \beta_4\,, & y_3 &= \beta_2\,,\\
x_4 &= -\alpha_1 + \beta_3 - \beta_4\,, & y_4 &= -\beta_1 + \beta_2\,.
\end{aligned}
$$

In the new coordinates we have $p_x = \alpha_2$, $P_y = \alpha_4$. Therefore, the Hamilton function H does not depend on the conjugate variables β_2, β_4. Thus, the number of degrees of freedom is lowered by two: we obtain a family of Hamiltonian systems with two degrees of freedom depending on two parameters α_2, α_4. The corresponding symplectic coordinates are the variables $\alpha_1, \alpha_3, \beta_1, \beta_3$. For $\alpha_2, \alpha_4 = 0$, the function F is an integral of the "reduced" system. As a result, the latter Hamiltonian system is completely integrable. In particular, the functions $\alpha_1, \alpha_3, \beta_1, \beta_3|_t$ can be determined by quadratures. Then the other "cyclic" coordinates $\beta_2(t), \beta_4(t)$ can be found from the equations

$$
\dot\beta_2 = \frac{\partial K}{\partial \alpha_2}\,, \quad \dot\beta_4 = \frac{\partial K}{\partial \alpha_4}\,; \qquad K(\alpha, \beta) = H(x, y)|_{\alpha, \beta}
$$

by taking the integrals.

5.5 Consider also Hamiltonian systems with two degrees of freedom whose Hamiltonian has the form

$$
H = \frac{1}{2}(p_x^2 + p_y^2) + \frac{1}{2}ax^2 + \frac{1}{2}by^2 - x^2y - \frac{\varepsilon}{3}y^3\,; \qquad a, b, \varepsilon = \text{const}\,.
$$

Such systems describe a point mass motion in the Euclidean plane $\mathbb{R}^2 = \{x, y\}$ in a force field with potential of degree three. For $a = b = -\varepsilon = 1$ we have the

Hénon–Heiles system mentioned above. The other known integrability cases are as follows.

1) $a = b, \varepsilon = 1$. The Hamilton equations are separated after changing to the variables $x - y, x + y$. This leads to the existence of an additional integral quadratic in the momenta (Aizawa and Saitô, 1972).
2) $\varepsilon = 6$, a and b are arbitrary. Then there is the following integral found by D. Greene:

$$x^4 + 4x^2y^2 + 4p_x(p_x y - p_y x) - 4ax^2 y + (4a - b)(p_x^2 + ax^2) .$$

As noticed by Trev, for $b = 4a$ the corresponding Hamilton equations are separable in parabolic coordinates.

3) $\varepsilon = 16,\ b = 16a$. The additional integral of degree four in the momenta was found by Hall [95]. For $b = 1$ it has the form

$$\frac{1}{4}p_x^4 + \left(\frac{x^2}{2} + 4x^2y\right)p_x^2 - \frac{4}{3}x^3 p_x p_y + \frac{x^4}{4} - \frac{4}{3}x^4 y - \frac{8}{9}x^6 - \frac{16}{3}x^4y^2 .$$

6 Isomorphisms of some Integrable Hamiltonian Systems

Local isomorphisms of non-degenerate completely integrable Hamiltonian systems, which were considered in the Introduction, can sometimes be extended to isomorphisms in the whole. Here are some relevant examples.

6.1 Consider the Jacobi problem on the motion of a point mass on an ellipsoid

$$ax^2 + by^2 + cz^2 = 1 .$$

Let x, y, z, p_x, p_y, p_z be the natural "excessive" canonical variables (§1.9, Chap. II). After the canonical change

$$p_1 = p_x/\sqrt{a}\,, \quad x_1 = x\sqrt{a}\,, \ \ldots$$

the Hamiltonian of the problem takes the form

$$H = \frac{abc\,G}{2F}\,, \qquad F = ax_1^2 + bx_2^2 + cx_3^2\,,$$

$$G = \frac{(p_2x_3 - p_3x_2)^2}{a} + \frac{(p_3x_1 - p_1x_3)^2}{b} + \frac{(p_1x_2 - p_2x_1)^2}{c} .$$

Recall (see §3.6, Chap. I) that when the area integral equals zero, the rotation of a rigid body in an axisymmetric force field is described by the Hamilton equations with the Hamiltonian

$$E = \frac{1}{2}\left\{\frac{(p_2x_3 - p_3x_2)^2}{I_1} + \frac{(p_3x_1 - p_1x_3)^2}{I_2} + \frac{(p_1x_2 - p_2x_1)^2}{I_3}\right\}$$
$$+ V(x_1, x_2, x_3) .$$

Here x_1, x_2, x_3 are the direction cosines of the unit vector along the symmetry axis of the field. The factor $f = x_1^2 + x_2^2 + x_3^2$, which does not contribute to the motion equation, is omitted.

Now put $V = \varepsilon(I_1 x_1^2 + I_2 x_2^2 + I_3 x_3^2)$. Then we come to the equations of the Brun problem which, according to Steklov's analogy, are equivalent to Clebsch's integrable case of the Kirchhoff equations. Putting $I_1 = a, I_2 = b, I_3 = c$ transforms the Hamiltonian E to $G/2 - \varepsilon F$. It is clear that the equations $\{E = 0\}$ and $\{H = abc\varepsilon\}$ define the same hypersurface in $\mathbb{R}^6 = \{p, x\}$. We are going to show that the trajectories of these two dynamical systems coincide. Indeed,

$$\dot{x} = \frac{\partial H}{\partial p} = \frac{abc}{2F}\frac{\partial G}{\partial p}, \quad \dot{p} = -\frac{\partial H}{\partial x} = -\frac{abc}{2}\left(\frac{\partial G}{\partial x}F^{-1} - G\frac{\partial F}{\partial x}F^{-2}\right), \tag{6.1}$$

$$\dot{x} = \frac{\partial E}{\partial p} = \frac{1}{2}\frac{\partial G}{\partial p}, \quad \dot{p} = -\frac{\partial E}{\partial x} = -\frac{1}{2}\frac{\partial G}{\partial x} + \varepsilon\frac{\partial F}{\partial x}. \tag{6.2}$$

Put $H = abc\varepsilon$ (i.e., $G/F = 2\varepsilon$). Then after the time change $D\,d\tau = (abc/F)\,dt$ the system (6.1) takes the form

$$\frac{dx}{d\tau} = \frac{1}{2}\frac{\partial G}{\partial p}, \quad \frac{dp}{d\tau} = -\frac{1}{2}\frac{\partial G}{\partial x} + \varepsilon\frac{\partial F}{\partial x}. \tag{6.3}$$

Since (6.2) and (6.3) are identical, the systems (6.1) and (6.2) have the same trajectories. In particular, the integrals of these system coincide.

Thus, the Jacobi problem is a special case of the Clebsch–Tisserand–Brun problem in rigid body dynamics.

6.2 Consider also the rotation of a rigid body in a force field with the potential

$$V = (a, \alpha) + (b, \beta) + (c, \gamma),$$

where a, b, c are constant vectors. This can be interpreted as the potential energy of a heavy charged magnetized body rotating in homogeneous gravitational, electric, and magnetic fields. The motion of the body is described by Eqs. (3.1–3.2) of Chap. II.

Bogoyavlensky proved the complete integrability of this problem for the case when the inertia tensor is spherical. He reduced the motion equations to the Neumann problem of a point mass motion on the three-dimensional sphere in a force field with a quadratic potential [24]. In order to show this we use the Euler identity

$$\begin{aligned}(\dot{\xi}^2 + \dot{\eta}^2 + \dot{\zeta}^2 + \dot{\chi}^2)(\xi^2 + \eta^2 + \zeta^2 + \chi^2) \\ = (\chi\dot{\xi} - \xi\dot{\chi} + \zeta\dot{\eta} - \eta\dot{\zeta})^2 + (\chi\dot{\eta} - \eta\dot{\chi} + \xi\dot{\zeta} - \zeta\dot{\xi})^2 \\ + (\chi\dot{\zeta} - \zeta\dot{\chi} + \eta\dot{\xi} - \xi\dot{\eta})^2 + (\xi\dot{\xi} + \eta\dot{\eta} + \zeta\dot{\zeta} + \chi\dot{\chi})^2.\end{aligned} \tag{6.4}$$

This follows from the multiplication rule for quaternions. Using this identity, Lagrange proved that any natural number can be represented as a sum of four squares. In view of (6.4) and the relations (3.6) in Chap. I the kinetic energy of the body can be written in the form

$$T = \frac{I}{2}(\omega_1^2 + \omega_2^2 + \omega_3^2) = 2I(\dot{\xi}^2 + \dot{\eta}^2 + \dot{\zeta}^2 + \dot{\chi}^2),$$

where I is the inertia tensor. This expression also describes the kinetic energy of a point mass motion on the three-dimensional sphere

$$\xi^2 + \eta^2 + \zeta^2 + \chi^2 = 1.$$

It remains to note that the direction cosines depend quadratically on the quaternion components ξ, η, ζ, χ (see (3.7), Chap. I).

6.3 Some Hamiltonian systems, whose equations have algebraic right-hand sides, turn out to be isomorphic under fractionally linear transformations with singularities. The first example follows from Volterra's investigations of the free rotation of the Zhukovsky gyrostat, which is a generalization of the Euler problem [232]. Recall that motion of the gyrostat is governed by the equations

$$I\dot{y} = (Iy + \lambda) \times y, \qquad \lambda \in \mathbb{R}^3. \tag{6.5}$$

We are going to show that this system can be reduced to the Euler equations and thus can be solved explicitly. Namely, Volterra represented Eqs. (6.5) in the remarkable form

$$\dot{y}_i = \frac{\partial(f_1, f_2)}{\partial(y_j, y_k)}, \qquad f_1 = \frac{(Iy, y)}{2\sqrt{I_1 I_2 I_3}}, \qquad f_2 = \frac{(Iy + \lambda)^2}{2\sqrt{I_1 I_2 I_3}}. \tag{6.6}$$

Here and below the indices i, j, k range over even permutations of the numbers $1, 2, 3$. The functions f_1, f_2 are integrals of (6.5). Put $f_1 = h_1$, $f_2 = h_2$, $h_1, h_2 =$ const. Introduce the homogeneous coordinates

$$(z_1, z_2, z_3, z_4) \in \mathbb{C}^4 \setminus \{0\}$$

such that

$$y_i = z_i/z_4, \qquad i = 1, 2, 3. \tag{6.7}$$

Then Eqs. (6.6) give rise to the following differential relations

$$z_4\dot{z}_i - z_i\dot{z}_4 = \frac{\partial(\varphi_1, \varphi_2)}{\partial(z_j, z_k)}, \tag{6.8}$$

$$\varphi_l(z) = z_4^2\left(f_l\left(\frac{z_1}{z_4}, \frac{z_2}{z_4}, \frac{z_3}{z_4}\right) - h_l\right), \qquad l = 1, 2.$$

Note that φ_1, φ_2 are quadratic forms of z_i and $\varphi_1 = 0, \varphi_2 = 0$. Using these conditions, from (6.8) we also deduce the relations

$$z_i\dot{z}_j - z_j\dot{z}_i = \frac{\partial(\varphi_1, \varphi_2)}{\partial(z_k, z_4)}. \tag{6.9}$$

According to a well-known theorem from algebra, under an appropriate non-degenerate linear transformation

$$z_r = \sum_{s=1}^{4} c_{es}\xi_s , \qquad c_{rs} \in \mathbb{C} \tag{6.10}$$

the quadratic forms φ_1, φ_2 can be diagonalized as follows

$$\varphi_1 = \frac{1}{2}\sum \mu_k \xi_k^2 , \quad \varphi_2 = \frac{1}{2}\sum \xi_k^2 .$$

In the coordinates y_i the change (6.10) defines a projective (fractionally linear) transformation $\mathbb{C}^3 \to \mathbb{C}^3$. Itconserves the form of the relations (6.8), (6.9), whereas their right-hand sides are divided by the Jacobian $\kappa = \det \|c_{rs}\|$. In the new variables ξ_k these relations take the form

$$\begin{aligned} \xi_4\dot{\xi}_i - \xi_i\dot{\xi}_4 &= (\mu_k - \mu_j)\xi_j\xi_k/\kappa , \\ \xi_i\dot{\xi}_j - \xi_j\dot{\xi}_i &= (\mu_k - \mu_4)\xi_j\xi_4/\kappa . \end{aligned} \tag{6.11}$$

These are exactly the Euler equations describing free motion of a rigid body written in terms of homogeneous coordinates ξ_i such that $y_i = \xi_i/\xi_4$ are components of the angular momentum of the body. The coefficients $\mu_1^{-1}, \mu_2^{-1}, \mu_3^{-1}$ are proportional to the principal moments of inertia of the body. We note that they depend not only on I and λ, but also on the integral constants h_1, h_2. The isomorphism between the two problems is the composition of two projective transformations with one affine. In the real phase space this is an isomorphism with singularities which corresponds to the different topology of invariant curves of these two problems.

6.4 Recently some other unexpected isomorphisms of solved integrable problems of dynamics have been discovered. For example, Heine and Khorozov [96] found a fractionally linear transformation which takes the equations and integrals of Kovalevskaya's problem to those of the Clebsch case mentioned above. An analogous isomorphism between the Goryachev–Chaplygin case and the three particle Toda system was discovered in [19]. These results were obtained by thorough investigation of the complex geometry of algebraic completely integrable Hamiltonian systems (see [4]).

7 Separation of Variables

7.1 The most simple and effective way of explicitly solving Hamiltonian systems is the method of separation of variables. According to Jacobi, the problem of solving the canonical equations

$$\dot{p} = -\partial H/\partial q , \quad \dot{q} = -\partial H/\partial p ; \qquad (p, q) \in \mathbb{R}^{2n} \tag{7.1}$$

is reduced to searching for a complete integral of the Hamilton–Jacobi equation

$$\frac{\partial V}{\partial t} + H\left(\frac{\partial V}{\partial q}, q, t\right) = 0 . \tag{7.2}$$

A complete integral is an n-parameter family of solutions $V(t, q, x)$ that satisfies the non-degeneracy condition

$$\det \| \partial^2 V/\partial q \partial x \| \neq 0 .$$

If the Hamiltonian H does not depend explicitly on time, then by the substitution $V(q, t) = -Kt + W(q)$ Eq. (7.2) is reduced to the form

$$H(\partial W/\partial q, q) = K(x) . \tag{7.3}$$

Since

$$\det \| \partial^2 W/\partial q \partial x \| = \det \| \partial^2 V/\partial q \partial x \| \neq 0 ,$$

the function $W(q, x)$, being a "complete" integral of (7.3), can be regarded as the generating function of the canonical transformation $(p, q) \to (y, x)$:

$$y = \partial W/\partial x , \quad p = \partial W/\partial q .$$

In the new canonical coordinates x, y the function $H(p, q) = K(x)$ does not depend on y and, therefore, the Hamilton equations are immediately integrated:

$$x = x_0 , \quad y = y_0 + \omega(x_0)t , \quad \omega(x) = \partial K/\partial x .$$

We emphasize that the function K in (7.3) is not determined and to specify it uniquely one must impose supplementary conditions. The usual choice is $K(x_1, \ldots, x_n) = x_n$: the trajectories of the system with this Hamiltonian are straight lines in $\mathbb{R}^2 = \{x, y\}$.

7.2 If (7.3) possesses a complete integral of the form

$$W(q, x) = \sum_{k=1}^{n} W_k(q_k, x_1, \ldots, x_n) ,$$

then we say that the variables $q_1, \ldots, q_n$ are *separable*.

Here are examples of Hamiltonians for which equation (7.3) can be solved by separation of variables:

$$(a)\ H = f_n(f_{n-1}(\cdots f_2(f_1(p_1, q_1), p_2, q_2) \ldots, p_{n-1}, q_{n-1}), p_n, q_n) ,$$

$$(b)\ H = \sum f_s(p_s, q_s) / \sum g_s(p_s, q_s) .$$

For case (a) we may put

$$W = W_1(q_1, x_1) + W_2(q_2, x_1, x_2) + \cdots + W_n(q_n, x_{n-1}, x_n) ,$$

where the functions W_k satisfy the equations

$$f_k\left(x_{k-1}, \frac{\partial W}{\partial q_k}, q_k\right) = x_k, \qquad 2 \le k \le n,$$
$$f_1\left(\frac{\partial W_1}{\partial q_1}, q_1\right) = x_1 .$$

Since W_k depends only on q_k, and $x_1, \dots, x_n$ are parameters, these equations may be regarded as ordinary differential equations, which can be easily integrated.

For case (b) we assume that $K = x_0$ and seek a complete integral of the equation

$$\sum_k \left[x_0 g_k\left(\frac{\partial W}{\partial q_k}, q_k\right) - f_k\left(\frac{\partial W}{\partial q_k}, q_k\right)\right] = 0$$

in the form

$$\sum W_k(q_k, x_0, x_k),$$

where W_k, as a function of q_k, satisfies the ordinary differential equation

$$x_0 g_k\left(\frac{\dot{W}_k}{\dot{q}_k}, q_k\right) - f_k\left(\frac{\dot{W}_k}{\dot{q}_k}, q_k\right) = x_k, \qquad \sum x_k = 0 .$$

Here x_0 and any $n-1$ variables of $x_1, \dots, x_n$ may be viewed as independent parameters.

We emphasize that we do not intend to find all solutions of (7.3); it is sufficient to know at least one n-parameter family of such solutions.

Note that the cases (a) and (b) may occur simultaneously, and, besides, more complicated kinds of separation of variables are possible. As an example, we recall Stäckel's result (1895). Let Φ be the determinant of the matrix $\|\varphi_{ij}(q_j)\|$ $(1 \le i, j \le n)$, and Φ_{ij} be the algebraic complement (cofactor) of φ_{ij}. Suppose that in the symplectic coordinates $p_1, \dots, p_n, q_1, \dots, q_n$ the Hamilton function has the following form

$$H(p, q) = \sum_{s=1}^{n} \Phi_{1s}(q) f_s(p_s, q_s)/\Phi(p, q) . \tag{7.4}$$

Then the Hamilton equations can be integrated. Putting $K(x) = x_1$, we write Eq. (7.3) as follows

$$\sum \Phi_{1m}\left[\sum_k x_k \varphi_{km}(q_m) - f_m\left(\frac{\partial W}{\partial q_m}, q_m\right)\right] = 0 .$$

Its complete integral can be found as a sum

$$W(q, x) = \sum_m W_m(q_m, x_1, \dots, x_n) ,$$

where W_k, as a function of q_k, satisfies the equation

$$f_m\left(\frac{dW_m}{dq_m}, q_m\right) = \sum_k x_k \varphi_{km}(q_m) .$$

Stäckel's Hamiltonian systems contain Liouville's systems as a special case: the latter are described by Hamiltonians

$$\frac{1}{2\sum_{i=1}^n A_i} \sum_{j=1}^n \left[\frac{p_j^2}{B_j} + C_j\right] . \tag{7.5}$$

The functions A_i, B_i, C_i depend only on the coordinate q_i, and $\sum B_i, \sum C_i$ differ from zero. Liouville's systems are often encountered in applications.

7.3 The problem on separation of variables was quite popular in the last century. Its urgency can be seen from the following observation: if a Hamiltonian system with the Hamiltonian $H = \sum g^{ij}(q) p_i p_j / 2$ is solvable by separation of variables, then in the Laplace–Beltrami equation

$$\Delta f = -\frac{1}{\sqrt{g}} \sum_{i,j} \frac{\partial}{\partial q_i} \left(\sqrt{g} g^{ij} \frac{\partial f}{\partial q_j}\right) = 0 , \qquad g = \det \| g_{ij} \|$$

the variables q are also separable (see [177]).

Levi-Civita found a criterion for a system with a Hamilton function $H(p, q)$ to be solved by separation of variables in given symplectic coordinates. Namely, the function H must satisfy the following system of equations

$$\begin{gathered}\frac{\partial H}{\partial p_j}\frac{\partial H}{\partial p_k}\frac{\partial^2 H}{\partial q_j \partial q_k} - \frac{\partial H}{\partial p_j}\frac{\partial H}{\partial q_k}\frac{\partial^2 H}{\partial q_j \partial p_k} \\ - \frac{\partial H}{\partial q_j}\frac{\partial H}{\partial p_k}\frac{\partial^2 H}{\partial p_j \partial q_k} + \frac{\partial H}{\partial q_j}\frac{\partial H}{\partial q_k}\frac{\partial^2 H}{\partial p_j \partial p_k} = 0 , \\ 1 \le j \le k \le n .\end{gathered} \tag{7.6}$$

For $n = 2$ and $n = 3$ the solutions of (7.6) were studied by Morera and Dal Acqua (see the survey in [200]).

If variables p, q of a system with the Hamiltonian $H(p, q)$ are not separable, it does not imply that this system cannot be solved by separation of variables at all. It may happen that after an appropriate change $p, q \to y, x$ we arrive at separable canonical variables x, y. The problem of the existence of separable variables in a Hamiltonian system is much more complicated.

Let H have the "natural" form $T + V$, and the canonical change $p, q \to y, x$ be an extension of the "contact" transformation

$$q = f(x), \qquad y = \left(\frac{\partial f}{\partial x}\right)^T p .$$

If in certain new symplectic coordinates x, y the original Hamiltonian system is solvable by separation of variables, then it possesses a complete set of involutive integrals quadratic in the momenta (see §7.4). The possibility of separation of variables in systems with quadratic integrals is discussed in [198]. The existence problem for a complete set of polynomial integrals of Hamiltonian systems will be considered in Chap. VII.

Note that if arbitrary canonical transformations are acceptable, then any completely integrable system is solvable by separation of variables. To see this it is sufficient to pass to angle–action variables. In such a general formulation, the existence problem of separable canonical variables is equivalent to that of a complete set of involutive integrals.

7.4 Let $W(q,x)$ be a complete integral of Eq. (7.3). Put $p = \partial W/\partial q$. Since $\det \| \partial^2 W \partial q \partial x \| \neq 0$, one can express x (at least locally) in terms of p, q. Put also $x_1 = F_1(p,q), \ \dots \ , x_n = F_n(p,q)$. It is easy to show that the functions $F_1, \dots, F_n$ are independent and their Poisson brackets equal zero.

To find a complete set of commuting integrals in a system with separable variables it is not necessary to write down the complete solution of Eq. (7.3). For case (a) in §7.2 such integrals are represented by the functions

$$F_1 = f_1(p_1, q_1)\,, \quad F_2 = f_2(f_1(p_1, q_1), p_2, q_2)\,, \ \dots \ , \ F_n = H\,,$$

and for case (b)

$$F_0 = H\,, \quad F_s = f_s(p_s, q_s) - H g_s(p_s, q_s)\,, \qquad 1 \leq s \leq n\,.$$

The latter functions, being in involution, are not all independent in view of the relation $F_1 + \cdots + F_n = 0$. A complete set of independent integrals is obtained from $F_1, \dots, f_n$ by rejecting one of the functions F_k $(k \neq 1)$.

When solving the Goryachev–Chaplygin problem (§5), we used case (b) of separation of symplectic coordinates. The additional integral in this problem is of third (but not second) degree in the momenta (cf. §6.3), because the transformation to the special canonical coordinates is notan extension of a coordinate transformation in the configuration space.

In Stäckel's Hamiltonian system (7.4), the functions

$$F_k = \sum_s \frac{\Phi_{ks} f_s}{\Phi}\,, \qquad k = 1, \dots, n$$

form a complete set of integrals in involution.

Thus, if a Hamiltonian system is solvable by the Hamilton–Jacobi method after separation of variables, then one can immediately write down $l = \dim M/2$ independent involutive integrals.

7.5 Many important problems of Hamiltonian mechanics were solved by using *elliptic coordinates* in $\mathbb{R}^n$ (or their degenerations). These coordinates were introduced and applied by Jacobi [105]. Euler was aware of them for the case $n = 2$.

Let $0 < a_1 < a_2 < \cdots < a_n$ be distinct positive numbers. For each $x = (x_1, \ldots, x_n) \in \mathbb{R}^n$ the equation

$$f(\lambda) = \sum_s \frac{x_s^2}{a_s - \lambda} = 1$$

has n real roots $\lambda_1, \ldots, \lambda_n$ which separate $a_1, \ldots, a_n$ (Fig. 9). These roots serve as curvilinear coordinates in $\mathbb{R}^n$ and are called Jacobi's elliptic coordinates.

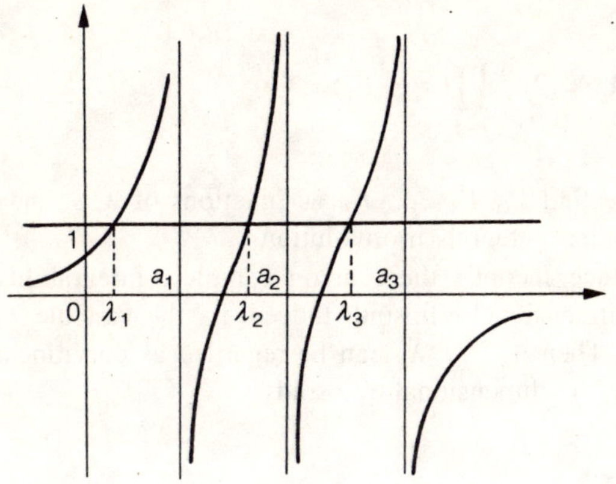

Fig. 9.

One can show that

$$x_i^2 = \prod_{s=1}^{n} (a_i - \lambda_s) \Big/ \prod_{\substack{s=1 \\ s \neq i}}^{n} (a_i - a_s) . \tag{7.7}$$

Using this formula one can easily derive the relation $4 \sum \dot{x}_s^2 = \sum M_s \dot{\lambda}_s^2$, where

$$M_s = \prod_{i \neq s} (\lambda_i - \lambda_s) \Big/ \prod (a_i - \lambda_s) . \tag{7.8}$$

The reader may notice the duality of the expressions (7.7) and (7.8).

Now we pass to the symplectic coordinates λ_s, $\mu_s = M_s \dot{\lambda}_s / 4$. Then the kinetic energy of free motion of a point mass in $\mathbb{R}^n$ takes the form

$$H = \frac{1}{2} \sum \dot{x}_s^2 = 2 \sum \mu_s^2 / M_s(\lambda) . \tag{7.9}$$

Here it is not immediately clear how the variables λ, μ could be separated. However, we can resort to the following formula found by Jacobi: the sum

$$\sum_{s=1}^{n} \frac{\lambda_s^m}{\prod_{i \neq s} (\lambda_s - \lambda_i)}$$

vanishes for $m < n - 1$ and equals 1 for $m = n - 1$. This enables us to represent (7.9) in the form

$$\sum_{s=1} \frac{\sum_{m=0}^{n-1} F_m \lambda_s^m}{\prod_{i \neq s}(\lambda_s - \lambda_i)} = 2 \sum_{s=1} \frac{\mu_s^2 \prod_j (\lambda_s - a_j)}{\prod_{i \neq s}(\lambda_s - \lambda_i)} .$$

Here $F_{n-1} = H$, while $F_0, F_1, \ldots, F_{n-2}$ are still arbitrary. Now the variables λ, μ separate; we can put

$$\sum_{m=0}^{n-1} F_m \lambda_s^m = 2\mu_s^2 \prod_{j-1}^{n} (\lambda_s - a_j) .$$

From this system of equations we find $F_0, F_1, \ldots, F_n$ as functions of λ, μ; they provide a complete set of independent integrals in involution.

From this result one can deduce Jacobi's theorem on complete integrability of a geodesic motion on a multidimensional ellipsoid. Indeed, fix the variable λ_1 by putting, for example, $\lambda_1 = 0$. Then $\lambda_2, \ldots, \lambda_n$ can be regarded as curvilinear orthogonal coordinates on the $(n - 1)$-dimensional ellipsoid

$$\sum_{s=1}^{n} x_s^2 / a_s^2 = 1 .$$

The Hamiltonian of the Jacobi problem is given by formula (7.9) in which we must put $\lambda_1 = 0$, $\mu_1 = 0$. The separation of the variables $\lambda_2, \ldots, \lambda_n, \mu_2, \ldots, \mu_n$ is performed according to the scheme above. Note that for the two-dimensional ellipsoid the Hamiltonian takes the form (7.5): for $n = 3$ we obtain a Liouville Hamiltonian system. Upon fixing the value of one of the variables $\lambda_2, \ldots, \lambda_n$ we may, by the same method, establish the complete integrability of the problem of geodesics on multidimensional hyperboloids of arbitrary types. The analysis of the qualitative behavior of geodesics on a two-dimensional ellipsoid, which is based on Jacobi's formulas, can be found in [12]. Jacobi showed that the problem of free motion on an ellipsoid remains integrable in the presence of an elastic force which is directed to the center of the ellipsoid and proportional to the distance between the center and the moving point.

As another example of the application of elliptic coordinates, we consider the planar motion of a point mass in the attraction field of two fixed centers. This problem was integrated by Euler in 1760. Let (x_1, x_2) be Cartesian coordinates in the plane of motion, and $(0, -c)$, $(0, c)$ the coordinates of the two attracting centers $(c > 0)$. Pass to elliptic coordinates in this plane by assuming that $a_2 - a_1 = 2c$. This means, in particular, that for each fixed λ the equation

$$\frac{x_1^2}{a_1 - \lambda} + \frac{x_2^2}{a_2 - \lambda} = 1$$

defines a conic with foci at these fixed centers. In the symplectic coordinates λ, μ the Hamiltonian of this problem is

$$H = 2\frac{(a_1-\lambda_1)(a_2-\lambda_1)}{\lambda_2-\lambda_1}\mu_1^2 + 2\frac{(a_1-\lambda_2)(a_2-\lambda_2)}{\lambda_1-\lambda_2}\mu_2^2 + V(\lambda_1, \lambda_2) , \qquad (7.10)$$

V being the interaction potential. Let r_1, r_2 be the distances from the moving mass to the attracting centers. Using (7.7) for $n = 2$, it is easy to obtain

$$r_1^2 = (x_2+c)^2 + x_1^2 = (\sqrt{a_2+\lambda_1} + \sqrt{a_2+\lambda_2})^2 ,$$
$$r_2^2 = (x_2-c)^2 + x_1^2 = (\sqrt{a_2+\lambda_1} - \sqrt{a_2+\lambda_2})^2 .$$

Therefore

$$V = \frac{\gamma_1}{r_1} + \frac{\gamma_2}{r_2} = \frac{\gamma_1 r_2 + \gamma_2 r_1}{r_1 r_2}$$
$$= \frac{(\gamma_1+\gamma_2)\sqrt{a_2+\lambda_1} - (\gamma_1-\gamma_2)\sqrt{a_2+\lambda_2}}{\lambda_1-\lambda_2} . \qquad (7.11)$$

As a result, the variables λ_1, μ_1 and λ_2, μ_2 separate and hence the problem of two fixed centers is integrable. Lagrange showed that the integrability is preserved if an elastic force directed to the midpoint of the segment joining the two fixed centers acts on the moving point. A qualitative investigation of the problem of two fixed centers can be found in Charlier's book [57]. To conclude, taking (7.11) into account, we note that the Hamiltonian (7.10) has the form of the Liouville system (7.5).

7.6 In a number of cases (when, for example, some of $a_1, \ldots, a_n$ coincide) Jacobi's elliptic coordinates degenerate. Studying such cases is useful, because this may give us new systems with separable variables.

Suppose that $n = 3$: this case is most important in practice. There are 10 different types of degeneration of the elliptic coordinates; the customary Cartesian coordinates in $\mathbb{R}^3$ are among them (see [177, Chap. 5]). We indicate two interesting cases.

Let $\mathbb{R}^3 = \{x, y, z\}$. Jacobi's elliptic coordinates are defined by the equation

$$\frac{x^2}{a-\lambda} + \frac{y^2}{b-\lambda} + \frac{z^2}{c-\lambda} = 1 , \qquad a \ge b \ge c \ge 0 . \qquad (7.12)$$

We put

$$a = \alpha d + d^2 , \quad b = \beta d + d^2 , \quad c = 2d^2 , \qquad \lambda = \mu d + d^2 ,$$

and displace the origin to the point $(0, 0, d)$; so that $x = x'$, $y = y'$, $z = z' - d$. Then rewrite (7.12) in terms of the new coordinates and let d tend to infinity. As a limit, we obtain the family of surfaces

$$\frac{(x')^2}{\alpha-\mu} + \frac{(y')^2}{\beta-\mu} - \mu + 2z' . \qquad (7.13)$$

In fact, Eq. (7.13) defines three different families of paraboloids. Through each point in $\mathbb{R}^3$ there pass three surfaces of these families intersecting orthogonally. As $d \to \infty$, Jacobi's elliptic coordinates transform to the new *parabolic* coordinates μ_1, μ_2, μ_3.

We mention two important problems which are solved by separation of variables under a transformation to parabolic coordinates.

1) Kepler's problem in a homogeneous force field, i.e., the motion of a point mass under the gravitational attraction of a fixed center and another force of constant magnitude and direction. This problem was solved by Lagrange in 1766. Its alternative interpretation is a hydrogen atom in a homogeneous electric field.
2) The motion of a point mass on the paraboloid (7.13) with vertical axis (the paraboloid pendulum). This problem was solved by Painlevé (1895). A detailed analysis of the motion of the paraboloid pendulum was performed by Chaplygin [54].

We mention another limit case of the elliptic coordinates, when the parameters $a > b > c > 0$ tend to each other (for example, $a \to b + 0$, $c \to b - 0$). In the limit, Eq. (7.12) has one single root and a double root $\lambda = -b$. The family of ellipsoids transforms to a family of concentric spheres having their center at the origin, whereas the hyperboloids turn into elliptic cones. As a result, curvilinear orthogonal coordinates in $\mathbb{R}^3$ arise, which are called *spheroconical* coordinates (for details see, for example, [177, Chap. 5]).

Using spheroconical coordinates, Neumann solved the problem of motion of a point mass on a sphere in $\mathbb{R}^3$ in a force field with a potential quadratic in x, y, z (1859). This problem is also integrable in the multi-dimensional case (see [186]).

8 The Heisenberg Representation

This section is devoted to an effective method of solving Hamiltonian systems, which is based on their representation in Heisenberg's form (equivalent terminology is the Lax representation, the L-A pair, the isospectral deformation method).

8.1 The Heisenberg representation of a system of differential equations

$$\dot{x} = v(x, t), \qquad x \in M^n \tag{8.1}$$

is defined by square matrices L and A, whose entries are smooth (complex-valued) functions of x, t, such that the following matrix identity holds

$$\dot{L} = [A, L], \qquad [A, L] = AL - LA. \tag{8.2}$$

The solutions of (8.1) obviously satisfy Eq. (8.2). To rule out trivial cases ($L = 0$ for example), we introduce the notion of exact representation when all solutions of (8.2) satisfy the system (8.1).

Differential equations of the form (8.2) seem to have first achieved widespread acceptance in quantum mechanics in connection with Heisenberg's approach in which the observables vary with time, whereas the configurations are constant.

Theorem 1 [159]. *The eigenvalues of the matrix L are integrals of the system* (8.1).

The proof is based on the following property: given finite-dimensional matrices X and Y, $\mathrm{tr}XY = \mathrm{tr}YX$, where tr is the trace operator. Now calculate the derivative

$$\begin{aligned}(L^k)^{\cdot} &= \dot{L}L^{k-1} + L\dot{L}L^{k-2} + \cdots + L^{k-1}\dot{L} \\ &= AL^k - LAL^{k-1} + \cdots + L^{k-1}AL - L^kA\,.\end{aligned}$$

Using the relation

$$L^kAL^{s+1} - L^{k+1}AL^s = (L^kAL^s)L - L(L^kAL^s)\,,$$

we find that $\mathrm{tr}(L^k)^{\cdot} = (\mathrm{tr}L^k)^{\cdot} = 0$. Therefore, the traces of powers of L are integrals of the system (8.1). To complete the proof, we recall that coefficients of the characteristic equation $|L - \lambda I_n| = 0$ are uniquely determined by $\mathrm{tr}L^k$, $1 \le k \le n$.

Theorem 1 is well-known in quantum mechanics in connection with the change from Heisenberg's description of motion to that proposed by Schrödinger. It should be noted that the independence of integrals obtained by this method must be checked separately for each specific system.

8.2 As a simple example of Heisenberg's representation we again consider the Euler equations

$$\dot{m} = m \times \omega\,, \qquad m = I\omega\,, \tag{8.3}$$

describing the evolution of the angular momentum of a free rigid body.

To each vector a of three-dimensional Euclidean space with the Cartesian coordinates (a_1, a_2, a_3) there corresponds the skew-symmetric matrix

$$A = \begin{pmatrix} 0 & -a_3 & a_2 \\ a_3 & 0 & -a_1 \\ -a_2 & a_1 & 0 \end{pmatrix}.$$

Under this correspondence, a vector product in $\mathbb{R}^3$ is represented by a matrix commutator. Therefore, Eq. (8.3) can be written as the matrix commutation equation $\dot{M} = [\Omega, M]$. We see that in this case Heisenberg's representation is exact. The traces of M, M^2, M^3 equal $0, -2(m, m), 0$ respectively.

This observation can be extended to the Poincaré equations (Chap. I, §2):

$$\begin{aligned}\frac{d}{dt}\left(\frac{\partial \mathcal{L}}{\partial \omega_s}\right) &= \sum_{i,j} c_{is}^{j}\frac{\partial \mathcal{L}}{\partial \omega_j}\omega_i + v_s(\mathcal{L})\,, \\ \frac{d}{dt}x_s &= \sum_{\alpha} \omega_\alpha v_\alpha(x_s)\,, \qquad 1 \le s \le n\,.\end{aligned} \tag{8.4}$$

Here $\mathcal{L}(\omega, x)$ is the Lagrange function, c_{is}^{j} the structure constants of the Lie algebra g, and $v_1, \ldots, v_n$ the basis of left-invariant vector fields on the corresponding Lie group G. Putting

$$m_s = \partial\mathcal{L}/\partial\omega_s \, ,$$

we introduce the matrices L and A with the entries

$$L_{ks} = \sum_\alpha c_{ks}^\alpha m_\alpha \, , \quad A_k^s = \sum_\alpha c_{k\alpha}^s \omega_\alpha \, .$$

Theorem 2. *The solutions of the system* (8.4) *satisfy the matrix equation*

$$\dot{L} = [A, L] + B \, , \qquad B_{ij} = [v_i, v_j](\mathcal{L}) \, . \tag{8.5}$$

To obtain (8.5) we multiply the first equation in (8.4) by c_{ik}^s, sum up in s, and use the Jacobi identity for the structure constants of the Lie algebra.

If the Lagrangian $\mathcal{L}$ is left-invariant (when $v_s(\mathcal{L}) = 0$)), it depends only on the variables ω, and the matrix B in (8.5) vanishes. In this case, the Poincaré equations turn out to be a closed system on the algebra g and the pair L, A gives their Heisenberg representation. The latter is not always exact: for an Abelian group G we have $c_{ij}^k = 0$, and Eq. (8.5) degenerates to a trivial identity. However, Heisenberg's representation is exact for a simple Lie algebra (as in the Euler problem).

8.3 Let $M = D \times \mathbb{T}^m$, where D is a domain in $\mathbb{R}^k = \{I_1, \dots, I_k\}$, and $\mathbb{T}^m$ the m-dimensional torus with the angle coordinates $\varphi_1, \dots, \varphi_m$.

Proposition. *The system*

$$\dot{I}_1 = \dots = \dot{I}_k \, , \quad \dot{\varphi}_1 = \omega_1(I) \, , \; \dots \, , \; \dot{\varphi}_m = \omega_m(I) \tag{8.6}$$

admits an exact Heisenberg representation (8.2) *such that the entries of L and A are single-valued functions in $D \times \mathbb{T}^m$ and the eigenvalues of L are $0, I_1, \dots, I_k$.*

Corollary. *Each completely integrable Hamiltonian system admits a Heisenberg representation in a neighborhood of its invariant tori.*

For the most part of solved Hamiltonian systems their exact Heisenberg representation is defined in the whole phase space (see the surveys [72] and [73]).

To prove this proposition we consider the box-diagonal matrices

$$L = \begin{pmatrix} I_1 & & & & & \\ & \ddots & & & & \\ & & I_k & & & \\ & & & L_1 & & \\ & & & & \ddots & \\ & & & & & L_m \end{pmatrix} , \quad A = \begin{pmatrix} 0 & & & & & \\ & \ddots & & & & \\ & & 0 & & & \\ & & & A_1 & & \\ & & & & \ddots & \\ & & & & & A_m \end{pmatrix} ,$$

with the boxes

$$L_s = \begin{pmatrix} \Psi_s & \Psi_s \\ -\Psi_s & -\Psi_s \end{pmatrix}, \quad A_s = \begin{pmatrix} 0 & i\omega_s/2 \\ -i\omega_s/2 & 0 \end{pmatrix},$$
$$\Psi_s = \exp(i\varphi_s), \quad 1 \le s \le m .$$

One can check easily that the solutions of the system (8.6) satisfy the matrix Eq. (8.2). The eigenvalues of L are the numbers $I_1, \ldots, I_k$, and 0 (with multiplicity $2m$).

8.4 Here are several examples of Heisenberg's representation in the theory of interacting particles.

Consider a closed Toda lattice, i.e., n particles on the linewith coordinates $x_1, \ldots, x_n$, whose motion is governed by the system

$$\ddot{x}_i = -\partial V/\partial x_i, \qquad i = 1, \ldots, n, \tag{8.7}$$
$$V = \sum_{k=1}^{n} \exp(x_k - x_{k+1}), \qquad x_{n+1} = x_1 .$$

Put

$$L = \begin{pmatrix} b_1 & a_1 & 0 & \cdots & 0 & a_n \\ a_1 & b_2 & a_2 & \cdots & 0 & 0 \\ \vdots & \vdots & \vdots & & \vdots & \vdots \\ a_n & 0 & 0 & \cdots & a_{n-1} & b_n \end{pmatrix},$$
$$A = \begin{pmatrix} 0 & a_1 & 0 & \cdots & 0 & a_n \\ -a_1 & 0 & 0 & \cdots & 0 & 0 \\ \vdots & \vdots & \vdots & & \vdots & \vdots \\ -a_n & 0 & 0 & \cdots & -a_{n-1} & 0 \end{pmatrix}, \tag{8.8}$$

where $2a_k = \exp[(x_k - x_{k+1})/2]$, $2b_k = -\dot{x}_k$. As shown by Hénon, Flaschka, and Manakov (1974), Eqs. (8.7) admit the representation (8.2), (8.8). It follows from Theorem 1 that n eigenvalues of L are first integrals of the system (8.7), which, as one can show, are independent and commute. A complete set of independent commuting integrals can also be obtained as the traces of the matrices $L, L^2, \ldots, L^n$. For example,

$$\mathrm{tr} L = -\sum \dot{x}_k/2 .$$

These integrals are polynomials in the velocities $\dot{x}_1, \ldots, \dot{x}_n$ with the coefficients being sums of finite numbers of exponents of the linear expressions $\sum c_k x_k$ ($c_k =$ const).

Now we consider conditions under which a Hamiltonian system with the Hamiltonian

$$H = \frac{1}{2}\sum_{i=1}^{n} y_i^2 + \sum_{i \le j} V(x_i - x_j)$$

can be represented in the matrix form (8.2). Following Moser, we seek the representation (8.2) with the matrices

$$L = \begin{pmatrix} y_1 & \alpha(x_1 - x_2) & \cdots & \alpha(x_1 - x_n) \\ \alpha(x_2 - x_1) & y_2 & \cdots & \alpha(x_2 - x_1) \\ \vdots & \vdots & & \vdots \\ \alpha(x_n - x_1) & \alpha(x_n - x_2) & \cdots & y_n \end{pmatrix},$$

$$A = \begin{pmatrix} \sum\limits_{i \neq j} V(x_j - x_1) & \beta(x_1 - x_2) & \cdots & \beta(x_1 - x_n) \\ \beta(x_2 - x_1) & \sum\limits_{j \neq 2} V(x_j - x_2) & \cdots & \beta(x_2 - x_n) \\ \vdots & \vdots & & \vdots \\ \beta(x_n - x_1) & \beta(x_n - x_2) & \cdots & \sum\limits_{j \neq n} V(x_j - x_n) \end{pmatrix}.$$

The function α is supposed to be odd. Thus, the Heisenberg representation is possible if and only if $\beta = \alpha'$ and

$$\alpha'(y)\alpha(z) - \alpha(y)\alpha'(z) = \alpha(y + z)[V(y) - V(z)] .$$

This equation was solved by Calogero, Ol'shanetsky, Perelomov, Pidkujko, and Stepin (1978). It turns out that

$$V = -\alpha''/2\alpha , \tag{8.9}$$

whereas $\alpha(z)$ coincides with one of the following functions

$$\begin{array}{llll} \text{I} & z^{-1} , & & \\ \text{II} & a \operatorname{cotanh}(az), & a \sinh^{-1}(az) , & \\ \text{III} & a \operatorname{cotan}(az), & a \sin^{-1}(az) , & \\ \text{IV} & a\dfrac{\operatorname{cn}(az)}{\operatorname{sn}(az)} , & a\dfrac{\operatorname{dn}(az)}{\operatorname{sn}(az)} , & \dfrac{a}{\operatorname{sn}(az)} . \end{array}$$

Here sn, cn, and dn are Jacobi's elliptic functions. The replacement of a by ia transforms functions of type II to those of type III, while functions of types II and III turn into those of type I as $\alpha \to 0$.

Let the function α be of type IV and $a = 1$. Then (8.9) gives rise to Weierstrass's $\wp$-function:

$$\wp(z) = \frac{1}{z^2} + \sum{}' \left[\frac{1}{(z - 2m\omega - 2n\omega')^2} - \frac{1}{(2m\omega + 2n\omega')^2} \right] .$$

Here n, m range over all integers not vanishing simultaneously. This is an elliptic function with periods $(2\omega, 2\omega')$. The invariants of the $\wp$-function

$$g_2 = 60 \sum{}' (2m\omega + 2n\omega')^{-4} , \quad g_3 = 140 \sum{}' (2m\omega + 2n\omega')^{-6}$$

are related to the modulus of Jacobi's functions as follows

$$g_2 = \frac{4}{3}(k^4 - k^2 + 1)\,, \quad g_3 = \frac{4}{27}(2k^6 - 21k^4 - 21k^2 + 2)\,.$$

9 Algebraically Integrable Systems

9.1 In many solved problems of Hamiltonian mechanics the solutions can be extended to meromorphic functions on the complex time plane. Moreover, the general solution is expressed in terms of Abelian functions[2]. In old textbooks on mechanics the main attention was given to integrable problems solvable in terms of the elliptic function of time (see, for example, [233]). Before we proceed to general definitions we consider as an example the Euler equations describing free rotation of a rigid body around a fixed point:

$$\begin{aligned} I_1\dot{\omega}_1 + (I_3 - I_2)\omega_2\omega_3 &= 0\,, \\ I_2\dot{\omega}_2 + (I_1 - I_3)\omega_1\omega_3 &= 0\,, \\ I_3\dot{\omega}_3 + (I_2 - I_1)\omega_1\omega_2 &= 0\,. \end{aligned} \tag{9.1}$$

These have the two polynomial integrals

$$I_1\omega_1^2 + I_2\omega_2^2 + I_3\omega_3^2 = \mathcal{J}\mu^2\,, \quad I_1^2\omega_1^2 + I_2^2\omega_2^2 + I_3^2\omega_3^2 = \mathcal{J}^2\mu^2\,, \tag{9.2}$$

where $\mathcal{J}$, μ are integral constants.

In typical situation the surfaces (9.2) (as real surfaces in $\mathbb{R}^3 = \{\omega\}$) intersect in two closed curves. Now consider the complexification of $\mathbb{R}^3$ by regarding ω_k as complex variables. It turns out that the system of algebraic equations (9.2) defines an elliptic curve in $\mathbb{C}^3$ with a few points removed. To show this we write down the general solution of (9.1) in terms of Jacobi's elliptic functions:

$$\begin{aligned} &\omega_1 = \varepsilon\mu\sqrt{\frac{\mathcal{J}(\mathcal{J} - I_3)}{I_1(I_1 - I_3)}}\,\mathrm{cn}\tau\,, \quad \omega_2 = \varepsilon'\mu\sqrt{\frac{\mathcal{J}(\mathcal{J} - I_2)}{I_2(I_2 - I_3)}}\,\mathrm{sn}\tau\,, \\ &\omega_3 = \varepsilon''\mu\sqrt{\frac{\mathcal{J}(I_3 - \mathcal{J})}{I_3(I_1 - I_3)}}\,\mathrm{dn}\tau\,, \quad \tau = \mu\sqrt{\frac{\mathcal{J}(I_1 - \mathcal{J})(I_2 - I_3)}{I_1 I_2 I_3}}(t - t_0)\,, \\ &k^2 = \frac{(I_1 - I_2 - 2)(\mathcal{J} - I_3)}{(I_2 - I_3)(I_1 - \mathcal{J})}\,. \end{aligned} \tag{9.3}$$

Here ε, ε', ε'' equal ± 1; k is the modulus of the elliptic functions. Expressions (9.3) give the representation of the complex curve (9.2) in parameteric form. Since the elliptic functions are double-periodic, the parameter τ ranges over a two-dimensional

2 "During World War I, when I was a student, the Abelian functions (following the traditions established by Jacobi) were uncontestably reputed as a peak of mathematics. Each of us felt ambitious to advance in this field by himself. And what do we see now? The young generation is hardly familiar with the Abelian functions at all." Klein F. Lectures on development of mathematics in XIX century.

real torus. However, the curve (9.2) is not isomorphic (as a real variety) to such a torus because the elliptic functions (9.3) have poles.

It is well-known that between any two elliptic functions f_1 and f_2 with the same periods there exists a relation of the form $\Psi(f_1, f_2) = 0$, where Ψ is a polynomial of two variables. For example, the Weierstrass function $\mathfrak{P}$ and its derivative $\mathfrak{P}'$, which obviously have the same periods, are connected by the algebraic equation

$$(\mathfrak{P}')^2 - 4\mathfrak{P}^3 + g_2\mathfrak{P} + \mathfrak{g}_3 = 0 ,$$

where g_2, g_3 are the invariants of the function $\mathfrak{P}$. More generally, any m elliptic functions ($m \geq 2$) with the same periods are connected by $m-1$ algebraic equations. An example is given by Eqs. (9.2) for elliptic functions (9.3). The Euler Eqs. (9.1) are Hamiltonian (§1.2): the symplectic structure is defined by the Lie–Poisson bracket

$$\{I_1\omega_1, I_2\omega_2\} = I_3\omega_3 , \ \ldots$$

with the kinetic energy as Hamiltonian. However, this bracket is degenerate: the square of the angular momentum $F = \sum(I_i\omega_i)^2$ commutes with all functions on the algebra $so(3)$ (such functions are called the Casimir functions). As mentioned in §2, Chap. I, the restriction of this bracket to the integral surface $F = \text{const} > 0$ is non-degenerate.

In the sense of the geometrical version of the Liouville theorem on completely integrable systems (§4), Eqs. (9.2) define exactly one-dimensional invariant tori and the angular coordinate $u = \tau/2K \ (\mathrm{mod}\, 2\pi)$ on these tori varies uniformly with the time (K being a complete elliptic integral). According to (9.3), ω_k are elliptic functions of the angular coordinate u.

9.2 Here we present a brief survey of the theory of Abelian functions required in the sequel (for details see, for example, [71, 172]). Consider the m-dimensional complex space $\mathbb{C}^m$ with the coordinates $(z_1, \ldots, z_m)$. Recall that the function $F(z)$ is called meromorphic if it can be represented as a ratio $f(z)/g(z)$, where f, g are entire functions, i.e., convergent power series in the whole space $\mathbb{C}^>$. If $g(a) = 0$ and $f(a) \neq 0$, then the point $a \in \mathbb{C}^m$ is called a pole of the function F. Note that for $m \geq 1$ poles as well as zeros are not isolated points. If $g(a) = f(a) = 0$ and there exists no limit of $F(z)$ for $z \to a$, then the point $z = a$ is called an indeterminate point of the meromorphic function F. Indeterminate points can exist only if $m \geq 2$ (for $m = 2$ such points are isolated). Here is a simple example: the rational function $F = z_1/z_2$ has an indeterminate point at the origin.

An Abelian function F is a meromorphic function in $\mathbb{C}^m$ with $2m$ linearly independent (over the reals) periods $\omega^{(1)}, \ldots, \omega^{(2m)}$: $F(z + \omega^{(j)}) = F(z)$ for all $z \in \mathbb{C}^m$ and $j = 1, \ldots, 2m$.

Abelian functions of one complex variable are exactly the elliptic functions. According to the Weierstrass–Poincaré theorem, between any $m+1$ Abelian functions with the same periods there exists an algebraic relation.

The period matrix of an Abelian function

$$W = (\omega^{(1)} \cdots \omega^{(2m)}) = \begin{pmatrix} \omega_1^{(1)} & \cdots & \omega_1^{(2m)} \\ \vdots & & \vdots \\ \omega_m^{(1)} & \cdots & \omega_m^{(2m)} \end{pmatrix}$$

is called the Riemannian matrix. By no means every $n \times 2m$ matrix is Riemannian. The matrix W is Riemannian if and only if there exists a skew-symmetric non-degenerate integer $2n \times 2m$ matrix N such that
1) $WNW^t = 0$,
2) the matrix $iWN\overline{W^t}$ determines a positive definite Hermitian form.

Let Γ be the lattice in $\mathbb{C}^m$ generated by the vectors $\omega^{(j)}$; all the lattice vectors have the form

$$k_1\omega^{(1)} + \cdots + k_{2m}\omega^{(2m)}, \qquad k_j \in \mathbb{Z}. \tag{9.4}$$

It is obvious that Γ is a group by addition. It is useful to introduce the quotient space $\mathbb{C}^m/\Gamma$ by identifying points in $\mathbb{C}^m$ that differ by the vectors (9.4).This space is a $2m$-dimensional torus $\mathbb{T}^{2m}$ which is called an Abelian torus. We can say that Abelian functions are meromorphic functions on $\mathbb{T}^{2m}$.

Each compact Riemann surface of genus m is naturally connected with a field of Abelian functions of m complex variables. Recall that the Riemann surface X is a two-dimensional manifold covered by complex charts in such a way that a transformation from one chart to another is a holomorphic mapping. The simplest example of a compact Riemann surface is a two-dimensional torus—the quotient $\mathbb{C}^m/\Gamma$. Its genus equals 1.

Riemann surfaces of algebraic functions are often encountered in applications. An important example is the function $w(z)$ defined by the equations

$$w^2 = P_{(2m+1)}(z), \qquad w^2 = P_{(2m+2)}(z), \tag{9.5}$$

where P_k are polynomials of degree k with simple roots. The function w is double-valued: after circling twice around the zeros of the polynomial $P_k(z)$ it takes its original value. Therefore, the Riemann surface of the algebraic function $w = \sqrt{P(z)}$ is a two-sheeted covering of the complex plane $\mathbb{C} = \{F\}$; the zeros of the polynomial $P(z)$ are the branch points of this covering. Riemann himself was aware of the structure of such surfaces: after natural compactification they turn out to be diffeomorphic to a sphere with m handles (the Riemann surface of genus m). From the classical point of view, Riemann's ideas were described in the book [89]; for the modern explanation see [82].

Let $z = x + iy$ be a local coordinate on the Riemann surface X. We refer to 1-forms $\varphi = a\,dx + b\,dy = a\,dz + b\,dy = \alpha\,dz + \beta\overline{z}$ as simply differentials. If in a neighborhood of any point of X the differential φ has the form $f(z)\,dz$, f being a holomorphic function, then φ is called a holomorphic differential (the Abelian differential of the first kind). Holomorphic differentials span a linear space whose dimension coincides with the genus of the surface X. For example, if X is defined by the first Eq. (9.5), then the differentials

$$\varphi_k = \frac{z^k}{\sqrt{P_{(2m+1)}(z)}} dz \,, \qquad 1 \le k \le m$$

are holomorphic and linearly independent.

On each Riemann surface of genus m one can choose $2m$ closed curves (cycles) $a_1, \ldots, a_m, b_1, \ldots, b_m$ such that when dissected along these cycles the surface unfolds in a polygon with $4m$ sides (for $m = 2$ see Fig. 10).

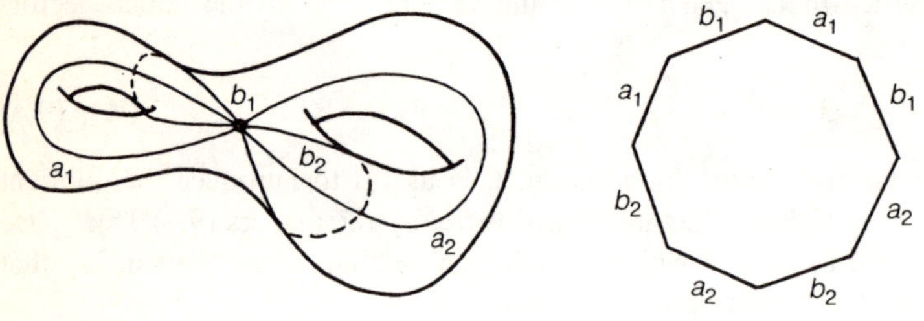

Fig. 10.

Let $\varphi_1, \ldots, \varphi_m$ be a basis of holomorphic differentials on X. Since these differentials are closed ($d\varphi_k = 0$), their periods are correctly defined:

$$\oint_{a_i} \varphi_k = \omega_k^i \,, \quad \oint_{b_i} \varphi_k = \omega_k^{i+m} \,. \tag{9.6}$$

By the Stokes theorem, the values of the integrals (9.6) do not change under continuous deformations of the loops a_i, b_i. The matrix of periods of holomorphic differentials

$$(\omega_k^{(i)}, \omega_k^{(i+m)}) \,, \qquad 1 \le i, k \le m \tag{9.7}$$

turns out to be Riemannian. Therefore, as mentioned above, each Riemann surface of genus m is naturally connected with a field of Abelian functions of m complex variables. The Abelian torus $\mathbb{T}^{2m} = \mathbb{C}^{2m}/\Gamma$ is called the Jacobi variety (Jacobian) Jac(X) of the Riemann surface X.

One of the key points of the theory of Riemann surfaces is the Jacobi inversion problem: for a given point $\zeta = (\zeta_1, \ldots, \zeta_m) \in \text{Jac}(X)$ find m points $z_1, \ldots, z_m$ on the surface X such that

$$\sum_{k=1}^{m} \int_{z_0}^{z_k} \varphi_j \equiv \zeta_j \,, \qquad 1 \le j \le m \,. \tag{9.8}$$

Here z_0 is a fixed point of X and $\equiv$ denotes equivalence up to addition by the lattice vectors (9.4). Changing to other path from z_0 to z_k leads to the extra integral

$$\oint_\gamma \varphi_j ,$$

where γ is a cycle on X. The cycle γ can be represented as a linear combination of the basis cycles a_i, b_i with integer coefficients:

$$\gamma = \sum p_i a_i + \sum q_k b_k .$$

Then the left side of (9.8) will change by

$$\oint_\gamma \varphi_j = \sum p_i \omega_j^i + \sum q_k \omega_j^{k+m} ,$$

which is exactly the j-th component of some lattice vector. Thus, the Jacobi inversion problem is correctly defined.

Consider the following Jacobi theorem: let f be an arbitrary meromorphic function on X. Then any rational symmetric function of $f(z_1), \ldots, f(z_m)$ is an Abelian function of $\zeta_1, \ldots, \zeta_m$, i.e., a meromorphic function on Jac(X).

As an example of an application of this theorem we consider Eqs. (5.2), to which the Chaplygin problem in rigid body dynamics is reduced. We represent these equations in the form

$$\begin{aligned} \int_{p_0}^{p_1} \frac{dz}{\sqrt{\Phi(z)}} + \int_{p_0}^{p_2} \frac{dz}{\sqrt{\Phi(z)}} &= \zeta_1 , \\ \int_{p_0}^{p_1} \frac{z\,dz}{\sqrt{\Phi(z)}} + \int_{p_0}^{p_2} \frac{z\,dz}{\sqrt{\Phi(z)}} &= \zeta_2 , \end{aligned} \tag{9.9}$$

where $p_0, c_k = \text{const}$ and

$$\zeta_1 = c_1 , \quad \zeta_2 = c_2 + t . \tag{9.10}$$

Since Φ is a polynomial of the sixth degree, the corresponding Riemann surface has genus 2.

In the space $\mathbb{C}^2 = \{\zeta_1, \zeta_2\}$ these equations determine a complex line, whereas on the Abelian torus (of complex dimension 2) we have a complex "winding".

According to the Jacobi theorem, any symmetrical function of p_1, p_2 is an Abelian function of ζ_1, ζ_2. In view of (9.10), these functions are meromorphic functions of the complex time t. In particular, the angular velocity component $\omega_3 = -(p_1 + p_2)/I_3$ is a single-valued meromorphic function on $\mathbb{C} = \{\approx\}$. One may show that ω_1^2, ω_2^2 also enjoy this property. However, ω_1, ω_2 themselves have algebraic branch points.

Equations of the type (9.9) often appear as a result of solving problems of classical mechanics. Examples are the integrable cases in the rigid body dynamics found by Kovalevskaya, Clebsch, Steklov, and Lyapunov (see §5). Besides, unlike the Goryachev–Chaplygin problem, in these cases the phase variables are single-valued functions on the Jacobian of a Riemann surface of genus 2.

9.3 Now we are ready to give the general definition of an *algebraic integrable* Hamiltonian system, which was formulated by Adler and van Moerbeke [4].

Suppose that in n-dimensional space with the Cartesian coordinates $z_1, \dots, z_n$ the Poisson bracket $\{\cdot, \cdot\}$ (generically degenerate) is defined such that z_i, z_j is a polynomial in $\mathbb{R}^n = \{z\}$. An example is the Lie–Poisson bracket. Let $F_1, \dots, F_k$ be the polynomial Casimir functions which commute with all the coordinates z_i. Suppose that the restriction of this bracket on the level surface $M_c = \{z : F_i(z) = c_i\}$ is non-degenerate.

Consider the Hamiltonian system

$$\dot{z}_i = \{z_i, H\} = f_i(z) , \qquad 1 \leq i \leq n , \tag{9.11}$$

where the Hamiltonian H and the functions f_i are polynomials in $\mathbb{R}^n$. This system can be regarded as an analytic system of differential equations in $\mathbb{C}^n$.

The Hamilton Eqs. (9.11) are called *algebraic completely integrable* provided the following conditions are fulfilled:

1) Apart from the Casimir functions $F_1, \dots, F_k$, Eqs. (9.2) have also $m = (n-k)/2$ polynomial integrals $F_{k+1} = H, \dots, F_{k+2}, \dots, F_{k+m}$, where the functions $F_1, \dots, F_{k+m}$ are independent almost everywhere. For almost all real constants $c_1, \dots, c_{k+m}$ the set

$$\{z \in \mathbb{R}^n : F_i(z) = c_i , \quad 1 \leq i \leq k+m\} \tag{9.12}$$

is compact; therefore, by the geometrical Liouville theorem (Theorem 1, §4), the varieties (9.12) are diffeomorphic to m-dimensional tori, and the phase flow is a quasi-periodic motion on them.

2) For almost all complex constants $c_1, \dots, c_{k+m}$ there exists an Abelian torus $\mathbb{T}^{2m}$ with the natural complex coordinates $\zeta_1, \dots, \zeta_{k+m}$ and n Abelian functions $z_i = z_i(\zeta_1, \dots, \zeta_{k+m})$ parameterizing the non-compact invariant manifolds $A_c = \{z \in C_n : F_i(z) = c_i, \ 1 \leq i \leq k+m\}$. The mapping $\mathbb{T}^{2m} \to A_c$ defined by the function $z(h)$ is one-to-one everywhere except several closed surfaces on $\mathbb{T}^{2m}$ of complex codimension 1.

3) On A_c the phase flow of the complexified system (9.11) takes the form

$$\dot{\zeta}_i = \mu_i = \text{const} , \qquad 1 \leq i \leq m .$$

The simplest example of an algebraic completely integrable system is the Euler problem considered in §9.1. More intricate examples are the integrable cases found by Kovalevskaya, Clebsch, Steklov, and Lyapunov.

Consider also the closed Toda lattice whose dynamics is governed by the Hamilton equations with the Hamiltonian

$$H = \frac{1}{2} \sum y_i^2 + e^{x_1 - x_2} + e^{x_2 - x_3} + \dots + e^{x_n - x_1} .$$

Like the total energy, the total momentum $\sum y_i$ is also conserved in this system. Let the barycenter of this system be fixed, i.e.,

$$\sum y_i = 0 . \tag{9.13}$$

We introduce new coordinates u, v as follows

$$v_k = \exp(x_k - x_{k+1}), \quad u_k = y_k - y_{k+1}; \qquad 1 \le k \le n. \tag{9.14}$$

Here $x_{n+1} = x_1$ and $y_{n+1} = y_1$. These are "excessive" coordinates since

$$F_1 = v_1 \cdots v_n = 1, \quad F_2 = u_1 + \cdots + u_n = 0.$$

The Poisson brackets $\{v_i, v_j\}, \{u_i, u_j\}$ equal zero while the brackets $\{v_i, u_j\}$ are linear combinations of v_k. In the new variables the dynamical equations have the form

$$\dot{v}_k = v_k u_k, \quad \dot{u}_k = v_{k-1} - 2v_k + v_{k+1}. \tag{9.15}$$

These can be rewritten as the Hamilton equations

$$\dot{v}_k = \{v_k, H\}, \quad \dot{u}_k = \{u_k, H\}, \tag{9.16}$$

where $H = T + \sum v_k$, and the kinetic energy T is expressed in terms of u_k by using the relations

$$y_1 = [(n-1)u_1 + (n-2)u_2 + \cdots + u_{n-1}]/n, \ \ldots. \tag{9.17}$$

The other relations are found by a cyclic permutation of indices $1, \ldots, n$. The relations (9.17) can be deduced from the second group of Eqs. (9.14) and Eq. (9.13).

The system (9.16) has the Hamiltonian form (9.11) with F_1, F_2 as Casimir functions. As shown in [3], the Hamiltonian system (9.16) is algebraic completely integrable. In particular, the momenta y_k and the exponents v_k are meromorphic functions of the complex time.

9.4 If the Hamiltonian system (9.11) is algebraic completely integrable, then almost all its solutions are meromorphic functions of time. More exactly, this system admits the solutions as the asymptotic expansions

$$z_j(t) = t^{-k_j}(z_j^{(0)} + z_j^{(1)}t + \cdots + z_j^{(p)}t^p + \cdots), \qquad 1 \le j \le n \tag{9.18}$$

with integer exponents $k_j \ge 0$, $\sum k_j \ge 1$. The coefficients of the expansion depend on $n-1$ free parameters:

$$z_j^{(p)} = z_j^{(p)}(\alpha_1, \ldots, \alpha_{n-1}). \tag{9.19}$$

One extra free parameter appears as a result of the change $t \to t - t_0$. In practice, the coefficients (9.19) are, as a rule, rational functions on a certain $(n-1)$-dimensional algebraic manifold.

Note that Eqs. (9.11) admit several essentially different families of meromorphic solutions of the type (9.18). This can be seen easily from the following simple example. Consider the Hamilton equations

$$\dot{x} = \frac{\partial H}{\partial y}, \quad \dot{y} = -\frac{\partial H}{\partial x}; \qquad H = \frac{y^2}{2} + f_{n+1}(x), \tag{9.20}$$

where $f_{n+1} = -ax^{n+1} + bx^n$ $(a \neq 0)$ is a polynomial with constant coefficients. We seek formally meromorphic solutions of the form

$$x = \frac{X_{-\alpha}}{t^\alpha} + \frac{X_{-\alpha+1}}{t^{\alpha-1}} + \cdots, \quad y = \frac{Y_{-\beta}}{t^\beta} + \frac{Y_{-\beta+1}}{t^{\beta-1}} + \cdots, \tag{9.21}$$

where α, β are integer nonnegative numbers such that $\alpha + \beta \geq 1$. The coefficients $X_{-\alpha}, \ldots, X_{-\beta}, \ldots$ $(X_{-\alpha} \neq 0, X_{-\beta} \neq 0)$ can be complex numbers. We are going to find a "complete" solution of (9.20); in this case the coefficients of the expansion (9.21) must depend on a "modulus"—an arbitrary parameter.

We shall refer to the number k of different one-parameter families of the meromorphic solutions (9.21) as the Kovalevskaya number.Such numbers were introduced in [157]. It turns out that if $n = -1, 0, 1$ or $n \geq 4$ then $k = 0$ and $k = 1$ for $n = 2$, $k = 2$ for $n = 3$.

Indeed, substituting the Laurent series (9.21) in (9.20) and matching the coefficients of higher degrees of $1/t$ we arrive at the linear relations

$$\beta = \alpha + 1\,, \qquad \beta + 1 = n\alpha\,.$$

If $n = -1$ or $n = 0$, then $\beta < 0$, whereas for $n = 1$ this system is not consistent. Besides, $\alpha = 2/(n-1)$ must be an integer. Therefore, $n \leq 3$. As a result, if n differs from 2 or 3, then the Hamiltonian system has no solutions. As a matter of fact, for $n \leq 4$ all the solutions are entire functions on $\mathbb{C} = \{\approx\}$, while for $n \geq 4$ the general solution is multivalued.

Let $n = 2$. Then

$$X_{-2} = 6/a\,, \quad X_{-3} = -12/a\,.$$

Substituting the series (9.21) into the left and right sides of the Hamilton equations and matching the coefficients of the same degrees of t, we obtain an infinite sequence of algebraic relations. The pairs $X_\lambda, X_{-\lambda}$ can be found from these relations by a recursion procedure. For $\lambda \neq 4$ these relations have a unique solution, while for $\lambda = 4$ all the relations degenerate to the same equation $4X_4 = Y_3$. Thus, the coefficient X_4 (or X_3) can be regarded as an arbitrary parameter (a modulus) and, thereby, for $n = 2$ the Kovalevskaya number equals 1.

If $n = 3$, then $\alpha = 1, \beta = 2$ and

$$X_{-1} = \pm\sqrt{2/a}\,, \quad X_{-2} = \mp\sqrt{2/a}\,.$$

For each of the two possible sign choices in Eq. (9.20) we have a one-parameter family of its meromorphic solutions. In both options the coefficient X_3, for example, can be regarded as arbitrary parameter. These two families are different, since they have different coefficients at higher degrees of $1/t$. Therefore, $k = 2$.

Note that for $n = 2, 3$ the general solution of (9.21) is expressed in terms of elliptic functions of time. In addition, in the first case the function $x(t)$ has a single pole in the parallelogram of periods of the elliptic functions, while in the second case this function has two simple poles, whose residues have different signs. Therefore, since the solutions are periodic, for $n = 2$ there exists only one family of meromorphic solutions, while for $n = 3$ there are exactly two such solutions.

The ideas considered above can be generalized to an arbitrary system of differential equations in $\mathbb{C}^n = \{z\}$ with polynomial right-hand sides. Substituting formal Laurent series (9.18) in the equations and comparing the coefficients of the same degrees of t, we find the restrictions on the multiplicities k_j of the poles, and then obtain an infinite sequence of polynomial equations in the coefficients $z_j^{(p)}$ of the Laurent series. Each of these equations contains only a finite number of unknown coefficients. The set of all these polynomial equations constitutes a certain algebraic variety in the infinite-dimensional space of the coefficients $z_j^{(p)}$. The dimension of the variety is not greater than $n - 1$, since the system of differential equations is autonomous. The Kovalevskaya number k of a system of polynomial equations is the number of connected components that have dimension $n - 1$. The Kovalevskaya numbers are the simplest topological invariants of an analytical system of differential equations. One may also consider more refined invariants of the algebraic variety constructed above (the homology groups, for example). Note that the complex dimension of some of its connected components can be greater than 1.

If $k = 0$, then the general solution of the original system of differential equations cannot be meromorphic. In particular, in this case the Hamiltonian system (9.11) is not algebraic completely integrable. This provides grounds for Kovalevskaya's method of identification of a system of differential equations as an algebraic completely integrable system. She applied this method first to rigid body motion around a fixed point. It turned out that $k \neq 0$ only in the integrable cases found by Euler, Lagrange, and Kovalevskaya. This method is successfully used in searching for new integrable problems of classical mechanics and mathematical physics.

9.5 Now consider the differential equations with polynomial right-hand sides

$$\dot{z}_i = f_i(z_1, \ldots, z_n)\,, \qquad 1 \leq i \leq n\,, \tag{9.22}$$

which are invariant under the similarity transformation

$$t \to t/\alpha\,, \quad z_1 \to \alpha^{g_1} z_1\,, \ \ldots\,, \ z_n \to \alpha^{g_n} z_n\,, \tag{9.23}$$

where g_j are positive integers. Equations (9.22) are invariant under such a transformation if and only if

$$f_i(\alpha^{g_1} z_1, \ldots, \alpha^{g_n} z_n) = \alpha^{g_i+1} f_i(z_1, \ldots, z_n)\,. \tag{9.24}$$

For example, let f_i be homogeneous polynomials of degree $m > 1$; then we can put $g_1 = \cdots = g_n = g$ in (9.23). In view of (9.24), this implies that $g = 1/(m-1)$ which is integer only if $m = 2$. Thus, equations with quadratic right-hand sides admit the similarity transformation of the form (9.23). Important examples are the Euler–Poincaré equations on Lie algebras. Equations (9.15) present a more complex example: they admit the group action

$$t \to \frac{t}{\alpha}\,, \quad u_k \to \alpha u_k\,, \quad v_k \to \alpha^2 v_k\,.$$

A similar example is given by the Euler–Poisson equations describing rigid body rotation around a fixed point.

For the "quasi-homogeneous" Eqs. (9.22) the problem of the uniqueness of the general solution can be solved completely. We reproduce here the investigation of Eqs. (9.22) made by Yoshida [237], who used Kovalevskaya's method. First note that the system (9.22) admits the particular solutions

$$z_1 = c_1 t^{-g_1}\,, \ \ldots\,, \ z_n = c_n t^{-g_n}\,, \tag{9.25}$$

where the constants c_k satisfy the system of algebraic equations

$$f_i(c_1, \ldots, c_n) = -g_i c_i\,, \qquad 1 \le i \le n\,.$$

These, as a rule, have non-zero complex roots.

We find the general solution of (9.22) in the form

$$z_i = (c_i + x_i)t^{-g_i}\,, \qquad 1 \le i \le n\,.$$

Now write the equations for the new variables x_i. First, we have

$$\begin{aligned} \frac{d}{dt}(x_i t^{-g_i}) &= \frac{d}{dt}(z_i - c_i t^{-g_i}) \\ &= f_i((c_1 + x_1)\, t^{-g_1}, \ldots, (c_n + x_n)t^{-g_n}) \\ &\quad - f_i(c_1 t^{-g_1}, \ldots, c_n t^{-g_n}) \\ &= \sum_{j=1}^{n} \frac{\partial f_i}{\partial z_j}\bigg|_{z=ct^{-g}} x_j t^{-g_j} \\ &\quad + \sum_{|m|\ge 2} \frac{\partial^{m_1+\cdots+m_n} f_i}{\partial^{m_1} z_1 \cdots \partial^{m_n} z_n}\bigg|_{z=ct^{-g}} x_1^{m_1} \cdots x_n^{m_n} t^{-(m,g)}\,, \end{aligned} \tag{9.26}$$

where $|m| = \sum m_j$, $(m, g) = m_1 g_1 + \cdots + m_n g_n$. Then, differentiating (9.24) by z_j and putting $\alpha = 1/t$, $z_i = c_i$, we obtain

$$t^{-g_j} \frac{\partial f_i}{\partial z_j}(ct^{-g}) = t^{-g_i - 1} \frac{\partial f_i}{\partial z_j}(c)\,.$$

More generally,

$$t^{-\sum m_j g_j} \frac{\partial^{m_1+\cdots+m_n} f_i}{\partial^{m_1} z_1 \cdots \partial^{m_n} z_n}(ct^{-g}) = t^{-g_i - 1} K^{(i)}_{m_1 \cdots m_n}\,,$$

$$K^{(i)}_{m_1 \ldots m_n} = \frac{\partial^{m_1+\cdots+m_n} f_i}{\partial^{m_1} z_1 \cdots \partial^{m_n} z_n}(c)\,.$$

As a result, Eqs. (9.26) take on the form

$$t\dot{x}_i = \sum_{j=1}^{n} K_{ij} x_j + \sum_{m=2}^{\infty} K^i_{m_1 \cdots m_n} x_1^{m_1} \cdots x_n^{m_n}\,, \tag{9.27}$$

$$K_{ij} = \frac{\partial f_i}{\partial z_j}(c) + g_j \delta_{ij}\,,$$

where δ_{ij} is the Kronecker tensor. Let $\rho_1, \ldots, \rho_n$ be the eigenvalues of the matrix K. Recall that by an appropriate similarity transformation $K \to C^{-1}KC$ the matrix K is reduced to the diagonal matrix $\mathrm{diag}[\rho_1, \ldots, \rho_n]$ if and only if all its eigenvalues have prime elementary divisors.

Theorem (Lyapunov). *If the solutions of* (9.22) *are single-valued functions of complex time, then*
1) $\rho_i \in \mathbb{Z}$ *for all* i,
2) $\rho_1, \ldots, \rho_n$ *have prime elementary divisors.*

To prove this we note that after the time change $\tau = \ln t$ Eqs. (9.27) become autonomous ($dx/d\tau = t\dot{x}$) and, therefore, we can apply the well-known Lyapunov theorem which asserts that the solutions can be expanded by small parameters $\varepsilon_1, \ldots, \varepsilon_n$ in a convergent series

$$x_i(\tau) = \sum_{|m|=1}^{\infty} X^{(i)}_{m_1 \cdots m_n} \varepsilon_1^{m_1} \cdots \varepsilon_n^{m_n} e^{(m_1\rho_1 + \cdots + m_n\rho_n)\tau} ,$$

where $X^{(i)}_m$ are polynomials of τ with constant coefficients (see [167, Chap.2]). Going back to the original time t, we obtain the following expansions

$$x_i(t) = \sum_{m=1}^{\infty} X^{(i)}_{m_1 \cdots m_n} \varepsilon_1^{m_1} \cdots \varepsilon_n^{m_n} t^{m_1\rho_1 + \cdots + m_n\rho_n} , \tag{9.28}$$

where $X^{(i)}_m$ are polynomials of $\ln \tau$.

Let γ be a continuous curve in a bounded domain in $\mathbb{C}$ that does not contain the point $t = 0$. According to [167, Chap.2], if t belongs to a small neighborhood of the curve γ, then the series (9.28) are convergent for all $\varepsilon = (\varepsilon_1, \ldots, \varepsilon_n)$ in a small neighborhood of zero in $\mathbb{C}^n = \{\varepsilon\}$. If γ has self-intersections (for example, when it is a cycle in $\mathbb{C}$) then (9.28) determines an analytic continuation of the functions $x_i(t)$ along γ. It is obvious that if there are non-integer numbers among $\rho_1, \ldots, \rho_n$, then the functions $x_i(t)$, together with $z_i(t)$, branch after circling around the point $t = 0$. If all the ρ's are integer, but the matrix K is not diagonalizable, then $x_i(t), z_i(t)$ branch as well: in this case the polynomials $X^{(i)}_m$, $|m| = 1$ contain nontrivial terms with $\ln t$. This establishes the theorem.

The matrix K and the condition for its eigenvalues to be integers appeared in Kovalevskaya's work on the dynamics of a heavy rigid body [122]. Yoshida proposed calling the numbers $\rho_1, \ldots, \rho_n$ *the Kovalevskaya exponents*. If the solutions(9.28) are meromorphic and the series (9.28) are infinite, then $\rho_i \geq 0$. Kovalevskaya's investigations were complemented and improved by Lyapunov [165], who showed that the solutions of the Euler–Poisson equations branch in all cases, except the integrable problems found by Euler, Lagrange and Kovalevskaya.

10 Perturbation Theory

10.1 Let $M = D \times \mathbb{T}^n$, $\mathbb{T}^n = \{\varphi \bmod 2\pi\}$, where D is a domain in $\mathbb{R}^n = \{I\}$, be endowed with the standard symplectic structure, and let $H(I, \varphi, \varepsilon) : M \times (-\varepsilon_0, \varepsilon) \to \mathbb{R}$ be an analytic function such that $H(I, \varphi, 0) = H_0(I)$. The canonical equations with the Hamiltonian H_0 can be immediately integrated:

$$\dot{I} = -\frac{\partial H_0}{\partial \varphi}, \quad \dot{\varphi} = \frac{\partial H_0}{\partial I}; \qquad I = I_0, \quad \varphi = \varphi^0 + \omega(I_0)t .$$

According to Poinçaré, the investigation of the system

$$\dot{I} = -\frac{\partial H}{\partial \varphi}, \quad \dot{\varphi} = \frac{\partial H}{\partial I}; \qquad H = H_0(I) + \varepsilon H_1(I, \varphi) + \cdots \tag{10.1}$$

for small values of ε is a basic problem of dynamics.

The idea of the classical perturbation theory is the following: to find a canonical transformation

$$(I, \varphi) \to (\mathcal{J}, \psi), \qquad I = -\frac{\partial S}{\partial \varphi}, \quad \psi = \frac{\partial S}{\partial \mathcal{J}};$$
$$S(\mathcal{J}, \varphi, \varepsilon) = S_0 + \varepsilon S_1 + \cdots ,$$

depending on ε analytically, such that
1) $S_0 = \mathcal{J}\varphi$ (the transformation is close to identity),
2) the functions $S_k(\mathcal{J}, \varphi)$ are periodic in φ with period 2π for all $k \geq 1$,
3) in the new variables $H = K(\mathcal{J}, \varepsilon)$.

Therefore, if any function $f(I, \varphi, \varepsilon)$ which is 2π-periodic in φ is expressed in terms of $\mathcal{J}, \psi$, then it is also 2π-periodic in ψ.

Given such a transformation, the Hamilton Eqs. (10.1) can be integrated completely. The functions $\mathcal{J}_s = \mathcal{J}_s(I, \varphi, \varepsilon)$, $\mathcal{J}_s(I, \varphi, 0) = I_s$ $(1 \leq s \leq n)$ form a complete set of independent integrals in involution.

10.2 The function $S(\mathcal{J}, \varphi)$ satisfies the equation

$$\left(\frac{\partial H_0}{\partial \mathcal{J}}, \frac{\partial S_1}{\partial \varphi} \right) + H_1(\mathcal{J}, \varphi) = K_1(\mathcal{J}) , \tag{10.2}$$

where $K_1(\mathcal{J})$ is to be determined later. We expand the perturbation H_1 in a multiple Fourier series:

$$H_1 = \sum_{m \in \mathbb{Z}^n} H_m(\mathcal{J}) \exp i(m, \varphi) .$$

If Eq. (10.2) has a solution which is periodic in φ, then

$$K_1(\mathcal{J}) = \frac{1}{(2\pi)^n} \int_{\mathbb{T}^n} H_1(\mathcal{J}, \varphi)\, d^n\varphi .$$

Let

$$S_1 = \sum_{m \neq 0} S_m(\mathcal{J}) \exp i(m, \varphi) \ .$$

Then

$$S_m(\mathcal{J}) = \frac{H_m(\mathcal{J})}{i(m, \omega(\mathcal{J}))} \ . \tag{10.3}$$

In the sequel we use the *secular set* $B \subset D$, i.e., the set of points $\mathcal{J} \in D$ such that

$$\sum_{m \neq 0} \left| \frac{H_m(\mathcal{J})}{(m, \omega(\mathcal{J}))} \right|^2 = \infty \ .$$

In particular, in the set B there are points $\mathcal{J} \in D$ such that $(m, \omega(\mathcal{J})) = 0$, $m \neq 0$, and $H_m(\mathcal{J}) \neq 0$[3]. According to Bessel's inequality

$$\sum_m S_m^2 < \infty \ ,$$

the generating function S_1 is not defined on the set $B \times \mathbb{T}^n \subset D \times \mathbb{T}^n$.

As a matter of fact, the secular set is the set of those tori of the unperturbed integrable problem which are destroyed under a perturbation of order ε. In the typical situation, B is everywhere dense in D. This is connected with a well-known difficulty, the appearance of "small denominators" (or small divisors) which make it impossible to construct not only converging, but even formal series of the classical perturbation theory.

It is well-known from the theory of Diophantine approximations that for almost all frequencies $\omega = (\omega_1, \ldots, \omega_n) \in \mathbb{R}^n$ the estimate

$$|(m, \omega)| \geq \mu / |m|^{n+1} \tag{10.4}$$

is valid for all $m \in \mathbb{Z}^n$ $(m \neq 0)$, where μ is a positive constant depending on ω. Let $\omega = \partial H_0 / \partial \mathcal{J}$, and the Hamiltonian H_0 be non-degenerate. Then the estimate (10.4) holds for almost all $\mathcal{J}$. The Fourier coefficients of an analytic function diminish exponentially fast as $|m|$ grows. Let $\mathcal{J}$ be such that inequality (10.4) holds for all $|m|$. According to (10.3), in this case the numbers $S_m(\mathcal{J})$ also diminish exponentially fast as $|m| \to \infty$. Hence, for such values of $\mathcal{J}$ the function $S_1(\mathcal{J}, \varphi)$ is analytic and periodic in φ. One can show in a similar way that if (10.4) is fulfilled, then the analytic functions $S_k(\mathcal{J}, \varphi)$, $k \geq 1$ are well defined. However, to ensure the convergence of the series

$$\sum_{k \geq 1} S_k(\mathcal{J}, \varphi) \varepsilon^k \ , \tag{10.5}$$

additional conditions are needed.

---

3 According to the terminology of celestial mechanics, the Fourier coefficient $H_{m_1 \cdots m_n}(\mathcal{J})$ is called secular if $m_1 \omega(\mathcal{J}) + \cdots + m_n \omega_n(\mathcal{J}) = 0$ for some $\mathcal{J}_1, \ldots, \mathcal{J}_n$.

Introduce the set

$$\Omega_\mu = \{\omega \in \Omega : |(m, \omega)| > \mu/|m|^{n+1}, \ m \neq 0\} ,$$

where Ω is a certain bounded domain in the space of frequencies $\omega = (\omega_1, \ldots, \omega_n)$. One can show that for small μ the Lebesgue measure of the complement of Ω_μ with respect to Ω does not exceed a number of order μ.

Let Λ_μ be a preimage of the set Ω_μ under the mapping

$$\mathcal{J} \to \partial H_0/\partial \mathcal{J} .$$

Kolmogorov [117] proved that for small ε the series (10.5) converges for all $\mathcal{J}$ from the set Λ_μ with the value of μ fixed. The proof of Kolmogorov's theorem relies upon the procedure of successive approximations of Newton's type, which was first proposed by Newcomb in celestial mechanics. A direct proof of the convergence of the series (10.5) based on estimating coefficients is still unknown.

The relations

$$I_k = \partial S/\partial \varphi_k , \qquad 1 \leq k \leq n ,$$
$$S(\mathcal{J}, \varphi, \varepsilon) = \sum_{k \geq 0} S_k(\mathcal{J}, \varphi)\varepsilon^k , \qquad \mathcal{J} \in \Lambda_\mu \tag{10.6}$$

define n-dimensional invariant tori of the perturbed Hamiltonian system with "strongly" incommensurable frequencies. These tori are called the Kolmogorov tori; they depend on ε analytically. The Kolmogorov tori are n-dimensional Lagrange manifolds, since the covector field $I = \partial S/\partial \varphi$ is potential (see §2).

Various versions of Kolmogorov's theorem on conservation of quasi-periodic motions were obtained by Arnol'd and Moser. For a survey of KAM (Kolmogorov–Arnol'd–Moser) theory see the book [14, Chap. 5].

Theorem 1. *Suppose that equations* (10.1) *possess n analytic integrals*

$$F_i : D \times \mathbb{T}^n \times (-\kappa, \kappa) \to \mathbb{R} ,$$

such that
1) *for all ε the functions $F_1, \ldots, F_n$ are in involution,*
2) $F_i(I, \varphi, 0) = f_i(I)$, $1 \leq k \leq n$,
3) *the Jacobian*

$$\frac{\partial(f_1, \ldots, f_n)}{\partial(I_1, \ldots, I_n)}$$

does not vanish in the domain D. Then on the set $G \times \mathbb{T}^n \times (-\alpha, \alpha)$ with $(\alpha \ll 1)$, G being a compact subdomain in D, there exists an analytic generating function $S(\mathcal{J}, \varphi, \varepsilon)$ which satisfies conditions 1)–3) *in* §10.1.

If Eq. (10.1) have integrals formally analytic in ε (formal series in degrees of ε with coefficients analytic in $D \times \mathbb{T}^n$), which satisfy the conditions of the theorem, then one can (at least formally) construct series of perturbation theory defined for $(\mathcal{J}, \varphi) \in D \times \mathbb{T}^n$. Let us prove this.

Let $F_s(I, \varphi, \varepsilon) = f_s(I) + \sum \varepsilon^k F_{sk}(I, \varphi)$. Consider the system of equations

$$F_s\left(\frac{\partial S}{\partial \varphi}, \varphi, \varepsilon\right) = f_s(\mathcal{J}) + \sum_{k\geq 1} \varepsilon^k f_{sk}(\mathcal{J}) , \qquad 1 \leq s \leq n ,$$

with the analytic functions $f_{sk} : D \to \mathbb{R}$ to be determined later. For $\varepsilon = 0$ relations (10.6) hold provided $S_0 = \mathcal{J}\varphi$. Since $F_s(I, \varphi, 0) = f_s(I)$ and the Jacobian in condition 3) of Theorem 1 does not vanish, for given f_{sk} we can define the formal series

$$I(\varphi, \varepsilon) = \frac{\partial S}{\partial \varphi} = \mathcal{J} + \varepsilon \frac{\partial S_1}{\partial \varphi} + \cdots , \tag{10.7}$$

which satisfies (10.6). We claim that the differential form

$$I(\varphi, \varepsilon)\, d\varphi = \frac{\partial S}{\partial \varphi}\, d\varphi$$

is exact. To prove this we need the following simple

Lemma. *Let the system of equations*

$$F_s(p, q) = c_s , \qquad 1 \leq s \leq n$$

be defined in $\mathbb{R}^{2n} = \{p, q\}$, *and* $p_s = f_s(q, c_1, \ldots, c_n)$ *be its solution. If the functions* $F_1, \ldots, F_n$ *are in involution (with respect to the standard symplectic structure in* $\mathbb{R}^{2n}$*), then the form* $\sum f_s(q, c)\, dq_s$ *is a complete differential.*

Proof. The functions $G_s = p_s - f_s(q, F_1, \ldots, F_n)$ are, obviously, constant. Since $F_1, \ldots, F_n$ commute, we have

$$\{G_s, G_m\} = \frac{\partial F_m}{\partial q_s} - \frac{\partial f_s}{\partial q_m} = 0 .$$

Q.E.D.

For arbitrary choice of $f_{sk}(\mathcal{J})$, the functions $S_k(\mathcal{J}, \varphi)$ are multivalued on $\mathbb{T}^n$. This can be avoided by an appropriate choice of f_{sk}. First, let $k = 1$. From (10.6) it follows that

$$\left(\frac{\partial f_s}{\partial \mathcal{J}}, \frac{\partial S_1}{\partial \varphi}\right) = f_{s1}(\mathcal{J}) - F_{s1}(\mathcal{J}, \varphi) . \tag{10.8}$$

If we put

$$f_{s1} = \frac{1}{(2\pi)^n} \int\limits_{\mathbb{T}^n} F_{s1}(\mathcal{J}, \varphi)\, d^n\varphi ,$$

then from (10.8) we obtain the periodic solution S_1. For $k \geq 1$, to specify S_k and f_{sk} we have an equation of the form (10.8), whose right-hand side includes the known functions S_m and f_{sm} $(m < k)$.

In the new canonical variables $\mathcal{J}, \psi$ the functions $F_1, \ldots, F_n$ depend only on $\mathcal{J}$ and ε. Since the latter are first integrals of the Hamiltonian system (10.1) and they are independent, the same holds for $\mathcal{J}_1, \dots, \mathcal{J}_n$. Therefore, the Hamiltonian H does not depend on the angles ψ:

$$\frac{\partial H}{\partial \psi} = -\dot{\mathcal{J}} = 0 .$$

Theorem 1 is proved.

11 Normal Forms

11.1 Consider a Hamiltonian system

$$\dot{z} = J\frac{\partial H}{\partial z} , \qquad z = (p, q) \in \mathbb{R}^{2n}$$

in a neighborhood of the point $z = 0$. Let a real analytical function H be represented by a converging power series in z with the leading terms of second order: $H = \sum_{k\geq 2} H_k$. The point $z = 0$ is, obviously, an equilibrium position.

The linearized system

$$\dot{z} = J\frac{\partial H_2}{\partial z}$$

has four types of eigenvalues: real pairs $(-a, a)$, $a \neq 0$, purely imaginary pairs $(-ib, ib)$, $b \neq 0$, quadruples $(\pm a \pm b)$ $a \neq 0$, $b \neq 0$, and multiple zeros (see, for example, [14, Chap. 6]). In the first and third cases the equilibrium position $z = 0$ is stable automatically. Consider the case when the eigenvalues of the linearized system are imaginary and distinct. Then, as we know ([217]), there exists a linear canonical change of coordinates which reduces the quadratic form H_2 to

$$\frac{1}{2}\sum \alpha_s(x_s^2 + y_s^2) . \tag{11.1}$$

The eigenvalues are exactly the numbers $\pm\alpha_1, \dots, \pm\alpha_n$.

Theorem 1 (Birkhoff). *If $\alpha_1, \dots, \alpha_n$ are independentover the field of rational numbers, then there exists a formal canonical transformationx, $y \to \xi, \eta$ defined by formal power series*

$$x = u(\xi, \eta) = \xi + \cdots , \qquad y = v(\xi, \eta) = \eta + \cdots , \tag{11.2}$$

which takes $H(x, y)$ to a Hamiltonian $K(\rho)$ which is a formal power series in $\rho_s = \xi_s^2 + \eta_s^2$.

If the series (11.2) converge, then the equations with the Hamiltonian H can be easily integrated. Indeed, the functions $\rho_1, \dots, \rho_n$, being converging power series in x, y, form a complete set of independent integrals in involution. From the canonical equations

$$\dot{\xi}_s = \Omega_s\eta_s , \quad \dot{\eta}_s = -\Omega_s\xi_s , \qquad \Omega_s = 2\,\partial K/\partial\rho_s$$

it follows that $\xi_s(t)$ and $\eta_s(t)$ are linear combinations of $\sin \Omega_s t$ and $\cos \Omega_s t$. Hence, the original coordinates x, y are quasi-periodic functions of time with the frequencies $\Omega_1, \dots, \Omega_n$. In particular, the equilibrium position $z = 0$ is stable.

Corollary. *If $\alpha_1, \dots, \alpha_n$ are rationally incommensurable, then the Hamilton equations admit n independent formal integrals in involution of the following form*

$$\rho_s = x_s^2 + y_s^2 + \text{terms of order } \geq 3 \, , \qquad 1 \leq s \leq n \, . \tag{11.3}$$

The series (11.3) are obtained from the functions $\xi_s^2 + \eta_s^2$ by the formal canonical change (11.2). Every formally analytic integral of the Hamilton equations is a power series in the integrals (11.3). Indeed, in the new variables ξ, η such an integral depends only on $\rho_1, \dots, \rho_n$.

Theorem 2. *If a system with the Hamiltonian $H = \sum_{k \geq 2} H_k$ possesses n analytic integrals in involution*

$$G_m = \frac{1}{2} \sum \kappa_{ms}(x_s^2 + y_s^2) + \sum_{k \geq 2} G_{mk} \, ,$$

and, besides, $\det \|\kappa_{ms}\| \neq 0$, *then the Birkhoff transformation* (11.2) *converges.*

This argument shows why (following Birkhoff) we refer to a Hamiltonian system with a converging Birkhoff transformation as integrable. Postponing discussion of convergence until Chap. VI, we notice that the Birkhoff series are, as a rule, divergent. Theorem 2 was first proved by Rüssmann for $n = 2$ [208], and afterwards by Vey for the multidimensional case [231].

The Rüssmann–Vey theorem was generalized by Ito in [103].

Theorem 3. *Suppose that the numbers $\alpha_1, \dots, \alpha_n$ are rationally incommensurable, and in a neighborhood of the point $z = 0$ the Hamiltonian system has n analytic integrals $G_1 = H, G_2, \dots, G_n$ independent almost everywhere. Then there exists an analytic canonical transformation $z = \Phi(\zeta)$, $\zeta = (\xi, \eta)$ such that $\Phi(0) = 0$ and in the new variables ζ the integrals $G_1, \dots, G_n$ are analytic functions of $\xi_s^2 + \eta_s^2$ $(1 \leq s \leq n)$.*

We emphasize that in Ito's theorem the quadratic forms of the Maclaurin series $G_2, \dots, G_n$ may be degenerate or even absent. Besides, the functions G_k are not supposed to be in involution because any analytic integrals of a Hamiltonian system are automatically in involution provided the eigenvalues $\alpha_1, \dots, \alpha_n$ are non-resonant. For this reason, in Theorem 2 we can omit the involution condition as well.

Theorem 1 admits a generalization to the case when the numbers $\alpha_1, \dots, \alpha_n$ are rationally dependent. Consider all vectors $j = (j_1, \dots, j_n)$ with integer components for which $(j, \alpha) = 0$. These vectors form a free Abelian group Γ of a certain rank r. If $\alpha_1, \dots, \alpha_n$ are independent, then, obviously, $r = 0$.

Now perform a formal canonical change $x, y \to \xi, \eta$ of the form (11.2). In the new variables ξ, η the Hamiltonian $H(x, y)$ is represented by a formal power series $K(\xi, \eta)$. Introduce the complex variables $\zeta_s = \xi_s + i\eta_s$ $\bar{\zeta}_s = \xi_s - i\eta_s$ and expand K in a series in the products

$$\zeta^k \bar{\zeta}^l = \prod_{s=1}^{n} \zeta_s^{k_s} \bar{\zeta}_s^{l_s} .$$

We say that the formal series $K(\xi, \eta)$ has a normal form if its expansion contains only the terms $\zeta_s^k \bar{\zeta}_s^l$ with $(k - l) \in \Gamma$. In particular, if $\alpha_1, \ldots, \alpha_n$ are independent, then in the normal form of the Hamiltonian there are only the following terms

$$\zeta^k \bar{\zeta}^l = (\xi_1^2 + \eta_1^2)^{k_1} \cdots (\xi_n^2 + \eta_n^2)^{k_n} .$$

The series $K(\xi, \eta)$ has a normal form if and only if $\mathfrak{D}(K) = 0$, where

$$\mathfrak{D} = \sum_{s=1}^{n} \alpha_s \left(\xi_s \frac{\partial}{\partial \eta_s} - \eta_s \frac{\partial}{\partial \xi_s} \right) .$$

The proof can be easily deduced from the formula

$$\mathfrak{D}(\zeta^k \bar{\zeta}^l) = i(\alpha, k - l)\, \zeta^k \bar{\zeta}^l .$$

Theorem 4. *There exists a formal canonical transformation of the form* (11.2) *such that the original Hamiltonian $H(x, y)$ transforms to a normal form, i.e.,* $\mathfrak{D}(K) = 0$.

The proof can be found in [182]. For $\Gamma = \{0\}$ this theorem coincides with the Birkhoff theorem.

Let us show that in the considered case one can indicate $n - r$ commuting independent formal integrals of the type

$$G = \frac{1}{2} \sum \beta_s (\xi_s^2 + \eta_s^2) ,$$

where the vector $\beta = (\beta_1, \ldots, \beta_n)$ is orthogonal to all vectors from the group Γ. Indeed,

$$\dot{G} = \sum \beta_s \left(\xi_s \frac{\partial K}{\partial \eta_s} - \eta_s \frac{\partial K}{\partial \xi_s} \right) = 0 ,$$

if $\beta \perp \Gamma$. Since rank $\Gamma = r$, one can find $n - r$ linearly independent vectors β.

As an example, consider the Hénon–Heiles system whose Hamiltonian is

$$H = \frac{1}{2}(y_1^2 + y_2^2 + x_1^2 + x_2^2) + 2x_1^2 x_2 - \frac{2}{3} x_2^3 .$$

Here $n = 2$ and $\alpha_1 = \alpha_2 = 1$. The group Γ is defined by the condition $j_1 + j_2 = 0$; rank $\Gamma = 1$. To find an integral independent of H, we put $\beta_1 = \beta_2 = 1$. Then

$$G = (\xi_1^2 + \eta_1^2 + \xi_2^2 + \eta_2^2)/2 .$$

After transforming H to normal form according to Theorem 3, we obtain the integral

$$K = \frac{1}{2}(\xi_1^2 + \eta_1^2 + \xi_2^2 + \eta_2^2) + \cdots ,$$

which has the same leading terms as G. Taking into account terms of degree ≥ 3, one can show by direct computations that the functions K and G are indeed independent. For the discussion of results of the numerical investigations performed by Hénon and Heiles in connection with construction of formal integrals see Gustavson [94] and Moser [182].

11.2 The procedure of normalization of a Hamiltonian system in a neighborhood of a stable equilibrium position is closely related to the classical scheme of perturbation theory. Indeed, we may introduce a small parameter ε by the substitution $x \to \varepsilon x$, $y \to \varepsilon y$ and then change to the polar coordinates I, φ by using the formulas

$$x_s = \sqrt{2I_s} \sin \varphi_s , \quad y_s = \sqrt{2I_s} \cos \varphi_s .$$

As a result, we obtain a Hamiltonian system

$$\dot{I}_s = -\frac{\partial H}{\partial \varphi_s} , \quad \dot{\varphi}_s = \frac{\partial H}{\partial I_s}$$

with the Hamiltonian

$$H = \sum_{m \geq 0} \varepsilon^m H_m^*(I, \varphi) , \qquad H_0^* = \sum \alpha_s I_s , \quad H_m^* = H_{m+2}(x, y)|_{I,\varphi} ,$$

which is 2π-periodic in φ. If the frequencies α_s are rationally independent, then there exists a formal power series of the classical perturbation theory which corresponds exactly to the Birkhoff transformation. In this way one can deduce the Rüssmann–Vey theorem from Theorem 1 in §10.

11.3 In applications the function H usually depends on several parameters $\varepsilon \in D$ (D is a domain in $\mathbb{R}^m$). We suppose that the function $H(z, \varepsilon)$ is analytic in z, ε and $\partial H(0, \varepsilon)/\partial z = 0$ for all ε. If for all ε the eigenvalues of the linearized system are purely imaginary and distinct, then by an appropriate linear symplectic transformation analytic in ε the form H_2 can be reduced to the "normal" form (11.1). The coefficients α_s are obviously analytic in ε.

The following theorem slightly strengthens the Rüssmann and Vey result

Theorem 5 [135]. *Suppose we are given n integrals in involution*

$$G_k(x, y, \varepsilon) = \frac{1}{2} \sum x_{ks}(\varepsilon)(x_s^2 + y_s^2) + \sum_{j \geq 3} G_{kj}(x, y, \varepsilon) ,$$

which are analytic in ε and such that $\det \|x_{ks}(\varepsilon)\| \neq 0$ *for all $\varepsilon \in D$. Then there exists an analytic canonical transformation $x, y \to \xi, \eta$, also analytic in ε, which takes $H(x, y, \varepsilon)$ to the Hamiltonian $K(\rho_1, \ldots, \rho_n, \varepsilon)$, $\rho_s = \xi_s^2 + \eta_s^2$.*

If the series $\sum G_{kj}$ are formal (not necessarily convergent), then one can find a formal canonical transformation which "normalizes" the Hamiltonian H. In particular, the Birkhoff transformation exists also for rationally dependent frequencies $\alpha_1, \ldots, \alpha_n$ provided the conditions of Theorem 5 are fulfilled.

An analog of Theorem 4 for Hamiltonian systems depending on a parameter is given in [123]. The reduction of Hamiltonian systems depending on a parameter and having integrals with degenerate quadratic terms to the Birkhoff normal form is considered in [197]. Let $n = 2$ and the coefficients α_1, α_2 in the quadratic form of the Hamiltonian (11.1), as functions of ε, satisfy the condition

$$m_1\alpha_1 + m_2\alpha_2 \not\equiv 0$$

for all integer m_1, m_2 which do not vanish simultaneously. As proved in [197], if a Hamiltonian system admits a formal integral

$$F = F_q + F_{q+1} + \cdots, \qquad q \geq 2,$$

which is analytic in ε and such that the homogeneous forms F_q and H_2 are functionally independent for all ε, then there exists a normalizing Birkhoff transformation depending on ε analytically.

Apparently, an analog of Theorem 3 is valid for Hamiltonian systems depending on a parameter.

11.4 In the general case, when not all eigenvalues $\pm\lambda_1, \ldots, \pm\lambda_n$ are purely imaginary, the Hamilton equations can be reduced to a Birkhoff normal form as well. For a detailed discussion of these questions see, for example, [217]. The Hamilton equations generally have invariant asymptotic surfaces Σ completely filled by trajectories which tend to equilibrium positions as $t \to \pm\infty$. It turns out that a Birkhoff transformation can be defined by divergent power series, but these series converge at points on Σ.

Consider more closely the case when the eigenvalues $\pm\lambda_1, \ldots, \pm\lambda_n$ are real and differ from zero. The equilibrium position is obviously stable. Without loss of generality we may assume that all $\lambda_k > 0$. Suppose that all of them are distinct. Then in appropriate canonical coordinates x, y the Hamilton function is reduced to the following form

$$H = \sum_{k\geq 2} H_k, \qquad H_2 = \sum_{j=1}^{n} \lambda_j x_j y_j. \tag{11.4}$$

Using the general arguments given in §2 of Chap. II, we are going to find an n-dimensional invariant asymptotic surface Σ in the form

$$y = \partial S/\partial x, \tag{11.5}$$

where the function $S(x)$ satisfies the Hamilton–Jacobi equation

$$H\left(x, \frac{\partial S}{\partial x}\right) = 0. \tag{11.6}$$

We have put zero in the right-hand side of this equation because the equilibrium position $x = y = 0$ belongs to Σ. Now we search for S as a series in homogeneous forms of $x_1, \ldots, x_n$:

$$S = \sum_{k \geq 2} S_k(x) . \tag{11.7}$$

Taking (11.4) into account, we obtain the following chain of equations for successive determination of $S_2, \ldots, S_k, \ldots$:

$$\begin{aligned} &\sum \lambda_j x_j \partial S_2/\partial x_j = 0 , \\ &\sum \lambda_j x_j \partial S_k/\partial x_j = W_k , \qquad k \geq 3 . \end{aligned} \tag{11.8}$$

Here W_k are some functions uniquely determined by H_r and S_r with $r < k$. Since $\lambda_k > 0$, we have $S_2 \equiv 0$. Now let S_k contain a term of the type

$$s\, x_1^{\alpha_1} \cdots x_n^{\alpha_n} , \qquad \sum \alpha_j = k .$$

Then, from (11.8) we obtain

$$s = w/(\lambda_1 \alpha_1 + \cdots + \lambda_n \alpha_n) , \tag{11.9}$$

where w is the coefficient of $x_1^{\alpha_1} \cdots x_n^{\alpha_n}$ in the homogeneous form W_k. Thus, the power series for S is uniquely determined. The denominators $\lambda_1 \alpha_1 + \cdots + \lambda_n \alpha_n$ in (11.9) are isolated from zero, and, therefore, nothing prohibits the convergence of the series (11.7) in the analytic case. For reversible systems as well as for some systems with gyroscopic forces, convergence follows from the papers of Bohl and Poincaré (see also [45,46]). One may show that if the Hamilton function is infinitely differentiable, then the corresponding Hamilton–Jacobi equation has a solution of the class $C^\infty(\mathbb{R}^n = \{x\})$ with the critical point $x = 0$ (see [37]).

Thus, the Hamiltonian system with the Hamiltonian (11.4) has an n-dimensional invariant surface Σ, and $x_1, \ldots, x_n$ can be regarded as local coordinates on Σ. As a result, we obtain the following system with an equilibrium position at $x = 0$:

$$\dot{x} = \left. \frac{\partial H}{\partial y} \right|_{y = \partial S/\partial x} ,$$

or, in explicit form,

$$\dot{x}_j = \lambda_j x_j + \cdots , \qquad 1 \leq j \leq n . \tag{11.10}$$

Here dots denote terms of order ≥ 2. Since $\lambda_j > 0$, all solutions of (11.10) tend to the point $x = 0$ as $t \to -\infty$. To obtain the asymptotic surface with the trajectories tending to the equilibrium position as $t \to +\infty$, it is sufficient to permute the variables x and y.

Consider reversible systems with the Hamiltonian

$$H = \frac{1}{2} \sum a_{ij}(x) y_i y_j + V(x) ,$$

which are often encountered in applications. Let $x = 0$ be a non-degenerate local minimum of the potential energy V, and, in addition, $V(0) = 0$. Then the Hamilton–Jacobi equation

$$\frac{1}{2}\sum a_{ij}\frac{\partial S}{\partial x_i}\frac{\partial S}{\partial x_j} + V = 0$$

possesses two solutions $\pm S$ defined in a neighborhood of the point $x = 0$. The corresponding system (11.10) takes the form of a gradient system

$$\dot{x}_j = \sum_{k=1}^{n} a_{ik}(x)\frac{\partial S}{\partial x_k}, \qquad 1 \leq j \leq n .$$

According to the Birkhoff theorem, there is a formal canonical transformation $x, y \to \xi, \eta$ of the form

$$\xi_k = -\frac{\partial S}{\partial \eta_k}, \quad y_k = \frac{\partial S}{\partial x_k},$$
$$S(x, \eta) = S_2 + S_3 + \cdots, \qquad S_2 = \sum_{k=1}^{n} x_k \eta_k , \tag{11.11}$$

such that in the new coordinates the Hamilton function is a power series K in $w_1 = \xi_1\eta_1, \ldots, w_n = \xi_n\eta_n$ with the leading terms $\sum \lambda_s w_s$. It is clear that the generating function S satisfies the equation

$$H\left(x, \frac{\partial S}{\partial x}\right) = K(w) , \qquad w = \frac{\partial S}{\partial \eta}\eta . \tag{11.12}$$

One of the asymptotic surfaces is defined by the equations $\eta_1 = \cdots = \eta_n = 0$. Namely, substituting these values into (11.11) and (11.12), we obtain the known relations (11.5) and (11.6) defining this surface. In particular, the normalizing Birkhoff transformation (11.11) converges for $\eta = 0$. The same holds for the other asymptotic surface.

The detailed analysis of the convergence of normalizing transformations (not only for Hamiltonian systems) can be found in Bryuno's book [44].

11.5 Reducing to a normal form can be performed not only in a neighborhood of an equilibrium position, but, for example, in the vicinity of periodic trajectories. All the arguments above hold (with necessary modifications) in the latter case as well.

III Topological and Geometrical Obstructions to Complete Integrability

We start the analysis of the reasons for nonintegrable behavior of Hamiltonian systems with a discussion of relatively recently discovered "rough" *topological obstructions to integrability*. In [130] it was proved that a closed analytic surface with genus ≥ 2 cannot be the configuration space of an integrable analytic system. The reason is the existence of an infinite number of unstable periodic orbits on which the integrals are dependent. This result (unnoticed earlier due to the preference for local study of dynamical systems) was generalized in several directions. The proof of nonintegrability is based on variational methods and subtle results from the theory of singularities of analytic mappings.

1 Topology of the Configuration Space of an Integrable System

1.1 Consider a reversible mechanical system with two degrees of freedom (see Chap. I, §1). Suppose that the configuration space M is a compact oriented analytic surface. The topological structure of such surfaces is well known: they are spheres with a certain number κ of handles attached. The number κ is the genus of the surface.

The motion of the reversible system is governed by the Hamiltonian equations on the phase space T^*M. The cotangent bundle T^*M has a natural structure of a four-dimensional analytic manifold. We assume that the function $H : T^*M \to \mathbb{R}$ is everywhere analytic. Since $H = T(p, q) + V(q)$ and the kinetic energy $T(q, p)$ is a quadratic form in $p \in T_q^*M$ for all $q \in M$, the functions T and V are analytic on T^*M and M respectively. Solutions of the canonical equations

$$\dot{p} = -\frac{\partial H}{\partial q}, \quad \dot{q} = \frac{\partial H}{\partial p} \tag{1.1}$$

are analytic mappings from $\mathbb{R} = \{t\}$ into T^*M. On each trajectory the total energy $H = T + V$ is obviously constant.

Theorem 1 [130]. *If the genus of the surface M is not equal to 0 or 1 (i.e., M is not homeomorphic to the sphere S^2 or the torus $\mathbb{T}^2$), then the system* (1.1) *does not have an analytic first integral on T^*M that is independent of the energy integral.*

Recall that analytic functions are regarded as independent if they are independent at some point (then they are independent almost everywhere).

There are numerous well-known examples of integrable systems with configuration spaces which are homeomorphic to S^2 or $\mathbb{T}^2$ (for example, the inertial motion of a particle on the standard sphere or torus).

In the infinitely differentiable case, Theorem 1 is, in general, not true: for every smooth surface M there exists a smooth "natural" Hamiltonian $H = T+V$ such that the Hamiltonian system (1.1) has an additional smooth integral independent of H (more precisely, not everywhere dependent on H). Indeed, consider the standard sphere S^2 in $\mathbb{R}^3$. Suppose that the surface M is obtained from S^2 by attaching an arbitrary number of handles to some small domain $N \subset S^2$. Let H be the Hamiltonian function describing the inertial motion ($V \equiv 0$) of a particle on the imbedded surface $M \subset \mathbb{R}^3$. Outside the small domain N, the particle moves along the great circles of the sphere. Hence, the phase space T^*M contains an invariant subset which is diffeomorphic to the Cartesian product $D \times \mathbb{T}^2$ foliated by two-dimensional invariant tori. Points of the domain D "enumerate" these tori. Let $f : D \to \mathbb{R}$ be a smooth function that vanishes outside some subdomain G contained in D. The function f defines a smooth function F on $D \times \mathbb{T}^2$ which is constant on the invariant tori in $D \times \mathbb{T}^2$. We put $F = 0$ outside the set $D \times \mathbb{T}^2$. Then F is a smooth function on T^*M. Obviously, F is an integral of the canonical system (1.1) and for appropriate choice of f the functions F and H are not everywhere dependent.

1.2 Theorem 1 is a consequence of a stronger result establishing non-integrability of the equations of motion for fixed sufficiently large values of the total energy. The precise statement is as follows. For any value $h > \max_M V$, the energy level $\Sigma_h = \{z \in T^*M : T + V = h\}$ is a three-dimensional analytic manifold. It has a natural structure of a fiber bundle with base M and fiber S^1. The local coordinates on Σ_h are q and ϕ, where q are coordinates on M, and ϕ is an angular variable on the fiber

$$S^1_q = \{p \in T^*_qM : T(p, q) + V(q) = h\} ,$$

which is a circle in the cotangent plane. Since the Hamiltonian vector field v_H is tangent to Σ_h, we obtain a system of analytic differential equations on Σ_h.

Theorem 2. *If the genus of the surface M is not equal to 0 or 1, then for all $h > \max V$ the flow on Σ_h does not possess a non-constant analytic integral.*

1.3 For a C^∞ Hamiltonian system, the assumptions of Theorems 1 and 2 imply the nonexistence of nontrivial smooth integrals satisfying certain additional properties.

Theorem 3. *If the genus of the smooth surface M is not equal to 0 or 1, then for all $h > \max V$ the phase flow on Σ_h does not have a smooth first integral $f(q, \phi) : \Sigma_h \to \mathbb{R}$ such that:*

a) *f has a finite number of critical values,*

b) *the points $q \in M$ such that the sets $\{f(q, \phi) = c\}$ are finite or coincide with the whole fiber S^1_q are dense in M.*

If f is analytic, then conditions a) and b) are always satisfied. Moreover, condition b) holds for all $q \in M$. Statement a) is nontrivial; the proof is contained in [220].

More generally, if a compact oriented surface M is not homeomorphic to the sphere or the torus, then the equations of motion do not have an additional smooth integral that is analytic on the cotangent plane T_q^*M for all $q \in M$, and has a finite number of critical values. A well-known example of such integrals are integrals that are polynomial in momenta. The number of distinct critical values of a smooth function on a compact manifold is finite if, for example, all critical points are isolated, or the critical points form non-degenerate critical manifolds.

The examples of §1.1 do not contradict Theorem 3: condition b) is obviously false for points $q \in M$ that are sufficiently remote from the "singular" domain N.

1.4 Theorems 1–3 also hold in the case of non-orientable compact surfaces, if we exclude the projective plane $\mathbb{R}P^2$ and the Klein bottle K^2. Indeed, the standard regular two-sheet covering $N \to M$, where N is an oriented surface, induces a natural mechanical system on N. If the system on M has an additional integral, then the same holds for the system on N. Now it is sufficient to recall that if M is not homeomorphic to $\mathbb{R}P^2$ or K^2, then the genus of the surface N is greater than 1.

1.5 According to the Maupertuis principle, the trajectories of the mechanical system with energy $h > \max_M V$ are geodesics of the Riemannian metric

$$(ds)^2 = 2(h - V)T\,(dt)^2$$

on M. Let k be the Gauss curvature of the Maupertuis metric ds. By the Gauss–Bonnet formula,

$$\frac{1}{2\pi}\int_M k\,d\sigma = \chi(M)\,,$$

where $\chi(M)$ is the Euler characteristic of the compact surface M. If the genus of M is greater than one, then $\chi(M) < 0$. Hence, the average curvature is negative. If the curvature is negative everywhere, then the dynamical system on Σ_h is Anosov's system. Thus, it is ergodic on Σ_h [9]. This result holds also in the multidimensional case (if the sectional curvature is negative for all two-dimensional planes). In this case, the differential equations on Σ_h do not admit continuous non-constant integrals, since almost all the trajectories are dense in Σ_h. Of course, if the curvature is negative on average, this does not mean that it is everywhere negative.

1.6 A generalization of Theorems 1 and 2 to the multidimensional case is contained in Taimanov's papers [226–227].

First recall the definition of Betti numbers. Let M be a smooth compact n-dimensional manifold. Consider the real vector space of all differential k-forms on M. The set of exterior derivatives of all $(k-1)$-forms on M is a subspace in the vector space of all closed k-forms. Identifying closed forms which differ by a differential, define the factor space

$$(\text{closed forms})/(\text{differentials}) = H^k(M, \mathbb{R}) .$$

The dimension of the space $H^k(M, \mathbb{R})$ is called the k-th Betti number. It is usually denoted by $b_k(M)$, or simply b_k. For example, if M is an n-dimensional torus, then $b_k = \binom{n}{k}$ (see, for example, [75]) The Poincaré duality theorem yields $b_k = b_{n-k}$. In particular, for connected manifolds we always have $b_0 = b_n = 1$. The Euler characteristic is expressed in terms of Betti numbers by the formula:

$$\chi(M) = \sum_{j\geq 0}(-1)^j b_j .$$

Theorem 4. *Suppose that the configuration space of a natural system with n degrees of freedom is a connected analytic manifold M^n, and the Hamiltonian function is analytic on the phase space. If this system has n independent analytic integrals, then*

$$b_k(M^n) \leq \binom{n}{k} , \qquad k = 1, \ldots, n . \tag{1.2}$$

If $b_1(M^n) = n$, then the inequalities in (1.2) *are replaced by equalities.*

In particular, for an integrable system we have $b_1(M) \leq n$. This was conjectured by the author in [14]. For the case of two-dimensional oriented surfaces, we have $b_1 = 2\kappa$, where κ is the genus of the surface. Hence (1.2) coincides with the inequality $\kappa \leq 1$. Thus, Theorem 2 is a particular case of Theorem 4.

In [226–227] topological obstructions to the integrability in terms of the fundamental group of the manifold are also obtained. For integrable systems, the fundamental group must have no Abelian subgroups of finite index.

The assumption for the integral to be analytic in Theorem 4 can be weakened. We call an integrable Hamiltonian system on the energy surface $\Sigma_h = \{H = h\}$ geometrically simple if:

1) Σ_h contains a closed g_H^t-invariant subset Γ such that the complement $\Sigma_h \setminus \Gamma$ is dense in Σ_h and has a finite number of connected components fibered by n-dimensional tori over $(n-1)$-dimensional balls;
2) for every point $x \in \Sigma_h$ and any neighborhood W of x there exists a neighborhood $U \subset W$ of x such that the set $U \cap (\Sigma_h \setminus \Gamma)$ has a finite number of connected components.

It turns out that if a Hamiltonian system on the surface Σ_h, where $h > \max V$, is integrable and geometrically simple, then inequalities (1.2) are satisfied. Analytically integrable systems are geometrically simple (see [226]).

2 Proof of Nonintegrability Theorems

2.1 First we shall prove Theorem 3 which implies Theorems 1–2. The proof is based on certain simple facts from algebraic topology (see, for example, [75, 213]).

Proof of Theorem 3. Let ds be the Maupertuis metric on M. We fix a point $q \in M$ satisfying condition b). Since (M, ds) is a smooth two-dimensional compact oriented Riemannian manifold, by Gaidukov's theorem [83], for every nontrivial free homotopy class of closed curves in M there exists a geodesic semi-trajectory Γ, emanating from the point q, which asymptotically approaches some closed geodesic from the given homotopy class. We shall call such geodesic semi-trajectories Γ_q-*geodesics*.

Suppose that the Hamiltonian system has an infinitely differentiable first integral $F(q, \phi)$ on Σ_h. Each of its regular levels is a union of a finite number of two-dimensional invariant tori. Consider a circle S_q^1 in the cotangent plane $p \in T_q^*M$ consisting of all vectors p such that $T(p, q) + V(q) = h$. To each vector $p \in S_q^1$ there corresponds a motion $q(t)$ with the initial conditions $q(0) = q$, $p(0) = p$. The function F is constant on the trajectory $(q(t), p(t))$. We call the momentum p critical, if the corresponding value of the integral F is critical. Let us show that there exist an infinite number of critical momenta.

If the number of critical momenta is finite, then the circle S_q^1 breaks into a finite number of open sectors $\Delta_1, \dots, \Delta_n$ such that any momentum $p \in \Delta_i$, $i = 1, \dots, n$ is non-critical. For each $p \in \Delta_i$ let T_p^2 be the invariant torus containing the solution $q(t)$, $p(t)$. Since none of the values of the function F for $p \in \Delta_i$ are critical, the natural map

$$f_i : \Delta_i \times \mathbb{T}^2 \to D_i = \bigcup_{p \in \Delta_i} T_p^2$$

is continuous. Let $\pi : T^*M \to M$ be the projection of the cotangent bundle T^*M onto M. Denote $X_i = \pi(D_i)$. The continuous map $\pi \circ f_i : \Delta_i \times \mathbb{T}^2 \to X_i$ induces a homomorphism of the homology groups $g_i : H_1(\Delta_i \times \mathbb{T}^2) \to H_1(X_i)$. Since $X_i \subset M$, there exists a natural homomorphism $\phi_i : H_1(X_i) \to H_1(M)$. We denote by G_i the subgroup of the group $H_1(M)$ which is the image of the group $H_1(\Delta_i \times \mathbb{T}^2)$ under the homomorphism $\phi_i \circ g_i : H_1(\Delta_i \times \mathbb{T}^2) \to H_1(M)$. The elements of the group $H_1(M)$ are homology classes of cycles, and every class contains a cycle represented by a closed curve. Freely homotopic cycles are, obviously, homologous. The Γ_q-geodesics corresponding to the non-critical initial momenta are, of course, closed. For certain critical initial momenta, the Γ_q-geodesics may be not closed. Such geodesics are "windings" on certain cycles γ generating one-dimensional subgroups $\{n\gamma : n \in \mathbb{Z}\} \subset H_1(M)$. By the assumption, the number of critical momenta is finite. Hence, the number of such subgroups is also finite. We denote them by $N_1, \dots, N_m$. If an element $\alpha \in H_1(M)$ does not belong to the union $N_1 \cup \dots \cup N_m$, then the homology class α contains at least one closed Γ_q-geodesic. Since Γ_q-geodesics are the images of certain closed curves in the domains $\Delta_1 \times \mathbb{T}^2, \dots, \Delta_n \times \mathbb{T}^n$ under the maps $\pi \circ f_i$, the set $H_1(M) \setminus \left(\bigcup N_i\right)$ is covered by the subgroups $G_1, \dots, G_n$. Since $H_1(\Delta_i \times \mathbb{T}^2) \cong H_1(\mathbb{T}^2) \cong \mathbb{Z}^2$, the rank of the Abelian subgroups G_i does not exceed 2. It is well known that $H_1(M) \cong \mathbb{Z}^{2\kappa}$, where κ is the genus of the surface M. By the assumption, M is not homeomorphic to the sphere or the torus. Hence, $2\kappa \geq 4$. Thus, it is impossible to cover $H_1(M)$ by a finite number of one-dimensional and two-dimensional Abelian subgroups. This contradiction proves that the number of critical momenta is infinite.

Condition a) implies that the number of distinct critical values of the function $F : \Sigma_h \to \mathbb{R}$ is finite. Hence, for a fixed point $q \in M$, the function $F(q, \phi)$, $\phi \in S^1_q$, takes the same value at infinitely many points. By condition b), the function $F(q, \phi)$ is constant on S^1_q (i.e., F does not depend on ϕ). Since the surface M is compact and connected, any two of its points can be joined by a minimal geodesic. The function F is constant along any motion. Thus, it takes the same value at all points $q \in M$ satisfying condition b). Since the set of such points is everywhere dense in M, the continuous function F is constant. The theorem is proved.

2.2 In the case of inertial motion ($V \equiv 0$), Kolokol'tsov found another proof of Theorem 1 based on the introduction of a complex-analytic structure on M [120]. The idea of this proof goes back to Birkhoff [23, Chap. 2].

By a well known result from geometry, there exist local coordinates q_1, q_2 on M such that the metric takes the isothermic form:

$$ds^2 = \lambda(q_1, q_2)(dq_1^2 + dq_2^2)$$

(see, for example, [74, Chap. 2]). The coordinates q_1, q_2 are called *conformal* or *isothermic*. Let p_1, p_2 be the canonical momenta conjugate to q_1, q_2. In the variables p, q, the Hamiltonian of the inertial motion takes the form:

$$H = \frac{1}{2}\Lambda(q_1, q_2)(p_1^2 + p_2^2) .$$

Introducing the local complex variable $z = q_1 + iq_2$, we obtain

$$ds^2 = \lambda(z, \bar{z})\, dz\, d\bar{z} .$$

Let $z = f(w)$ be a holomorphic function. Then

$$dz = f'\, dw , \qquad d\bar{z} = \bar{f}'\, d\bar{w} , \qquad ds^2 = \lambda |f'|^2\, dw\, d\bar{w} .$$

Hence, the conformal form of the Riemannian metric is invariant under holomorphic coordinate transformations. We introduce a complex structure on M such that in some local complex chart the Riemannian metric is of the conformal type. Then this holds for all complex charts on M.

In §1.1 we already mentioned that for the case of inertial motion the existence problem for analytic integrals is reduced to the problem of the existence of an integral F_n that is a homogeneous polynomial in momenta. In explicit form,

$$F_n = \sum_{k+l=n} f_{k,l}(q_1, q_2) p_1^k p_2^l .$$

Lemma 1. *The function*

$$R_n = (f_{n,0} - f_{n-2,2} + f_{n-4,4} - \cdots) + i(f_{n-1,1} - f_{n-3,3} + \cdots) = \sum_{k+l=n} i^l f_{k,l}$$

is a holomorphic function of $z = q_1 + iq_2$.

This result was obtained by Birkhoff [23, Chap. 2] for $n \leq 2$. To prove Lemma 1, we calculate the Poisson bracket of the functions F_n and H:

$$\{F_n, H\} = \sum \left(\frac{\partial f_{k,l}}{\partial q_1} \Lambda p_1^{k+1} p_2^l + \frac{\partial f_{k,l}}{\partial q_2} \Lambda p_1^k p_2^{l+1} \right) - \sum f_{k,l} \frac{k}{2} \frac{\partial \Lambda}{\partial q_1} (p_1^2 + p_2^2) p_1^{k-1} p_2^l - \sum f_{k,l} \frac{l}{2} \frac{\partial \Lambda}{\partial q_2} (p_1^2 + p_2^2) p_1^k p_2^{l-1} \equiv 0 .$$

Let $p_1 = 1$ and $p_2 = i$. Then $p_1^2 + p_2^2 = 0$ and hence,

$$\frac{\partial}{\partial q_1} \sum f_{k,l} i^l + i \frac{\partial}{\partial q_2} \sum f_{k,l} i^l = 0 .$$

This equation coincides with the Cauchy–Riemann condition for the function $R_n(z)$ to be holomorphic.

Lemma 2. *Let* $z \to z(w)$ *be a holomorphic coordinate transformation. Then in the new variables the corresponding function* $\widetilde{R}_n(w)$ *is given by the equation* $R_n(z) = \widetilde{R}_n(w(z))(w'(z))^{-n}$.

Proof. Let $w = Q_1 + iQ_2$ and let P_1, P_2 be the new canonical momenta. Since

$$p_1\, dq_1 + p_2\, dq_2 = P_1\, dQ_1 + P_2\, dQ_2 ,$$

by the Cauchy–Riemann condition we get

$$p_1 = P_1 Q_1' + P_2 Q_2' , \qquad p_2 = P_1 \frac{\partial Q_1}{\partial q_2} + P_2 \frac{\partial Q_2}{\partial q_2} = -P_1 Q_2' + P_2 Q_1' ,$$

where primes denote derivatives with respect to q_1. Hence,

$$F_n(p, q) = \sum f_{k,l}(w)(P_1 Q_1' + P_2 Q_2')^k (-P_1 Q_2' + P_2 Q_1')^l .$$

Substituting $P_1 = 1$ and $P_2 = i$ into this equality, we obtain

$$\tilde{R}_n(w) = \sum f_{k,l}(w(z))(Q_1' + iQ_2')^k (-Q_2' + iQ_1')^l = \left[\sum f_{k,l}(z) i^l \right] (Q_1' + iQ_2')^n = R_n(z)(w')^n .$$

The lemma is proved.

Lemma 3. $R_n(z) \equiv 0$.

Proof. Suppose that $R_n(z) \not\equiv 0$. Then, according to Lemma 2, the differential form

$$(dz)^n/R_n(z) = (dw)^n/\widetilde{R}_n(w) \tag{2.1}$$

is invariant under holomorphic coordinate transformations. For $n = 1$, the form (1.1) is an ordinary Abelian differential, since the function R_1^{-1} is meromorphic on the Riemann surface M. If $n > 1$, then the form (2.1) is called an n-differential.

It is well known (see, for example, [102, Chap. 9]) that for any Abelian differential on a compact Riemannian surface M of genus κ the difference between the number of zeros and the number of poles equals $2\kappa - 2$. For an n-differential, this difference equals $2n(\kappa - 1)$ (the proof is the same as in the classical case $n = 1$). Since $R_n(z)$ is locally holomorphic, the n-differential (2.1) has no zeros. Hence, the number of poles is equal to $2n(1 - \kappa)$. Since the genus $\kappa > 1$, this number is negative. We have obtained a contradiction.

Lemma 4. *If $R_n \equiv 0$, then $F_n = HF_{n-2}$.*

Proof. Let

$$F_{n-2} = a_{n-2,0}p_1^{n-2} + a_{n-3,1}p_1^{n-3}p_2 + \cdots + a_{0,n-2}p_2^{n-2}\,.$$

The equality $F_n = HF_{n-2}$ holds if and only if the system of algebraic equations

$$\begin{array}{cc} a_{n-2,0} = f_{n,0}\,, & a_{n-3,1} = f_{n-1,1}\,, \\ a_{n-2,0} + a_{n-3,1} = f_{n-2,2}\,, & a_{n-3,1} + a_{n-5,3} = f_{n-3,3}\,, \\ \ldots\quad\ldots\quad\ldots & \ldots\quad\ldots\quad\ldots \end{array}$$

has a solution. The solution exists if

$$f_{n,0} - f_{n-2,2} + \cdots = f_{n-1,1} - f_{n-3,3} + \cdots = 0\,.$$

The lemma is proved.

To complete the proof of the theorem, it remains to apply induction by n. Thus, if there exists a polynomial integral of degree $2k$ or $2k + 1$, then there exists an integral of degree 0 or 2 respectively. The first case is trivial. In the second case, $F_1 = ap_1 + bp_2$. Lemma 3 yields $R_1 = a + ib \equiv 0$. Hence, $F_1 \equiv 0$. It follows that $F_n = 0$ for odd n, and $F_n = cH^{n/2}$ for even n.

Note that the proof of the nonexistence of a polynomial integral does not need the coefficients to be analytic. It is sufficient to assume that they are of the class $C^1(M)$. One might think that the proven result is more general than Theorem 1. However, as proved by Bolotin in [33], if M, T and V are analytic, then every polynomial integral is an analytic function on T^*M.

2.3 The proof of Theorem 4 is based on more complicated topological techniques and is not given here (see [226–227]). In contrast to the proof in §3.1, this proof does not provide additional qualitative information on the dynamical phenomena obstructing integrability. However, using Gromov's results, Paternain [199] proved that under similar assumptions the *topological entropy* is positive.

3 Geometrical Obstructions to Integrability

3.1 Another possible way to generalize Theorem 1 is to consider regions with geodesically convex boundary. Let M be a two-dimensional analytic surface and M' a compact submanifold with boundary. Denote by Σ' the set of points of the three-dimensional manifold $\Sigma = \{H = h\}$ that are mapped into the points of M' by the projection $\pi : TM \to M$. We call the manifold M' geodesically convex if for any pair of close points on the boundary $\partial M'$ the shortest geodesic joining them lies completely in M'.

Theorem 1. *Let M' be a geodesically convex submanifold with negative Euler characteristic. Then the geodesic flow in Σ has no non-constant analytic integrals. Moreover, there are no analytic integrals in any neighborhood of the set Σ' in Σ.*

If $\partial M' = 0$, we again obtain Theorem 1 of §3.1. Theorem 1 of the present section was first proved by the author under the assumption that rank $H_1(M', \mathbb{Z}) > 2$. Later, Bolotin [31] have replaced it by a weaker condition $\chi(M) < 0$, where χ is the Euler characteristic.

3.2 The proof of Theorem 1 is similar to the method used in [130]. It uses the fact that in every free homotopy class of closed curves in M' there exists an unstable closed geodesic. The existence of closed geodesics (without considering stability) in manifolds with convex boundary was established in the classical works of Whittaker [233] and Birkhoff [23]. Instead of the homology group, which was used in the proof of nonintegrability in the case of an empty boundary $\partial M'$, the proof of Theorem 1 is based on other topological invariants (see [31]).

3.3 Theorem 1 implies several interesting results about nonintegrability conditions for geodesic flows on a sphere and a torus. They were communicated to the author by S. Bolotin.

Corollary 1. *Suppose that there is a null-homotopic closed geodesic on the two-dimensional torus $\mathbb{T}^2$ with an analytic metric. Then the geodesic flow on $\mathbb{T}^2$ has no non-constant analytic integrals.*

Proof. Let γ be a contractible closed geodesic on $\mathbb{T}^2$. First, suppose that γ has no self-intersections. Then γ divides $\mathbb{T}^2$ into two geodesically convex regions. One of them is homeomorphic to the disk, and the Euler characteristic of the other is negative. Hence, the nonexistence of an integral is a consequence of Theorem 1.

Now consider the general case when the geodesic γ has self-intersections. Represent the torus as the factor space of $\mathbb{R}^2$ by the integer lattice $\mathbb{Z}^2$. The metric on $\mathbb{T}^2$ induces a metric on $\mathbb{R}^2$, and the geodesic γ is covered by a geodesic γ' on $\mathbb{R}^2$. Since γ is null-homotopic in $\mathbb{T}^2$, the geodesic γ' is closed. Hence, γ' lies entirely in some square in $\mathbb{R}^2$ with integer vertices. Let p be the length of the side of the square. Consider the new torus $\mathbb{T}^2_*$ obtained by factorization of $\mathbb{R}^2$ over the lattice $p\mathbb{Z}^2$. Clearly, if the geodesic flow on $\mathbb{T}^2$ has an additional analytic integral, then the same holds for the geodesic flow on the new torus $\mathbb{T}^2_*$. The geodesic γ' is lifted to a geodesic γ_* on $\mathbb{T}^2_*$. One of the connected components of $\mathbb{T}^2_* \setminus \gamma_*$ is homeomorphic to the two-dimensional torus with a disk removed. This component is, of course, geodesically convex and its Euler characteristic is negative. The proof is complete.

Of course, far from every metric on a two-dimensional torus has a contractible closed geodesic. However, in certain cases the existence of such geodesics can be established with the help of simple variational arguments (see Fig. 11).

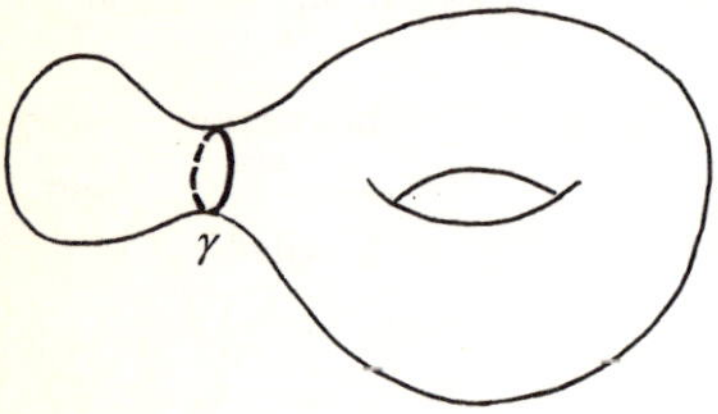

Fig. 11.

Now consider the case when M is homeomorphic to the two-dimensional sphere S^2. According to the famous Lusternik–Shnirelman theorem, on S^2 there always exist three closed geodesics γ_i without self-intersections (see, for example, [75]). It turns out that integrability of the geodesic flow depends on their mutual disposition.

Corollary 2. *Suppose that the geodesics γ_1, γ_2, γ_3 do not intersect and each of them can be contracted to a point without intersecting the other two. Then the geodesic flow on S^2 has no additional analytic integral.*

Indeed, the curves γ_i divide the sphere into several geodesically convex components. One of them has negative Euler characteristic (Fig. 12).

3.4 Following Bolotin, we apply these general results to systems with Newtonian type potentials. Let M be the configuration space of a natural system with two degrees of freedom, and let $V : M \to \mathbb{R}$ be the potential energy. We do not assume that the surface M is orientable. We say that V is a potential of Newtonian type if V is an analytic function everywhere except at a finite number of points $z_1, \ldots, z_n$, and if, in the conformal coordinates z (with respect to the metric defined by the

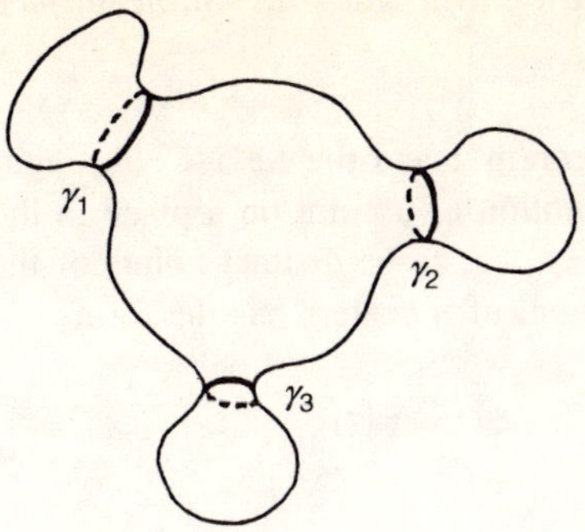

Fig. 12.

kinetic energy), with the origin at the singular point z_s, V has the form

$$V = -f(z)/|z|,$$

where the function f is analytic in a neighborhood of the point z_s and $f(0) > 0$.

Theorem 2 [32]. *Let M be compact, and suppose that the potential V has $n > 2\chi(M)$ singular points. Then for*

$$h > \sup_M V$$

there are no non-constant analytic integrals on the energy level surface $\Sigma_h = \{H = h\}$.

For $n = 0$, we obtain Theorem 1 in §3.1. The condition $n > 2\chi(M)$ is violated only if
a) M is a sphere and $n \leq 4$,
b) M is a projective plane and $n \leq 2$,
c) M is a torus or Klein bottle and $n = 0$.

If M is non-compact, then we need auxiliary conditions on the behavior of the kinetic energy at infinity. Assume that $\chi(M) \neq -\infty$; then we know from topology that M can be transformed into a compact surface M^* by adding a finite number of points z_k^* at infinity. Let $D_k \subset M^*$ be certain neighborhoods of the points z_k^* which are diffeomorphic to a disk. The additional assumption is that any closed curve in D_k, embracing the point z_k^*, cannot be contracted to the point z_k^* in the class of curves with bounded length (in the metric determined by the kinetic energy). We also assume that $\sup_M V < \infty$.

Theorem 3 [32]. *Let M be non-compact and the kinetic energy satisfy the above condition at infinity. Suppose that the potential energy has $n > 2\chi(M)$ singular* *points. Then for* $h > \sup_M V$ *the system has no analytic integrals on the surface* $\Sigma_h = \{H = h\}$.

Theorems 2 and 3 are proved by using Theorem 1 and the Levi–Civita regularization. As an illustration, we consider the motion of a point on a plane in the gravitational field of n fixed centers [25]. Let $z_1, \dots, z_n$ be distinct points of the complex plane $\mathbb{C}$. The Hamiltonian of the problem of n centers has the form

$$H(p, z) = \frac{1}{2}|p|^2 + V(z) , \qquad p \in \mathbb{C} , \quad z \in \mathbb{C} \setminus \{z_1, \dots, z_n\} ,$$

where

$$V(z) = -\sum_{k=1}^{n} \mu_k |z - z_k|^{-1} , \qquad \mu_k > 0 ,$$

is the gravitational potential. Here $M = \mathbb{C}$, $\chi(M) = 1$, the kinetic energy (given by the Euclidean metric in the plane) satisfies the necessary conditions at infinity, and the potential $V < 0$. Hence, by Theorem 3, for $n > 2$ the problem is nonintegrable on the energy level surface $H = h > 0$. Notice that the cases $n = 1$ and $n = 2$ correspond to the classical integrable Kepler and Euler problems.

We sketch the proof. Let Λ be the Riemann surface of the function $\sqrt{(z - z_1) \cdots (z - z_n)}$, and let $\pi : \Lambda \to \mathbb{C}$ be the natural projection. It can be shown that the Levi–Civita regularization (the transition from the configuration space $M = \mathbb{C}$ to the Riemann surface Λ) reduces the phase flow on the energy surface $H = h > 0$ to a geodesic flow on M with a complete metric. Let D be a disk on the complex plane $\mathbb{C}$ with a sufficiently large radius. Then the set $\Lambda' = \pi^{-1}(D)$ is compact, geodesically convex, and homotopically equivalent to Λ. By the Riemann–Hurwitz formula we have $\chi(M) = 2 - n < 0$ if $n > 2$.

Let us prove that Λ' is geodesically convex. Consider a motion $z(t)$ with energy $h > 0$ and suppose that $z(0) \in \partial D$ and the vector $\dot{z}(0)$ is tangent to ∂D. By Lagrange's formula,

$$\frac{d^2}{dt^2} \frac{|z|^2}{2} = |p|^2 - (z, V_z') = 2(h - V) + V > 0 .$$

Hence, for small t the trajectory $z(t)$ lies outside D. Thus, the set Λ' is geodesically convex. Now it remains to use Theorem 1.

It turns out that the conditions of Theorems 2 and 3 cannot be weakened. Let M be a two-dimensional surface, T the kinetic energy on M, and $z_1, \dots, z_n$ points of M. We fix the value h of the total energy. Then we have

Theorem 4 [33]. *There exists a Newtonian type potential* V *with singularities at the points* $z_1, \dots, z_n$ *such that the Hamiltonian system with Hamiltonian* $H = T + V$ *has on* $\{H = h\}$ *an additional analytic integral which is quadratic in momenta. Moreover,*

1) *if* M *is not compact, then* $V < h$;
2) *if* M *is compact and* $n > \chi(M)$, *then* $V < h$ *everywhere except at a finite number of points*;
3) *if* M *is compact and* $n \leq 2\chi(M)$, *then* $\sup_M V < h$;
4) *if* M *is not compact,* T *is Euclidean at infinity, and* $n \leq 2\chi(M)$, *then* $\sup_M V < h$.

The condition that the Riemannian metric T is Euclidean at infinity implies the following. Let M be obtained from a compact surface M^* by removing a finite number of points z_k^* at infinity. Then each point z_k^* has a punctured neighborhood with a conformal coordinate $z \in \mathbb{C}$, $1 < |z| < \infty$, such that the metric T tends to the Euclidean metric $|\dot{z}|^2/2$ as $z \to \infty$.

4 Systems with Gyroscopic Forces

4.1 Now let M be a two-dimensional oriented Riemannian manifold and ϕ the area 2-form on M. Clearly, any form of gyroscopic forces can be written as $f = \lambda\phi$, where λ is a function on M. We say that the form $f = \lambda\phi$ preserves its sign if $\lambda \geq 0$ ($\lambda \leq 0$) everywhere in M. The latter is evidently true if $f \equiv 0$ (i.e., the system is reversible).

Following Birkhoff, consider the existence problem for conditional polynomial integralsfor a system with gyroscopic forces. Recall that a first integral defined on a fixed level surface of the energy integral is called a conditional polynomial integral if it can be continued to a function on T^*M which is a polynomial in the momenta with single-valued coefficients on M.

Again we assume that the Newtonian type potential V has n singular points on M. We assume that all the objects are analytic.

Theorem 1 [33]. *Let* M *be compact and* $n > 2\chi(M)$. *Then for* $h > \sup_M V$ *on the energy level* $H = h$ *there are no non-constant conditional integrals that are polynomial in the momenta.*

If $n = 0$ and $f \equiv 0$, then this is again Theorem 1 of §1. It turns out that if the form f satisfies the inequality

$$\iint_D f + \oint_{\partial D} 2\sqrt{(h - V)T}\, dt > 0\,, \tag{4.1}$$

for any domain $D \subset M$ with a non-empty boundary, then the polynomial integrals in Theorem 1 can be replaced by analytic integrals [32]. If the form of the gyroscopic forces is exact, then condition (4.1) certainly holds when $h > \sup_M H_0$, where $H_0 = H(p, q)|_{p=0}$. For $f \equiv 0$, we obtain Theorem 2 of §3.3.

Theorem 2 [33]. *Let* M *be non-compact. Suppose that the kinetic energy is Euclidean at infinity and the potential* V *has* $n > 2\chi(M)$ *singular points* $z_1, \ldots, z_n$

of Newtonian type. Then there are no integrals on $T^*(M \setminus \{z_1, \ldots, z_n\})$ *that are polynomial in the momenta and independent of the Hamiltonian function* H.

In fact, if the form f is exact and the condition $h > \sup_M H_0$ is satisfied, then a stronger result holds: for $n > 2\chi(M)$ the system has no even analytic non-constant integrals on the surface $H = h$ [32]. With $f \equiv 0$, we again obtain Theorem 3 of §3.3.

4.2 As an example, consider the plane circular restricted problem of many bodies. Suppose that n points $z_1, \ldots, z_n$ are fixed in the plane M which rotates about a fixed point O with constant angular velocity ω (the vector ω is orthogonal to M). Let a point z with unit mass move on M in the gravitational field of the gravity centers $z_1, \ldots, z_n$. Here

$$L = \frac{1}{2}|\dot{z}|^2 + (\omega \times z, \dot{z}) - V\,,$$

where

$$V = -\sum_{k=1}^{n} \frac{\mu_k}{|z - z_k|} - \frac{\omega^2 |z|^2}{2}\,, \qquad \mu_k > 0\,.$$

The restricted many-body problem is integrable for $n = 1$ and all ω (Kepler's problem), and also for $n = 2$ and $\omega = 0$ (Euler's problem of two fixed centers). It turns out that for $n > 2$ and all ω this problem has no additional analytic integrals.

In the present case, we have

$$H_0 = V + \frac{|\omega \times z|^2}{2} = -\sum_{k=1}^{n} \frac{\mu_k}{|z - z_k|}$$

and $f = 2|\omega|\phi$, where ϕ is the standard area form on the plane M. Since $\chi(M) = 1$, for $n > 2$ and $h > 0$ the restricted $(n+1)$-body problem has no analytic integrals on the energy surface $H = T + V = h$.

For $n = 2$ and $\omega \neq 0$ (the restricted three-body problem), similar results have not been proved. Moreover, there is the well-known Chazy hypothesis on the integrability of the three-body problem for positive values of energy (see [8]). This hypothesis is connected with a more general idea: in a particle scattering problem on a non-compact configuration space the data at infinity (for example, the momenta of the particles) are candidates for the role of first integrals. However, realization of this approach encounters fundamental difficulties connected with determining the domain of definition and smoothness properties of the "scatter integrals". One of the difficulties is the possibility of capture in the problem of many interacting particles.

For the restricted three-body problem, weaker results on nonintegrability are known. Poincaré proved that there are no additional integrals that are analytic with respect to the masses μ_1 and μ_2 of the "heavy" particles [202]. Llibre and Simo [164], applying Alexeev's method of quasi-random trajectories, proved the nonexistence of additional analytic integrals in the case when the mass of one of the particles is small. Apart from these results, there is Siegel's theorem on the absence of new algebraic integrals [214]. This theorem was proved by the Bruns method. Apparently, the restricted three-body problem admits no first integrals that are polynomial in the momenta and independent of the energy integral.

4.3 For $n = 2\chi(M)$, the structure of the gyroscopic forces is of crucial importance in the problem of integrability of a Hamiltonian system.

Theorem 3 [33]. *Let M be compact, and suppose that the Newtonian type potential has $2\chi(M)$ singular points. If*

$$\iint_M f \neq 0\,, \tag{4.2}$$

then there exist no conditional polynomial integrals on the energy level $H = h$, where $h > \sup_M V$.

Let us show that under the conditions of Theorem 3 analytic first integrals can exist. Consider the motion of a charged particle on the flat torus $\mathbb{T}^2 = \{x, y \bmod 2\pi\}$ in a constant magnetic field. The dynamics is governed by the equations

$$\ddot{x} + \alpha\dot{y} = 0, \quad \ddot{y} - \alpha\dot{x} = 0\,, \qquad \alpha = \text{const}\,. \tag{4.3}$$

For this system, $\chi(M) = 0$ and the potential has no singularities. The form of gyroscopic forces is $-\alpha\, dx \wedge dy$. Hence, for $\alpha \neq 0$ condition (4.2) is satisfied. It follows that the system (4.3) admits no polynomial integrals, independent of the energy integral, with coefficients that are single-valued on the torus $\mathbb{T}^2$. The obvious linear integrals $\dot{x} + \alpha y$ and $\dot{y} - \alpha x$ are multivalued functions on the phase space $TM = \mathbb{R}^2 \times \mathbb{T}^2$. The function $\sin(y + \dot{x}/\alpha)$ is a single-valued analytic integral, but it is not polynomial in $\dot{x}$.

In the non-compact case, it is impossible to represent the conditions for the existence of an additional polynomial integral in the form of purely topological restrictions.

The proof of Theorem 3 is given in §7.1 under the assumption that M is a torus with a flat metric.

5 Generic Integrals

5.1 An alternative approach to the study of topological obstructions to the complete integrability of Hamiltonian systems was suggested by Fomenko [79]. He linked the existence of generic additional integrals with the topology of level surfaces of the energy integral and the number of stable closed trajectories.

We turn to exact statements. Let M^4 be the phase space of a Hamiltonian system with two degrees of freedom, and H the Hamiltonian function. We consider a fixed compact three-dimensional non-singular level surface of the Hamiltonian H. Suppose that the Hamiltonian system has a smooth integral f on Σ. We call

f a *Morse integral* if its critical points form a union of nondegenerate critical manifolds N. This means that the Hessian $d^2 f$ is nondegenerate on subspaces that are transverse to these submanifolds. Analysis of concrete integrable problems of classical mechanics shows that in the majority of known cases additional integrals are of the Morse type. Non-critical level surfaces of the function f are always orientable (we use the fact that the symplectic manifold M is oriented, and the level surface Σ is regular). The critical manifolds N may be non-orientable. Therefore, it is natural to call the integral f orientable, if all the critical manifolds are orientable. It can be shown that the integral f can always be made orientable by passing to a suitable two-sheeted covering over Σ.

We call a closed trajectory of the Hamiltonian system on Σ orientable, if its tubular neighborhood in Σ is foliated by two-dimensional concentric invariant tori (non-critical level surfaces of the function f). Clearly, if f has a strict local maximum or minimum at all points of a closed trajectory, then this trajectory is stable. The example of inertial motion on a flat two-dimensional torus shows that not every completely integrable system has stable periodic trajectories.

Theorem 1 [79]. *Assume that the Hamiltonian system has an orientable Morse integral f on the level surface Σ. If the homology group $H_1(\Sigma, \mathbb{Z})$ is finite or the rank of the fundamental group $\pi_1(\Sigma)$ is equal to one, then the Hamiltonian system has on Σ at least two stable closed trajectories. Moreover, f has a strict maximum or minimum on each of these trajectories.*

For inertial motion on the flat torus $\mathbb{T}^2$, we have $\Sigma = \mathbb{T}^3$ and hence $\operatorname{rank} \pi_1(\Sigma) = 3$. We give an example of a Hamiltonian system which has exactly two stable closed trajectories on its energy surfaces Σ. Consider the biharmonic oscillator governed by the equations

$$\ddot{x}_1 + \omega_1^2 x_1 = 0\,, \quad \ddot{x}_2 + \omega_2^2 x_2 = 0\,; \qquad \omega_1/\omega_2 \notin \mathbb{Q}\,.$$

For $h > 0$ the energy surface

$$\dot{x}_1^2 + \dot{x}_2^2 + \omega_1^2 x_1^2 + \omega_2^2 x_2^2 = 2h$$

is diffeomorphic to the three-dimensional sphere. Hence, $H_1(\Sigma, \mathbb{Z}) = 0$. For all $h > 0$ there are exactly two stable periodic solutions

$$x_1 = \frac{\sqrt{2h}}{\omega_1} \sin \omega_1 t\,, \quad x_2 = 0\,,$$

$$x_1 = 0, \quad x_2 = \frac{\sqrt{2h}}{\omega_2} \sin \omega_2 t\,.$$

5.2 Theorem 1 provides certain obstructions to integrability.

Corollary. *Assume that the group $H_1(\Sigma, \mathbb{Z})$ is finite and the Hamiltonian system has no stable periodic orbits on the energy surface Σ. Then the system has no additional Morse integrals on Σ.*

We point out one application of this result. Consider inertial motion with fixed (say, unit) velocity on an n-dimensional closed Riemannian manifold M. Let Σ^{2n-1} be the unit tangent bundle of M. The Riemannian metric (kinetic energy of the system) defines a dynamical system on Σ which is usually called the *geodesic flow*. In other words, the geodesic flow is the restriction of the Hamiltonian system with Hamiltonian $H = T$ on T^*M to the invariant surface $\Sigma = \{H = 1\}$. The Maupertuis principle reduces the motion in an arbitrary potential to a geodesic flow.

For example, the results of Anosov, Klingenberg and Takens imply that in the set of all geodesic flows on smooth Riemannian manifolds there is an open dense set of flows without stable periodic trajectories (see [53]). Thus, a generic flow has no stable periodic trajectories. Consider geodesic flows on the two-dimensional sphere. In this case we have $M = T^*S^2$, $\Sigma = SO(3)$ and $H_1(\Sigma, \mathbb{Z}) = \mathbb{Z}_2$. Therefore, a generic geodesic flow on the two-dimensional sphere has no additional Morse integralon non-singular energy surfaces.

5.3 Fomenko [80] has studied in detail the structure of three-dimensional energy level manifolds of integrable systems and found topological invariants which make it possible to distinguish non-isomorphic systems.

6 Topological Obstructions to the Existence of Linear Integrals

6.1 It is well known that the existence of integrals that are linear in momenta is closely related to the existence of symmetry groups acting on the configuration space (see §6 of the Introduction). It turns out that the existence of linear integrals imposes restrictions not only on the Riemannian metric (kinetic energy) and the potential of the force field, but also on the topology of the configuration space.

Let $H = T + V$ be the Hamiltonian of a reversible system,and $F = F_1 + F_0$ a linear integral (F_k is a homogeneous form of degree k in the momenta). Clearly,

$$\{H, F\} = \{T, F_1\} + \{T, F_0\} + \{V, F_1\} .$$

The terms in the right-hand side are homogeneous forms in momenta of degree 2, 1 and 0 respectively. Since the functions H and F commute, all these forms are equal to zero. Hence, $F_0 \equiv \text{const}$, and the function F_1 is an integral of the reversible system. Moreover, F_1 is an integral of the system with the Hamiltonian $H = T$.

Thus, linear integrals of the reversible system on T^*M are homogeneous in momenta. Hence, they can be represented in the form $F = y \cdot v(x)$, where y is the canonical momentum, and v a vector field on M. In the Lagrangian formalism, linear integrals are represented as

$$F = \frac{\partial T}{\partial \dot{x}} \cdot v = \langle \dot{x}, v \rangle , \tag{6.1}$$

where $\langle \cdot, \cdot \rangle$ is the Riemannian metric on M given by the kinetic energy. The vector field v defines a dynamical system on M. Let g_v^t be its phase flow:

$$\frac{d}{d\alpha} g^\alpha(x) = v(g^\alpha(x)) . \tag{6.2}$$

The theorem on local straightening of trajectories implies that in a small neighborhood of a non-critical point of the vector field v, the system of Eqs. (6.2) can be transformed to the following form:

$$\frac{dx_1}{d\alpha} = 1 , \qquad \frac{dx_k}{d\alpha} = 0 \quad \text{for} \quad k \geq 2 .$$

In these coordinates, $F = \partial T / \partial \dot{x}_1$ and the functions T and V do not depend on x_1. The phase flow g_v^α is reduced to the family of shifts

$$x_1 \to x_1 + \alpha , \quad x_2 \to x_2 , \ \dots , \ x_n \to x_n .$$

In particular, the functions T and V are invariant under the action of the group g_v^t. This is the converse of Noether's theorem for natural systems. The vector field v is called a *symmetry field*. In Riemannian geometry, it is usually called a *Killing field*. Since the function F is an integral of the geodesic flow, the group g_v^α transforms geodesics into geodesics. Since g^α conserves distances between any points of M, the group g^α is a one-parameter isometry group of the Riemannian manifold M.

The Killing fields on M form a Lie algebra of dimension $\leq n(n+1)/2$, and the equality holds only for manifolds of constant curvature. For example, on the two-dimensional sphere $|x| = \text{const}$ in the Euclidean space $\mathbb{R}^3 = \{x\}$, for any $e \in \mathbb{R}^3$ the field $e \times x$ is a Killing field. The corresponding Lie algebra is isomorphic to so(3).

6.2 If there exist several symmetry fields $v_1, \dots, v_k$, then the equations of motion have the same number of linear integrals

$$F_1 = y \cdot v_1 , \ \dots , \ F_k = y \cdot v_k .$$

These functions are independent and in involution if and only if the vector fields $v_1, \dots, v_k$ are independent and commute. This follows from the equality

$$\{F_i, F_j\} = y \cdot [v_i, v_j] .$$

Theorem 1 [1]. *Let M be a compact connected oriented even-dimensional manifold. If a natural Hamiltonian system on T^*M has $k \geq (\dim M)/2$ independent linear integrals, pairwise in involution, then $\chi(M) \geq 0$.*

In particular, if $\dim M = 2$, then linear integrals can exist only when M is a sphere or a torus. The proof of Theorem 1 is based on Kobayashi's result on commutative isometry groups of Riemannian manifolds [112]. Since M is compact, all Killing fields $v_1, \dots, v_k$ are complete: their phase flows g_i^α are defined for all α. Each group g_i^α is isomorphic to the circle $\mathbb{T}^1$ (when the orbits are closed), or the line $\mathbb{R}$ (when the orbits are non-compact). Since the Killing fields $v_1, \dots, v_k$

are independent and commute, the compact Lie isometry group

$$G = \mathbb{T}^m \times \mathbb{R}^{k-m}\,, \qquad 0 \le m \le k\,,$$

acts effectively on the Riemannian manifold M. An action is effective if only the unit element keeps all points of M fixed. However, if $\chi(M) < 0$, then $\dim G < n/2$ [112].

6.3 For $n = 2$, we can give a transparent proof of Theorem 1. It is based on the properties of the index of a vector field. Recall that the index of an isolated critical point x_* of a field v is the number of full rotations of the vector $v(x)$ (in the counter-clockwise direction) as x winds round x_* once.

If $\chi(M) < 0$, then the symmetry field v has at least one critical point. Since the phase flow g_v^α is an isometry group of the surface M, all critical points x_k are isolated and elliptic. In particular, $\mathrm{ind}_v(x_k) = 1$. By the Poincaré formula, the Euler characteristic

$$\chi(M) = \sum \mathrm{ind}_v(x_k)$$

is positive, which is a contradiction.

Formula (6.3) shows that on the sphere ($\chi = 2$) a symmetry field has exactly two critical points, while on the torus ($\chi = 0$) there are none.

6.4 In [1] Abrarov studied a more general problem on the existence of k involutive conditional integrals, linear in momenta. Such functions are integrals of the geodesic flow of the Maupertuis metric $(ds)^2 = 2(h - V)T(dt)^2$. In the interior of the region $B_h = \{x : V(x) \le h\}$, the metric ds is Riemannian, but for $h < \max_M V$ it has singularities. The length of curves lying on the boundary of B_h is zero, while the lengths of curves that are not contained in B_h are imaginary.

Theorem 2 [1]. *Let h be a regular value of the energy. Suppose that the equations of motion have $k \ge (\dim M)/2$ linear involutive conditional integrals on $\Sigma_h = \{H = h\}$. Then $\chi(B_h) \ge 0$.*

If $h > \chi(M)$, then Theorem 2 follows from Theorem 1. Suppose that the boundary of the region B_h is non-empty. We double B_h by identifying the boundaries of two copies of B_h and obtain a manifold $2B_h$ without boundary. Then the k-dimensional Lie group of isometries acts on $2B_h$ (see §6.2). To complete the proof, it is sufficient to apply the following formula from topology

$$\chi(2B) = 2\chi(B)\,.$$

In fact, the manifold $2B_h$ is not Riemannian, since $ds = 0$ on the boundary of B_h. To avoid this difficulty, we need to step back from the boundary to a small distance in the Maupertuis metric, and only then carry out the doubling. The details can be found in [1].

7 Topology of the Configuration Space of Reversible Systems with Nontrivial Symmetry Groups

7.1 Many results of this chapter can be generalized to the case of topological obstructions to the existence of nontrivial symmetry groups.

Consider a reversible system with two degrees of freedom. Let M be the compact configuration space, T the kinetic energy, and V the potential energy. All objects are assumed to be analytic.

Let h be the total energy of the system. Suppose that $h > \max_M V$. We denote by v the restriction of the Hamiltonian vector field with Hamiltonian $H = T + V$ to the three-dimensional analytic regular surface $\Sigma_h = \{H = h\}$. Let u be an analytic symmetry fieldon Σ_h: $[u, v] \equiv 0$.

Theorem 1. *If the genus of M is greater than one, then $u = \lambda v$, where $\lambda =$ const.*

Since every integral independent of the energy integral generates a nontrivial symmetry group, this result generalizes Theorems 1 and 2 of §3.1. In particular, Theorem 1 implies the nonexistence of multivalued analytic integrals. The main step in the proof of Theorem 1 is to establish that the vectors u, v are linearly dependent at all points of Σ_h. Since $v \neq 0$, we have $u = \lambda v$. It follows (see Chap. II, §3) that λ is an integral of the Hamiltonian system on Σ_h. The function λ is analytic and the genus of M is greater than one. Hence, by Theorem 1 of §1, we obtain $\lambda \equiv$ const.

7.2 Consider the simpler case when $V \equiv$ const. Then the trajectories of the system are geodesics in M. Introducing conformal coordinates q_1, q_2, we may assume that (see §2.2)

$$H = \Lambda(q_1, q_2)(p_1^2 + p_2^2)/2 .$$

The operator of differentiation with respect to the Hamiltonian vector field v becomes

$$L_v = \Lambda p_1 \frac{\partial}{\partial q_1} + \Lambda p_2 \frac{\partial}{\partial q_2} - \frac{\partial \Lambda}{\partial q_1} \frac{(p_1^2 + p_2^2)}{2} \frac{\partial}{\partial p_1} - \frac{\partial \Lambda}{\partial q_2} \frac{(p_1^2 + p_2^2)}{2} \frac{\partial}{\partial p_2} . \tag{7.1}$$

Let

$$L_u = Q_1 \frac{\partial}{q_1} + Q_2 \frac{\partial}{\partial q_2} + P_1 \frac{\partial}{\partial p_1} + P_2 \frac{\partial}{\partial p_2} \tag{7.2}$$

be the operator of differentiation with respect to the symmetry field u. Suppose that the functions Q_j and P_j are analytic in the momenta p_1 and p_2.

Theorem 2. *If the genus of the surface M is greater than one, then $u = \lambda v$, where λ is an analytic function of H.*

Strictly speaking, this result does not follow from Theorem 1, since we do not assume that the field u is analytic in the coordinates q_1, q_2. The proof of Theorem 2 is based on Birkhoff's method outlined in §2.2.

Let us expand the functions Q_j and P_j into series in homogeneous forms of the momenta:

$$Q_j = \sum_{m\geq 0} Q_j^{(m)}, \quad P_j = \sum_{m\geq 0} P_j^{(m)}.$$

Lemma 1. *If the operators* (7.1) *and* (7.2) *commute, then the operator* (7.1) *commutes with each of the operators*

$$L^{(0)} = P_1^{(0)} \frac{\partial}{\partial p_1} + P_2^{(0)} \frac{\partial}{\partial p_2}, \tag{7.3}$$

$$L^{(m)} = Q_1^{(m-1)} \frac{\partial}{\partial q_1} + Q_2^{(m-1)} \frac{\partial}{\partial q_2} + P_1^{(m)} \frac{\partial}{\partial p_1} + P_2^{(m)} \frac{\partial}{\partial p_2}, \quad m \geq 1. \tag{7.4}$$

This obvious result makes it possible to reduce the problem of analytic symmetry fields to the problem of symmetry fields with homogeneous polynomial components.

Lemma 2. *Suppose that the operators* (7.1) *and* (7.3) *commute. Then* $P_1^{(0)} = P_2^{(0)} \equiv 0$.

Of course, the statement of Lemma 2 holds for arbitrary genus of the surface M.

Now we shall assume that $m \geq 1$. Let $p_1 = 1$ and $p_2 = i$. Then P_1 and P_2 become complex functions of q_1 and q_2. We denote them by P_1^* and P_2^*.

Lemma 3. *The function* $R = \Lambda(P_1^* + iP_2^*)$ *is holomorphic in* $z = q_1 + iq_2$.

Proof. Calculate the commutator $[L_u, L_v]$ and equate to zero the coefficients of $\partial/\partial p_1$ and $\partial/\partial p_2$. For $p_1 = 1$ and $p_2 = i$, we obtain the equations

$$\begin{aligned} &\Lambda \frac{\partial P_1^*}{\partial q_1} + \frac{\partial \Lambda}{\partial q_1} P_1^* + i\left(\Lambda \frac{\partial P_1^*}{\partial q_2} + \frac{\partial \Lambda}{\partial q_1} P_2^*\right) = 0, \\ &i\left(\Lambda \frac{\partial P_2^*}{\partial q_2} + \frac{\partial \Lambda}{\partial q_2} P_2^*\right) + \Lambda \frac{\partial P_2^*}{\partial q_1} + \frac{\partial \Lambda}{\partial q_2} P_1^* = 0. \end{aligned} \tag{7.5}$$

From (7.5) it follows that

$$\frac{\partial}{\partial q_1} \Lambda(P_1^* + iP_2^*) + i \frac{\partial}{\partial q_2} \Lambda(P_1^* + iP_2^*) = 0.$$

This is the criterion for the function $\Lambda(P_1^* + iP_2^*) = 0$ to be holomorphic. The lemma is proved.

Lemma 4. $R(z) \equiv 0$.

Indeed, let $z \to w(z)$ be a holomorphic coordinate transformation. According to Lemma 2 of §2, the transformation law for the function R is as follows

$$R(z) = \widetilde{R}(w(z))(w'(z))^{-m} ,$$

where m is the degree of the homogeneous polynomial $\Lambda(P_1 + iP_2)$.

Since the genus of the surface M is greater than one, we can apply Lemma 3 of §2. Hence, $R \equiv 0$. The lemma is proved.

Therefore, $P_1 + iP_2 = 0$. By (7.5), we obtain two equations

$$\frac{\partial P_1^*}{\partial q_1} + i\frac{\partial P_1^*}{\partial q_2} = 0 , \qquad \frac{\partial P_2^*}{\partial q_1} + i\frac{\partial P_2^*}{\partial q_2} = 0 .$$

Thus, P_1^* and P_2^* are locally holomorphic functions of $z = q_1 + iq_2$. Hence, they satisfy the statement of Lemma 4: $P_1^* = P_2^* = 0$. Lemma 4 of §2 implies that $P_1 = HP_1'$ and $P_2 = HP_2'$, where H is the energy integral.

Now we will show that the polynomials Q_j are also entirely divisible by the polynomial H. Let Q_j^* be the value of the function Q_j for $p_1 = 1$, $p_2 = i$.

Lemma 5. *The function* $S = Q_1^* + iQ_2^*$ *is holomorphic in* $z = q_1 + iq_2$.

Proof. We calculate the commutator $[L_u, L_v]$ and equate to zero the coefficients of $\partial/\partial q_1$ and $\partial/\partial q_2$. Put $p_1 = 1$ and $p_2 = i$. Since $P_1^* = P_2^* = 0$, we obtain the equations

$$\begin{aligned} Q_1^*\frac{\partial \Lambda}{\partial q_1} + Q_2^*\frac{\partial \Lambda}{\partial q_2} - \Lambda\frac{\partial Q_1^*}{\partial q_1} - i\Lambda\frac{\partial Q_1^*}{\partial q_2} &= 0 , \\ iQ_1^*\frac{\partial \Lambda}{\partial q_1} + iQ_2^*\frac{\partial \Lambda}{\partial q_2} - \Lambda\frac{\partial Q_2^*}{\partial q_1} - i\Lambda\frac{\partial Q_2^*}{\partial q_2} &= 0 . \end{aligned} \tag{7.6}$$

From (7.6) it follows that

$$\frac{\partial}{\partial q_1}(Q_1^* + iQ_2^*) + i\frac{\partial}{\partial q_2}(Q_1^* + iQ_2^*) = 0 .$$

Hence, $S = Q_1^* + iQ_2^*$ is a locally holomorphic function of $z = q_1 + iq_2$. This proves Lemma 5.

Lemma 4 yields $S \equiv 0$. Using (7.6), we obtain two equations

$$\frac{\partial}{\partial q_1}\frac{Q_1^*}{\Lambda} + i\frac{\partial}{\partial q_2}\frac{Q_1^*}{\Lambda} = 0 , \qquad \frac{\partial}{\partial q_1}\frac{Q_2^*}{\Lambda} + i\frac{\partial}{\partial q_2}\frac{Q_2^*}{\Lambda} = 0 .$$

Thus, Q_1^*/Λ and Q_2^*/Λ are locally holomorphic functions of z. Lemma 4 implies that they are identically zero. By Lemma 4 of §2, we get $Q_j = HQ_j'$.

We proved that $u = Hu'$. Since the multiplier H is an integral of the equations of motion, u' is also a symmetry field, but its degree in the momenta is $m - 2$, where m is the degree of the field u. By induction in m, we reduce the problem to the case of symmetry fields with degrees $m = 0$ or $m = 1$.

The case $m = 0$ is covered by Lemma 2. If $m = 1$, then evidently $Q_1 = Q_2 \equiv 0$. Let $P_j = \xi_j p_1 + \eta_j p_2$. Then, as we have shown, $P_j^* = \xi_j + i\eta_j \equiv 0$. Hence, $\xi_j + i\eta_j = 0$. Thus, for $m = 1$ every symmetry fieldis identically zero. The proof of Theorem 2 is completed.

7.3 Topological obstructions to the existence of nontrivial symmetry fields of reversible systems were first obtained in the author's paper [149] (Theorem 2). Theorem 2 was formulated in [149] as a conjecture. It was proved in [36] by means of detailed analysis of doubly asymptotic orbits to periodic trajectories from various homotopy classes. More precisely, it was proved that under the assumptions of Theorem 1 of §1 there exists a hyperbolic orbit with transverse asymptotic surfaces. This implies, in particular, the stochastization of the phase flow and, as a consequence, the nonexistence of additional integrals and symmetry groups (see Chap. V).

We mention also Katok's paper [108], where the positivity of the topological entropywas established for geodesic flow on a closed two-dimensional surface with negative Euler characteristic. Positiveness of the topological entropy implies complicated behavior of the phase trajectories (see, for example, [118]). However, to translate this important result into the language of integrability theory of Hamiltonian systems further work is needed. Note that §1 contains an example of a geodesic flow on a sphere with $g > 1$ handles attached which admits a non-constant smooth integral. In this case, the entropy is supported by a small region of the phase space. It would be interesting to investigate the connection between the existence of nontrivial symmetry fields and positiveness of the topological entropy.

IV Nonintegrability of Hamiltonian Systems Close to Integrable Ones

We begin the consideration of the analytic obstacles to integrability with the analysis of Poincaré's "basic problem of dynamics". This problem deals with the Hamiltonian system whose Hamiltonian has the form

$$H = H_0(y_1, \ldots, y_n) + \varepsilon H_1(x_1, \ldots, x_n, y_1, \ldots, y_n) + \cdots .$$

The canonical coordinates $x \bmod 2\pi$, y are the action–angle variables for the "unperturbed" system with the Hamiltonian H_0. Following Poincaré, we consider questions of the existence in this system of additional integrals and nontrivial symmetry fields in the form of power series in the small parameter ε. The classical scheme of perturbation theory described in §10 of Chap. II is essential here. The destruction of a large number of resonant tori of the unperturbed problem for small values of $\varepsilon \neq 0$ turns out to prevent integrability of the Hamiltonian system.

1 The Poincaré Method

1.1 Consider an analytic system of differential equations

$$\dot{y}_j = \varepsilon \Phi_j + \ldots, \quad \dot{x}_k = \omega_k + \varepsilon \Psi_k + \ldots, \quad 1 \leq j \leq m, \ 1 \leq k \leq n. \tag{1.1}$$

Here ω_k depend only on the "slow" variables $y \in \mathbb{R}$, the variables x are angular (the right-hand sides of (1.1) are periodic in all x_k with period 2π; in other words $x \in \mathbb{T}^n$), ε is a small parameter. Dots denote terms of order ≥ 2 in ε.

Evidently, Hamiltonian equations that are close to integrable ones have the form (1.1). In this case $m = n$, and the frequencies ω_k are equal to $\partial H_0/\partial y_k$. Equations (1.1) are frequently encountered in applications. One can find many examples, for instance, in [14, Chap. 5].

Suppose that the system (1.1) has s integrals in the form of power series in ε:

$$H^{(i)} = H_0^{(i)}(x, y) + \varepsilon H_1^{(i)}(x, y) + \ldots, \qquad 1 \leq i \leq s. \tag{1.2}$$

The functions $H_r^{(i)}$, $r \geq 0$ are assumed to be analytic in x, y and 2π-periodic in the coordinates x. Thus, the integrals (1.2) are "single-valued" functions on the phase space of the system (1.1). Hamiltonian systems always admit an integral of the form (1.2); it is the energy integral.

Consider the problem of the existence in the system (1.1) of an additional integral in the form of a series

$$F = F_0(x, y) + \varepsilon F_1(x, y) + \cdots \tag{1.3}$$

with analytic coefficients F_r, $r \geq 0$ which are 2π-periodic in x. It is natural to assume the integrals (1.2–1.3) to be independent.

Now we extend the formulation of the problem and search for integrals which are formal power series (1.3). Some explanations are needed concerning these series. We regard the formal series $\sum f_i \varepsilon^i$ as equal to zero if all $f_i \equiv 0$. The series (1.3) is a formal integral of the system (1.1) if the formal series

$$\dot{F} = \sum \frac{\partial F_0}{\partial x_k} \omega_k + \varepsilon \left(\sum \frac{\partial F_0}{\partial y_j} \Phi_j + \sum \frac{\partial F_1}{\partial x_k} \omega_k \right) + \cdots \tag{1.4}$$

vanishes. The formal series (1.2) and (1.3) are regarded as independent if at least one minor of $(s+1)$-st order of the Jacobian matrix

$$\frac{\partial(H^{(1)}, \ldots, H^{(s)}, F)}{\partial(x, y)}$$

considered as a formal power series in ε does not vanish.

For $\varepsilon = 0$ we have the integrable unperturbed system

$$\dot{y}_j = 0\,, \quad \dot{x}_k = \omega_k(y)\,.$$

Its phase space $\mathbb{R}^m \times \mathbb{T}^n = \{y, x \bmod 2\pi\}$ is fibrated by n-dimensional tori $\{x, y : y = y_0, x \bmod 2\pi\}$ filled with quasi-periodic trajectories. Coordinates y "enumerate" these tori. We call the unperturbed system non-degenerate, if the equality $\sum \alpha_k \omega_k(y) \equiv 0$ with integer numbers α_k holds only for $\alpha_k = 0$. Non-resonant tori are dense in the phase space of the unperturbed system. For example, if $m = n$ and the Jacobian

$$\frac{\partial(\omega_1, \ldots, \omega_m)}{\partial(y_1, \ldots, y_m)}$$

does not vanish, then the system is non-degenerate. In particular, for Hamiltonian systems the inequality

$$\det \|\partial^2 H_0 / \partial y^2\| \neq 0$$

is a sufficient condition for non-degeneracy. We expand the functions Φ_j in multiple Fourier series:

$$\Phi_j = \sum \Phi^j_\alpha(y) \exp[i(\alpha, x)]\,, \qquad \alpha \in \mathbb{Z}^n\,. \tag{1.5}$$

The *Poincaré set* $\mathbf{P}_s \subset \mathbb{R}^m = \{y\}$ plays an important role in further analysis. This set is an analog of the secular set from §10, Chap. II. By definition, the Poincaré set consists of points $y \in \mathbb{R}^m$ for which there exist $m - s$ linearly independent integer vectors $\alpha, \alpha', \ldots \in \mathbb{Z}^n$ such that

1) $(\alpha, \omega(y)) = (\alpha', \omega(y)) = \ldots = 0$,
2) the vectors $\Phi_\alpha = (\Phi^1_\alpha, \ldots, \Phi^m_\alpha)$, $\Phi_{\alpha'} = (\Phi^1_{\alpha'}, \ldots, \Phi^m_{\alpha'}), \ldots$ are linearly independent.

In a typical situation the set $\mathbf{P}_s$ is dense in $\mathbb{R}^m$.

We denote by $C^\omega(V)$ the class of functions analytic in the domain $V \subset \mathbb{R}^u$. We call $M \subset V$ a *key* set (or a uniqueness set) for the class $C^\omega(V)$, if any analytic function vanishing on M vanishes identically in V. Thus, if analytic functions coincide on M, then they coincide everywhere in V. For example, a subset of an interval $\Delta \subset \mathbb{R}$ is a key set for the class $C^\omega(\Delta)$ if and only if it has a limit point inside Δ. The sufficiency of this condition is evident, and necessity follows from the Weierstrass theorem on infinite products. Note that if M is a uniqueness set for the class $C^p(V)$ $(0 \leq p \leq \infty)$, then M is dense in V.

Theorem 1. *Suppose that the unperturbed system is non-degenerate. Let the Jacobian matrix*

$$\frac{\partial(H_0^{(1)}, \ldots, H_0^{(s)})}{\partial(y_1, \ldots, y_m)} \tag{1.6}$$

have maximal rank at a point $y^0 \in \mathbb{R}^m$, and for any neighborhood U of the point y^0 let the Poincaré set $\mathbf{P}_s$ be a key set for the class $C^\omega(U)$. Then the system (1.1) *has no formal integral* (1.3) *with analytic coefficients which is independent of the integrals* (1.2).

The proof of Theorem 1 is based on the following

Lemma 1. *Let the functions $F_r(x, y)$ be continuously differentiable and let the series*

$$\sum_{r=0}^{\infty} F_r(x, y)\varepsilon^r$$

be a formal integral of the system (1.1). *If the system* (1.1) *is non-degenerate for $\varepsilon = 0$, then*
1) *the function F_0 does not depend on x,*
2) *the rank of the Jacobian matrix*

$$\frac{\partial(F_0, H_0^{(1)}, \ldots, H_0^{(s)})}{\partial(y_1, \ldots, y_m)} \tag{1.7}$$

does not exceed s at all points $y \in \mathbf{P}_s$.

Proof. Equating the right-hand side of relation (1.4) to zero, we obtain the chain of equations

$$\sum \frac{\partial F_0}{\partial x_k}\omega_k = 0\,, \quad \sum \frac{\partial F_0}{\partial y_j}\Phi_j + \sum \frac{\partial F_1}{\partial x_k}\omega_k = 0\,, \quad \ldots\,. \tag{1.8}$$

Then we multiply the first equation by $\exp[-i(\alpha, x)]$, $\alpha \in \mathbb{Z}^n$ and take the average of both parts over the n-dimensional torus $\mathbb{T}^n$:

$$\frac{1}{(2\pi)^n}\int_{\mathbb{T}^n}\left(\frac{\partial F_0}{\partial x}, \omega\right)e^{-i(\alpha, x)}\,dx = 0\,.$$

Integrating this relation by parts, we obtain the equality

$$(\alpha, \omega) f_\alpha = 0\,, \qquad \alpha \in \mathbb{Z}^n\,, \tag{1.9}$$

where

$$f_\alpha = \frac{1}{(2\pi)^n} \int_{\mathbb{T}^n} F_0 e^{-i(\alpha, x)}\, dx$$

is a coefficient in the Fourier expansion of the function $F_0 = F_0(x)$.

The function $(\alpha, \omega(y))$ is analytic, and for all $\alpha \neq 0$ it does not vanish identically. Hence, $(\alpha, \omega) \neq 0$ on a dense subset of $\mathbb{R}^m = \{y\}$. According to (1.9), $f_\alpha(y) \equiv 0$ for all $\alpha \neq 0$. Thus the function F_0 does not depend on x.

Now multiply the second equation by $\exp[-i(\alpha, x)]$, take its average over the torus $\mathbb{T}^n$, and integrate it by parts. As a result, we obtain the chain of relations

$$i(\alpha, \omega) F_\alpha^{(1)} + \left(\frac{\partial F_0}{\partial y}, \Phi_\alpha \right) = 0\,, \qquad \alpha \in \mathbb{Z}^n\,,$$

where $F_\alpha^{(1)}$ are Fourier coefficients of the function F_1.

Let $y \in \mathbf{P}_s$. Then

$$\left(\frac{\partial F_0}{\partial y}, \Phi_\alpha \right) = \left(\frac{\partial F_0}{\partial y}, \Phi'_\alpha \right) = \ldots = 0\,.$$

Consequently, the gradient of the function F_0 lies in the s-dimensional plane Π, orthogonal to the linearly independent vectors $\Phi_\alpha, \Phi_{\alpha'}, \ldots$. Since the functions (1.2) are integrals of the system (1.1), the gradients of the functions $H_0^{(1)}, \ldots, H_0^{(s)}$ lie in the same plane Π. Thus, the functions $F_0, H_0^{(1)}, \ldots, H_0^{(s)}$ are dependent at the points $y \in \mathbf{P}_s$: the rank of the Jacobian matrix (1.7) drops. The lemma is proved.

Proof of Theorem 1. According to Lemma 1, the functions $H_0^{(1)}, \ldots, H_0^{(s)}$ depend only on the variables y. Consequently, $s \leq m$. Since these functions are independent at the point y^0, we can assume that in some small neighborhood of the point y^0 the functions $H_0^{(r)}$ $(1 \leq r \leq s)$ can be regarded as a part of the local coordinates $z_1, \ldots, z_m$: $z_r = H_0^{(r)}$, $r \leq s$. Consequently, $F_0 = G_0(z_1, \ldots, z_m)$, where G_0 is analytic.

According to Lemma 1, the rank of the Jacobian matrix (1.7) does not exceed s at all points of the Poincaré set $\mathbf{P}_s$. Since all the minors of this matrix are analytic functions of x, y and $\mathbf{P}_s$ is a key set for the class of analytic functions, the functions $F_0, H_0^{(1)}, \ldots, H_0^{(s)}$ are dependent everywhere. In the new variables $\{z_j\}$ the matrix

(1.7) has the following form:

$$\left\| \begin{matrix} \partial G_0/\partial z_1 & \partial G_0/\partial z_2 & \dots & \partial G_0/\partial z_s & \dots & \partial G_0/\partial z_m \\ 1 & 0 & \dots & 0 & \dots & 0 \\ 0 & 1 & \dots & 0 & \dots & 0 \\ \vdots & \vdots & \ddots & \vdots & \ddots & \vdots \\ 0 & 0 & \dots & 1 & \dots & 0 \end{matrix} \right\| .$$

Since its rank equals s, the function G_0 does not depend on $z_{s+1}, \dots, z_m$. Thus, in the domain $U \times \mathbb{T}^n \subset \mathbb{R}^m \times \mathbb{T}^n$ (U is a small neighborhood of y^0) we have the equality

$$F_0 = G_0(H_0^{(1)}, \dots, H_0^{(s)}) .$$

Now consider the power series

$$F' = [F - G_0(H_0^{(1)}, \dots, H_0^{(s)})]/\varepsilon = F'_0 + \varepsilon F'_1 + \cdots . \tag{1.10}$$

The coefficients $F'_0, F'_1, \dots$ are analytic functions in the direct product $U \times \mathbb{T}^n$. It is clear that the series (1.10) is a formal integral of Eqs. (1.1). According to Lemma 1, the function F'_0 does not depend on the angular variables x and the functions $F'_0, H_0^{(1)}, \dots, H_0^{(s)}$ are dependent at all points of the set $\mathbf{P}_s \cap U$. According to the assumption of Theorem 1, $\mathbf{P}_s \cap U$ is a key set for the class of functions $C^\omega(U)$. Consequently, these functions are dependent at all points of the neighborhood U. Thus, in the domain U the relation

$$F'_0 = G_1(H_0^{(1)}, \dots, H_0^{(s)})$$

holds, where G_1 is some analytic function. The power series

$$F'' = [F' - G_1(H_0^{(1)}, \dots, H_0^{(s)})]/\varepsilon = F''_0 + \varepsilon F''_1 + \cdots \tag{1.11}$$

is a formal integral of Eqs. (1.1), and we can apply this procedure again.

According to (1.10),

$$F = G_0(H^{(1)}, \dots, H^{(s)}) + \varepsilon F' ,$$

and equality (1.11) yields the relation

$$F' = G_1(H^{(1)}, \dots, H^{(s)}) + \varepsilon F'' .$$

Consequently,

$$F = G_0(H) + \varepsilon G_1(H) + \varepsilon^2 F'' .$$

Applying Lemma 1 repeatedly, we arrive at the formal expansion

$$F = G_0 + \varepsilon G_1 + \varepsilon^2 G_2 + \cdots ,$$

where all coefficients G_r are functions of the integrals $H^{(1)}, \dots, H^{(s)}$. Consequently, any $(s+1)$-order minor of the Jacobian matrix of integrals (1.2) and (1.3) vanishes as a formal power series in ε. Theorem 1 is completely proved.

1.2 The following version of Theorem 1 is true.

Theorem 2. *Suppose that the system* (1.1) *is non-degenerate for* $\varepsilon = 0$, *the rank of the Jacobian matrix* (1.6) *equals* s *in the domain* $D \subset \mathbb{R}^m = \{y\}$, *and the set* $\mathbf{P}_s$ *is dense in* D. *Then the system* (1.1) *has no formal integral* (1.3) *with infinitely differentiable coefficients which is independent of the integrals* (1.2).

This theorem can be easily proved by the method of §1.1. Let $s = 0$ and let the set $\mathbf{P}_0$ be dense in D. Then the system (1.1) has no formal integral (1.3) with non-constant coefficients that are continuously differentiable in the domain $D \times \mathbb{T}^n \subset \mathbb{R}^m \times \mathbb{T}^n$. Indeed, according to Lemma 1, the function F_0 depends only on y and $dF_0 = 0$ at the points of the set $\mathbf{P}_0$. Since $\mathbf{P}_0$ is everywhere dense, $dF_0 \equiv 0$. Hence, $F_0 \equiv \text{const}$. In a similar way, one can prove that the other terms of the series (1.3) are also constant.

1.3 We apply the results of §1.1 and §1.2 to the Hamiltonian equations with Hamiltonian

$$H = H_0(y_1, \ldots, y_n) + \varepsilon H_1(x_1, \ldots, x_n, y_1, \ldots, y_n) + \cdots . \tag{1.12}$$

Evidently, these equations have the form (1.1). Expand the perturbing function H_1 in a multiple Fourier series:

$$H_1 = \sum H_\alpha(y) e^{i(\alpha, x)}, \qquad \alpha \in \mathbb{Z}^n .$$

Since

$$\dot{y}_j = -\varepsilon \frac{\partial H_1}{\partial x_j} - \cdots ,$$

we have $\Phi_j = -\partial H_1/\partial x_j$. Thus, the vectors Φ_α defined in §1.1 are equal to $-i\alpha H_\alpha$. As we already noted in §1.1, the frequencies ω_k coincide with the partial derivatives $\partial H_0/\partial y_k$. Consider the existence problem of a formal integral which is additional to the energy integral H and represented by the series (1.3) with coefficients analytic in the domain $D \times \mathbb{T}^n$. According to §1.1, it is necessary to put $s = 1$. In our case the Poincaré set $\mathbf{P}_1$ is defined as the set of points $y \in D$ for which there exist $n - 1$ linearly independent vectors $\alpha, \alpha', \ldots \in \mathbb{Z}^n$ such that
1) $(\alpha, \omega(y)) = (\alpha', \omega(y)) = \ldots = 0$,
2) $H_\alpha(y) \neq 0, \; H_{\alpha'}(y) \neq 0, \ldots$.

Theorem 3. *Suppose that the unperturbed Hamiltonian system is non-degenerate*:

$$\det \|\partial^2 H_0/(\partial y_i \partial y_j)\| \neq 0 \quad \text{in the domain } D .$$

Let $y \in D$ *be a non-critical point of the function* H_0 *and suppose that for any neighborhood* U *of* y *the set* $\mathbf{P}_1 \cap U$ *is a key set for the class of functions* $C^\omega(U)$. *Then the Hamiltonian equations with Hamiltonian* (1.12) *have no integral independent of the function H which is a formal power series*

$$\sum_{r\geq 0} F_r(x, y)\varepsilon^r \tag{1.13}$$

with coefficients analytic in the domain $D \times \mathbb{T}^n$.

This theorem is a consequence of Theorem 1 from §1.1. One can easily deduce from Theorem 2 the following

Theorem 4. *Let the function* H_0 *be non-degenerate in the domain D and suppose that the Poincaré set* $\mathbf{P}_1$ *is dense in D. Then the Hamiltonian equations have no formal integral independent of H of the type* (1.13) *with infinitely differentiable coefficients* $F_r : D \times \mathbb{T}^n \to \mathbb{R}$.

For the Poincaré sets $\mathbf{P}_s$ the inclusions $\mathbf{P}_0 \subset \mathbf{P}_1 \subset \mathbf{P}_2 \subset \ldots \subset \mathbf{P}_{n-1} \subset \mathbf{B}$ hold, where $\mathbf{B}$ is the secular set introduced in §10 of Chap. II. It is clear that in general the set $\mathbf{P}_0$ consists of isolated points.

Let $\mathbf{P}_{n-1}$ be a key set for the class $C^\omega(D)$ and suppose that the Hamiltonian system with Hamiltonian (1.12) has n formal integrals

$$F_0^{(1)} + \varepsilon F_1^{(1)} + \cdots , \; \ldots , \quad F_0^{(n)} + \varepsilon F_1^{(n)} + \cdots \tag{1.14}$$

with analytic coefficients 2π-periodic in x. Then the functions $F_0^{(1)}, \ldots, F_0^{(n)}$ are dependent at all points of $D \times \mathbb{T}^n$. This result is a simple consequence of Lemma 1. Perhaps the key property of the set $\mathbf{P}_{n-1}$ implies that the formal integrals (1.14) are dependent. However, this is not proven yet.

1.4 Now consider a system of non-autonomous canonical equations

$$\dot{y} = -\frac{\partial H}{\partial x}, \quad \dot{x} = \frac{\partial H}{\partial y}; \qquad H = H_0(y) + \varepsilon H_1(y, x, t) + \cdots . \tag{1.15}$$

The Hamiltonian function H is assumed to be analytic and 2π-periodic in x and t.

Equations (1.15) appear, for example, when we study an autonomous Hamiltonian system, if one of the angular coordinates x is taken as the new time. Let, for example, $\partial H/\partial y_1 \neq 0$. Then (at least locally) we can solve the equation $H(y, x, \varepsilon) = h$ with respect to y_1:

$$y_1 = -K(y_2, \ldots, y_n, x_2, \ldots, x_n, \tau, \varepsilon, h), \quad \tau = x_1 .$$

Since $\dot{x}_1 \neq 0$, the solutions $y_s(t), x_s(t)$ $(s \geq 2)$ of the original equations can be considered as functions of τ. According to Whittaker's theorem, the functions $y_s(t), x_s(t)$ $(s \geq 2)$ satisfy the canonical equations

$$\frac{dy_s}{d\tau} = -\frac{\partial K}{\partial x_s}, \quad \frac{dx_s}{d\tau} = \frac{\partial K}{\partial y_s} . \tag{1.16}$$

Indeed,

$$\frac{dy_s}{d\tau} = \frac{dy_s}{dx_1} = \frac{\dot{y}_s}{\dot{x}_1} = -\left(\frac{\partial H}{\partial x_s}\right) \Big/ \left(\frac{\partial H}{\partial y_1}\right) .$$

Differentiating the identity $H(-K, y_2, \ldots, y_n, \tau, x_2, \ldots, x_n, \varepsilon) = h$ with respect to x_s, we obtain:

$$\frac{\partial H}{\partial x_s} - \frac{\partial H}{\partial y_1}\frac{\partial K}{\partial x_s} = 0 .$$

Consequently,

$$\frac{dy_s}{d\tau} = -\frac{\partial K}{\partial x_s} , \qquad 2 \leq s \leq n .$$

The second half of the canonical equations is derived in the same way. Equations (1.16) have exactly the form (1.15).

Again it is useful to introduce the Poincaré set $\mathbf{P}_*$ (an analog of the set $\mathbf{P}_1$) as a set of points $y \in D$, satisfying the following conditions:

1) there exist n linearly independent vectors $k_s \in \mathbb{Z}^n$ and n integer numbers m_s such that $(k_s, \omega(y)) + m_s = 0$, $1 \leq s \leq n$,
2) the Fourier coefficients $H_{k_s, m_s}(y)$ of the expansion of the perturbing function

$$H_1 = \sum H_{km}(y) e^{((k,\alpha)+mt)}$$

do not vanish.

Note that if (1.15) are the Whittaker equations, obtained from the autonomous Hamiltonian equations with Hamiltonian (1.12) by reducting the order, then the Poincaré set $\mathbf{P}_*$ of the reduced system can be obtained as follows. It is the projection of the intersection of the original system's Poincaré set with the level surface $H_0(y_1, \ldots, y_n) = h$ to the plane $\mathbb{R}^{n-1} = \{y_2, \ldots, y_n\}$.

Theorem 5. *If the function F_0 is non-degenerate in the domain D, and $\mathbf{P}_*$ is a key set for the class $C^\omega(D)$, then Eqs.* (1.15) *have no formal integral*

$$\sum_{s \geq 0} F_s(y, x, t)\varepsilon^s$$

with analytic coefficients $F_s : D \times \mathbb{T}^{n+1} \to \mathbb{R}$.

The proof of Theorem 5 is based on successive applications of an auxiliary assertion, analogous to Lemma 1 from §1.1.

Lemma 2. *Let $F_s : D \times \mathbb{T}^{n+1} \to \mathbb{R}$ be continuously differentiable functions and let the series $\sum F_s\varepsilon^s$ be a formal integral of the system* (1.15) *with nondegenerate Hamiltonian function H_0. Then*

1) *$F_0(y, x, t)$ does not depend on x and t,*
2) *$dF_0 = 0$ on the set $\mathbf{P}_*$.*

If the Poincaré set $\mathbf{P}_*$ is dense in the domain D, then, obviously, the Eqs. (1.15) have no formal integral with continuously differentiable coefficients.

1.5 As mentioned in the Introduction, the formulation of the integrability problem for perturbed Hamiltonian systems belongs to Poincaré ([202]; [203, Chap. 5]). He proved the absence of "single-valued" integrals, independent of the energy integral and analytic in the phase variables and the parameter ε, under the assumption that the set $\mathbf{P}_1$ is dense. In [124] it is proved that the assumption of the density of the set $\mathbf{P}_1$ can be weakened: it is sufficient for $\mathbf{P}_1$ to be a key set for the class of analytic functions. In a talk on the I. G. Petrovsky seminar [129] we extended the Poincaré method to the case when integrals are sought as formal power series in ε with analytic or smooth coefficients (Theorems 3 and 4). An extension of the Poincaré method to Hamiltonian systems with periodic Hamiltonians is contained in [124]. A generalization of the Poincaré results to non-Hamiltonian systems of standard form (1.1) (Theorems 1 and 2) seems not to have been discussed in the literature.

We indicate one unsolved problem. Is it true that under the assumptions of Theorems 3–5 Hamiltonian systems have no single-valued integrals of an appropriate smoothness for small fixed values of the parameter $\varepsilon \neq 0$? In connection with this problem it is interesting to note that for small values of ε Hamiltonian systems (1.15) with one and a half degrees of freedom ($n = 1$) always have non-constant continuous integrals. This follows from the Kolmogorov theorem on the conservation of quasi-periodic motions (see [14], Chap. V and also §10 of Chap. II): for small values of ε the Kolmogorov tori form a perfect, nowhere dense set. For $n = 1$ these tori "divide" the phase space into connected pieces. The integral in question is the "Cantor staircase": a continuous function equal to a constant in the connected splits between the Kolmogorov tori.

If $n > 1$, then the complement to the set of Kolmogorov tori is connected and thus it is impossible to construct a non-constant Cantor staircase: since this complement is dense in the phase space of the perturbed system, any continuous function that is constant on the complement is constant everywhere in the phase space. In particular, there appears, in principle, the possibility of the existence of a trajectory that is dense in the connected gap between the Kolmogorov tori. Perhaps such a situation is actually typical (see the discussion, for example, in [11]). This would imply the nonexistence of non-constant continuous integrals of perturbed completely integrable Hamiltonian systems.

2 Applications of the Poincaré Method

2.1 We turn to the restricted three-body problem considered in §5 of Chap. I. First, suppose that the mass μ of Jupiter vanishes. Then in a fixed frame the asteroid revolves along a Kepler orbit about the Sun having unit mass. Let the orbits be ellipses. Then it is convenient to pass from Cartesian coordinates to the canonical Delaunay elements L, G, l, g: if a and e are the major semiaxis and the orbit eccentricity, then $L = \sqrt{a}$, $G = \sqrt{a(1-e^2)}$, G is the perihelion longitude, and l is the eccentric anomaly (an angle defining the asteroid location on the orbit). In the new coordinates the equations of the asteroid motion turn out to be canonical with the Hamiltonian function $F_0 = -1/(2L^2)$. If $\mu \neq 0$, then the Hamiltonian F can be expanded in power series in μ: $F = F_0 + \mu F_1 + \cdots$. Since in the fixed coordinate system, connected with the Sun and Jupiter, the Kepler orbits rotate with unit angular velocity, the Hamiltonian function F depends on L, G, l and $g-t$. We set $y_1 = L$, $y_2 = G$, $x_1 = l$, $x_2 = g - t$ and $H = F - G$. Now the function H depends only on x, y. Moreover, it is 2π-periodic in the angular variables x_1 and x_2. As a result, we represent the equations of asteroid motion in the form of the Hamiltonian system:

$$\dot{y}_i = -\frac{\partial H}{\partial x_i}, \quad \dot{x}_i = \frac{\partial H}{\partial y_i}; \quad H = H_0 + \mu H_1 + \cdots, \; H_0 = \frac{1}{2y_1^2} + y_2 . \tag{2.1}$$

The expansion of the perturbing function in a multiple trigonometric series in the angles x_1 and x_2 was studied by Leverrier (see, for example, [57]). It has the following form:

$$H_1 = \sum_{u=-\infty}^{\infty} \sum_{v=-\infty}^{\infty} H_{uv} \cos[ux_1 - v(x_1 + x_2)] .$$

The coefficients H_{uv}, depending on y_1 and y_2, generally do not vanish.

The Poincaré set of this problem consists of the lines $u/y_1^3 - v = 0$, $H_{uv} \neq 0$, parallel to the y_2 axis. It is dense in the half-plane $y_1 > 0$. Nevertheless, we cannot apply Theorem 4 on the absence of additional analytic integrals directly because the unperturbed problem is degenerate: $\det \|\partial^2 H_0/\partial y^2\| = 0$. This difficulty can be overcome. Indeed, the canonical equations with the Hamiltonians H and $\exp H$ have identical trajectories (although their solutions differ). Consequently, these equations are integrable or nonintegrable simultaneously. It remains to note that

$$\exp H = \exp H_0 + \mu(\exp H_0)H_1 + \cdots, \qquad \det \|\partial^2(\exp H_0)/(\partial y^2)\| \neq 0 .$$

As a result, we find that the equations of the restricted three-body problem in the form (2.1) have no integral $\Phi = \sum \Phi_s \mu^s$ independent of the function H with smooth coefficients on the set $D \times \mathbb{T}^2 = \{y, x\}$. Here D is an arbitrary domain in the half-plane $y_1 > 0$.

We can apply Whittaker's procedure of reducing the order to the autonomous Hamiltonian system (2.1). Fixing the energy constant $h < 0$, we solve the equation $H(x, y, \mu) = h$ with respect to y_2. We find that

$$-y_2 = K(y_1, x_1, x_2, h, \mu) = K_0 + \mu K_1 + \cdots, \qquad K_0 = -y_1^{-2}/2 .$$

Consider the variable $x_2 = \tau$ as a new time. The functions $x_1 = x(\tau)$ and $y_1 = y(\tau)$ satisfy the Whittaker equations

$$\frac{dy}{d\tau} = -\frac{\partial K}{\partial x}, \quad \frac{dx}{d\tau} = \frac{\partial K}{\partial y} . \tag{2.2}$$

The Poincaré set for these equations is dense on the half-line $y > 0$. Since the unperturbed system is non-degenerate ($d^2K_0/dy^2 \neq 0$), all the conditions of Theorem 5, §1 are satisfied. Thus, we conclude that for all negative values of energy h Eqs. (2.2) have no integral $\sum \Psi_s \mu^s$ with continuously differentiable coefficients in the domain $\Delta \times \mathbb{T}^2 = \{y, x, \tau\}$, where Δ is an arbitrary interval on the half-line $y > 0$.

2.2 "Passons à un autre problème; celui du mouvement d'un corps pesant autour d'un point fixe.

... On peut donc se demander si, dans ce problème, les considérations exposées dans ce Chapitre s'opposent à l'éxistence d'une intégrale uniforme autre que celles des forces vives et des aires." (H. Poincaré, [203], §86).

The symmetry group consisting of rotations of the body about a vertical line generates the linear integral $(I\omega, \gamma)$ (the projection of the angular momentum to the vertical is constant). Fix this constant and reduce the number of degrees of freedom to two. Thus, on the four-dimensional integral levels $M_c = \{\omega, \gamma : (I\omega, \gamma) = c,\ (\gamma, \gamma) = 1\}$ a Hamiltonian system with two degrees of freedom arises. The Hamiltonian function of this system (total energy of the body restricted to the level M_c) equals $H_0 + \varepsilon H_1$. Here H_0 is the kinetic energy (the Hamiltonian function of the integrable Euler problem of the inertial motion of the body), and εH is the potential energy of the body in a homogeneous gravitational field. The parameter ε is the product of the weight of the body and the distance from the center of mass to the suspension point. We regard ε as a small parameter. This is equivalent to the investigation of rapid rotations of the body in a moderate force field.

We can define the action–angle variables $y, x \bmod 2\pi$ in the unperturbed integrable Euler problem. The formulas of the transition from the special canonical variables L, G, l, g to the action–angle variables y, x can be found, for example, in [16]. In the new variables $H = H_0(y) + \varepsilon H_1(y, x)$. The action variables y_1, y_2 belong to the domain $\Delta = \{|y_1| \leq y_2, y_2 \geq 0\}$. The Hamiltonian $H_0(y_1, y_2)$ is a homogeneous function of degree 2. It is analytic in each of four connected subdomains of Δ. The domain Δ is broken in these subdomains by the three lines π_1, π_2, and $y_1 = 0$. The equation of the lines π_1 and π_2 is $2H_0/y_2^2 = I_2^{-1}$. They are symmetric in the vertical axis, tend to the line $y_1 = 0$ as $I_2 \to I_1$ and to the pair of lines $|y_1| = y_2$ as $I_2 \to I_3$ (recall that I_1, I_2, and I_3 are the principal moments inertia of the body and $I_1 \geq I_2 \geq I_3$). In fact, the expansion of the perturbing function H_1 in a multiple Fourier series in the angular variables x_1 and x_2 is contained in one Jacobi's paper:

$$\sum H_{m,1} e^{i(mx_1 + x_2)} + \sum H_{m,-1} e^{i(mx_1 - x_2)} + \sum H_{m,0} e^{imx_1}. \tag{2.3}$$

Jacobi indicated the expansions of the direction cosines of a freely rotating body in trigonometric series in $\omega_1 t$ and $\omega_2 t$ (ω_1 and ω_2 depend only on integration constants). Since the perturbing function is reduced to a linear combination of the direction cosines, after the change of $\omega_1 t$ and $\omega_2 t$ to the angles x_1 and x_2 the Jacobi expansions yield exactly the Fourier series (2.3). If the body is not symmetric, then the coefficients $H_{m,\pm 1}$ do not vanish for sufficiently large values of $|m|$ (see [131], Chap. III). This implies, in particular, that the Poincaré set **P** and the secular set **B** coincide in this problem. If the principal moments of inertia satisfy the inequality $I_1 > I_2 > I_3$, then the secular set consists of an infinite number of lines passing through the point $y = 0$ and accumulating near the pair of lines π_1 and π_2. One can show that the function H_0 is non-degenerate in the domain Δ.

If the function H_0 were analytic in the domain Δ, one could apply the results of §1, since the points y^0 lying on the lines π_1 and π_2 satisfy the conditions of Theorem 3. We overcome the difficulty connected with analytic singularities of the Hamiltonian function written in terms of the action–angle variables by considering the problem of an additional integral which is analytic on all integral level M_c. The two-dimensional tori of the unperturbed Euler problem correspond to the points $y \in$ **B**. These tori accumulate near the separatrices of the unstable stationary rotations about the mean inertia axis. The union of these resonant tori forms a key set for the class of analytic functions on the level surface M_c. Let $F = \sum F_r \varepsilon^r$ be an additional integral and let all the functions F_r be analytic on M_c. With the help of the Poincaré method one can prove that the functions F_0 and H_0 are dependent on the resonant tori of the Euler problem. These tori are "enumerated" by the action variables $y \in$ **B**. Taking into account the key property of this set of tori, we conclude that H_0 and F_0 are dependent everywhere on M_c. Now it is not difficult to deduce by induction that the integrals H and F are dependent (one can find the details in [131]). Thus, we obtain

Theorem 1 [125]. *If a heavy rigid body is not dynamically symmetric, then the equations of its rotation have no independent of the function $H_0 + \varepsilon H_1$ formal integral $\sum F_r \varepsilon^r$ with coefficients analytic on the level M_c.*

This theorem gives a negative answer to the question raised by Poincaré in [203, §86].

Going back to the original Hamiltonian system with three degrees of freedom, we see that the equations of rotation of a heavy non-symmetric rigid body with a fixed point have no additional integral which is a formal power series in ε with single-valued coefficients analytic in all phase space, and which commutes with the area integral. Taking into account the well known connection between integrals which are formal in ε and polynomial integrals of reversible systems (see §1 of Chap. II), we obtain the following result: in the non-symmetric case there are no additional integrals which, commute with the area integral, in the form of polynomials in momenta with coefficients analytic on the group $SO(3)$.

2.3 As one more example consider the perturbed motion of the Lagrange top (see §5 of Chap. II). More precisely, we deal with rotation of a heavy dynamically symmetric ($I_1 = I_2$) rigid body whose centre of mass is displaced slightly from the axis of dynamical symmetry. Let r_1, r_2, r_3 be the coordinates of the center of mass with respect to the inertia axes. Fixing the value $r_3 \neq 0$, we define a small parameter

$$\varepsilon = \sqrt{r_1^2 + r_2^2}/r_3 \, .$$

For $\varepsilon = 0$ we have a completely integrable Lagrange problem. We pose the problem of the existence of an additional integral which is a formal power series in ε, commuting with the area integral. This problem is solved following the method given in [125]. First, we reduce the order of the Hamiltonian system using the area integral, and then perform the transition to the action-angle variables $x_i \bmod 2\pi$, y_i $(i = 1, 2)$ of the unperturbed problem. Written in these variables the expansion of the perturbing function H_1 in a multiple Fourier series has the form (2.3) (where it is necessary to put $H_{m,0} = 0$). If

$$r_3 \neq 0 \, , \quad I_1 \neq I_3 \, , \tag{2.4}$$

then the Fourier coefficients $H_{m,\pm 1}$ do not vanish. As was shown in [210], under conditions (2.4) the secular set consists of an infinite number of closed curves surrounding the point y_0. The closure of the set **B** consists of the single point y_0. Now one would like to use Theorem 3, but the Hamiltonian function H_0 has a singularity for $y = y_0$. Consequently, Theorem 3 is not applicable directly. This difficulty can be overcome in the same way as in [125]: since the Hamiltonian function $H_0 + \varepsilon H_1$ is analytic in the phase space $T^*SO(3) = \mathbb{R}^3 \times SO(3)$, it is natural to consider the problem of the existence of a commuting with the area integral additional integral which is a formal power series in ε with coefficients analytic in $\mathbb{R}^3 \times SO(3)$.

It turns out, that under the conditions (2.4) such an integral does not exist [210]. Note that for $I_1 = I_3$ the perturbed problem is completely integrable (it is the Lagrange problem again); for $r_3 = 0$ there exist the integrable problems of Kovalevskaya ($I_1 = 2I_3$) and Goryachev–Chaplygin ($I_1 = 4I_3$ and the area integral constant vanishes). The question of the existence of an additional integral under the condition $r_3 = 0$ is much more complex: here the set **B** does not possess the key property.

3 Symmetry Groups

3.1 Following [143], we apply the Poincaré method to the existence problem of symmetry groups in systems of differential Eqs. (1.1). We consider the symmetries generated by a system of the following form:

$$\begin{gathered} y_j' = Y_j^0 + \varepsilon Y_j^1 + \cdots \, , \qquad x_k' = X_k^0 + \varepsilon X_k^1 + \cdots \, , \\ 1 \leq j \leq m \, , \quad 1 \leq k \leq n \, . \end{gathered} \tag{3.1}$$

The coefficients Y_j^r and X_k^s are assumed to be 2π-periodic in the coordinates $x_1, \ldots, x_n$.

We consider only the "non-degenerate" case, when the following conditions hold:
1) $n \geq m$ and the rank of the matrix $\|\partial\omega_k/\partial y_j\|$ equals m almost everywhere,
2) if $\sum \omega_k(y)\alpha_k \equiv 0$ with some integer α_k, then all $\alpha_k = 0$.

For example, for $m = 1$ these conditions hold if the curve $y \to \omega(y)$ is regular and intersects the resonance surfaces $\sum \omega_k \alpha_k = 0$ ($\alpha_k \in \mathbb{Z}$) transversely. If $m = n$, then the non-degeneracy conditions are reduced to the single one:

$$\det \|\partial\omega_k/\partial y_j\| \neq 0$$

almost everywhere. Note that in §1 the definition of a non-degenerate system contains only the second condition.

All the functions encountered below are assumed to be analytic.

First, we put $\varepsilon = 0$ and find all symmetry fields of the unperturbed integrable system. It is not difficult to show that for $\varepsilon = 0$ the condition for the vector fields (1.1) and (3.1) to commute is reduced to the set of equalities

$$\sum \frac{\partial Y_j^0}{\partial x_l}\omega_l = 0\,, \qquad 1 \leq j \leq m\,, \tag{3.2}$$

$$\sum \frac{\partial \omega_k}{\partial y_j} Y_j^0 = \sum \frac{\partial X_k^0}{\partial x_l}\omega_l\,, \qquad 1 \leq k \leq n\,. \tag{3.3}$$

Lemma 1. *If the unperturbed system is non-degenerate, then* $Y_j^0 \equiv 0$, *and the functions* X_k^0 *do not depend on* x.

Proof. Solve Eqs. (3.2) using the Fourier method. We put

$$Y_j^0 = \sum \zeta_\alpha(y)\exp[i(\alpha, x)]\,.$$

Then from (3.2) we find that $(\alpha, \omega(y))\zeta_\alpha \equiv 0$. Since the unperturbed system is non-degenerate and there are no zero divisors in the ring of analytic functions, $\zeta_\alpha = 0$ for all $\alpha \neq 0$. Consequently, the functions Y_j^0 depend only on y. In particular, the left-hand sides of equalities (3.3) do not contain the coordinates $x_1, \ldots, x_n$. Taking the average of both parts of (3.3) over $\mathbb{T}^n = \{x \bmod 2\pi\}$, we obtain the relation

$$\sum \frac{\partial \omega_k}{\partial y_j} Y_j^0 = 0\,. \tag{3.4}$$

According to the definition of non-degeneracy,

$$\operatorname{rank} \|\partial\omega_k/(\partial y_j)\| = m \leq n\,.$$

Thus, (3.4) implies the equalities $Y_j^0 \equiv 0$. We apply the Fourier method to the equations

$$\sum \frac{\partial X_k^0}{\partial x_l}\omega_l = 0$$

and find that the functions X_k^0 depend only on the slow variables y. The lemma is proved.

We put

$$Y_j^1 = \sum g_\alpha^j(y) \exp[i(\alpha, x)] .$$

From the condition for the vector fields (1.1) and (3.1) to commute we derive the following equalities in the first approximation in ε:

$$(\alpha, \omega) g_\alpha = (\alpha, X^0) \Phi_\alpha , \qquad \alpha \in \mathbb{Z}^n . \tag{3.5}$$

Here X^0, Φ_α, g_α are vectors with components X_k^0, Φ_α^j, g_α^j respectively; and Φ_α^j are Fourier coefficients of the function Φ^j (see §1). We use Lemma 1 when deriving (3.5).

Lemma 2. *Suppose that in some open domain $D \subset \mathbb{R}^m = \{y\}$ the frequency vector $\omega \neq 0$ and $\mathbf{P}_1 \cap D$ is a key set. Then there exists a function ξ_0 analytic in the domain D such that $X^0 = \xi_0 \omega$.*

Indeed, according to (3.5), the vectors X^0 and ω are linearly dependent at the points of the Poincaré set $\mathbf{P}_1$. Since $\mathbf{P}_1$ is a key set, X^0 and ω are dependent at all points of the domain D: $\mu X_0 = \lambda \omega$, $\mu^2 + \lambda^2 \neq 0$. Since $\omega \neq 0$, we have $\mu \neq 0$. Consequently, $X_0 = \xi_0 \omega$, and ξ_0 is an analytic function. The lemma is proved.

We substitute the vector field $\xi_0 \omega$ instead of X_0 in (3.5) and use the inequality $(\alpha, \omega(y)) \not\equiv 0$. Then $g_\alpha = \xi_0 \Phi_\alpha$ for all $\alpha \neq 0$. We put

$$\Psi_k = \sum \psi_\alpha^k(y) \exp[i(\alpha, x)] , \quad X_k^1 = \sum \eta_\alpha^k(y) \exp[i(\alpha, x)] .$$

Apart from (3.5), the condition for the systems (1.1) and (3.1) to commute implies in the first approximation in ε the following chain of relations:

$$i(\alpha, \omega) \eta_\alpha^k = (\alpha, X^0) \psi_\alpha^k + \sum \frac{\partial \omega_k}{\partial y_j} g_\alpha^j - \sum \frac{\partial X_k^0}{\partial y_j} \Phi_\alpha^j , \qquad \alpha \in \mathbb{Z}^n . \tag{3.6}$$

Putting in (3.6) $y \in \mathbf{P}_1$, $X_k^0 = \xi_0 \omega_k$, and using the inequality $\omega \neq 0$, we obtain the equality

$$\sum \frac{\partial \xi_0}{\partial y_j} \Phi_\alpha^j = 0 . \tag{3.7}$$

Lemma 3. *Let the Poincaré set $\mathbf{P}_0$ be a key set for $C^\omega(D)$. Then $\xi_0 = \mathrm{const}$.*

Indeed, since $\mathbf{P}_0 \subset \mathbf{P}_1$, the relation (3.7) holds at all points $y \in \mathbf{P}_0$. Since for $y \in \mathbf{P}_0$ there exist m linearly independent vectors $\Phi_\alpha, \Phi_{\alpha'}, \Phi_{\alpha''}, \ldots$ orthogonal to $\partial \xi_0 / \partial y$, we have $d\xi_0 = 0$. Consequently, $\xi_0 = \mathrm{const}$.

Theorem 1. *Suppose that in the domain $D \subset \mathbb{R}^m = \{y\}$ the unperturbed system is non-degenerate, $\omega \neq 0$ and $\mathbf{P}_0$ is a key set for $C^\omega(D)$. Then for $x, y \in \mathbb{T}^n \times D$ the vector field* (3.1) *differs from the field* (1.1) *by the multiplier $\xi_\varepsilon = \xi_0 + \varepsilon\xi_1 + \cdots$, $\xi_r = \text{const}$ $(r \geq 0)$.*

Proof. Let v_ε be the vector field (1.1) and let u_ε be the vector field (3.1). According to Lemmas 1–3, $u_0 = \xi_0 v_0$, where $\xi_0 = \text{const}$. Consequently, the vector field $w_\varepsilon = (u_\varepsilon - \xi_0 v_\varepsilon)/\varepsilon$ is also an analytic symmetry field. Again it follows from Lemmas 1–3 that $w_0 = \xi_1 v_0$ ($\xi_1 = \text{const}$), etc. As a result we obtain the equality $u_\varepsilon = \xi_\varepsilon v_\varepsilon$, where $\xi_\varepsilon = \xi_0 + \varepsilon\xi_1 + \cdots$.

3.2 Now consider the case when the system (1.1) has the integral

$$H = H_0(x, y) + \varepsilon H_1(x, y) + \cdots$$

with analytic coefficients 2π-periodic in x. Then according to the results of §1 the Poincaré set $\mathbf{P}_0$ cannot possess the key property. In the non-degenerate case the function H_0 does not depend on x (see Lemma 1, §1).

Theorem 2. *Suppose that the unperturbed system is non-degenerate. Let $dH_0 \neq 0$, $\omega \neq 0$ at some point $y^0 \in \mathbb{R}^m$ and let the Poincaré set $\mathbf{P}_1$ be a key set for the class of functions $C^\omega(U)$ for a neighborhood U of the point y_0. Then for $(x, y) \in \mathbb{T}^n \times D$ the vector field* (3.1) *differs from the field* (1.1) *by a multiplier $\xi_\varepsilon = \xi_0 + \varepsilon\xi_1 + \cdots$, where ξ_r are analytic functions of H.*

Proof. According to Lemmas 1 and 2, $u_0 = \xi_0 v_0$, where ξ_0 is an analytic function of y. Since $[u_0, v_0] = 0$, ξ_0 is an integral of the unperturbed system (1.1) (see §3, Chap. II). According to Lemma 1, §1, the functions ξ_0 and H_0 are dependent at the points of $\mathbf{P}_1 \cap D$. Then the functions ξ_0 and H_0 are dependent everywhere by the key property of this set. Since there are no critical points of the function H_0 in a small domain, according to the implicit function theorem, in this domain we have $\xi_0 = \Phi_0(H_0)$, where Φ_0 is some analytic function (see §1.1). Consequently, the vector field

$$w_\varepsilon = (u_0 - \Phi_0(H)v_\varepsilon)/\varepsilon$$

is again an analytic symmetry field. Analogously $w_0 = \Phi_1(H_0)v_0$, and so on. As a result, we obtain the equality $u_\varepsilon = \Phi(H, \varepsilon)v_\varepsilon$, where $\Phi = \Phi_0 + \varepsilon\Phi_1 + \cdots$. The theorem is proved.

Let us make several remarks.

1) Under the assumptions of Theorem 2 the system (1.1) does not admit nontrivial symmetry fields $u_\varepsilon = u_0 + \varepsilon u_1 + \cdots$ with single-valued coefficients analytic in the domain $D \times \mathbb{T}^n$.

2) We nowhere assumed the convergence of the power series in ε, introduced above. Thus, under the assumptions of Theorems 1 and 2, the system (1.1) does not admit symmetry fields which are formal power series in ε.
3) If the Poincaré set $\mathbf{P}_1$ is dense in D, then one can claim the absence of a symmetry field $u_\varepsilon = u_0 + \varepsilon u_1 + \cdots$ with coefficients which are smooth in the domain $D \times \mathbb{T}^n$.

3.3 Now we apply the results of §3.2 to Poincaré's "basic problem of dynamics". Consider the Hamiltonian system

$$\dot{x}_k = \frac{\partial H}{\partial y_k}, \quad \dot{y}_k = -\frac{\partial H}{\partial x_k}; \qquad 1 \leq k \leq n \tag{3.8}$$

with Hamiltonian $H = H_0(y) + \varepsilon H_1(x, y) + \cdots$ which is analytic and 2π-periodic in x. Let v_ε be the vector field (3.8), and u_ε the symmetry field (3.1).

Theorem 3. *Suppose that y^0 is a non-critical point of the function H_0 and at this point*

$$\det \|\partial^2 H_0/(\partial y^2)\| \neq 0\,. \tag{3.9}$$

Suppose also that in any small neighborhood U of the point y_0 the Poincaré set $\mathbf{P}_1$ is a key set for the class of functions $C^\omega(U)$. Then in the domain $U \times \mathbb{T}^n \subset \mathbb{R}^n \times \mathbb{T}^n$ the equality $u_\varepsilon = \Phi(H, \varepsilon) v_\varepsilon$ holds, where Φ is some analytic function.

Indeed, according to (3.9) the unperturbed system (3.8) is non-degenerate. The frequency vector $\omega = \partial H_0/\partial y$ does not vanish at the point $y = y^0$. It remains to use Theorem 2.

This result improves Theorem 3, §1. Recall that this theorem guarantees the absence of an additional analytic integral independent of the energy integral. Indeed, as it is noted in §3 of Chap. II, any integral of a Hamiltonian system generates a Hamiltonian symmetry field. Moreover, symmetry fields may correspond to multivalued integrals (recall that a multivalued function on M is defined as a closed 1-form ϕ: locally $\phi = dF$, where F is a function on M). This remark and Theorem 3 imply

Corollary. *Under the assumptions of Theorem* 3 *the Hamiltonian system* (3.8) *does not have a formal integral independent of H of the form*

$$\sum F_r(x, y)\varepsilon^r$$

with coefficients multivalued in $\mathbb{T}^n \times \mathbb{R}^n$.

3.4 Theorem 3 is applicable to many problems of Hamiltonian mechanics. For example, the plane restricted circular three-body problem does not admit a nontrivial symmetry group in the form of a power series in the small parameter μ with infinitely differentiable coefficients (cf. §2.1; the parameter μ equals the ratio of the mass of Jupiter to the mass of the Sun).

As one more example, consider the rotation of a heavy non-symmetric rigid body with the centre of mass close to the fixed point (see §2.2). This system with three degrees of freedom has two Hamiltonian symmetry fields v_ε and u_ε: the field v_ε corresponds to the energy integral, while the field u_ε corresponds to the integral of angular momentum (actually the field u_ε does not depend on the Poincaré small parameter). Using the results of §2.2 and Theorem 3, one can show that the Hamiltonian equations of this problem do not admit another symmetry field w_ε independent of the fields u_ε and v_ε, commuting with the field u_ε ($[u_\varepsilon, v_\varepsilon] = 0$), and analytic in the phase variables and in the parameter ε. We note one of the consequences of this result: the equations of rotation of a non-symmetric rigid body, whose centre of mass does not coincide with the suspension point, do not admit an additional integral, polynomial in the velocities, with coefficients which are multivalued on the group $SO(3)$. (The integral is assumed to be independent of the energy and the angular momentum integrals and to commute with the latter.) Note that there exist non-single-valued functions on the group $SO(3)$, because $SO(3)$ is not simply connected. These functions are actually two-valued: they take one of two possible values when returning to the neighborhood of an initial point on $SO(3)$.

4 Reversible Systems with a Torus as the Configuration Space

4.1 In this section we study Hamiltonian systems of the following form:

$$\dot{x}_k = \frac{\partial H}{\partial y_k}, \quad \dot{y}_k = -\frac{\partial H}{\partial x_k}; \qquad 1 \le k \le n, \tag{4.1}$$
$$H = H_0(y) + \varepsilon H_1(x).$$

The function H_0 is a non-degenerate quadratic form in the momenta $y_1, \dots, y_n$ with constant coefficients:

$$H_0 = \frac{1}{2} \sum_{i,j=1}^{n} a_{ij} y_i y_j, \qquad \det \|a_{ij}\| \neq 0. \tag{4.2}$$

The function H_1 is assumed to be analytic on the n-dimensional torus $\mathbb{T}^n = \{x_1, \dots, x_n \bmod 2\pi\}$. Certainly, the Hamiltonian system (4.1) is a particular case of the systems discussed in the introduction to this chapter. While this system preserves the essential features of the general case, it is simpler to analyze from the technical point of view.

If the form H_0 is positive definite, Eqs. (4.1) describe the dynamics of a reversible mechanical system on the n-dimensional torus $\mathbb{T}^n = \{x\}$ with kinetic energy H_0 and a small potential εH_1.

We introduce some notation. Let $\xi = (\xi_1, \dots, \xi_n)$ and $\eta = (\eta_1, \dots, \eta_n)$ be vectors in $\mathbb{R}^n$. We put

$$(\xi, \eta) = \sum_{i=1}^{n} \xi_i \eta_i \,, \quad \langle \xi, \eta \rangle = \sum_{i,j=1}^{n} a_{ij} \xi_i \eta_j \,.$$

Hence, the relation (4.2) can be written in the brief form:

$$H_0 = \langle y, y \rangle / 2 \,.$$

In our future analysis an important role will be played by the Fourier expansion of the perturbing function

$$H_1 = \sum h_m e^{i(m,x)} \,, \qquad h_m = \text{const} \,, \quad m \in \mathbb{Z}^n \,.$$

Following Poincaré, we pose the question of the existence in the system (4.1) of a complete set of independent integrals in the form of power series $\sum F_k(x, y)\varepsilon^k$ with analytic coefficients 2π-periodic in x. We emphasize that there is no reason at all to demand convergence of these power series. Using the non-degeneracy of the unperturbed problem with the Hamiltonian H_0, one can show (see below) that the formal integrals, which form the complete set of integrals, are pairwise in involution.

As was shown in §10 of Chap. II, the existence of a complete set of independent formal integrals is closely related to the possibility of realizing the classical scheme of perturbation theory. We give some explicit expressions to be used later.

The basic idea of perturbation theory is finding a canonical transformation $y, x \bmod 2\pi \to u, v \bmod 2\pi$, depending on ε, such that written in the new variables the Hamiltonian $H_0 + \varepsilon H_1$ has the following form:

$$K_0(u) + \varepsilon K_1(u) + \varepsilon^2 K_2(u) + \cdots \,.$$

If this transformation can be found, then the initial system of Hamiltonian equations is easily integrated. We search for the canonical transformation in the form

$$y_i = \frac{\partial S}{\partial x_i} \,, \quad v_i = \frac{\partial S}{\partial u_i} \,; \qquad 1 \le i \le n \,,$$
$$S = S_0(u, x) + \varepsilon S_1(u, x) + \cdots \,.$$

Put $S_0 = \sum u_i x_i$. Then, for $\varepsilon = 0$, we have the identity transformation. The generating function S satisfies the Hamilton–Jacobi equation

$$H_0(\partial S / \partial x) + \varepsilon H_1(x) = K_0(u) + \varepsilon K_1(u) + \cdots \,. \tag{4.3}$$

Expand the left-hand side of this equation in a power series in ε and equate coefficients of equal powers of ε. We obtain an infinite chain of equations for finding successively $S_1, S_2, \dots$:

$$\sum_k \frac{\partial H_0}{\partial u_k}\frac{\partial S_1}{\partial x_k} + H_1(x) = K_1(u)\,,$$

$$\sum_k \frac{\partial H_0}{\partial u_k}\frac{\partial S_2}{\partial x_k} + \frac{1}{2}\sum_{i,j} a_{ij}\frac{\partial S_1}{\partial x_i}\frac{\partial S_1}{\partial x_j} = K_2(u)\,,$$

$$\dots \quad \dots \quad \dots \quad \dots \quad \dots \quad \dots \tag{4.4}$$

$$\sum_k \frac{\partial H_0}{\partial u_k}\frac{\partial S_m}{\partial x_k} + \frac{1}{2}\sum_{i,j} a_{ij}\sum_{p+q=m}\frac{\partial S_p}{\partial x_i}\frac{\partial S_q}{\partial x_j} = K_m(u)\,,$$

$$\dots \quad \dots \quad \dots \quad \dots \quad \dots \quad \dots$$

It is not difficult to show that Eqs. (4.4) have a unique (formal) solution $S_1, S_2, \dots$ in the form of trigonometric series in $x_1, \dots, x_n$ with zero coefficients S_m^0:

$$S_m = \sum{}' S_m^\tau(u)e^{i(\tau,x)}\,, \qquad \tau \in \mathbb{Z}^m \setminus \{0\}\,.$$

Consider the first equation of the system (4.4) and solve it by the Fourier method. Using (4.3), we obtain

$$K_1 = h_0\,; \quad S_1^\tau = \frac{ih_\tau}{(\omega, \tau)}\,, \qquad \tau \neq 0\,, \tag{4.5}$$

where $\omega = (\omega_1, \dots, \omega_n)$, $\omega_s = \partial H_0/\partial u_s$. It is clear that $(\omega, \tau) = \langle u, \tau\rangle$. From (4.5) we see that the function S_1 is not defined at points of $\mathbb{R}^n = \{u\}$ which lie on the hyperplanes

$$\langle \tau, u\rangle = 0\,, \quad h_\tau \neq 0\,.$$

We call the set of all these points the first order Poincaré set and denote it by $\mathbf{P}^1$ (in §1 it was denoted by $\mathbf{P}_1$).

On solving the second equation of the system (1.4) by the Fourier method, we formally obtain $S_2^k = ih'_k/\langle u, k\rangle$, where

$$h'_k = \frac{1}{2}\sum_{\tau+\sigma=k}\frac{\langle \tau, \sigma\rangle h_\tau h_\sigma}{\langle u, \tau\rangle\langle u, \sigma\rangle}\,, \qquad k \neq 0\,. \tag{4.6}$$

The other equations of the system (4.4) are solved in the same way.

4.2 If the Poincaré set is sufficiently large, then the corresponding system cannot be completely integrable:

Theorem 1_1**.** *Suppose that the Poincaré set consists of an infinite number of distinct hyperplanes. Then the system* (4.1) *does not have n formal integrals* $\sum F_k\varepsilon^k$ *with analytic coefficients* $F_k : \mathbb{R}^n \times \mathbb{T}^n \to \mathbb{R}$ *which are independent for* $\varepsilon = 0$.

Theorem 1_1 is a simple consequence of the following statement: if the system (4.1) has n integrals

$$F_0^{(1)}(x,y)+\varepsilon F_1^{(1)}(x,y)+\cdots\,,\ \ldots\,,\ F_0^{(n)}(x,y)+\varepsilon F_1^{(n)}(x,y)+\cdots\,,$$

then

1) the functions $F_0^{(1)},\ldots,F_0^{(n)}$ are independent of x,

2) the Jacobian $\dfrac{\partial(F_0^{(1)},\ldots,F_0^{(n)})}{\partial(y_1,\ldots,y_n)}\neq 0$ at the points of the set $\mathbf{P}^1$.

This statement easily follows from Lemma 1, §1.

Now we prove Theorem 1_1. First, note that the union of an infinite number of distinct hyperplanes in $\mathbb{R}^n$, which pass through the origin, is a key set for the class of analytic functions in $\mathbb{R}^n$. Indeed, let Π be one of the "limit" hypersurfaces of the set $\mathbf{P}^1$, and let the point a not lie on Π. Take a line l which intersects Π transversely at a point distinct from the origin (see Fig. 13). The line l intersects an infinite number of distinct hyperplanes of $\mathbf{P}^1$. Let f be an analytic function in $\mathbb{R}^n$ which vanishes on the set $\mathbf{P}^1$. It is clear that the restriction of f to the line l is an analytic function which has a non-isolated zero. Consequently, $f|_l\equiv 0$; in particular, $f(a)=0$. To complete the proof we use the analyticity of the Jacobian of the functions $F_0^{(1)},\ldots,F_0^{(n)}$.

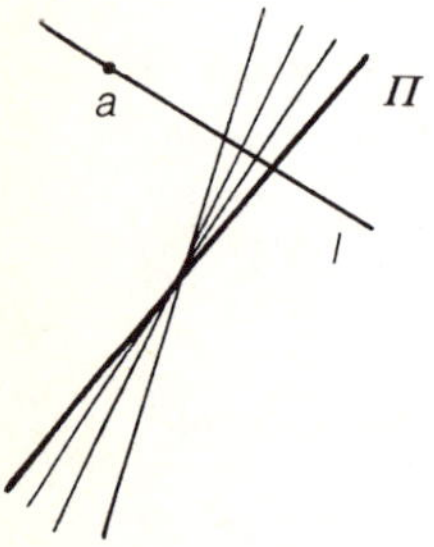

Fig. 13.

Assume that the conditions of Theorem 1_1 are valid. Then there is no complete set of independent integrals of the form $\sum F_k(x,y)\varepsilon^k$, whose coefficients F_k are analytic in the direct product $D\times\mathbb{T}^n$. (Here D is any open domain in $\mathbb{R}^n=\{y\}$ which has a non-empty intersection with $\overline{\mathbf{P}}^1\setminus\mathbf{P}^1$; the bar denotes the closure of a set.)

Theorem $1_1'$. *Let the conditions of Theorem 1_1 hold. Suppose that the Hamiltonian system (4.1) has $n-1$ analytic integrals*

$$F^{(1)}=F_0^{(1)}+\varepsilon F_1^{(1)}+\cdots\,,\ \ldots\,,\ F^{(n-1)}=F_0^{(n-1)}+\varepsilon F_1^{(n-1)}+\cdots\,,\qquad(4.7)$$

where the functions $F_0^{(1)}, \ldots, F_0^{(n-1)}$ *are independent at some point of* $(\overline{\mathbf{P}}^1 \setminus \mathbf{P}^1) \times \mathbb{T}^n$. *Then the Hamiltonian equations have no integral independent of the functions* $F^{(1)}, \ldots, F^{(n-1)}$ *and represented as a formal series* $\sum F_s \varepsilon^s$ *with coefficients being analytic and single-valued in* $\mathbb{R}^n \times \mathbb{T}^n$.

This theorem is a consequence of Theorem 3, §1 and the key property of the Poincaré set $\mathbf{P}^1$.

The Hamiltonian function $H_0 + \varepsilon H_1$ can be among the integrals (2.1). Note that the only critical point of the function H_0 is the point $y = 0$. Hence, it follows, in particular, that for $n = 2$ Theorem 1_1 can be strengthened by replacing the condition that functions H_0 and F_0 are independent by the weaker condition that the series $H_0 + \varepsilon H_1$ and $\sum F_s \varepsilon^s$ are independent.

4.3 Assume that the condition of Theorem 1_1 does not hold. Then there is a solution S_1 of the first equation of the system (4.4) which is analytic in $(\mathbb{R}^n \setminus \mathbf{P}^1) \times \mathbb{T}^n$. In particular, the functions (4.6) are defined and analytic in the domain $\mathbb{R}^n \setminus \mathbf{P}^1$. In this situation a second order Poincaré set $\mathbf{P}^2$ can be defined as the set of hyperplanes $\langle k, u \rangle = 0$, $k \neq 0$ on which the corresponding functions h'_k do not vanish.

Theorem 1_2. *If the set* $\mathbf{P}^2$ *consists of an infinite number of distinct hyperplanes, then the system* (4.1) *does not have n formal integrals* $\sum F_s \varepsilon^s$ *with analytic coefficients* $F_s : \mathbb{R}^n \times \mathbb{T}^n \to \mathbb{R}$ *which are independent for* $\varepsilon = 0$.

The proof follows the same lines as that of Theorem 1_1. We can also prove Theorem $1'_2$, obtained from Theorem $1'_1$ by replacing $\mathbf{P}^1$ by $\mathbf{P}^2$.

Notice an interesting particular case when $h'_k \equiv 0$ for all $k \neq 0$. Then $S_2 \equiv 0$. From the system (4.4) we obtain by induction that $S_3 = S_4 = \cdots \equiv 0$. In this case Eqs. (4.1) are completely integrable. If the points of the set $\{m \in \mathbb{T}^n : h_m \neq 0\}$ lie on orthogonal (in the Euclidean metric $\langle\ ,\ \rangle$) straight lines passing through the origin, then obviously $h'_k = 0$ for $k \neq 0$. It will be shown below that in this case the canonical variables can be separated.

The Poincaré sets of higher orders are defined recursively. Indeed, suppose that there exist solutions $S_1, S_2, \ldots, S_{p-1}$ of the first $p - 1$ equations of the system (4.4), and these solutions are analytic in $(\mathbb{R}^n \setminus (\mathbf{P}^1 \cup \ldots \cup \mathbf{P}^{p-1})) \times \mathbb{T}^n$. Then the p-th order Poincaré set $\mathbf{P}^p$ is well defined, and Theorems 1_p and $1'_p$ hold. If the perturbing function H_1 is a trigonometric polynomial, then each of the sets $\mathbf{P}^p$ consists of only a finite number of distinct hyperplanes (i.e., $\overline{\mathbf{P}}^p = \mathbf{P}^p$) so that Theorems 1_p $(p = 1, 2, \ldots)$ lead to no conclusions about the integrability of the Hamiltonian system (4.1). Similar situations are often encountered in analysis. For instance, there are series whose convergence or divergence cannot be established by an infinite number of logarithmic tests.

If the Poincaré sets of all orders are defined, we can put

$$\mathbf{P}^\infty = \bigcup_{p=1}^{\infty} \mathbf{P}^p .$$

It turns out that Theorems 1_∞ and $1'_\infty$ are valid. Admittedly, these theorems are not constructive: to check if their conditions hold, we have to perform an infinite number of steps of perturbation theory.

The conditions of Theorems 1_∞ and $1'_\infty$ can be made effective in the case of trigonometric polynomials.

5 A Criterion for Integrability in the Case when the Potential is a Trigonometric Polynomial

5.1 Let the function $H_1 : \mathbb{T}^n \to \mathbb{R}$ be a trigonometric polynomial. Then the set

$$\Delta = \{m \in \mathbb{Z}^n : h_m \neq 0\}$$

is obviously finite. As in the general case, it is supposed to be invariant under the reflection $m \to -m$. We assume that $H_1 \neq$ const. Hence, Δ contains at least two elements.

Theorem 1 [155]. *Assume that the quadratic form H_0 is positive definite. Then the Hamiltonian system with Hamilton function $H_0 + \varepsilon H_1$ has a complete set of first integrals formally analytic in ε, independent for $\varepsilon = 0$ if and only if the points of the set Δ are located on $d \leq n$ straight lines which intersect orthogonally (in the metric $\langle\ ,\ \rangle$) at the origin.*

The proof of sufficiency is easy. For, let $l_1, \ldots, l_d$ be the straight lines in $\mathbb{R}^n$ referred to in the theorem. Denote by $k_i \neq 0$ the point of the set $\mathbb{Z}^n \cap l_i$ which is closest to the origin. We supplement $k_1, \ldots, k_d$ by integer-valued vectors $k_{d+1}, \ldots, k_n$ up to a basis in $\mathbb{R}^n$. Now perform a linear transformation $x' = Mx$ with the non-degenerate integer-valued matrix

$$M = \left\| \begin{matrix} k_1 \\ \vdots \\ k_n \end{matrix} \right\| ,$$

and extend it to a canonical transformation $x, y \to x', y'$ by putting $y' = (M^T)^{-1} y$. In the new variables x', y' the Hamiltonian function $H_0 + \varepsilon H_1$ takes the form

$$\frac{1}{2}\Big(\sum_{i=1}^{d} a'_{ii}(y'_i)^2 + \sum_{j=1}^{n}\sum_{i>d} a'_{ij} y'_i y'_j\Big) + \varepsilon \sum_{i=1}^{d} f_i(x'_i) , \qquad (5.1)$$

where $a'_{ij} =$ const, and f_i are 2π-periodic trigonometric polynomials. The variables x', y' can obviously be separated. Hence, the Hamilton system with Hamiltonian (5.1) has the following set of n independent integrals

$$F_i = \frac{1}{2}\Big(a'_{ii}(y'_i)^2 + y'_i \sum_{s>d} a'_{is} y'_s\Big) + \varepsilon f_i(x'_i) , \qquad 1 \leq i \leq d ,$$

$$F_j = y'_j , \qquad j > d .$$

Returning to the old variables x, y, we obtain the set of integrals linear in ε (or independent of ε) whose coefficients are analytic functions in $\mathbb{R}^n \times \mathbb{T}^n = \{y, x \bmod 2\pi\}$.

Now we put $\varepsilon = 1$ and consider the system with Hamiltonian $H_0 + H_1$. It has been already noted in §1 of Chap. II that if the system with Hamiltonian $H_0 + H_1$ has n integrals, which are polynomial in the momenta and have independent leading homogeneous forms, then the system with Hamiltonian $H_0 + \varepsilon H_1$ has n first integrals, analytic in ε, which are independent for $\varepsilon = 0$. On the other hand, Ziglin [247] showed that if the system in question has r independent polynomial integrals, then it has r integrals of the same form with independent leading homogeneous forms. With the help of these remarks one can derive from Theorem 1 the following interesting

Corollary 1. *If the system with Hamiltonian $H_0 + H_1$ has n independent polynomial integrals, then it has n independent commuting polynomial integrals of degree not greater than two.*

Remark. It seems that this assertion also holds in the more general case when the potential energy H_1 is any analytic function in $\mathbb{T}^n = \{x \bmod 2\pi\}$ (and not merely a trigonometric polynomial). Earlier Bialy [22] proved this conjecture in the special case when $n = 2$ and the Hamiltonian system has an additional polynomial integral of degree not greater than four. Notice that the existence problem of an additional polynomial integral of a given degree is much simpler than that for an integral in the form of a polynomial whose degree is not fixed in advance.

Consider k orthogonal straight lines in $(\mathbb{R}^n, \langle\ ,\ \rangle)$ which pass through the origin O. Take on each line two points equidistant from and on opposite sides of $O \in \mathbb{R}^n$. We call the convex hull of these $2k$ points a k-dimensional rhomboid. The number of l-dimensional faces of the k-dimensional rhomboid is $2^{l+1} C_k^{l+1}$; in particular this polyhedron has precisely $2k$ vertices and 2^k faces. It can be shown that the k-dimensional rhomboid is a convex polyhedron, dual to a k-dimensional parallelepiped.

Corollary 2. *Assume that the Hamiltonian system with Hamiltonian $H_0 + H_1$ has n independent polynomial integrals. Then the convex hull of the set Δ is a k-dimensional rhomboid ($k \leq n$); on the boundary of this rhomboid there are no points of Δ other than vertices.*

As an example, consider the system with the potential

$$H_1 = \sum_{i<j} f(x_i - x_j)\,, \tag{5.2}$$

where the even function $f(\cdot)$ is a non-constant 2π-periodic trigonometric polynomial (a potential of pairwise interaction). One can show that the convex hull of the set Δ in this problem is an $(n-1)$-dimensional polyhedron with $2C_n^2$ vertices.

Since $2C_n^2 > 2(n-1)$ for $n > 2$, for $n \geq 3$ the system with potential (5.2) does not possess a complete set of polynomial integrals. This conclusion does not depend on the form of the Euclidean metric $\langle\ ,\ \rangle$.

Adler and van Moerbeke [3] considered a particular case of this problem, when the metric $\langle\ ,\ \rangle$ in $\mathbb{R}^n$ is standard and $f = \cos(\cdot)$. This is the classical version of the Gross–Neveu system which is well known in theoretical physics. The Hamiltonian function has the following form:

$$H = \frac{1}{2}\sum y_s^2 + \sum_{i<j} \cos(x_i - x_j) .$$

This function can be regarded as the Hamiltonian of the problem of n points on a circle, joined by elastic springs. By using the Kovalevskaya method, it was shown [3] that for $n = 3$ and $n = 4$, for almost all initial conditions, the variables y_s and $\exp(ix_s)$ are not meromorphic functions of complex time. In particular, the Gross–Neveu system is not algebraically integrable. In connection with this result it is worth remarking that an algebraically nonintegrable system can be completely integrable (see §9 of Chap. II).

5.2 We introduce in $\mathbb{Z}^n$ the standard lexicographic order relation, denoted henceforth by $\prec$. We say that $\sigma \prec \delta$ if, for the least subscript s such that $\sigma_s \neq \delta_s$, the inequality $\sigma_s < \delta_s$ holds. We say that $\sigma \preceq \delta$ if $\sigma \prec \delta$ or $\sigma = \delta$.

Definition. Let α be the greatest element of Δ and β element of the set $\Delta \setminus \{\alpha\}$ which is the greatest linearly independent of α. We call the vector α a vertex of Δ, and the vector β a vertex of Δ adjacent to α.

Leaving aside the trivial case of integrability when all the points of Δ lie on a single straight line passing through the origin, we shall assume henceforth that an adjacent vertex always exists.

The proof of Theorem 1 is based on the following assertion which is interesting in itself.

Theorem 2 [155]. *Let α and β be vertices of Δ, and*

$$m\langle\alpha, \alpha\rangle + 2\langle\alpha, \beta\rangle \neq 0 \tag{5.3}$$

for all integers $m \geq 0$. Then the Hamiltonian system with Hamiltonian $H_0 + \varepsilon H_1$ does not have n formally analytic integrals independent for $\varepsilon = 0$.

We emphasize that for the validity of Theorem 2 only the non-degeneracy of the quadratic form H_0 is needed. Theorem 2 is proved with the help of perturbation theory. The unattached coefficients of the integrals (the functions $F_0^{(s)}$) turn out to be independent of the angular coordinates x. Moreover, they are dependent on the hyperplanes $\langle y, m\alpha+\beta\rangle = 0$. Since the coefficients $F_0^{(s)}$ are analytic and the vectors α and β are linearly independent, the functions $F_0^{(s)}$ are dependent everywhere on

$\mathbb{R}^n = \{y\}$. The points $y \in \mathbb{R}^n$ lying on the hyperplane $\langle y, m\alpha + \beta\rangle = 0$ correspond to resonant tori of the unperturbed integrable problem which are destroyed on the m-th step of perturbation theory.

Theorem 3. *Let α and β satisfy the conditions of Theorem 2. Suppose that the Hamiltonian system with Hamiltonian $H_0 + \varepsilon H_1$ has $n-1$ analytic integrals*

$$F_0^{(1)} + \varepsilon F_1^{(1)} + \dots, \;\dots,\; F_0^{(n-1)} + \varepsilon F_1^{(n-1)} + \dots,$$

where the functions $F_0^{(1)}, \dots, F_0^{(n-1)}$ are independent at some point of $\Gamma \times \mathbb{T}^n$, and Γ is the hyperplane $\langle \alpha, y\rangle = 0$. Then the Hamiltonian system has no integral independent of the functions

$$F_0^{(s)} + \varepsilon F_1^{(s)} + \dots, \qquad 1 \le s \le n-1,$$

in the form of a formal series $\sum F_r(y, x)\varepsilon^r$ with coefficients which are single-valued and analytic in $\mathbb{R}^n \times \mathbb{T}^n$.

One can deduce from Theorem 3 the following

Corollary 3. *Let the vectors $\alpha, \beta \in \Delta$ satisfy the conditions of Theorem 2 and Γ be the hyperplane $\langle \alpha, y\rangle = 0$. Suppose that the system with Hamiltonian $H_0 + H_1$ has $n-1$ polynomial integrals $F^{(1)}, \dots, F^{(n-1)}$ whose leading homogeneous forms are independent at some point of $\Gamma \times \mathbb{T}^n \subset \mathbb{R}^n \times \mathbb{T}^n$. Then the Hamiltonian system has no additional integral, independent of the functions $F^{(1)}, \dots, F^{(n-1)}$.*

The remainder of this section contains the proof of Theorems 1–3.

5.3 Basic Lemma. Let the vertices α and β of the set Δ satisfy condition (5.3). Then the set $\mathbf{P}^k$ contains the hyperplane $\langle k\alpha + \beta, y\rangle = 0$. In particular the secular set $\mathbf{P}^\infty$ consists of an infinite number of distinct hyperplanes and its closure contains the hyperplane $\langle \alpha, y\rangle = 0$.

Let

$$S = S_0 + \varepsilon S_1 + \varepsilon^2 S_2 + \cdots$$

be the generating function of a canonical transformation of perturbation theory (see §3). We put

$$S_m = \sum{}' S_m^\tau(y) e^{i(\tau, x)}, \qquad \tau \in \mathbb{Z}^n, \quad \tau \neq 0.$$

According to (4.5),

$$S_1^\tau = \frac{ih_\tau}{(\omega, \tau)}, \qquad \tau \neq 0. \tag{5.4}$$

The Fourier coefficients S^{τ}_m, $m = 2, 3, \dots$ can be found with the help of the following inductive formula:

$$S^{\tau}_m = \frac{1}{2i(\omega, \tau)} \sum_{u+v=m, \sigma+\delta=\tau} \langle \sigma, \delta \rangle S^{\sigma}_u S^{\delta}_v \,. \tag{5.5}$$

This formula is a consequence of relations (4.4). It is clear that S^{τ}_r can be represented as fractions whose denominators are products of expressions of the form (ω, τ).

The definition of lexicographic order implies inequalities $\alpha \succ 0$ and $\alpha \succ \gamma$ for all $\gamma \in \Delta \setminus \{\alpha\}$.

Lemma 1. *The functions $S^{\tau}_r \equiv 0$ for all $\tau \succ r\alpha$.*

The proof is performed using induction in r. For $r = 1$ the lemma is a consequence of relation (5.4) and the definition of the vertex α. Suppose that the lemma holds for all $r \leq m$. The function S^{τ}_{r+1} is calculated with the help of the equality (5.5). Let $\tau \succ (r+1)\alpha$. Now we show that there is a multiplier S^{τ}_w with $\tau \succ w\alpha$, $w \leq r$ in any term of the right-hand side of (5.5). Indeed, if $\sigma \preceq u\alpha$ and $\delta \preceq v\alpha$, then $\sigma + \delta \preceq (u+v)\alpha = (r+1)\alpha \prec \tau$. But this inequality contradicts the condition of the summation $\sigma + \delta = \tau$. The multipliers S^{τ}_w vanish due to the assumption of the induction. The lemma is proved.

Lemma 2.

$$S^{m\alpha}_m = \frac{\langle \alpha, \alpha \rangle}{2im(\omega, \alpha)} \sum_{u+v=m} uv S^{u\alpha}_u S^{v\alpha}_v \,. \tag{5.6}$$

Proof. The expression (5.6) is deduced from (5.5) by putting $\tau = m\alpha$. We consider only non-zero terms in the right-hand side of (5.5). According to Lemma 1 the relations $\sigma \preceq u\alpha$, $\delta \preceq v\alpha$, and $\sigma + \delta = m\alpha = (u+v)\alpha$ hold. Consequently $\sigma = u\alpha$ and $\delta = v\alpha$. The lemma is proved.

Lemma 3.

$$S^{m\alpha}_m = K_m \left(\frac{\langle \alpha, \alpha \rangle}{i(\omega, \alpha)} \right)^{m-1} \left(S^{\alpha}_1 \right)^m \,, \tag{5.7}$$

where

$$K_1 = 1 \,, \quad K_m = \sum_{u+v=m} \frac{uvK_u K_v}{2m} \,. \tag{5.8}$$

We prove this lemma by induction in m. Formulas (5.7) and (5.4) coincide for $m = 1$. Suppose that Lemma 3 is valid for $m \leq r$. Then

$$S_{r+1}^{(r+1)\alpha} = \frac{\langle\alpha,\alpha\rangle \sum_{u+v=r+1} uvK_uK_v}{2i(r+1)(\omega,\alpha)} \left(\frac{\langle\alpha,\alpha\rangle}{i(\omega,\alpha)}\right)^{u+v-2} \left(S_1^\alpha\right)^{u+v}$$
$$= K_{r+1}\left(\frac{\langle\alpha,\alpha\rangle}{i(\omega,\alpha)}\right)^{r} \left(S_1^\alpha\right)^{r+1} .$$

Lemma 4. *Suppose that the vectors τ and α are linearly independent, and $(m+1)\alpha+\beta \prec \tau \prec m\alpha$. Then $S_m^\tau \equiv 0$.*

The validity of Lemma 4 for $m = 1$ follows from the definition of the vertices α and β. Suppose that it holds for all $m \leq r$. Use formula (5.5) for $m = r+1$. By the assumption of the induction and Lemma 1, the product $S_u^\sigma S_v^\delta$ may differ from zero only in the following cases:

1) the vectors α, σ, δ are pairwise linearly dependent,
2) $\sigma \preceq u\alpha$ and $\delta \preceq (v-1)\alpha+\beta$, or $\sigma \preceq (u-1)\alpha+\beta$ and $\delta \preceq v\alpha$.

In the first case the vector τ is obviously parallel to α, and in the second one the inequality $\tau = \sigma+\delta \prec (u+v-1)\alpha+\beta = r\alpha+\beta$ holds. The lemma is proved.

Lemma 5.

$$S_{m+1}^{m\alpha+\beta} = \frac{1}{i(\omega, m\alpha+\beta)} \sum_{u+v=m,\, u>0,\, v\geq 0} \langle u\alpha, v\alpha+\beta\rangle S_u^{u\alpha} S_{v+1}^{v\alpha+\beta} . \qquad (5.9)$$

Relation (5.9) is derived from (5.5) with the help of Lemma 4. First we note that either $\sigma \preceq (u-1)\alpha+\beta$ or $\delta \preceq (v-1)\alpha+\beta$ (otherwise the vectors σ, δ, and $\sigma+\delta$ are pairwise dependent). If $\sigma \prec u\alpha$ and $\delta \prec v\alpha$, then we obtain the following contradiction: $m\alpha+\beta = \sigma+\delta \prec (u+v)\alpha+\beta = m\alpha+\beta$. Consequently, according to Lemma 4, in formula (5.5) one must take into account only the following pairs of vectors σ and δ: 1) $\sigma = u\alpha$, $\delta = (v-1)\alpha+\beta$ and 2) $\sigma = (u-1)\alpha+\beta$, $\delta = v\alpha$. It remains to use the symmetry of formula (5.5) in σ and δ. The lemma is proved.

We transform relation (5.9) as follows:

$$i(\omega, m\alpha+\beta) S_{m+1}^{m\alpha+\beta} =$$
$$\sum_{u+v=m,\, u>0,\, v\geq 0} \frac{\langle\alpha, v\alpha+\beta\rangle}{i(\omega, v\alpha+\beta)} u S_u^{u\alpha} i(\omega, v\alpha+\beta) S_{v+1}^{v\alpha+\beta} . \qquad (5.10)$$

Introduce the following notations:

$$\lambda_m = mS_m^{m\alpha}, \quad \mu_{m+1} = i(\omega, v\alpha+\beta) S_{m+1}^{m\alpha+\beta}, \quad l_v = \frac{\langle\alpha, v\alpha+\beta\rangle}{i(\omega, v\alpha+\beta)} .$$

Then (5.10) can be written in the following form:

$$\mu_{m+1} = \sum_{u+v=m, u>0, v\geq 0} l_v \lambda_u \mu_{v+1} .$$

Lemma 6.

$$\mu_{m+1} = a_m \lambda_1^m \mu_1 , \tag{5.11}$$

where

$$a_0 = 1 , \quad a_m = \sum_{u+v=m, u>0, v\geq 0} u K_u h^{u-1} a_v l_v , \quad h = \frac{\langle \alpha, \alpha \rangle}{i(\omega, \alpha)} .$$

The proof is performed using (5.7) and induction in m.
We put $uK_u = r_u$. Then (5.8) yields:

$$r_1 = 1 , \quad r_m = \sum_{u+v=m, u>0, v>0} \frac{r_u r_v}{2} . \tag{5.12}$$

Using the new notation we obtain

$$a_m = \sum_{u+v=m, u>0, v\geq 0} r_u h^{u-1} a_v l_v . \tag{5.13}$$

Lemma 7.

$$1 - \sqrt{1-2z} = \sum_{n=1}^{\infty} r_n z^n .$$

Corollary.

$$r_m = (2m-3)!!/m!! , \quad m > 1 .$$

Proof of Lemma 7. The relation (5.12) implies that the power series

$$f(z) = \sum_{n=1}^{\infty} r_n z^n$$

satisfies the equation $f^2 - 2f + 2z = 0$. Since $f(0) = 0$, we have $f(z) = 1 - \sqrt{1-2z}$. The lemma is proved.

From (5.13) we obtain successively:

$$a_1 = r_1 l_0 , \quad a_2 = r_2 h l_0 + r_1^2 l_0 l_1 ,$$
$$a_3 = r_3 h^2 l_0 + r_1 r_2 h l_0 l_1 + r_1 r_2 h l_0 l_2 + r_1^3 l_0 l_1 l_2 , \ \ldots$$

Lemma 8. *For $m \geq 1$*

$$a_m = \sum_{0=j_0<j_1<\ldots<j_k<m} r_{j_1-j_0} r_{j_2-j_1} \cdots r_{m-j_k} h^{m-k-1} l_{j_0} l_{j_1} \cdots l_{j_k} \,. \tag{5.14}$$

Formula (5.14) can be derived from (5.13) by induction in m.

Now we analyze of the structure of the Poincaré set. Since the vectors α and β are linearly independent, the hyperplanes $(\omega, \alpha) = 0$ and $\Gamma_m = \{y : (\omega, m\alpha+\beta) = 0\}$ do not coincide. In accordance with Lemma 1, the functions S^{σ}_{τ} are analytic almost everywhere on Γ_m for $r < m + 1$. In order to understand whether the hyperplane is in the set $\mathbf{P}^{m+1}$, it is necessary to analyze the inequality $\mu_{m+1} \neq 0$. Now we use relation (5.11). In this relation $\lambda_1 = S^{\alpha}_1$, $\mu_1 = i(\omega, \beta) S^{\beta}_1$. According to (5.4) and the definition of the vertices α and β, the coefficients S^{α}_1 and S^{β}_1 do not vanish. The vectors α and β are not collinear, consequently $(\omega, \beta) \not\equiv 0$ on the hyperplane Γ_m. Thus, $\mu_{m+1} \neq 0$ if and only if $a_m \neq 0$. Consider two cases: $\langle \alpha, \alpha \rangle = 0$ and $\langle \alpha, \alpha \rangle \neq 0$. In the first case $h = 0$ and (according to Lemma 8) $a_m = l_0 l_1 \ldots l_{m-1}$. According to assumption (5.3), we have $\langle \alpha, \beta \rangle \neq 0$, consequently $l_s \neq 0$ for all s. As a result, we have $a_m \neq 0$. To handle the second case, we introduce the number $\lambda = \langle \alpha, \beta \rangle / \langle \alpha, \alpha \rangle$. At the points of the hyperplane Γ_m the equality $(\omega, \beta) = -m(\omega, \alpha)$ holds, hence

$$l_v = \frac{\lambda + v}{v - m} h \,.$$

Since in the case under consideration $h \neq 0$, relation (5.14) implies that $a_m = 0$ if and only if λ is a root of the polynomial

$$P_m(x) = \sum_{0=j_0<j_1<\ldots<j_k<m} r_{j_1-j_0} \cdots r_{m-j_k} \frac{(x + j_0) \cdots (x + j_k)}{(j_0 - m) \cdots (j_k - m)} \,. \tag{5.15}$$

Lemma 9.

$$P_m = \frac{(-1)^m}{m!} x \left(x + \frac{1}{2}\right) \cdots \left(x + \frac{m-1}{2}\right) . \tag{5.16}$$

In order to prove Lemma 9, we consider the new polynomials

$$Q_n(y) = \left. \frac{P_n(x)}{-x} \right|_{x=y-n} , \quad Q_0 = -\frac{1}{y} \,. \tag{5.17}$$

Lemma 10. *The following recurrent relation holds:*

$$mQ_m = \sum_{u+v=m,\, u>0,\, v\geq 0} r_u (v - y) Q_v \,. \tag{5.18}$$

Proof. We make the change $m - j_l = i_{k-l+1}$ in (5.15). As a result we obtain:

$$P_m(x) = -\frac{x}{m} r_m + \sum_{0<i_1<\ldots<i_k<m} r_{m-i_k} r_{i_k - i_{k-1}} \cdots r_{i_1} \frac{x}{-m} \frac{x+m-i_k}{-i_k} \cdots \frac{x+m-i_1}{-i_1} .$$

Summing up with respect to i_k separately, we obtain:

$$P_m(x) = \frac{x}{-m} \sum_{i_k=0}^{m-1} P_{i_k}(x) r_{m-i_k} , \quad P_0 \equiv 1 .$$

This relation can be rewritten in the following form:

$$m \frac{P_m}{x} = \sum_{k=0}^{m-1} (k - m - x) \frac{P_k}{x+m-k} r_{m-k} .$$

We put $x + m = y$ and $P_n = (n - y) Q_n$, and obtain the equality

$$m Q_m(y) = \sum_{k=0}^{m-1} (k - y) r_{m-k} Q_k(y) , \quad Q_0 = -\frac{1}{y},$$

which is equivalent to (5.18).

Lemma 11. *The following identity holds*

$$\sum_{n=0}^{\infty} Q_n z^n = -\frac{1}{y} \left(\frac{1+\sqrt{1-2z}}{2} \right)^{2y} . \tag{5.19}$$

Proof. We put

$$g(z) = \sum_{n=0}^{\infty} Q_n z^n .$$

Relation (5.18) yields the following differential equation for the function g:

$$z \frac{dg}{dz} = \left(z \frac{dg}{dz} - y g \right) f .$$

Here f is the function defined in Lemma 7. We solve this differential equation with the initial condition $g(0) = -1/y$ and obtain:

$$g(z) = -\frac{1}{y} \left(\frac{1+\sqrt{1-2z}}{2} \right)^{2y} .$$

The lemma is proved.

Function (5.19) is analytic for small z. To find its Maclaurin series, we put $g(z) = F(\phi^{-1}(z))$, where

$$F(z) = -\frac{1}{y}\left(\frac{1+z}{2}\right)^{2y}, \quad \phi = \frac{1-z^2}{2}.$$

Since $\phi'(1) \neq 0$, we can use the Bürmann–Lagrange theorem [102]:

$$g(z) = g(0) + \sum_{m=1}^{\infty} \frac{z^m}{m!} \frac{d^{m-1}}{dz^{m-1}}\bigg|_{z=1} [F'(z)\Psi^m(z)], \tag{5.20}$$

where

$$\Psi = \frac{z-1}{\phi(z)} = -\frac{2}{1+z}.$$

Using formulas (5.19) and (5.20), we get:

$$m!Q^m = \left(\frac{2m-1}{2} - y\right)\left(\frac{2m-2}{2} - y\right)\cdots\left(\frac{m+1}{2} - y\right).$$

Returning to the old variable x, we use the relation (5.17) and obtain relation (5.16) for the polynomial $P_m(x)$. Lemma 9 is proved.

Now we continue analyzing the Poincaré set. Lemma 9 shows us that $a_m \equiv 0$ on the hyperplane Γ_m if and only if $\lambda = \langle\alpha, \beta\rangle / \langle\alpha, \alpha\rangle$ coincides with one of the following numbers: $0, -1/2, -1, \ldots, -(m-1)/2$. According to the assumption (5.3), $\lambda \neq -m/2$ for all integer $m \geq 0$, consequently the hyperplane $\Gamma_m = \{y : \langle y, m\alpha + \beta\rangle = 0\}$ is contained in the set $\mathbf{P}^{m+1} \subset \mathbf{P}^\infty$. The hyperplanes Γ_m accumulate as $m \to \infty$ to the limit plane $\langle y, \alpha\rangle = 0$. The proof of the Basic Lemma is completed.

5.4 Now we prove Theorems 2 and 3. Let n analytic functions

$$F^{(k)} = \sum_{s=0}^{\infty} F_s^{(k)}(x, y)\varepsilon^s, \qquad 1 \leq k \leq n \tag{5.21}$$

be first integrals of the Hamiltonian system with Hamiltonian $H_0 + \varepsilon H_1$. All functions $F_s^{(k)}$ are 2π-periodic in the variables $x_1, \ldots, x_n$.

Lemma 12. *The functions $F_0^{(1)}, \ldots, F_0^{(n)}$ do not depend on the angular variables $x_1, \ldots, x_n$ and their Jacobian*

$$\frac{\partial(F_0^{(1)}, \ldots, F_0^{(n)})}{\partial(y_1, \ldots, y_n)} \tag{5.22}$$

vanishes on the set $\mathbf{P}^\infty$.

Theorem 2 follows immediately from the Basic Lemma and Lemma 12. Indeed, since the set $\mathbf{P}^\infty \subset \mathbb{R}^n$ consists of an infinite number of distinct hyperplanes passing through the origin, $\mathbf{P}^{\infty}$ is a key set for the class of analytic functions in $\mathbb{R}^n$. Consequently, by Lemma 12, the analytic function (5.22) vanishes identically. This, in turn, implies the dependence of integrals (5.21) for $\varepsilon = 0$.

The proof of Theorem 3 uses one more auxiliary construction by Poincaré [203, § 81].

Lemma 13. *Suppose that the functions $F_0^{(1)}, \dots, F_0^{(n-1)}$ are independent at some point $y_0 \in \Gamma$, and the Hamiltonian system has an additional formal integral $F = \sum F_s(x, y)\varepsilon^s$ independent of the functions $F^{(1)}, \dots, F^{(n-1)}$. Then there exists a neighborhood V of the point x_0 in $\mathbb{R}^n = \{y\}$ and a formal integral*

$$\Phi = \sum_{s=0}^{\infty} \Phi_s(y, x)\varepsilon^s$$

with coefficients analytic in $V \times \mathbb{T}^n$ such that the functions $F_0^{(1)}, \dots, F_0^{(n-1)}$ and Φ_0 are independent in $V \times \mathbb{T}^n$.

According to Lemma 12 the functions $F_0^{(1)}, \dots, F_0^{(n-1)}$ and Φ_0 depend only on the variables y. Here we do not give the proof of Lemma 13 since it repeats Poincaré's arguments presented in §1.1. The proof of Theorem 3 follows from Lemma 12, the Basic Lemma from §5.3, and the fact that the intersection $\mathbf{P}^{\infty} \cap V$ is a key set for the class of functions which are analytic in the domain V.

Lemma 12 generalizes the well-known Poincaré assertion on the dependence of the functions $F_0^{(1)}, \dots, F_0^{(n)}$ on the set $\mathbf{P}^1$ (see §1). The first part of the lemma on the dependence of the functions $F_0^{(s)}$ on the angular coordinates x has been already proved in §1.1. The second part is deduced from Theorem 1 of §10, Chap. II: if the Jacobian (5.22) does not vanish on the domain $D \subset \mathbb{R}^n$, then there exists a generating function

$$S = \sum_{r \geq 0} S_r(u, x)\varepsilon^r$$

of the classical perturbation theory, whose coefficients are analytic in the direct product $D \times \mathbb{T}^n$.

This enables us to derive Lemma 12 as follows. If the Jacobian (5.22) is non-zero at some point $y_0 \in \mathbf{P}^{\infty}$, then it is also non-zero in some neighborhood V of this point. Consequently, in $V \times \mathbb{T}^n$ we can (at least formally) construct a perturbation series in powers of ε with analytic coefficients. Then, by the definition of the Poincaré set $\mathbf{P}^{\infty}$, at least one of the functions S_r $(r = 1, 2, \dots)$ is not analytic at the points of $\{y_0\} \times \mathbb{T}^n \subset V \times \mathbb{T}^n$.

5.5 Now we prove Theorem 1. Perform the canonical transformation $x, y \to x', y'$ according to the formulas $y' = (B^T)^{-1}y$, $x' = Bx$, where B is an integer unimodular matrix. In the new variables the Hamiltonian $H_0 + H_1$ has an analogous form, and the set Δ is mapped to the set $\Delta' = \{m'\}$, $m' = (B^T)^{-1}m$. Since the integer vectors m are transformed in the same way as the momenta y, the validity of integrability condition (5.3) can be checked in the initial variables. Indeed, let a and b be vectors of $\mathbb{Z}^n$, a' and b' their images under the map $m \to (B^T)^{-1}m$. Then

$$\langle a', b'\rangle' = (BAB^T a', b') = (Aa, b) = \langle a, b\rangle \ .$$

First, we prove Corollary 2 of Theorem 1. Let $\mathscr{E}(\delta)$ be the convex hull of the set Δ. It is a convex polyhedron in $\mathbb{R}^n$.

Lemma 14. *Let α be a vertex of the polyhedron $\mathscr{E}(\delta)$, Γ an edge which is adjacent to it, and β the closest point to α of the set $\Delta \cap \Gamma$. Then there exists an integer unimodular matrix B such that under the transformation $m \to m' = (D^T)^{-1}m$ the points α and β are mapped to the vertices of the set Δ'.*

The proof is based on the recursive application of the following well-known algebraic statement: for any integer vector $k_1 = (k_1^1, \dots, k_1^m)$ with relatively prime coordinates there exist $m - 1$ integer vectors $k_2, \dots, k_m$ such that $\det \|k_j^l\| = \pm 1$. Let l be the greatest common divisor of the components of the vector $\alpha - \beta$, and B_1 an integer unimodular $n \times n$ matrix, whose bottom line consists of the components of the vector $(\alpha - \beta)/l$. The vector $(\alpha - \beta)/l$ is mapped to the vector $e_n = (0, \dots, 0, 1)^T$ under the transformation

$$m \to f(m) = (B_1^T)^{-1} m \ .$$

Now project the convex polyhedron $\mathscr{E}(\delta')$, $\Delta' = f(\Delta)$ to the hyperplane spanned by the basis vectors $e_1, \dots, e_{n-1}$. The edge $\Gamma' = f(\Gamma)$ is projected into a vertex of the obtained convex polyhedron. Consider an edge Λ adjacent to this vertex. With the help of a proper unimodular transformation the edge Λ can be made parallel to the $(n-1)$ coordinate axis. The matrix of this transformation may be assumed to have the form $\begin{Vmatrix} B_2 & 0 \\ 0 & 1 \end{Vmatrix}$. Repeat this operation $n-2$ more times. It is easy to check that the matrix

$$B_1 \begin{Vmatrix} B_2 & 0 \\ 0 & 1 \end{Vmatrix} \cdots \begin{Vmatrix} B_n & 0 & \cdots & 0 \\ 0 & 1 & \cdots & 0 \\ \vdots & \vdots & \ddots & \vdots \\ 0 & 0 & \cdots & 1 \end{Vmatrix}$$

is as desired. The lemma is proved.

Lemma 15. *Let α and β be neighboring vertices of the polyhedron $\mathscr{E}(\delta)$. If the Hamiltonian system is completely integrable, then the angle between the vectors α and β is not less than $\pi/2$.*

This assertion is a direct consequence of Lemma 14 and Theorem 2.

Lemma 16. *Suppose that a convex polyhedron in $(\mathbb{R}^n, \langle\ ,\ \rangle)$ is symmetric in the origin, and the angle between the position vectors of any two neighboring vertices is not less than $\pi/2$. Then this polyhedron is a rhomboid.*

We perform the proof using induction in the dimension of the polyhedron P. The lemma is obviously valid for $\dim M = 1$. Assume that Lemma 16 holds for $\dim M \leq m$. Let α be one of the vertices of an $(m+1)$-dimensional polyhedron, and Π_α the closed half-space in $\mathbb{R}^{m+1}$ whose boundary $\partial\Pi_\alpha$ passes through the origin and is orthogonal to the vector α. We also assume that the vector α does not belong to Π_α. By the condition of the lemma, all vertices of M, joined by edges with α, lie in Π_α. Actually all vertices of M except α are in Π_α. Indeed, suppose that there exists a vertex β which does not belong to Π_α. The convex polyhedron M is the union of the convex hull M_α of all vertices except α, and the convex hull R_α of one-dimensional edges of M which are adjacent to α. Obviously, the vertex β does not lie in R_α. The line segment Γ joining α and β belongs to the convex polyhedron M. The sets Γ and R_α have only one common point (the point α), since otherwise $\Gamma \subset R_\alpha$, and hence the point β can-not be a vertex of M. On the other hand, the line segment Γ does not lie completely in M_α, otherwise $M = M_\alpha$. We get a contradiction.

In a similar way, all vertices of M except $-\alpha$ are in $\Pi_{-\alpha}$. Thus, M is the convex hull of the points α and $-\alpha$, and of the remaining vertices of the polyhedron M which lie in $\partial\Pi_\alpha$. This hull is a rhomboid according to the assumption of the induction. The lemma is proved.

Now turn to the proof of Theorem 1.

Lemma 17. *Let P be a hyperplane in $\mathbb{R}^n$, and let points of the set $(P \cap \mathbb{Z}^n) \subset \mathbb{R}^n$ form a subgroup of $\mathbb{Z}$ of rank $n-1$. Then there exists an integer unimodular matrix B whose last $n-1$ columns (or rows) are vectors of $P \cap \mathbb{Z}^n$.*

The proof can be easily derived from well-known results on the construction of subgroups in $\mathbb{Z}^n$ (see, for example, [39, Chap. 7]).

Let α be one of the vertices of the rhomboid $\mathscr{E}(\delta)$, and Π_α the closed half-space introduced in the proof of Lemma 16. The intersection $\partial\Pi_\alpha \cap \mathbb{Z}^n$ is a subgroup of $\mathbb{Z}^n$ of rank $\dim \mathscr{E}(\delta) - 1$. Consider (if necessary) the complement of this subgroup to a subgroup of the rank $n-1$ which is formed by the intersection of a hyperplane in $\mathbb{R}^n$ with the set $\mathbb{Z}^n$ and does not contain the vector α. According to Lemma 17 there exists a matrix B whose last $n-1$ rows are vectors of this subgroup, and the first row (a vector of $\mathbb{Z}^n$) has a positive projection to α in the metric $\langle\, ,\,\rangle$. After the canonical change of coordinates $x \to Bx$, $y \to (B^T)^{-1}y$ we have:

a) the first coordinate of each vector $\tau \in \partial\Pi_\alpha \cap \mathbb{Z}^n$ vanishes,

b) the first coordinate of the vector α is positive,

c) the vector α is maximal element of Δ (with respect to the standard order relation $\prec$ in $\mathbb{Z}^n$).

d) if the vector τ is not in Π_α, then $0 \prec \tau$.

Lemma 18. *If the system with Hamiltonian $H_0 + \varepsilon H_1$ is completely integrable, then all points of the set $\Delta \setminus \Pi_\alpha$ lie on the line segment Γ which joins the points 0 and α.*

Assume the contrary. The vector α is the vertex of the set Δ due to the property c). Let β be the vertex of Δ adjacent to α. In accordance with our assumptions and the definition of the adjacent vertex, the vector β does not lie either in the half-space Π_α or on the line segment Γ. Since the vectors α and β are in one half-space $\mathbb{R}^n \setminus \Pi_\alpha$, the scalar product $\langle \alpha, \beta \rangle$ is positive. Consequently, condition (5.3) holds and, according to Theorem 2, the Hamiltonian system is nonintegrable. The obtained contradiction proves the lemma.

Applying Lemma 18 to all vertices of the rhomboid $\mathscr{E}(\delta)$, we prove the validity of Theorem 1.

6 Some Generalizations

6.1 The condition for the quadratic form $H_0 = (Ax, x)/2$ to be non-degenerate can be replaced by the weaker conditions:
i) $Am \neq 0$ for all integer vectors $m \neq 0$,
ii) the vectors $A\alpha$ and $A\beta$ are linearly independent.

Note that if $\det A = 0$, conditions i)–ii) can be satisfied simultaneously only for $n \geq 3$.

We give a simple example. If

$$A = \begin{Vmatrix} 0 & 1 & \sqrt{2} \\ 1 & 0 & 0 \\ \sqrt{2} & 0 & 0 \end{Vmatrix}, \qquad \alpha = (1, 0, 0)^T, \quad \beta = (1, -1, 0)^T,$$

then the matrix A is degenerate, but conditions i) and ii) are satisfied.

6.2 Note that Theorem 2 does not hold when the Fourier coefficients of the perturbing function H_1 depend on y. An instructive counter-example is

$$H = a^2 y_1^2 + ab y_1 y_2 + b^2 y_2^2 + \frac{\varepsilon}{a y_1 - b y_2} (\sin x_1 - \sin x_2) .$$

The Hamiltonian system with this Hamiltonian can be integrated by separation of variables: the analytic functions

$$F_1 = a^3 y_1^3 - a y_1 H + \varepsilon \sin x_1 , \quad F_2 = b^3 y_2^3 - b y_2 H + \varepsilon \sin x_2$$

form a complete set of independent integrals.

In this problem $\alpha = (1, 0)^T$, $\beta = (0, 1)^T$ so that inequality (5.3) takes the form: $b/a \neq -m/2$ for all integer $m \geq 0$. The "limiting" line $\langle \alpha, y \rangle = 2a y_1 + b y_2 = 0$ is not the same as the line $a y_1 - b y_2 = 0$ at points of which the Hamiltonian is not defined (cf. Theorem 3). However, integrability occurs for all (including irrational) values of the ratio b/a.

Let $\alpha', \alpha'', \ldots$ be elements of the set Δ located between vertices α and β (under the lexicographic ordering $\prec$ in $\mathbb{Z}^n$). Clearly, the vectors $\alpha', \alpha'', \ldots$ are parallel to α. Modifying the reasoning of §5, one can show that Theorems 2 and 3 hold when the coefficients $h_\alpha, h'_\alpha, h''_\alpha, \ldots, h_\beta$ are constant (then the remaining Fourier coefficients can be non-constant analytic functions of the variables $x_1, \ldots, x_n$).

6.3 If the perturbing function H_1 is not a trigonometric polynomial, then the existence problem for complementary integrals of a Hamiltonian system becomes much simpler: as a rule, the nonintegrability of the perturbed system can be established after a finite number of steps of perturbation theory.

Consider for definiteness the case of two degrees of freedom. Let $H = H_0 + \varepsilon H_1$, where

$$H_0 = \frac{1}{2}\sum_{j,k=1}^{2} a_{jk} y_j y_k , \quad H_1 = \sum_{m\in\mathbb{Z}^2} h_m e^{i(m,x)} , \qquad h_m = \text{const} .$$

Theorem 1. *Let $n = 2$ and let the Poincaré set $\mathbf{P}^1$ consist of just two straight lines. Then the Hamiltonian equations have a complementary formal integral if and only if these lines are orthogonal (in the metric $\langle\ ,\ \rangle$).*

Proof of sufficiency. Since $\mathbf{P}^1$ consists of two lines, we have $\tau = \lambda\tau_0$ and $\sigma = \mu\sigma_0$ in (4.6), where $\tau_0, \sigma_0 \in \mathbb{Z}^2$; λ and μ are integers. Vectors τ_0 and σ_0 are orthogonal, since the lines which form the set $\mathbf{P}^1$ are orthogonal. We show that the Hamiltonian system can be integrated by separation of variables. Indeed, let $\tau_0 = (\tau_1, \tau_2)$, $\sigma_0 = (\sigma_1, \sigma_2)$. We put

$$X_1 = (\tau_1 x_1 + \tau_2 x_2) , \quad X_2 = (\sigma_1 x_1 + \sigma_2 x_2) .$$

This homogeneous transformation of the angle coordinates is uniquely continued up to a homogeneous canonical transformation $x, y \to X, Y$. In the new variables

$$H = \frac{1}{2}(A_{11}Y_1^2 + 2A_{12}Y_1Y_2 + A_{22}Y_2^2) + \varepsilon(f(X_1) + g(X_2)) ,$$

where $A_{ij} = \text{const}$, and f, g are analytic 2π-periodic functions. Since $\langle\tau_0, \sigma_0\rangle = 0$, $A_{12} = 0$. Consequently the Hamiltonian system has two integrals linear in ε:

$$\frac{1}{2}A_{11}Y_1^2 + \varepsilon f(X_1) , \quad \frac{1}{2}A_{22}Y_2^2 + \varepsilon g(X_2) .$$

Proof of necessity. First, consider the case when the perturbing function H_1 is a trigonometric polynomial. The vectors $\alpha = \pm\lambda_*\tau_0$ and $\beta = \pm\mu_*\sigma_0$ (λ_* and μ_* are some positive numbers) can be taken as vertices of the set Δ. Let $\langle\alpha, \alpha\rangle \geq 0$ (the case $\langle\alpha, \alpha\rangle \leq 0$ can be considered in the same way). If $\langle\alpha, \beta\rangle \neq 0$, then without loss of generality we can assume $\langle\alpha, \beta\rangle > 0$ (otherwise replace β by $-\beta$). Thus, for all integer $m \geq 0$ the inequality $m\langle\alpha, \alpha\rangle + 2\langle\alpha, \beta\rangle > 0$ holds. Hence, if the Hamiltonian system has an additional integral, then (by Theorem 2, §5) $\langle\alpha, \beta\rangle = 0$. The last condition is obviously equivalent to the condition $\langle\tau_0, \sigma_0\rangle = 0$.

Now take the case when the function H_1 is not a polynomial. We use the results of §4. Since τ_0 and σ_0 are linearly independent, the numbers λ and μ are uniquely defined for any fixed $k = \lambda\tau_0 + \mu\sigma_0$. By the assumption, the function H_1 is not a polynomial; hence, among the numbers λ and μ infinitely many are distinct. If $\langle\tau_0, \sigma_0\rangle \neq 0$, it follows from (4.6) that $\mathbf{P}^2$ consists of an infinite number of distinct lines so that it is a key set for the class of functions analytic in $\mathbb{R}^2 = \{y_1, y_2\}$. Using Theorem 1, §4 we complete the proof of absence of a formal integral.

6.4 Theorem 1 can be extended (with certain refinements) to systems with $n > 2$ degrees of freedom. Assume that all the points of the set Δ lie on $l \leq n$ straight lines passing through the origin, their directing vectors being linearly independent. We can then claim that the Hamiltonian system with Hamiltonian $H_0 + \varepsilon H_1$ has n single-valued analytic integrals, which are independent for all sufficiently small ε, if and only if these l lines are pairwise orthogonal (in the metric $\langle\ ,\ \rangle$). The system is obviously integrable for $l = 1$.

As an example consider the system with the Hamiltonian

$$H = \frac{1}{2}\sum_{s=1}^{n} y_s^2 + \varepsilon[f(x_1 - x_2) + \cdots + f(x_{n-1} - x_n)]\,, \tag{6.1}$$

where f is a real analytic 2π-periodic function. This system describes the dynamics of an "aperiodic" chain of n particles on a straight line. Apart from the energy integral, the equations of motion have the following integral $y_1 + \cdots + y_n$: the total momentum of the system of interacting particles is conserved. It turns out that if $n > 2$ and $f \neq \text{const}$, then the system with Hamiltonian (6.1) does not have a complete set of independent integrals. Indeed, in this case $l = n - 1$ and the corresponding lines are defined by the vectors $(1, -1, 0, \ldots, 0)^T$ and $(0, \ldots, 0, 1, -1)^T$ which are not all pairwise orthogonal. If we "close" the chain by adding to Hamiltonian (6.1) the term $\varepsilon f(x_n - x_1)$, our above proposition is no longer applicable: $l = n$ lines are located in the hyperplane orthogonal to the vector $(1, 1, \ldots, 1)^T$. The problem of complete integrability of the "periodic" chain is considered in the next section.

6.5 Apparently Theorem 1 is valid in the case of a pseudo-Euclidean metric $\langle\ ,\ \rangle$ too. At least Corollary 2 of Theorem 1 holds. One can prove this using the arguments presented in §5.5. For $n = 2$ we have only one type of pseudo-Euclidean space. Here as usual we define a rhombus as a parallelogram with orthogonal diagonals. When neighboring vertices come close to one another, the rhombus degenerates into a line segment, situated on one of the isotropic straight lines.

6.6 In the case of two degrees of freedom the results of §5 can be strengthened. Indeed, there holds

Theorem 2. *The Hamiltonian system with Hamiltonian $H_0 + \varepsilon H_1$ (the form H_0 is positive definite) admits a nontrivial symmetry field*

$$u_\varepsilon = u^0(x, y) + \varepsilon u^1(x, y) + \cdots$$

with analytic coefficients u_r ($r \geq 0$), 2π*-periodic in* x_1 *and* x_2*, if and only if the points of the set* Δ *lie on one or two orthogonal straight lines, passing through the origin.*

In particular, validity of the conditions of Theorem 2 is equivalent to the existence for the Hamiltonian equations of an additional integral in the form of a series in ε with multivalued coefficients. The analogous statement holds for integrals of a reversible system with Hamiltonian $H_0 + H_1$, which are polynomials in the momenta y_1, y_2 with coefficients which are multivalued on the configuration space $\mathbb{T}^2 = \{x \bmod 2\pi\}$.

Theorem 2 can be proved by the method of §3 with the help of the results of §5 on the construction of the Poincaré set.

7 Systems of Interacting Particles

7.1 The dynamical behavior of n like interacting particles on a straight line is described by the Hamiltonian system with Hamiltonian

$$H = \frac{1}{2}\sum y_s^2 + \sum_{i<j} f(x_i - x_j) , \tag{7.1}$$

where $f(\cdot)$ is an even function (the potential of paired interactions). Moser [185] and Calogero [50] showed that the system with Hamiltonian (7.1) is completely integrable if f is the Weierstrass $\wp$-function (or its degenerate cases z^{-2}, $\sin^{-2} z$, $\sinh^{-2} z$). It was shown in [201] that this is the only case when there exists a polynomial integral of third degree which is independent of the integrals H and $P = \sum y_s$.

We shall consider analytic 2π-periodic potentials f. An example is the system of three points on a circle connected pairwise by elastic springs.

Theorem 1 [145]. *If* $f \neq$ const *and* $n > 2$*, then the system with Hamiltonian* (7.1) *does not have a complete set of n first integrals which are polynomial in the momenta with independent leading homogeneous forms.*

Remark. For $n = 3$ there is no supplementary integral which is a polynomial in the momenta and independent of the functions H and P.

It is worth emphasizing that the Weierstrass $\wp$-function has poles on the real axis.

7.2 Here is a sketch of the proof of Theorem 1 for the case $n = 3$. We pass to the inertial barycentric reference frame; in this frame the total momentum is zero.

The new canonical coordinates have the form

$$Y_1 = y_1 - y_2\,, \quad Y_2 = y_2 - y_3, \quad Y_3 = y_1 + y_2 + y_3\,,$$
$$X_1 = x_1 + x_3\,, \quad X_2 = -x_1 + x_2 + x_3\,, \quad X_3 = -x_2 + x_3\,.$$

We put $Y_3 = 0$, and perform the reduction to a system with two degrees of freedom whose Hamiltonian is

$$Y_1^2 - Y_1Y_2 + Y_2^2 + f(X_1) + f(X_2) + f(X_1 + X_2)\,. \tag{7.2}$$

First, we take the case when f is a trigonometric polynomial. Then the convex hull of Δ is a hexagon (see Fig. 14). In this case the absence of new integrals follows from Corollary 2 of Theorem 1, §5.

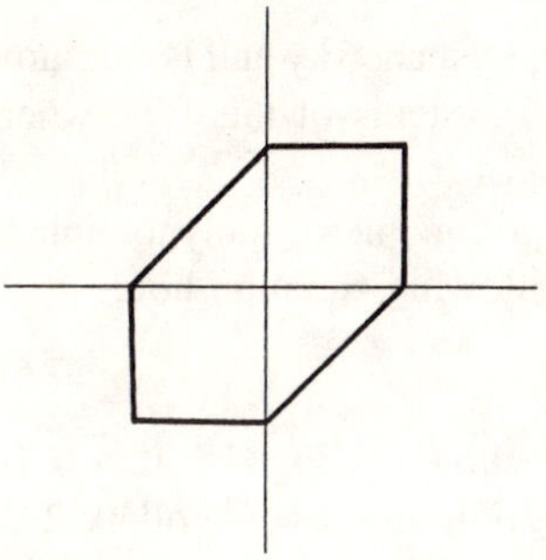

Fig. 14.

Now assume that f is not a polynomial. We use Theorem 1_2, §4. Let $k = (m, n) \in \mathbb{Z}^2$, where $m \neq 0$, $n \neq 0$ and $m \neq n$. The condition for the function h'_k to vanish on the line $\langle k, Y\rangle = 0$ can be written as

$$\frac{f_m}{m}\frac{f_n}{n} = \frac{\overline{f}_n}{n}\frac{f_{m-n}}{m-n} - \frac{\overline{f}_m}{m}\frac{\overline{f}_{m-n}}{m-n}\,. \tag{7.3}$$

Here f_s is the s-th Fourier coefficient of the function f; the bar denotes complex conjugation. For an even function f, obviously, $\overline{f}_s = f_s$.

Assume that $f_\lambda \neq 0$ for some $\lambda \neq 0$. If the Poincaré set $\mathbf{P}^2$ consists of a finite number of distinct straight lines, then equality (7.3) certainly holds for $n = \lambda$ and all sufficiently large m. We put $f_m/m = a_m$. Then

$$a_{m+\lambda} = \frac{a_m a_\lambda}{a_m + a_\lambda}\,. \tag{7.4}$$

Since f is not a polynomial, the coefficients $a_m \neq 0$ for some indefinitely large m. From (7.4) by induction we find

$$a_{m+s\lambda} = \frac{a_m a_\lambda}{s a_m + a_\lambda}\,. \tag{7.4'}$$

Since $a_m \neq 0$, we have

$$\lim_{s\to\infty} f_{m+s\lambda} = \lim_{s\to\infty} (m+s\lambda)a_{m+s\lambda} = f_\lambda \neq 0\,.$$

Consequently, the function f is not analytic (in fact, it is not even summable on the interval $[0, 2\pi]$). Hence, if $f \not\equiv$ const, the system with Hamiltonian (7.2) has no additional polynomial integral.

7.3 Using the results of §5 we can prove nonintegrability of some other familiar Hamiltonian systems. As an example, consider the system with two degrees of freedom with the Hamiltonian

$$H = \frac{1}{2}(y_1^2 + y_2^2) + \alpha[f(x_1 - x_2) + f(x_1 + x_2)] + \beta \sum f(x_i) + \gamma \sum f(2x_i)\,.$$

Here f is a periodic function; $\alpha, \beta, \gamma =$ const. In [194] Olshanetsky and Perelomov proved the complete integrability of multidimensional systems of this type when f is the Weierstrass $\mathcal{P}$-function (or a degenerate form of it).

First, we take the case when the potential f is a trigonometric polynomial. It turns out that integrability occurs if and only if the following equation holds

$$\alpha(\beta^2 + \gamma^2) = 0\,. \tag{7.5}$$

Indeed, for $\gamma \neq 0$, the convex hull of Δ is the square shown in Fig. 15. If $\alpha \neq 0$, the mid points of the sides of the square are points of Δ. In this case Corollary 2 of Theorem 1, §5 guarantees the absence of an additional integral, polynomial in the momenta, with periodic coefficients. Let $\gamma = 0$, while $\alpha\beta \neq 0$. Then the convex hull of Δ is the same as the inner square, see Fig. 15. If $\alpha \neq 0$, the mid-points of its sides belong to Δ and hence, Corollary 2 of Theorem 1, §5 is again applicable.

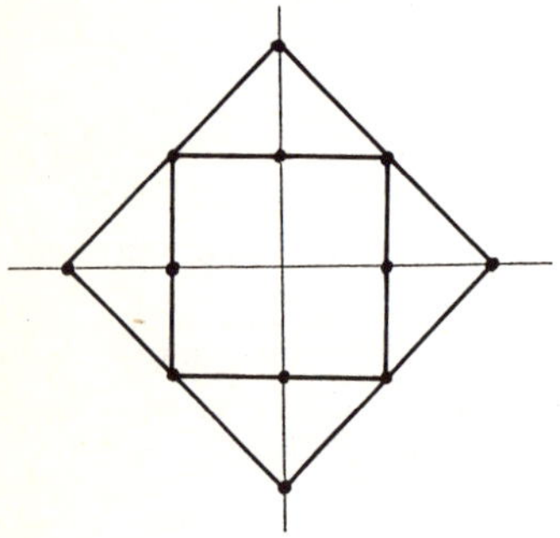

Fig. 15.

In the general case, when the potential is any even analytic function, the criterion for integrability is also equality (7.5). The proof is performed by the method of §2.

8 Birth of Isolated Periodic Solutions as an Obstacle to Integrability

8.1 Let the autonomous system

$$\dot{z} = v(z). \tag{8.1}$$

be defined on an $(m+1)$-dimensional manifold M^{m+1}, and let γ be a closed trajectory of a periodic solution of this system. We denote by $x \bmod 2\pi$ an angular coordinate on γ. The coordinate x is assumed to change uniformly in time: $\dot{x} = \omega = \text{const}$. The period equals $p = 2\pi/\omega$. A small neighborhood of γ in M^{m+1} is clearly diffeomorphic to the direct product of the circle $\mathbb{T}^1 = \{x \bmod 2\pi\}$ and some neighborhood of zero in m-dimensional space $\mathbb{R}^m = \{y\}$. The system (8.1), written in terms of the variables $x \bmod 2\pi, y_1, \ldots, y_m$, takes the following form:

$$\dot{x} = \omega + f(x, y)\,, \quad \dot{y} = Y(x, y)\,. \tag{8.2}$$

The right-hand sides of these equations are 2π-periodic in x, and $f(x, 0) = 0$, $Y(x, 0) = 0$. The periodic solution is defined by the simple relations

$$x = \omega t + x_0\,, \quad y = 0\,. \tag{8.3}$$

Introduce the square matrix of order m with periodic elements

$$\Omega(x) = \left.\frac{\partial Y}{\partial y}\right|_{y=0}\,,$$

and rewrite the system (8.2) in the more convenient form:

$$\dot{x} = \omega + f(x, y)\,, \quad \dot{y} = \Omega y + g(x, y)\,. \tag{8.4}$$

It is clear that $g = O(|y|^2)$.

According to the Lyapunov–Floquet theorem [64], one can reduce Ω to a constant matrix using a linear change of variables y with coefficients which are 2π-periodic in x. However, we shall not use this result henceforth.

The linear differential equations

$$\dot{u} = \Omega(x)u\,, \qquad x = \omega t + x_0 \tag{8.5}$$

are called variational equations of the periodic solution (8.3). Let $\Lambda(t)$ be a solution of the matrix equation

$$\dot{\Lambda} = \Omega(t)\Lambda \tag{8.6}$$

with the initial condition $\Lambda(0) = I$. The matrix $P = \Lambda(p)$ is called the monodromy matrix of the periodic solution (8.3), and the eigenvalues $\rho_1, \ldots, \rho_m$ of P are called multipliers of this solution. Let $\Lambda'(t)$ be a solution of (8.6) with the condition $\Lambda'(\tau) = I$. Then $\Lambda'(t) = \Lambda(t)\Lambda^{-1}(\tau)$, and $P_\tau = \Lambda'(\tau + p) = \Lambda(\tau + p)\Lambda^{-1}(\tau) = \Lambda(\tau)P\Lambda^{-1}(\tau)$. Consequently the eigenvalues of the matrices P_τ and P coincide. In particular, the multipliers of a periodic solution do not depend on x_0.

Since $\det P \neq 0$, we can put $P = \exp(pA)$. The eigenvalues $\alpha_1, \ldots, \alpha_m$ of the matrix A are called characteristic exponents of the periodic solution (8.3). They are connected with the multipliers by the relations $\rho_s = \exp(p\alpha_s)$ and are defined up to additive terms of the form $i\omega n$, $n \in \mathbb{Z}$.

A periodic solution is called *non-degenerate* if all its multipliers differ from 1.

Lemma 1. *In a small neighborhood of the trajectory γ of a non-degenerate periodic solution there is no periodic trajectory which does not coincide with γ and has a period close to that of γ.*

Indeed, consider an m-dimensional surface $\Pi = \{x, y : x = x_0\}$ which intersects γ transversely. Define the Poincaré map $F : \Pi \to \Pi$ in the following way. The trajectory of a solution of the system (8.2) with initial conditions $x(0) = x_0$, $y(0) = y_0$ ($|y_0|$ is small) intersects the surface Π at a point (x_0, y_1) after an interval of time which is close to p. We put $y_1 = F(y_0)$. It is clear that $F(0) = 0$ and

$$\partial F/\partial y|_{y=0} = P \, .$$

Periodic trajectories of the system (8.4) correspond to fixed points of the map F, but by the implicit function theorem the equations $F(y) = y$ have no nontrivial solutions for small $|y|$ due to the non-degeneracy of γ.

Remark. The multipliers of a periodic solution are often defined in a different way (see, e.g., [203, Chap. 3]). Let $z_0(t)$ be a p-periodic solution of the system (8.1). We put $z = z_0(t) + \delta z$ and linearize Eqs. (8.1) with respect to δz. As a result, we obtain a linear system with p-periodic coefficients:

$$(\delta z)^{\cdot} = A\delta z \, , \qquad A = \left.\frac{\partial v}{\partial z}\right|_{z_0(t)} .$$

Let $B(t)$ be the fundamental matrix of this system ($\dot{B} = AB$, $B(0) = I$), and $C = B(p)$. It is clear that $Cv_0 = v_0$, where $v_0 = v(z_0(0))$. Consequently, one of the eigenvalues of the matrix C is 1. It is easy to show that the other eigenvalues of C coincide with the multipliers of the periodic solution $z_0(\cdot)$.

8.2 The following theorem shows that if there exist first integrals independent at points of a periodic solution, then this solution is degenerate.

Theorem 1 (*Poincaré* [203]). *If Eqs.* (8.1) *admit k integrals which are independent at some point of the periodic trajectory γ, then at least k multipliers are equal to 1.*

Proof. Let $H(z)$ be an integral of the system (8.1). In the variables $x \bmod 2\pi$, y this function can be represented in the following form:

$$H = H_0(x) + (y, h(x)) + o(y) \, .$$

Here the function H_0 and the covector h are 2π-periodic in x. Differentiating this function along solutions of the system (8.4), we obtain:

$$\dot{H} = \frac{\partial H_0}{\partial x}\omega + \frac{\partial H_0}{\partial x} f + (\Omega y, h) + \left(y, \frac{\partial h}{\partial x}\right)\omega + o(y) = 0 \,.$$

Thus $H_0 = \text{const}$, and

$$\dot{h} = \frac{\partial h}{\partial x}\omega = -\Omega^T h \,. \tag{8.7}$$

The linear systems (8.5) and (8.7) are conjugate. Consequently,

$$(h(t), u(t)) = \text{const} \,. \tag{8.8}$$

Clearly, the function h is p-periodic in t, and $u(t+p) = Pu(t)$, where P is the monodromy matrix. According to (8.8),

$$(h(0), u(0)) = (h(p), u(p)) = (h(0), Pu(0)) = (P^T h(0), u(0)) \,.$$

Since $u(0)$ is an arbitrary vector, we have:

$$P^T h(0) = h(0) \,.$$

Therefore, $h(0)$ is an eigenvector of the matrix P with unit eigenvalue. According to the assumptions of the theorem, there are k such linearly independent vectors. Thus, at least k eigenvalues of the matrix P^T equal 1. It remains to note that the eigenvalues of the matrices P and P^T coincide.

Using Theorem 1, we may prove nonintegrability of dynamical systems in the following situation. Suppose that a set of non-degenerate periodic solutions of the analytic system (8.1) forms a key set for the class of analytic functions in M. Then the system (8.1) does not admit non-constant integrals which are analytic everywhere in M.

Theorem 2. *Suppose that the dynamical system* (8.1) *admits* $k+1$ *symmetry fields which are linearly independent at some point of a periodic trajectory* γ. *Then at least* k *multipliers equal 1.*

Indeed, let the system (8.4) admit a symmetry group which is the phase flow of the system

$$x' = V(x, y) \,, \quad y' = W(x, y) \,, \tag{8.9}$$

whose right-hand side is 2π-periodic in x. Calculate the commutator of the differential operators

$$(\omega + f)\frac{\partial}{\partial x} + (\Omega y + g)\frac{\partial}{\partial y} \quad \text{and} \quad V\frac{\partial}{\partial x} + W\frac{\partial}{\partial y} \,,$$

and put the coefficient of $\partial/\partial y$ equal to zero under the condition $y = 0$. As a result, we get the equality

$$\dot{w} = \frac{\partial w}{\partial x}\omega = \Omega w\ , \tag{8.10}$$

where $w(x) = W(x, 0)$. Equations (8.10) coincide with (8.5). Consequently

$$w(p) = Pw(0)\ .$$

Since the function $w(t)$ is p-periodic, $w(0)$ is an eigenvector of the monodromy matrix with eigenvalue 1.

If $w(0) = 0$, then the symmetry field is tangent to γ. Suppose that there exist $k + 1$ linearly independent symmetry fields. Then the matrix P has k linearly independent eigenvectors with eigenvalue 1. Hence, k multipliers of the periodic trajectory equal 1.

Corollary 1. *At all points of a non-degenerate periodic trajectory any symmetry field is linearly dependent on the field v.*

This assertion, indicated in [143], has a clear geometric meaning. Indeed, let γ be a non-degenerate closed trajectory. Then, according to Lemma 1 of §8.1, in a small neighborhood of γ the system (8.1) does not have other closed trajectories with close periods. If u is a symmetry field, then $g_u^\tau(\gamma)$ is a closed trajectory of the Eqs. (8.1). For small τ its period nearly coincides with the period of γ. Consequently, $g_u^\tau(\gamma) \equiv \gamma$ for all τ. Hence, vectors u and v are dependent at points of the trajectory γ.

Denote by Γ the union of all non-degenerate trajectories of the system (8.1).

Corollary 2. *Let M be a compact analytic manifold, v an analytic vector field on M. If Γ is a key set for the class $C^\omega(M)$, then any analytic symmetry field u of the system* (8.1) *is linearly dependent on v at all points of M. Moreover, if $v \neq 0$, then $u = \lambda v$, $\lambda =$* const.

Indeed, by Corollary 1 the vectors u and v are linearly dependent at all points of Γ. Now let Φ be an arbitrary analytic 2-form on M. Since $\Phi(u, v)$ is an analytic function on M which vanishes on Γ, and Γ is a key set, we have $\Phi(u, v) \equiv 0$. Now we use the following fact: given a 2-form Φ_0 at the point $z_0 \in M$, there exists an analytic differential form Φ_z on M coinciding with Φ_0 for $z = z_0$. This implies the dependence of the fields u and v at all points of M. The assumption that the form Φ_0 can be continued to the whole of M can be proved using the well-known fact that any compact analytic manifold M can be imbedded into the space $\mathbb{R}^k$. Let ϕ_0 be the restriction to $T_{z_0}M$ of the 2-form Φ_0, defined at the point $z_0 \in M \subset \mathbb{R}^k$. Evidently, the form ϕ_0 can be continued to an analytic form ϕ on all $\mathbb{R}^k$ (e.g., all coefficients of ϕ can be chosen constant). It remains to restrict the form ϕ to M.

If $v \neq 0$, then $u = \lambda(z)v$, where λ is an analytic function on M, and λ is an integral of (8.1) (see §3 of Chap. II). According to Theorem 1, $d\lambda = 0$ on the set Γ. Since Γ is a key set, $\lambda =$ const. Q.E.D.

Remark. Let M be a smooth manifold, v a smooth vector field on M. If the set Γ is dense on M, then any differentiable symmetry field u is linearly dependent on the field v at all points of M. Moreover, if $v \neq 0$, then $u = \lambda v$, $\lambda = \text{const}$. The proof is obvious.

As an example, consider a reversible mechanical system with compact con-ᵀuration space M, kinetic energy T and potential energy V. Let h be an energy ᴊnstant exceeding $\max_M V$. According to the Maupertuis principle, the projections to the manifold M of phase trajectories, which lie on the compact energy surface

$$\Sigma_h = \{x, \dot{x} : T + V = h\} ,$$

coincide with the geodesics of the metric

$$(dp)^2 = 2(h - V)T(dt)^2 .$$

Suppose that the Gauss curvature along any two-dimensional direction in the Riemann manifold (M, dp), is negative. Then the equations of motion on Σ_h generate an Anosov system [9]. We discussed this situation above in connection with the topological obstacles to integrability (§1 of Chap. III). The essential feature of Anosov systems is the exponential divergence of trajectories. In particular, all periodic trajectories are hyperbolic: the multipliers are real and differ from ± 1. It is known [9] that periodic trajectories are dense in Σ. Since hyperbolic trajectories are non-degenerate, this implies the absence of nontrivial differentiable symmetry fields defined on all Σ. In particular, the equations of motion do not admit multivalued integrals.

8.3 Suppose that the system (8.4) has k independent integrals $F_1, \ldots, F_k$ and l symmetry fields $u_1, \ldots, u_l$ linearly independent at points of a periodic trajectory γ. It follows from Theorems 1 and 2 that the characteristic equation $|P - \rho E| = 0$ has a root $\rho = 1$ of multiplicity $\geq \max(k, l - 1)$. Under some additional assumptions this estimate can be strengthened.

Theorem 3. *Suppose that the functions $F_1, \ldots, F_k$ are integrals of the dynamical systems*

$$z' = u_1 \ , \ \ldots \ , \ z' = u_l \ .$$

Then the root $\rho = 1$ of the characteristic polynomial $|P - \rho E|$ has multiplicity not less than $k + l - 1$.

Proof. We put

$$F_j = F_j^0(x) + (y, f_j(x)) + o(y) .$$

The function F_j^0 is constant (see §8.2). Represent the dynamical system $z' = u_r$ in the form (8.9):

$$x' = V_r(x, y) \ , \quad y' = W_r(x, y) \ . \tag{8.11}$$

Since the function F_j is a first integral of Eqs. (8.11), we have:

$$F'_j = (w_r, f_j) + O(y) ,$$

where $w_r(x) = W_r(x, 0)$. We put in this equality $y = 0$ and obtain the relations

$$(f_j, w_r) = 0 ; \qquad 1 \le j \le k , \quad 1 \le r \le l . \tag{8.12}$$

According to the results of §8.2,

$$Pw_r = w_r , \quad P^T f_j = f_j ,$$

where the covectors $\{f_j\}$ are linearly dependent, and in the set of vectors $\{w_r\}$ there are at least $l - 1$ linearly independent ones.

We reduce the monodromy matrix P to the Jordan form. It is clear that each linearly independent vector w defines a Jordan cell of the form

$$\begin{Vmatrix} 1 & 1 & 0 & 0 & 0 \\ 0 & 1 & 1 & 0 & 0 \\ & \ddots & \ddots & \ddots & \\ 0 & 0 & 0 & 1 & 1 \\ 0 & 0 & 0 & 0 & 1 \end{Vmatrix} .$$

Let the covector f correspond to the same Jordan cell. Then all except two components of w and f vanish. Let $w_\mu \ne 0$ and $f_\nu \ne 0$. Then the value $\nu - \mu + 1$ coincides with the dimension of the Jordan cell. In particular, $\nu \ge \mu$. According to (8.12), $\nu > \mu$. Thus, if the pair of vectors w, f corresponds to the same Jordan cell, then its dimension ≥ 2. Consequently, the multiplicity of the corresponding root $\rho = 1$ of the characteristic equation $|P - \rho E| = 0$ is not less than two. As a result, we conclude that the multiplicity of the unit multiplier is not less than $k + l - 1$. Q.E.D.

8.4 Now consider the case of Hamiltonian systems. Let γ be a closed trajectory of an autonomous Hamiltonian system with Hamiltonian H. Since $dH \ne 0$ at the points of γ, by Theorem 1 one of the multipliers equals 1. Thus, the periodic solutions of Hamiltonian systems are degenerate in the sense of the definition presented in §8.1. Suppose that the periodic trajectory γ lies on the energy surface $\Sigma = \{H = \text{const}\}$, and only one of its multipliers equals 1. It is easy to show that γ, considered as a periodic trajectory of the Hamiltonian dynamical system on Σ, is non-degenerate. In this case it is natural to call the periodic trajectory γ *isoenergetically non-degenerate*.

Theorem 4 (*Poincaré* [203]). *Suppose that there exist k integrals $F_1, \dots, F_k$ in involution and the functions $H, F_1, \dots, F_k$ are independent at some point of γ. Then at least $2k + 1$ multipliers of the periodic trajectory γ equal 1.*

Corollary. *All multipliers of periodic trajectories, which lie on invariant tori of a completely integrable Hamiltonian system, equal 1.*

Theorem 4 is a consequence of Theorem 3. Indeed, the Hamiltonian equations admit $k+1$ independent integrals $H, F_1, \dots, F_k$ and $k+1$ linearly independent symmetry fields: they are the Hamiltonian fields $v_H, v_{F_1}, \dots, v_{F_k}$. Since these integrals are pairwise in involution, the functions $H, F_1, \dots, F_k$ are integrals of the systems

$$\dot{z} = v_H \, , \quad \dot{z} = v_{F_1} \, , \ \dots \, , \quad \dot{z} = v_{F_k} \, .$$

According to Theorem 3, at least $2(k+1)-1 = 2k+1$ multipliers of the periodic trajectory γ equal 1.

The original proof of Theorem 4 given by Poincaré is based on another idea. Let $f(\rho) = |P - \rho E|$ be the characteristic polynomial of the monodromy matrix corresponding to a periodic solution of a Hamiltonian system with n degrees of freedom. We put $f(\rho) = (\rho - 1)g(\rho)$. According to the well-known Poincaré–Lyapunov theorem, the polynomial g is reflexive, i.e.

$$\rho^{2n-2} g(1/\rho) = g(\rho) \, .$$

Consequently, if the equation $g(\rho) = 0$ has a root $\rho = 1$, then its multiplicity is even and not less than 2. Thus, if the Hamiltonian equations admit an independent integral F, then there is a pair of unit multipliers. One of these multipliers equals 1 due to the existence of the nontrivial Hamiltonian symmetry field v_F.

8.5 Now we turn to the Hamiltonian system with Hamiltonian

$$H = H_0(y_1, \dots, y_n) + \varepsilon H_1(x_1, \dots, x_n, y_1, \dots, y_n) + o(\varepsilon) \, , \tag{8.13}$$

which is analytic in y, $x \bmod 2\pi$ and ε. First, we put $\varepsilon = 0$ and assume that for $y = y^0$ the frequencies

$$\omega_1 = \partial H_0/\partial y_1 \, , \ \dots \, , \ \omega_n = \partial H_0/\partial y_n \tag{8.14}$$

satisfy $n-1$ independent resonance relations of the form

$$k_1\omega_1 + \dots + k_n\omega_n = 0 \, , \qquad k_i \in \mathbb{Z} \, , \quad \sum |k_i| \neq 0 \, .$$

In other words, all the trajectories of the unperturbed system, lying on an n-dimensional invariant torus $\mathbb{T}^n_{y^0} = \{x \bmod 2\pi, y^0\}$, are closed. These periodic trajectories are certainly isoenergetically degenerate (see the corollary of Theorem 4), and as a rule they do not remain closed after the perturbation. Poincaré was the first to notice that in a typical situation the perturbed Hamiltonian system has for small values of $\varepsilon \neq 0$ some non-degenerate periodic solutions which are transformed as $\varepsilon \to 0$ to periodic solutions lying on the resonant torus $\mathbb{T}^n_{y^0}$.

Now we turn to precise statements. Since, by assumption, there are no equilibrium points on the torus $\mathbb{T}^n_{y^0}$, at least one frequency (8.14) does not vanish for $y = y^0$. For example, let $\omega_n \neq 0$. For $y = y^0$ the function

$$H_1(\omega_1 t + \lambda_1, \dots, \omega_{n-1} t + \lambda_{n-1}, \omega_n t, y_1^0, \dots, y_n^0)$$

is periodic in t with some period τ. It is clear that τ is a period of the unperturbed periodic solutions. We put

$$h(\lambda_1, \ldots, \lambda_{n-1}) = \frac{1}{\tau} \int_0^\tau H_1 \, dt .$$

This function is 2π-periodic in $\lambda_1, \ldots, \lambda_{n-1}$.

Theorem 5 (*Poincaré* [203]). *Suppose that for $y = y^0$*

$$\det \left\| \frac{\partial^2 H_0}{\partial y_i \partial y_j} \right\| \neq 0 \quad \text{and} \quad \det \left\| \begin{matrix} \omega_1 \ldots \omega_n & 0 \\ & \omega_1 \\ \dfrac{\partial^2 H_0}{\partial y_i \partial y_j} & \ldots \\ & \omega_n \end{matrix} \right\| \neq 0 . \tag{8.15}$$

Let λ_0 be a non-degenerate critical point of the function h:

$$dh(\lambda_0) = 0 , \quad \det \left\| \frac{\partial^2 h}{\partial \lambda_i \partial \lambda_j} \right\|_{\lambda = \lambda_0} \neq 0 .$$

Then for small values of $\varepsilon \neq 0$ there exists an isoenergetically non-degenerate τ-periodic solution of the perturbed Hamiltonian system; it depends analytically on ε, and for $\varepsilon = 0$ coincides with a periodic solution of the unperturbed system

$$y = y^0 , \ x_1 = \omega_1(y^0)t + \lambda_1^0 , \ \ldots$$
$$\ldots , \ x_{n-1} = \omega_{n-1}(y^0)t + \lambda_{n-1}^0 , \ x_n = \omega_n(y^0)t .$$

A detailed proof of this result, based on the implicit function theorem, is presented by Poincaré [203, Chap. 3–4]. The case of two degrees of freedom is considered in [131].

For $n = 2$ the second determinant (8.15) is equal to

$$-\left(\omega_1^2 \frac{\partial^2 H_0}{\partial y_2^2} - 2\omega_1 \omega_2 \frac{\partial^2 H_0}{\partial y_1 \partial y_2} + \omega_2^2 \frac{\partial^2 H_0}{\partial y_1^2} \right) . \tag{8.16}$$

Let H_0 be a positive definite quadratic form with constant coefficients:

$$H_0 = \langle y, y \rangle / 2 = \sum a_{ij} y_i y_j / 2 .$$

Then relation (8.16) takes the form

$$-\langle y, y \rangle \delta ,$$

where $\delta = \det \| a_{ij} \|$. Since for the periodic solution $y^0 \neq 0$, in this case the second condition (8.15) certainly holds.

Note that if the expression (8.16) does not vanish for $y = y^0$, then the curve $H_0(y_1, y_2) = H_0(y_1^0, y_2^0)$ on the plane $\mathbb{R}^2 = \{y_1, y_2\}$ does not have an inflection point at $y = y^0$.

Let

$$H_1 = \sum h_m(y) \exp[i(m, x)] , \qquad m \in \mathbb{Z}^n .$$

Denote by Λ the set of $m \in \mathbb{Z}^n$ for which $(m, \omega(y^0)) = 0$. We put

$$R(x) = \sum h_m(y^0) \exp[i(m, x)] , \qquad m \in \Lambda . \tag{8.17}$$

It is clear that

$$h(\lambda_1, \dots, \lambda_{n-1}) = R(\lambda_1, \dots, \lambda_{n-1}, 0) .$$

Since the function h is defined and smooth on the $(n-1)$-dimensional torus $\mathbb{T}^{n-1} = \{\lambda_1, \dots, \lambda_{n-1} \bmod 2\pi\}$, it has at least n distinct critical points. They may be degenerate. In a typical situation all the critical points of h are non-degenerate. It is well known from Morse theory that in this case the number of distinct critical points is not less than 2^{n-1}. For $n = 2$ they are the points of maximum and minimum of the function h on the circle $\mathbb{T}^{n-1}$.

8.6 The following assertion is a basis for proving nonintegrability of the perturbed equations (cf. Lemma 1, §1). Suppose that the function

$$F = F_0(x, y) + \varepsilon F_1(x, y) + \cdots , \qquad (x, y) \in \mathbb{T}^n \times \mathbb{R}^n$$

is an integral of the non-degenerate Hamiltonian system with Hamiltonian (8.13). Then

1) F_0 does not depend on x,
2) the functions H_0 and F_0 are dependent on the Poincaré set $\mathbf{P}_1$.

The first part of this assertion follows easily from the non-degeneracy of the unperturbed system. Using Theorem 5, we prove the dependence of functions H_0 and F_0 on the set K of unperturbed tori which satisfy the conditions of this theorem. It is clear that always $K \subset \mathbf{P}_1$. However, one can indicate examples when the sets K and $\mathbf{P}_1$ do not coincide. Certainly, in the general case K is dense in $\mathbb{R}^n = \{y\}$.

In order to prove the assertion in question, we consider non-degenerate periodic solutions $\gamma(\varepsilon)$ which are generated by the family of periodic solutions lying on the resonant torus $y^0 \in K$ (see Theorem 5). Since the solutions $\gamma(\varepsilon)$ are non-degenerate for $\varepsilon \neq 0$, the functions H and F are dependent at all points of the trajectories $\gamma(\varepsilon)$. Let ε tend to zero. Then the periodic solution $\gamma(\varepsilon)$ is transformed into one of the periodic solutions $\gamma(0)$ of the unperturbed problem which lies on the torus $y = y^0$, and the functions H and F become equal to H_0 and F_0. By continuity they are dependent at all points of the trajectory $\gamma(0)$. Consequently, the rank of the Jacobian matrix

$$\frac{\partial(H_0, F_0)}{\partial(x, y)}$$

equals one at the points $(x, y) \in \gamma(0)$. In particular, at these points the rank of the matrix

$$\frac{\partial(H_0, F_0)}{\partial(y_1, \dots, y_n)}$$

also equals one. To complete the proof, it remains to note that the functions H_0 and F_0 do not depend on x.

The same idea is applicable to the more general problem of the existence of a vector field u_ε commuting with the original Hamiltonian field v_ε: according to Theorem 2, the fields u_ε and v_ε are linearly dependent at the points of a non-degenerate periodic solution $\gamma(\varepsilon)$. Let ε tend to zero. Then we see that u_0 and v_0 are dependent at the points of the trajectory $\gamma(0)$. The field u_0 does not depend on x because of the non-degeneracy of the unperturbed system. Consequently, the fields u_0 and v_0 are dependent at all points $y \in K$.

Thus, the birth of a large number of isoenergetically non-degenerate periodic solutions in perturbed Hamiltonian systems prevents their integrability. Unfortunately, we cannot prove in the same way the nonexistence of integrals (or symmetry fields) which depend analytically on ε. The situation is that for small but fixed values of ε Theorem 5 guarantees the existence of only a finite number of isoenergetically non-degenerate periodic solutions. This is certainly not sufficient to prove nonintegrability of the perturbed system for small fixed values of $\varepsilon \neq 0$.

8.7 Following Poincaré [203, §75], we indicate asymptotic formulae for the characteristic exponents of τ-periodic solutions mentioned in Theorem 5. Recall (see §8.1) that the characteristic exponents α are connected with the multipliers ρ by the relation $\rho = \exp(\alpha\tau)$. We set

$$\alpha = \mu\sqrt{\varepsilon}\,. \tag{8.18}$$

The function μ turns out to be analytic in $\sqrt{\varepsilon}$:

$$\mu = \mu_0 + \mu_1\sqrt{\varepsilon} + \cdots\,.$$

The coefficient μ_0 is a root of the equation

$$\begin{vmatrix} -\mu_0\tau & \cdots & 0 & & & \\ \vdots & \ddots & \vdots & & B & \\ 0 & \cdots & -\mu_0\tau & & & \\ & & & \mu_0\tau & \cdots & 0 \\ & -A & & \vdots & \ddots & \vdots \\ & & & 0 & \cdots & -\mu_0\tau \end{vmatrix} = 0\,, \tag{8.19}$$

where the elements of the $n \times n$ matrices A and B equal respectively

$$\tau\frac{\partial^2 H_0}{\partial y_i \partial y_j}(y^0) \quad \text{and} \quad \tau\frac{\partial^2 R}{\partial x_i \partial x_j}(x^0)\,.$$

Here the function R is defined by equality (8.17), and $x^0 = (\lambda_1^0, \ldots, \lambda_{n-1}^0, 0)$.

Equation (8.19) can be represented in the simpler form:

$$|AB + \mu_0^2\tau^2 I| = |BA + \mu_0^2\tau^2 I| = 0\,. \tag{8.20}$$

By the definition of the function R, the vector $\omega^0 = (\omega_1(y^0), \ldots, \omega_n(y^0))^T$ is an eigenvector of the matrix B with zero eigenvalue. Consequently, $\det B = 0$, and $\mu = 0$ is a double root of the polynomial (8.20). This root corresponds to the unit multiplier of the perturbed periodic solution. The other roots are divided into pairs with opposite signs. This fact follows from the Poincaré–Lyapunov theorem on the reflexivity of the characteristic polynomial for multipliers of a periodic solution of a Hamiltonian system.

The matrices A and B are symmetric; but in a general case $(AB)^T = BA \neq AB$. Thus, the eigenvalues of the matrix AB (or BA) are not real in a typical situation. In particular, the set of the multipliers may contain subsets of the form $\rho, \overline{\rho}, \rho^{-1}, (\overline{\rho})^{-1}$ $(\operatorname{Re}\rho \neq 0,\ \operatorname{Im}\rho \neq 0)$.

Consider the particular case, when the matrix A is proportional to unity: $A = \kappa I$, $\kappa \in \mathbb{R}$. For definiteness, let $\kappa > 0$. Then the squares of the roots of the characteristic polynomial (8.20) (the numbers μ_0^2) are real. They are proportional to the non-zero eigenvalues of the matrix B. It is easy to see that these numbers coincide with the eigenvalues of the matrix

$$\left\| \frac{\partial^2 h}{\partial\lambda_i \partial\lambda_j} \right\|_{\lambda=\lambda_0}, \qquad 1 \leq i, j \leq n-1 . \tag{8.21}$$

In a typical situation all critical points of the function $h : \mathbb{T}^{n-1} \to \mathbb{R}$ are nondegenerate. Recall that the number of negative eigenvalues of the matrix (8.21) is called the index k of the function h at the critical point λ_0. According to (8.20), $2k$ multipliers of the perturbed periodic solution are real positive numbers. Half of them consist of numbers which are greater than one, and the other half consist of numbers less than one. The other multipliers lie on the unit circle. Thus, the index k can be called the instability degree of the periodic solution.

Let c_k be the number of critical points of index k. The following relations are known from Morse theory (see, e.g., [75, Chap. 2]):

$$c_k \geq C_{n-1}^k , \qquad 0 \leq k \leq n-1 , \tag{8.22}$$

$$\sum (-1)^k c_k = 0 . \tag{8.23}$$

Inequalities (8.22) yield estimates of the number of perturbed periodic trajectories of given instability degree. In particular, a periodic solution with real multipliers corresponds to a maximum of the function h.

For $n = 2$ there are two types of perturbed isoenergetically non-degenerate periodic solutions. For the first one $\mu_0^2 > 0$, and both multipliers are real. Such solutions are called hyperbolic. They are unstable. For the second type $\mu_0^2 < 0$ and both multipliers are on the unit circle. These solutions are called elliptic. They are stable in the linear approximation. Relation (8.23) shows that for small values of $\varepsilon \neq 0$ an even number of non-degenerate periodic solutions is generated by a family of periodic solutions lying on a resonant unperturbed torus. Moreover, half of these non-degenerate solutions consist of elliptic solutions, and the other half consist of hyperbolic ones. This statement follows from the fact that local minima and maxima of a periodic function of a single variable alternate. It is usually said that after the destruction of a resonant invariant torus, pairs of isolated periodic solutions appear.

According to the results of KAM-theory the trajectory of a typical elliptic periodic solution is surrounded by invariant tori. The hyperbolic periodic solution has two invariant surfaces (separatrices), filled by solutions which tend asymptotically to the periodic trajectory as $t \to +\infty$ or $t \to -\infty$. Distinct asymptotic surfaces may intersect, generating an entangled net. Properties of the asymptotic surfaces will be discussed in detail in the next chapter.

Fig. 16.

Trajectories of the perturbed problem are presented in Fig. 16. More precisely, on a fixed three-dimensional level of the energy integral a two-dimensional surface is taken. In Fig. 16 invariant curves of the Poincaré map are given. Non-degenerate periodic trajectories correspond to isolated points, and Kolmogorov tori correspond to closed curves which are close to concentric circles.

8.8 Theorem 5 can be generalized to Hamiltonian systems with Hamiltonian

$$H = H_0(u) + \varepsilon H_1(u) + \ldots + \varepsilon^{k-1} H_{k-1}(u) + \varepsilon^k H_k(u, v) + o(\varepsilon^k) . \qquad (8.24)$$

Such systems appear when a finite number of perturbation theory steps is performed in the Hamiltonian system with Hamiltonian function (8.13) (see e.g. §4).

Suppose that an invariant torus $\{u = u_0, v \bmod 2\pi\}$ of the unperturbed problem is filled by periodic trajectories. Consider the function h on an $(n-1)$-dimensional torus, where h is the average of $H_k(u^0, v)$ along the trajectories of the unperturbed problem. One can show that if conditions (8.15) hold with $u = u^0$ and critical points of the function h are non-degenerate, then for small $\varepsilon \neq 0$ the perturbed system has at least 2^{n-1} distinct non-degenerate periodic solutions of the same period which are analytic in ε. Their characteristic exponents have the asymptotics

$$\alpha = \alpha_0(\sqrt{\varepsilon})^k + o((\sqrt{\varepsilon})^k) , \qquad \alpha_0 \neq 0 .$$

Following [229], we indicate one of the possible applications of this result. Consider a system with two degrees of freedom with a Hamiltonian of the following form (see §4):

$$H = H_0(y) + \varepsilon H_1(x) , \qquad H_0 = \frac{1}{2}\sum a_{ij}y_iy_j , \quad a_{ij} = \text{const} . \tag{8.25}$$

If the perturbing function H_1 is a trigonometric polynomial, then, according to Theorem 5, there exists only a finite number of non-degenerate families of periodic solutions which are analytic in ε. One can find other such families by applying canonical transformations of perturbation theory.

Assume that the quadratic form H_0 is positive definite. We put

$$H_1 = \sum h_m \exp[i(m, x)] , \quad \Delta = \{m \in \mathbb{Z}^2 : h_m \neq 0\} .$$

Denote by $\langle\ ,\ \rangle$ the inner scalar product generated by the form H_0.

Theorem 6. *Let α and β be vertices of the set Δ which satisfy the conditions of Theorem 2 of §5, and $y^0 \neq 0$ be a point of $\mathbb{R}^2$ on one of the lines $\langle k\alpha + \beta, y\rangle = 0$, $k = 0, 1, 2, \ldots$, where the components of the integer vector $k\alpha + \beta$ are relatively prime. Then at least two periodic solutions on the resonant torus $y = y^0 \neq 0$ of the unperturbed problem transform after perturbation into isoenergetically non-degenerate periodic solutions with the same period.*

The idea of the proof is as follows. We perform the canonical transformation

$$x, y \to u, v , \qquad x_s = \frac{\partial S}{\partial y_s} , \quad v_s = \frac{\partial S}{\partial u_s} ; \qquad s = 1, 2 ,$$
$$S = S_0 + \varepsilon S_1 + \ldots + \varepsilon^{k-1} S_{k-1} ,$$

where $S_1, \ldots, S_{k-1}$ satisfy the first $k - 1$ equations of the infinite system (4.4). As a result, the Hamiltonian function $H_0(y) + \varepsilon H_1(x)$ takes the form (8.24). Suppose that the point $u^0 \neq 0$ lies on one of the lines $\langle u, k\alpha + \beta\rangle = 0$, $k = 0, 1, \ldots$, where the components of the vector $k\alpha + \beta \in \mathbb{Z}^2$ are relatively prime. We put

$$H_k = \sum h_m^*(u) \exp[i(m, v)] . \tag{8.26}$$

It is easy to show that there is a unique integer vector $\tau = (\tau_1, \tau_2)$, $\tau_2 > 0$ such that

a) $h_\tau^*(u^0) \neq 0$,

b) $\langle u^0, \tau\rangle = 0$.

Now we can use the generalized Theorem 5. Indeed, we find from the expansion (8.26) that

$$h(\lambda) = \sum_{\langle u^0, \tau\rangle = 0} h_\tau(u^0) e^{i\tau_2\lambda} = h_0 + h_\tau e^{i\tau_2\lambda} + \bar{h}_\tau e^{-i\tau_2\lambda} .$$

The derivative h' vanishes for at least two values of λ because of the periodicity of h. At these points $h'' \neq 0$. Indeed, otherwise

$$h_\tau e^{i\tau_2\lambda} - \overline{h}_\tau e^{-i\tau_2\lambda} = h_\tau e^{i\tau_2\lambda} + \overline{h}_\tau e^{-i\tau_2\lambda} = 0 .$$

Since $\exp(i\tau_2\lambda) \neq 0$, we have $h_\tau \overline{h}_\tau = 0$. This contradicts property a). Conditions (8.15) evidently hold because the form H_0 is positive definite (see §8.5). The theorem is proved.

Using Theorem 6 we find that if the convex hull of Δ is not a rhombus, then the system with Hamiltonian (8.25) has for $\varepsilon > 0$ an infinite set of distinct isoenergetically non-degenerate solutions with the same period (or energy). Unfortunately, the set of ε, for which there exist the solutions obtained on the k-th step of perturbation theory, decreases indefinitely as $k \to \infty$. Therefore, for each fixed $\varepsilon > 0$ we can guarantee the existence of a large but finite number of non-degenerate periodic solutions. This means that we cannot prove nonintegrability of the system (8.25) for small fixed $\varepsilon > 0$. However, we can prove that there is no family of first integrals analytic in ε and no nontrivial symmetry groups.

Consider as an example the motion of three particles of unit mass on a smooth unit circle which attract or repel elastically. Let $x_i \bmod 2\pi$ be the angular coordinates of these points, y_i the angular momenta. The Hamiltonian has the form (see §7)

$$H = \frac{1}{2}\sum y_i^2 + \varepsilon \sum_{i<j} \cos(x_i - x_j) . \qquad (8.27)$$

We reduce the number of degrees of freedom with the help of the angular momentum integral and obtain a Hamiltonian system with two degrees of freedom (see §7.2). The convex hull of the set Δ is presented in Fig. 14. Take two vectors $\alpha = (1, 0)$ and $\beta = (0, -1)$ as vertices. The components of the vector $k\alpha + \beta = (k, -1)$ are clearly relatively prime, and $\langle \alpha, \beta \rangle > 0$. Consequently, by Theorem 6 the system with Hamiltonian (8.27) is nonintegrable because an infinite number of families of isoenergetically non-degenerate long-periodic trajectories exist in the perturbed system.

9 Non-degenerate Invariant Tori

9.1 Suppose that the system of differential equations

$$\dot z = v(z) , \qquad z \in M^{m+k} \qquad (9.1)$$

has a k-dimensional invariant torus $\mathbb{T}^k$, filled by trajectories of quasi-periodic motions. In a small neighborhood of this torus one can introduce the coordinates $x_1 \bmod 2\pi, \dots, x_k \bmod 2\pi,\ y_1, \dots, y_m$ such that Eq. (9.1) takes the following form:

$$\dot x = \omega + f(x, y) , \quad \dot y = \Omega y + g(x, y) . \qquad (9.2)$$

Here $\omega = (\omega_1, \dots, \omega_k)$ is the constant frequency vector of quasi-periodic motions on $\mathbb{T}^k$; $f(x, 0) = 0$, and $g(x, y) = O(|y|^2)$. The invariant torus is defined by the equation $y = 0$. Elements of the $m \times m$ matrix Ω are 2π-periodic in $x_1, \dots, x_k$.

For $k = 1$ we have Eqs. (8.2). It was noted in §8.1 that in this case Ω can be reduced to a constant matrix by a periodic in x, linear change of the coordinates y (the Lyapunov–Floquet theorem). However, for $k > 1$ such a reduction is not always possible (these problems are discussed e.g. in [13]). We shall assume that the matrix Ω is constant. This is the case which appears in applications (see §10.1).

The linear equations

$$\dot{\xi} = \Omega\xi\,, \qquad \xi \in \mathbb{R}^m \tag{9.3}$$

are called variational. The role of characteristic exponents is played by the eigenvalues of the matrix Ω. As in the case $k = 1$, the existence of nontrivial integrals and symmetry groups in system (9.2) imposes essential restrictions on the spectrum of the matrix Ω.

Let H be a single-valued integral of the system (9.2). In a neighborhood of the invariant torus this function can be represented in the following form:

$$H = H_0(x) + (y, h(x)) + O(y)\,.$$

Here, H_0 (respectively h) is a function (respectively, a covector field) on the invariant torus $\mathbb{T}^k$.

Lemma 1. *Suppose that the frequencies $\omega_1, \dots, \omega_k$ are rationally incommensurable. Then $H_0 = \text{const}$, and the field h satisfies the equation*

$$\left(\frac{\partial h}{\partial x}, \omega\right) = -\Omega^T h\,. \tag{9.4}$$

The proof repeats the arguments of §8.2.

Assume that the system (9.2) admits a symmetry group—the phase flow of the system

$$x' = V(x, y)\,, \quad y' = W(x, y)\,,$$

whose right-hand side is 2π-periodic in the coordinates $x_1, \dots, x_k$.

Lemma 2. *The vector field $w(x) = W(x, 0)$ satisfies the equation*

$$\left(\frac{\partial w}{\partial x}, \omega\right) = \Omega h\,. \tag{9.5}$$

Lemma 2 is easily proved by the method of §8.2.

Let $x(t)$ be a quasi-periodic motion on a k-dimensional invariant torus. It is clear that the vector-function $w(x(t))$ satisfies the variational Eqs. (9.3). The linear systems (9.4) and (9.5) are conjugate: $(h, w) \equiv \text{const}$. Indeed, according to (9.4) and (9.5), the function $\phi = (h, w)$ satisfies the linear equation

$$\left(\frac{\partial \phi}{\partial x}, \omega\right) = 0 . \tag{9.6}$$

We put

$$\phi = \sum \phi_\lambda \exp[i(\lambda, x)] , \qquad \lambda \in \mathbb{Z}^k ,$$

and use Eq. (9.6):

$$\sum i\phi_\lambda(\lambda, \omega) \exp[i(\lambda, x)] = 0 .$$

Consequently, $(\lambda, \omega)\phi_\lambda = 0$. Since $(\lambda, \omega) \neq 0$ for $\lambda \neq 0$, all ϕ_λ except ϕ_0 vanish. Thus $\phi = \phi_0 = \text{const}$.

9.2 The existence of first integrals which are independent at points of an invariant torus imposes some restrictions on the structure of the matrix Ω.

Theorem 1. *Suppose that the system* (9.2) *admits r integrals which are independent at some point of the invariant torus* $\mathbb{T}^k$. *Then the matrix* Ω *has at least r eigenvalues of the form* $i(\lambda, \omega)$, $\lambda \in \mathbb{Z}^k$ *(with multiplicities)*.

Let $k = 1$. Then the monodromy matrix $P = \exp(2\pi\Omega/\omega)$ has the eigenvalue $\rho = 1$ of multiplicity $\geq r$. Consequently, Theorem 1 contains the Poincaré theorem (see §8.2) as a particular case.

Proof of Theorem 1. We expand the function $h(x)$ in a Fourier series:

$$h = \sum h_\lambda \exp[i(\lambda, x)] , \qquad \lambda \in \mathbb{Z}^k , \quad h_\lambda \in \mathbb{R}^m . \tag{9.7}$$

Substituting this series into Eq. (9.4), equating coefficients of identical harmonics, we obtain the chain of relations

$$\Omega^T h_\lambda = -i(\lambda, \omega) h_\lambda , \qquad \lambda \in \mathbb{Z}^k . \tag{9.8}$$

Consequently h_λ is an eigenvector of the matrix Ω^T with the eigenvalue $-i(\lambda, \omega)$.

Suppose that there are not more than $r - 1$ distinct non-zero eigenvectors h_λ which satisfy (9.8). Since the frequencies $\omega_1, \ldots, \omega_k$ are rationally incommensurable, the numbers (λ, ω) and (λ', ω) $(\lambda, \lambda' \in \mathbb{Z}^k)$ coincide only for $\lambda = \lambda'$. Hence, there are not more than $r - 1$ non-zero coefficients in the expansions (9.7). Thus, the covector fields $h(x)$ corresponding to r integrals of the system (9.2) are linearly dependent. Consequently, these integrals are dependent at all points of the invariant torus. We obtain a contradiction. In order to complete the proof it remains to note that the eigenvalues of the matrices Ω and Ω^T coincide.

Theorem 2. *Suppose that the system* (9.2) *admits* $s + k$ *symmetry fields which are independent at some point of the invariant torus* $\mathbb{T}^k$. *Then the matrix* Ω *has at least s eigenvalues of the form* $i(\lambda, \omega)$, $\lambda \in \mathbb{Z}^k$ *(with multiplicities)*.

The proof of this theorem is completely analogous to that of Theorem 1. For $k = 1$ we obtain Theorem 2 of §8.

9.3 Systems having first integrals as well as symmetry groups satisfy

Theorem 3. *Suppose that the system (9.2) admits r integrals $H_1, \dots, H_r$ and l symmetry fields $u_1, \dots, u_l$ which are independent at some point of the invariant torus $\mathbb{T}^k$. Moreover, let*

$$\left(\frac{\partial H_i}{\partial z}, u_j\right) = 0 , \qquad 1 \leq i \leq r , \quad 1 \leq j \leq l . \tag{9.9}$$

Then the matrix Ω has not less than $r + l - k$ eigenvalues of the form $i(\lambda, \omega)$, $\lambda \in \mathbb{Z}^k$ (with multiplicities).

For $k = 1$ this theorem coincides with Theorem 3 of §8. The proof is based on the same ideas.

Let us apply Theorem 3 to a Hamiltonian system with n degrees of freedom which has a non-resonant k-dimensional invariant torus. Suppose that this system admits r independent involutive integrals. Then the spectrum of the corresponding matrix Ω contains not less than $2r - k$ numbers of the form $i(\lambda, \omega)$, $\lambda \in \mathbb{Z}^k$. For $k = 1$ we obtain Theorem 4 of §8. The proof is based on the fact that the Hamiltonian vector fields

$$v_{H_1} , \ \dots , \ v_{H_r}$$

are symmetry fields satisfying the relations (9.9). To obtain the desired estimate, we use Theorem 3.

9.4 Consider an illustrative example. Let the $2(m + l)$-dimensional phase space be endowed with a symplectic structure

$$dY \wedge dX + dZ_- \wedge dZ_+ ,$$

where $X = (X_1, \dots, X_m) \bmod 2\pi$, $Y = (Y_1, \dots, Y_m)$, $Z_\pm = (Z_1^\pm, \dots, Z_l^\pm)$. Consider the Hamiltonian function

$$H = (\omega, Y) + (Y, \Gamma Y)/2 + (Z_-, \Omega Z_+) + G(X, Y, Z) . \tag{9.10}$$

Here ω is a non-resonant set of numbers $\omega_1, \dots, \omega_m$, Γ and Ω are constant matrices, the function G is 2π-periodic in X, and its expansion in Maclaurin series in the variables Y and Z begins with terms of third order. The Hamiltonian equations

$$\begin{aligned}
\dot{X} &= \omega + \Gamma Y + O_2(Y, Z) , \\
\dot{Y} &= O_3(Y, Z) , \\
\dot{Z}_+ &= \Omega Z_+ + O_2(Y, Z) , \\
\dot{Z}_- &= -\Omega^T Z_- + O_2(Y, Z) .
\end{aligned}$$

obviously have the m-dimensional invariant torus $\mathbb{T}^m = \{X, Y, Z : Y = Z = 0\}$ filled by trajectories of the quasi-periodic motions $X = \omega t + X_0$. According to (9.3), the variational equations have the form:

$$\dot{Y} = 0\,, \quad \dot{Z}_+ = \Omega Z_+\,, \quad \dot{Z}_- = -\Omega^T Z_-\,. \tag{9.11}$$

The spectrum of this linear system with constant coefficients contains m zeros, l eigenvalues of the matrix Ω, and l eigenvalues of the matrix $-\Omega^T$.

Suppose that the matrix Ω has no purely imaginary eigenvalues. Then, according to the results of §9.3, any $m+1$ integrals of the Hamiltonian system in question are dependent at the points of the m-dimensional torus $\{Y = Z = 0\}$.

For example, suppose that for all $\zeta \in \mathbb{C}^l$

$$\mathrm{Re}(\zeta, \Omega\bar{\zeta}) \geq \mu \|\zeta\|^2\,, \qquad \mu = \mathrm{const} > 0\,, \tag{9.12}$$

where the bar denotes complex conjugation. Let λ be an eigenvalue of the matrix Ω with the eigenvector $\xi \in \mathbb{C}^l$, $\Omega\xi = \lambda\xi$. Then

$$\mathrm{Re}(\xi, \Omega\bar{\xi}) = \mathrm{Re}(\xi, \overline{\lambda\xi}) = \mathrm{Re}\bar{\lambda}|\xi|^2 > 0\,.$$

Consequently all eigenvalues of the matrix Ω ($-\Omega^T$) are in the right (left) half-plane.

9.5 Consider the more general case: let the matrices Γ and Ω in the expansion of Hamiltonian (9.10) be 2π-periodic in $X_1, \ldots, X_m$. The variational equations have again the form (9.11). However, they cannot be reduced in general to a system with constant coefficients.

Lemma 3. *Suppose that the matrix Ω satisfies condition* (9.12) *for all X. Then any $m+1$ integrals of Hamilton equations with Hamiltonian* (9.10) *are dependent at all points of the m-dimensional non-resonant invariant torus $Y = 0$, $Z = 0$.*

Proof. In a neighborhood of this torus any function can be represented in the form:

$$F = F_0(X) + (y(X), Y) + (z_+(X), Z_+) + (z_-(X), Z_-) + O_2(Y, Z)\,.$$

The functions $F_0, y, z_\pm$ are 2π-periodic in X. If F is an integral of the Hamiltonian equations, then

$$\begin{aligned} &\left(\frac{\partial F_0}{\partial X}, \omega\right) = 0\,, \quad \left(\frac{\partial y}{\partial X}, \omega\right) = 0\,, \\ &\left(\frac{\partial z_+}{\partial X}, \omega\right) = -\Omega^T z_+\,, \quad \left(\frac{\partial z_-}{\partial X}, \omega\right) = \Omega z_-\,. \end{aligned} \tag{9.13}$$

Since the set of frequencies is non-resonant, $F_0 = \mathrm{const}$ and $y = \mathrm{const}$. The last two Eqs. (9.13) can be represented in the following form:

$$\dot{z}_+ = -\Omega^T z_+\,, \quad \dot{z}_- = \Omega z_-\,.$$

Here the dots denote the derivative along solutions of the system $\dot{X} = \omega$. According to (9.12), $z_\pm(t) \to 0$ as $t \to \pm\infty$. Since the trajectories $X = \omega t + X_0$ are dense on the m-dimensional torus $\mathbb{T}^m = \{X \bmod 2\pi\}$, $z_\pm(X) \equiv 0$.

We neglect an inessential constant term and get

$$F = (y_0, Y) + O_2(Y, Z) , \qquad y_0 = \text{const} .$$

It is clear that the integrals are independent at the points of the torus $\mathbb{T}^m$ if and only if the corresponding constant vectors y_0 are independent. Since $y_0 \in \mathbb{R}^m$, the number of independent integrals does not exceed m. The lemma is proved.

Note that if Eqs. (9.11) are reducible, then Lemma 3 follows from Theorem 3 of this section. The tori we dealt with in Lemma 3 may be called *hyperbolic*. They are a direct generalization of the hyperbolic periodic solutions of §8.

9.6 According to the Lyapunov–Floquet theorem, equations (9.11) are reducible for $m = 1$. Now we show that for $l = 1$ they are also reducible if the frequencies $\omega_1, \ldots, \omega_m$ are strongly incommensurable:

$$|(k, \omega)| \geq \alpha/|k|^\beta , \qquad \alpha, \beta = \text{const} > 0 \tag{9.14}$$

for all $k \in \mathbb{Z}^m$.

In this case the problem is reduced to the analysis of the equation

$$\dot{z} = \Omega(X)z , \qquad X \in \mathbb{T}^m . \tag{9.15}$$

Perform a invertible change $z \to \zeta$ using the relation $z = \zeta u$, where u is a positive function on $\mathbb{T}^m$. Assume that such a change transforms the Eq. (9.15) to the equation

$$\dot{\zeta} = c\zeta , \qquad c = \text{const} . \tag{9.16}$$

Then the function u satisfies the equation

$$\left(\frac{\partial u}{\partial X}, \omega\right) = (\Omega - c)u .$$

Since by assumption $u > 0$, we can put $u = \exp v$. Then

$$\left(\frac{\partial v}{\partial X}, \omega\right) = \Omega - c . \tag{9.17}$$

We set

$$c = \frac{1}{(2\pi)^m} \int_{\mathbb{T}^m} \Omega \, d^m X . \tag{9.18}$$

Since condition (9.14) holds, Eq. (9.17) has a smooth solution $v : \mathbb{T}^m \to \mathbb{R}$, provided the function Ω is smooth.

Thus, if (9.14) is valid, then (9.15) is transformed to (9.16); moreover, the constant c is defined by formula (9.18), and $u = \exp v$, where v is a solution of the homologic Eq. (9.17).

10 Birth of Hyperbolic Invariant Tori

10.1 Now turn again to Poincaré's "basic problem of dynamics". We consider Hamiltonian systems with real-analytic Hamiltonians

$$\begin{aligned} &H_0(y) + \varepsilon H_1(x, y) + o(\varepsilon) , \\ &y \in D \subset \mathbb{R}^n , \quad x \in \mathbb{T}^n , \qquad \varepsilon \in (-\varepsilon_0, \varepsilon_0) , \quad \varepsilon_0 > 0 . \end{aligned} \tag{10.1}$$

Suppose that for $y = y^0$ the frequencies of the unperturbed problem

$$\omega_1 = \partial H_0/\partial y_1 , \ \dots , \ \omega_n = \partial H_0/\partial y_n$$

are rationally commensurable. More precisely, there is a nontrivial subgroup Λ of the group $\mathbb{Z}^n$ such that $(\omega, \tau) = 0$ for all $\tau \in \Lambda$. Let $\operatorname{rank}\Lambda = l$ and $m = n - l$. According to the theory of Abelian groups, there exist n vectors $\tau'_1, \dots, \tau'_m, \tau_1, \dots, \tau_l$ of $\mathbb{Z}^n$ such that the matrix K_0, whose columns are formed by components of these vectors, is unimodular (its determinant equals one), and the vectors $\tau_1, \dots, \tau_l$ generate the group Λ.

We put

$$K = \|\tau_1, \dots, \tau_l\| , \quad K' = \|\tau'_1, \dots, \tau'_m\| .$$

The matrices K and K' have dimensions $n \times l$ and $n \times m$ respectively. It is clear that $\operatorname{rank}K = l$, $\operatorname{rank}K' = m$, and $K^T\omega = 0$.

The invariant torus $y = y^0$ of the unperturbed system is resonant. It is foliated by m-dimensional non-resonant invariant tori

$$\mathbb{T}^m(x^0, y^0) = \{x, y : y = y^0, K^T(x - x^0) \equiv 0 \bmod 2\pi\} . \tag{10.2}$$

It turns out that under certain conditions not all tori (10.2) are destroyed by the perturbation. Some of them just slightly deform. This result, obtained by Treshchev, is an extension of the Poincaré theorem on birth of periodic solutions (§8.5).

Now we give precise statements. We put

$$\begin{aligned} &\Pi = \left\| \frac{\partial^2 H_0}{\partial y^2}(y^0) \right\| , \\ &H_1(x, y^0) = \sum H^\tau \exp[i(\tau, x)] , \qquad \tau \in \mathbb{Z}^n . \end{aligned} \tag{10.3}$$

In order to take the average of the function H_1 along non-degenerate trajectories, we introduce new angular variables z according to the relation $z = K_0^T x$, and put $x = \omega t + x^0$. Then $z = \omega^* t + z^0$, where

$$\omega^* = K_0^T\omega , \quad z^0 = K_0^T x^0 .$$

Since $K_0 = \|K', K\|$ and $K^T\omega = 0$, we have $\omega^* = (\omega^*_1, \dots, \omega^*_m, 0, \dots, 0)^T$. It is clear that the new frequencies $\omega^*_1, \dots, \omega^*_m$ are rationally incommensurable, because otherwise the rank of the group Λ would be greater than l. We put $z^0 =$

$(0, \ldots, 0, \lambda_1, \ldots, \lambda_l)$ and perform the averaging with respect to time:

$$h(\lambda) = \lim_{T\to\infty} \frac{1}{T} \int_0^T H_1(\omega t + x^0, y^0)\, dt .$$

The function $h : \mathbb{T}^l \to \mathbb{R}$ can be obtained in a different way. We can eliminate from the Fourier series (10.3) all terms corresponding to the vectors τ which do not belong to the group Λ:

$$h(\lambda) = \sum H^{K\mu} \exp[i(\mu, \lambda)] , \qquad \mu \in \mathbb{Z}^l .$$

Let $dh(\lambda^0) = 0$ and

$$V = \left\| \frac{\partial^2 h}{\partial \lambda^2}(\lambda^0) \right\| .$$

According to Morse theory, in a typical situation the function h has not less than 2^l critical points.

Theorem 1 [230]. *Let the following conditions hold:*
1) $\det \Pi \neq 0$,
2) *the matrix* $VK^T\Pi K$ *has no eigenvalue which is a positive real number or zero,*
3) *the numbers* $\omega_1^*, \ldots, \omega_m^*$ *are strongly incommensurable:*

$$|\nu_1\omega_1^* + \cdots + \nu_m\omega_m^*| \geq c|\nu|^{-N} , \qquad c, N > 0 \tag{10.4}$$

for all integer $\nu_1, \ldots, \nu_m$ *which do not vanish simultaneously.*

Then for small $\varepsilon \geq 0$ *there exists a family of m-dimensional invariant tori* $\mathbb{T}^m_\varepsilon$ *of the Hamiltonian system with Hamiltonian* (10.1). *These tori are filled by trajectories of quasi-periodic motions with frequencies* $\omega_1^*, \ldots, \omega_m^*$, *and* $\mathbb{T}^m_0$ *coincides with the torus* $\mathbb{T}^m(x^0, y^0)$, *where* $K^T x^0 \equiv \lambda^0 \bmod 2\pi$.

Moreover, there exists an analytic in ε, x, y $(\varepsilon > 0)$, *canonical change of coordinates (the change is* C^∞*-smooth in* $\sqrt{\varepsilon}$ *for* $\varepsilon = 0$) :

$$(x; y) \to (X, Z_+; Y, Z_-) , \qquad dx \wedge dy = dX \wedge dY + dZ_+ \wedge dZ_- ,$$

$$X = (X_1, \ldots, X_m) \bmod 2\pi , \quad Y = (Y_1, \ldots, Y_m) , \quad Z_\pm = (Z_1^\pm, \ldots, Z_l^\pm) ,$$

such that in the new variables the Hamiltonian (10.1) *takes the form*

$$(\omega^*, Y) + \sqrt{\varepsilon}(Y, \Gamma(\varepsilon)Y)/2 + \sqrt{\varepsilon}(Z_-, \Omega(X, \varepsilon)Z_+) \\ + \sqrt{\varepsilon}F(X, Y, Z, \varepsilon) + \varepsilon G(X, Y, Z, \varepsilon),$$

where $\Gamma = \Gamma_0 + O(\sqrt{\varepsilon})$, $\det\Gamma_0 \neq 0$ *and* $\Omega = \Omega_0 + O(\sqrt{\varepsilon})$; *for all* $\zeta \in \mathbb{C}^l$ *the following inequalities hold:*

$$\operatorname{Re}(\zeta, \Omega_0\bar{\zeta}) \geq \mu|\zeta|^2 , \qquad \mu = \text{const} > 0 ,$$

and the expansion of the function F *(respectively* G*) in a Maclaurin series in* Z *(respectively in* Y *and* Z*) begins with terms of degree* ≥ 3. *The invariant tori* $\mathbb{T}^m_\varepsilon$ *have the form* $\{(X, Y, Z) : Y = 0,\ Z = 0\}$.

The proof of Theorem 1 is based on the ideas of KAM theory. According to §9, for small $\varepsilon > 0$ the invariant tori $\mathbb{T}_\varepsilon^m$ are hyperbolic. For $m = 1$ these tori transform to periodic solutions, and Theorem 1 becomes a particular case of the Poincaré theorem of §8.5. Indeed, condition 3) of Theorem 1 is certainly valid, and condition 1) coincides with the non-degeneracy condition for the unperturbed system. The non-degeneracy of the matrix $VK^T\Pi K$ is equivalent to the following two conditions: $\det V \neq 0$, and $\det(K^T\Pi K) \neq 0$. The first one is reduced to the non-degeneracy condition for the critical point of the function h, and the second is equivalent to the second inequality (8.15). Consequently, the Poincaré theorem is applicable.

We emphasize that all invariant tori which exist due to Theorem 1 are unstable. However, for $m = 1$ the perturbed system has by the Poincaré theorem (see §8.7) closed trajectories which are stable in the linear approximation. Moreover, the Poincaré theorem asserts that the family of invariant tori $\mathbb{T}_\varepsilon^1$ is analytic in ε for small $|\varepsilon|$. Theorem 1 implies analyticity in ε only for small positive ε.

Apparently, condition 2) of Theorem 1 is technical. The situation is as follows. After a small perturbation of the Hamiltonian function the isoenergetically non-degenerate periodic solutions do not disappear and transform to periodic solutions of the same period. Melnikov [174], Moser [181] and Graff [92] showed that hyperbolic reducible tori with strongly incommensurable sets of frequencies (condition (10.4)) survive perturbations of Hamiltonian equations. An analogous general result for non-hyperbolic tori was obtained by Eliasson [76], (see also Pöschel [207]). Note that the families of non-hyperbolic tori turn out to be Cantorian in ε. (Recall that the corresponding hyperbolic families are analytic in ε for $\varepsilon > 0$.) We believe that a non-hyperbolic version of Theorem 1 may be proved.

10.2 The case $m = n - 1$ is the most interesting from the point of view of the exact integration problem for the canonical equations with Hamiltonian (10.1). In this case formal expressions for the perturbed invariant tori and their stable and unstable manifolds have been obtained already by Poincaré in [203]. Here only one nontrivial relation

$$\sum k_i\omega_i = 0\,, \qquad k_i \in \mathbb{Z}$$

holds.

The unique angular variable λ (see §10.1) is defined by the equality

$$\lambda = \sum k_i x_i\,.$$

It is clear that

$$h(\lambda) = \sum H^{\mu k}\exp(i\mu\lambda)\,, \qquad \mu \in \mathbb{Z}\,.$$

The resonant invariant torus $\{y = y^0,\ x \bmod 2\pi\}$ of the unperturbed problem is foliated by a one-parameter family of $(n-1)$-dimensional non-resonant tori

$$\mathbb{T}^{n-1}(\lambda, y^0) = \{x, y : y = y^0,\ \sum k_i x_i = \lambda \bmod 2\pi\}\,. \tag{10.5}$$

We shall assume that the frequencies of quasi-periodic motions on the tori (10.5) satisfy condition (10.4).

In the case under consideration the matrix $K^T \Pi K$ is reduced to a single real number

$$\sum_{i,j=1}^{n} \frac{\partial^2 H_0}{\partial y_i \partial y_j}(y^0) k_i k_j \, . \tag{10.6}$$

Assume that the sum (10.6) does not vanish and all critical points of the function h are non-degenerate. In particular, the number of critical points is even, half of them consisting of local minima (where $h'' > 0$), and the other half of local maxima (where $h'' < 0$). Then, according to Theorem 1, half of the critical points of the function h correspond to $(n-1)$-dimensional invariant tori of the unperturbed problem, which for small values of the parameter $\varepsilon > 0$ transform to hyperbolic tori $\mathbb{T}_\varepsilon^{n-1}$ of the perturbed Hamiltonian equations.

Assume that the equations with Hamiltonian (10.1) have a complete set of integrals in the form of power series in ε:

$$F_1^0 + \varepsilon F_1^1 + \cdots \, , \; \ldots \, , \quad F_n^0 + \varepsilon F_n^1 + \cdots \, . \tag{10.7}$$

The coefficients in these series are assumed to be analytic in x, y and 2π-periodic in x. The functions $F_1^0, \ldots, F_n^0$ do not depend on the variables $x \bmod 2\pi$ due to the non-degeneracy of the unperturbed problem. Now we show that these functions are dependent at the point $y = y^0$. We use the arguments of §8.6. According to Lemma 3 from §9, integrals (10.7) are dependent on the torus $\mathbb{T}_\varepsilon^{n-1}$. Since these functions do not depend on the variables x, the Jacobian

$$\frac{\partial(F_1^0, \ldots, F_n^0)}{\partial(y_1, \ldots, y_n)}$$

vanishes for $y = y^0$. Q.E.D.

Thus, the birth of hyperbolic invariant $(n-1)$-dimensional tori in the perturbed Hamiltonian equations is an obstruction to complete integrability. Analogous reasoning shows that the birth of a large number of m-dimensional hyperbolic tori prevents the existence of $m+1$ independent integrals which are analytic in ε.

10.3 It can be shown that Theorem 1 holds for Hamiltonians of a more general form:

$$H_0(y) + \varepsilon H_1(y) + \cdots + \varepsilon^{s-1} H_{s-1}(y) + \varepsilon^s H_s(x, y) + o(\varepsilon^s) \, . \tag{10.8}$$

The role of perturbing function is played by H_s, and in order to calculate the matrix V, one must take the average of H_s along trajectories of the Hamiltonian system with Hamiltonian H_0.

The systems with Hamiltonians (10.8) appear naturally after a finite number of steps of the classical perturbation theory, applied to the system with the original

Hamiltonian (10.1). Using the generalized Theorem 1, we can find new hyperbolic tori in the perturbed problem.

We demonstrate this idea on the problem considered in §5. We consider the Hamiltonian system with Hamiltonian

$$H = H_0(y) + \varepsilon H_1(x) , \tag{10.9}$$

where

$$H_0 = \frac{1}{2}\sum a_{ij}y_iy_j$$

is a positive definite quadratic form with constant coefficients, and H_1 is a trigonometric polynomial in the angular variables $x_1, \ldots, x_n$. In this case the sum (10.6) certainly does not vanish. Since the Fourier expansion of H_1 has a finite number of terms, Theorem 1 guarantees the existence of a finite number of families of hyperbolic tori in the perturbed problem. This is not sufficient to prove its nonintegrability.

We shall denote by the brackets $\langle\ ,\ \rangle$ the scalar product defined by the metric H_0. Let Δ be the "spectrum" of the polynomial H_1 (see §5). Leaving aside the trivial integrable case when all the points of $\Delta \subset \mathbb{Z}^n$ lie on a single straight line, we shall assume that Δ contains at least two linearly independent elements. Hence we can introduce the vertex α of the set Δ and the adjacent vertex β (see §5). The vectors α and β are linearly independent.

Theorem 2 [230]. *Suppose that*
1) $2\langle\alpha, \beta\rangle/\langle\alpha, \alpha\rangle \notin -\mathbb{Z}_+$, $\qquad \mathbb{Z}_+ = \{0, 1, 2, \ldots\}$,
2) *for all integer* $j \geq 0$ *the components of the vectors* $j\alpha + \beta$ *are relatively prime.*

Then on each plane $\pi_j = \{y \in \mathbb{R}^n : \langle y, j\alpha + \beta\rangle = 0\}$ *there exists a subset* $W_j(\varepsilon_0)$ *such that for all* $y \in W_j(\varepsilon_0)$ *the Hamiltonian system with Hamiltonian* (10.9) *has for* $0 < \varepsilon < \varepsilon_0$ *hyperbolic invariant tori* $\mathbb{T}_y^{n-1}(\varepsilon)$ *depending smoothly on* $\varepsilon \geq 0$*. Moreover, for each* j_0 *there exists* $\varepsilon_0(j) > 0$ *such that the measure of* $W_j(\varepsilon_0(j))$ *is positive and the measure of the set*

$$\pi_j \bigcap_{\varepsilon>0} W_j(\varepsilon)$$

equals zero.

This result is a multidimensional analog of Theorem 6, §8; it is proved in a similar way. For example, put $j = 0$. Since the components of the vector β are relatively prime, the set Δ contains only one pair of non-zero vectors, parallel to β: they are β and $-\beta$. Thus, the function $h(\lambda)$ equals $c\cos(\lambda) + c_1$, where $c, c_1 = \text{const}$, and $c \neq 0$. It has exactly two non-degenerate critical points.

Consider an integer unimodular matrix K_0 corresponding to the subgroup $\beta\mathbb{Z} \subset \mathbb{Z}^n$ (see §10.1). The frequency vector for invariant $(n-1)$-dimensional tori of the unperturbed problem has the form

$$\omega^* = K_0^T A y , \qquad A = \|a_{ij}\| , \tag{10.10}$$

where $y \in \pi_0 = \{y \in \mathbb{R}^n : \langle \beta, y \rangle = 0\}$. Since the matrix $K_0^T A$ is non-degenerate, the set of $y \in \pi_0$ for which the frequencies (10.10) satisfy condition (10.4) for no $c, N > 0$ has zero measure.

Let $W(\varepsilon_0)$ be the set of points of π_0 for which there exists a hyperbolic invariant torus (see Theorem 1) for all $0 < \varepsilon < \varepsilon_0$. Since the measure of the set

$$\pi_0 \setminus \bigcup_{\varepsilon > 0} W(\varepsilon)$$

vanishes, for some $\varepsilon_0 > 0$ the measure of $W(\varepsilon_0)$ is positive.

Thus, Theorem 2 is proved for $j = 0$. To prove the theorem for $j \geq 0$, we must first perform j steps of the perturbation theory in order to reduce Hamiltonian (10.9) to the form (10.8), and then apply the generalized Theorem 1. All the details can be found in [230].

10.4 Theorem 2 can be applied to a number of problems of classical mechanics. As an example, take the Gross–Neveu system which describes the dynamics of interacting particles (see §6, Chap. I). The Hamiltonian function has the following form:

$$H = \frac{1}{2} \sum_{i=1}^{n} y_i^2 + \varepsilon \sum_{k<j} \cos(x_k - x_j) .$$

The set Δ consists of the vectors $e_k - e_j$, where $e_1 = (1, 0, \ldots, 0), \ldots, e_n = (0, \ldots, 0, 1)$. For $n \geq 3$ the vectors $(1, 0, \ldots, 0, -1)$ and $(1, 0, \ldots, -1, 0)$ are the vertices. It is clear that

$$j\alpha + \beta = (j + 1, 0, \ldots, -1, -j) .$$

The coordinates of these vectors are relatively prime for all $j \geq 0$. Consequently, according to Theorem 2, almost all points of the hyperplanes $\langle j\alpha + \beta, y \rangle = 0$ correspond to resonant n-dimensional tori which, when destroyed, give rise to hyperbolic $(n - 1)$-dimensional tori of the perturbed system. The existence of so many hyperbolic tori prevents complete integrability of the Gross–Neveu system.

11 Non-Autonomous Systems

11.1 Consider a non-autonomous system of canonical Hamilton equations

$$\dot{x}_i = \partial H / \partial y_i , \quad \dot{y}_i = -\partial H / \partial x_i ; \qquad 1 \leq i \leq n \tag{11.1}$$

with Hamiltonian $H(x, y, \phi)$ which is 2π-periodic in $\phi = \omega t$, $\omega = \text{const}$. Introduce two new conjugate canonical variables $\phi \bmod 2\pi$, ψ and the Hamiltonian function

$$\mathcal{H} = \omega\psi + H(x, y, \phi) .$$

The system (11.1) can be written in the form of an autonomous Hamiltonian system with $n+1$ degrees of freedom:

$$\begin{aligned} \dot{\phi} &= \partial \mathcal{H}/\partial \psi = \omega\,, \quad & \dot{\psi} &= -\partial \mathcal{H}/\partial \phi\,, \\ \dot{x} &= \partial \mathcal{H}/\partial y = \partial H/\partial y\,, \quad & \dot{y} &= -\partial \mathcal{H}/\partial x = -\partial H/\partial x\,. \end{aligned} \tag{11.2}$$

More precisely, projections of the trajectories of system (11.2), parameterized by the variable t, to the phase space of system (11.1) are solutions of Eqs. (11.1). Here we certainly assume that the constant $\phi_0 = \phi - \omega t$ vanishes.

Since $\partial \mathcal{H}/\partial \psi = \omega \neq 0$, one can apply the Whittaker procedure of reducing the order (see §1.4) to the autonomous Hamiltonian system (11.2). The number of degrees of freedom is reduced by one. As a result we obtain exactly the initial system (11.1).

Assume that Eqs. (11.1) admit a quasi-periodic solution

$$x = f(\omega t, \omega_1 t, \ldots, \omega_k t)\,, \quad y = g(\omega t, \omega_1 t, \ldots, \omega_k t)\,.$$

Here f and g are smooth functions on the $(k+1)$-dimensional torus $\mathbb{T}^{k+1} = \{\lambda, \lambda_1, \ldots, \lambda_k \bmod 2\pi\}$; $\omega, \omega_1, \ldots, \omega_k$ are incommensurable constant frequencies. We shall show, that the autonomous system (11.2) has an invariant $(k+1)$-dimensional torus with the same frequencies. Indeed, it is sufficient to put

$$\begin{aligned} &x = f(\lambda, \lambda_1, \ldots, \lambda_k)\,, \quad y = g(\lambda, \lambda_1, \ldots, \lambda_k)\,, \\ &\phi = \lambda\,, \quad \psi = -H(f, g, \lambda)/\omega\,, \qquad \dot{\lambda} = \omega\,,\ \dot{\lambda}_1 = \omega_1\,,\ \ldots\,,\ \dot{\lambda}_k = \omega_k\,. \end{aligned}$$

For $k=0$ we have a periodic solution.

Let a function $F(x, y, \phi)$, which is 2π-periodic in $\phi = \omega t$, be an integral of the system (11.1). Then it is also an integral of the system (11.2).

Assume that the Hamiltonian system (11.1) admits m involutive integrals $F_1, \ldots, F_m$ such that the rank of the Jacobian matrix

$$\frac{\partial(F_1, \ldots, F_m)}{\partial(x, y)}$$

equals m at some point of the trajectory of a periodic solution $t \to z(t)$. Then at least $2m+1$ multipliers of this solution equal one. The obvious proof uses Theorem 4, §8 and the relation

$$\operatorname{rank} \frac{\partial(\mathcal{H}, F_1, \ldots, F_m)}{\partial(\phi, \psi, x, y)} \geq m+1\,.$$

The analogous results are also valid for the invariant torus with $k \geq 1$.

11.2 Assume that a non-autonomous Hamiltonian system depends on a small parameter ε, and the system is completely integrable for $\varepsilon = 0$. In the action–angle variables $x \bmod 2\pi$, y of the unperturbed problem the Hamiltonian function has the following form:

$$H = H_0(y) + \varepsilon H_1(x, y, \phi) + o(\varepsilon)\,. \tag{11.3}$$

The function H is assumed to be 2π-periodic in ϕ (cf. §11.1).

Now pass to the autonomous system with Hamiltonian

$$\mathcal{H} = \mathcal{H}_0 + \varepsilon\mathcal{H}_1 + o(\varepsilon) , \qquad \mathcal{H}_0 = \omega\psi + H_0(y) , \quad \mathcal{H}_1 = H_1 . \tag{11.4}$$

In the perturbation theory for Hamiltonian systems the assumption of non-degeneracy of the unperturbed problem plays an essential role. Unfortunately, the Hessian of the function $\mathcal{H}_0$ with respect to the momenta ψ, y vanishes identically.

This difficulty can be overcome in the following way. Consider the system with Hamiltonian $\exp\mathcal{H}$. This system has the same trajectories, but the velocities of the motions along these trajectories differ by constant multipliers. In particular, periodic trajectories of the initial system are transformed by the change $\mathcal{H} \to \exp\mathcal{H}$ to periodic trajectories. It is clear that

$$\exp\mathcal{H} = \exp\mathcal{H}_0 + \varepsilon\mathcal{H}_1 \exp\mathcal{H}_0 + o(\varepsilon) . \tag{11.5}$$

Second derivatives of the new unperturbed Hamiltonian with respect to the momenta ψ, y are

$$\frac{\partial^2 \exp\mathcal{H}_0}{\partial y_i \partial y_j} = (a_{ij} + \omega_i\omega_j)e^{\mathcal{H}_0} ,$$

$$\frac{\partial^2 \exp\mathcal{H}_0}{\partial y_i \partial\psi} = \omega_i\omega e^{\mathcal{H}_0} , \qquad \frac{\partial^2 \exp\mathcal{H}_0}{\partial\psi^2} = \omega^2 e^{\mathcal{H}_0} ,$$

where

$$a_{ij} = \frac{\partial^2 H_0}{\partial y_i \partial y_j} , \qquad \omega_i = \frac{\partial H_0}{\partial y_i} .$$

The Hessian of the function $\exp\mathcal{H}_0$ with respect to y, ψ equals

$$(\exp\mathcal{H}_0)^{n+1}\omega^2 \det\|\partial^2 H_0/(\partial y_i \partial y_j)\| .$$

Consequently, if the function H_0 is non-degenerate with respect to the momenta y, then the unperturbed system (11.5) is also non-degenerate. This notation allows us to apply the results concerning perturbations of autonomous Hamiltonian systems to the system with Hamiltonian (11.5). Note that the replacement of a Hamiltonian by its exponent was first applied by Poincaré in the circular restricted three-body problem (see §2.1).

11.3 Assume that for $y = y_0$ the frequencies

$$\omega , \ \omega_1 = \partial H_0/\partial y_1 , \ \ldots , \ \omega_n = \partial H_0/\partial y_n$$

satisfy the single nontrivial resonance relation

$$k_0\omega + k_1\omega_1 + \cdots + k_n\omega_n = 0 .$$

Since $\omega \neq 0$, we have $\sum_{i\geq 1} |k_i| \neq 0$. It is natural to assume that the highest common divisor of the integer numbers $k_0, \ldots, k_n$ equals one.

We put $K = (k_0, k_1, \ldots, k_n)^T$. Let

$$H_1(x, y^0, \phi) = \sum H^\tau \exp[i(\tau_0\phi + \tau_1 x_1 + \cdots + \tau_n x_n)] , \qquad \tau \in \mathbb{Z}^{n+1} .$$

Introduce the 2π-periodic function of a single variable

$$h(\lambda) = \sum H^{K\mu} \exp(i\mu\lambda) , \qquad \mu \in \mathbb{Z} .$$

This function is the average of the perturbing function $H_1 : \mathbb{T}^{n+1} \to \mathbb{R}$ along the trajectories of the unperturbed problem (see §10.1).

Theorem 1. *Suppose that the function* (11.3) *is analytic in* x, y, ϕ, ε, *and for* $y = y^0$

1) $\det \|\partial^2 H_0/(\partial y_i \partial y_j)\| \neq 0$,

2) $\delta = \sum\limits_{i,j\geq 1} \dfrac{\partial^2 H_0}{\partial y_i \partial y_j} k_i k_j \neq 0$,

3) *for some positive constants* c *and* N *the relation*

$$|\omega_0 \nu_0 + \omega_1 \nu_1 + \cdots + \omega_n \nu_n| \geq c|\nu|^{-N}$$

holds for all integer $\nu_0, \ldots, \nu_n$ *which do not vanish simultaneously.*

Let $\lambda = \lambda^0$ *be a non-degenerate critical point of the function* h, *and* $h''(\lambda_0)\delta < 0$. *Then for small values of* $\varepsilon > 0$ *the perturbed system with Hamiltonian* (11.3) *has the family of quasi-periodic solutions*

$$x = \Phi + \sqrt{\varepsilon} f(\Phi, \phi, \sqrt{\varepsilon}) , \quad y = y^0 + \sqrt{\varepsilon} g(\Phi, \phi, \sqrt{\varepsilon}) , \qquad (11.6)$$

the functions f *and* g *being analytic and* 2π-*periodic in* $\Phi = (\Phi_1, \ldots, \Phi_n)$ *and* ϕ, *where*

$$\Phi_i = \omega_i t + \lambda_i^0 , \quad \phi = \omega t , \quad k_1 \lambda_1^0 + \cdots + k_n \lambda_n^0 \equiv \lambda^0 \bmod 2\pi .$$

In order to prove this theorem, we pass to an autonomous Hamiltonian system with $n+1$ degrees of freedom which is defined by the Hamiltonian (11.4), and then replace the Hamiltonian $\mathcal{H}$ by $\exp \mathcal{H}$. According to the results of §11.2, assumption 1) implies the non-degeneracy of the system with the Hamiltonian $\exp \mathcal{H}_0$. Let Π be the Hessian of the function $\exp \mathcal{H}_0$ with respect to the momenta y, ψ. It is easy to show that $K^T \Pi K$ coincides with the number δ from condition 2). According to the assumption $\delta \neq 0$. Now we can use Theorem 1, §10. Conditions 1) and 3) of this theorem are certainly valid. Since $h''(\lambda^0)\delta < 0$, condition 2) is also valid. Consequently, the perturbed Hamiltonian system with Hamiltonian (11.5) has for small values of $\varepsilon > 0$ an n-dimensional hyperbolic invariant torus, filled by trajectories of quasi-periodic motions. This torus is analytic in ε for $\varepsilon > 0$. For $\varepsilon = 0$ it coincides with the closure of the quasi-periodic trajectories

$$\phi = \omega t + \phi^0 , \quad x_i = \omega_i t + x_i^0 , \qquad \psi = \psi^0 , \quad y = y^0 ,$$
$$k_0 \phi^0 + \sum_{i\geq 1} k_i x_i^0 \equiv \lambda^0 \bmod 2\pi .$$

The constant ψ^0 is not essential, hence we can put $\psi^0 = 0$. By projecting the perturbed n-dimensional hyperbolic torus to the space of the variables x, y, ϕ, we obtain an n-dimensional torus filled by quasi-periodic trajectories with n incommensurable frequencies. Q.E.D.

In a typical situation (e.g., when h is a Morse function) there exists a critical point λ^0 of the function h with negative value of $h''(\lambda^0)\delta$. For $n = 1$ we obtain $2\pi/\omega$-periodic solutions of the perturbed problem. In fact, this case is considered in the classical Poincaré theorem on the birth of non-degenerate periodic solutions (Theorem 5, §8). Note two special properties. First, the perturbed solutions depend on ε analytically for all ε close to zero (not only for $\varepsilon > 0$, see Theorem 1). Second, for $\varepsilon > 0$ the perturbed periodic solutions of elliptic type correspond to critical points of the function h at which $h''(\lambda^0)\delta > 0$. These solutions are stable in the linear approximation. The conditions of their Lyapunov stability are indicated in [171].

In the variables $x \bmod 2\pi$, y, $\phi \bmod 2\pi$ the trajectories of quasi-periodic solutions (11.6) lie on n-dimensional hyperbolic tori, on which any n involutive single-valued integrals of the Hamiltonian system with Hamiltonian (11.3) are dependent. Hence the birth of a large amount of n-dimensional hyperbolic tori prevents the integrability of the perturbed problem.

11.4 Consider the particular case, when the Hamiltonian function has the form

$$H = H_0(y) + \varepsilon H_1(x, \phi) , \qquad \phi = \omega t , \quad \omega > 0 , \tag{11.7}$$

where $H_0 = \left(\sum a_{ij} y_i y_j\right)/2$ is a positive definite quadratic form with constant coefficients, and the perturbing function H_1 is a trigonometric polynomial in $x_1, \ldots, x_n, \phi$.

Following [110], we study the existence problem of a complete set of commuting integrals which are formal power series

$$\sum F_k(x, y, \phi)\varepsilon^k \tag{11.8}$$

with single-valued analytic coefficients defined in the direct product $\mathbb{T}^n_x \times \mathbb{R}^n_y \times \mathbb{T}^1_\phi$.

Let $\xi = (\xi_0, \xi_1, \ldots, \xi_n)$, $\eta = (\eta_0, \eta_1, \ldots, \eta_n)$ be two vectors of $\mathbb{R}^{n+1}$. We put

$$\langle \xi, \eta \rangle = \omega \xi_0 \eta_0 + \sum a_{ij} \xi_i \eta_j \, .$$

It is clear that the form $\langle \, , \, \rangle$ defines a scalar product in $\mathbb{R}^{n+1}$.

Assume that

$$H_1 = \sum h_\tau \exp[i(\tau_0 \phi + \tau_1 x_1 + \cdots + \tau_n x_n)] \, ,$$

and introduce the "spectrum" Δ of the function H_1 as the set of $\tau \in \mathbb{Z}^{n+1}$ for which $h_\tau \neq 0$. The set Δ is finite and invariant with respect to the involution $\tau \to -\tau$.

Theorem 2 [110]. *The Hamiltonian system with Hamiltonian* (11.7) *admits n commuting integrals of the form* (11.8), *which are independent for $\varepsilon = 0$ if and only if the points of the set Δ lie on n straight lines intersecting orthogonally* (*in the metric* $\langle \, , \, \rangle$) *at the origin.*

The sufficiency of the conditions of Theorem 2 is obvious: in this case variables in the Hamiltonian function separate,and the equations of motion have n integrals quadratic in the momenta. For example, under the conditions of Theorem 2, for $n = 1$ the function H_1 has the form $f(kx + l\phi)$, where $f(\cdot)$ is some trigonometric polynomial of a single variable, and the integer numbers k, l are relatively prime. The function

$$ly/k + H_0(y) + \varepsilon f(kx + l\phi)$$

is an integral for $k \neq 0$. The case $k = 0$ is trivial: the function H_0 is an integral of the equations of motion.

The necessity of the conditions of Theorem 2 is established in [110] by the method of [155] (see §5).

11.5 Suppose that the points of Δ do not lie on a single straight line (otherwise the system with Hamiltonian (11.7) is obviously completely integrable). Introduce the vertex α of the set Δ and the adjacent vertex β; the vectors α and β are linearly independent. We put $Y = (\omega, y_1, \dots, y_n)$.

Theorem 3. *Suppose that*
1) *the ratio* $2\langle\alpha, \beta\rangle / \langle\alpha, \alpha\rangle$ *does not belong to the set* $\{0, -1, -2, \dots\}$,
2) *for all integer* $j \geq 0$ *the components of each vector* $j\alpha + \beta$ *are relatively prime.*

Then on each plane $\pi_j = \{y \in \mathbb{R}^n : \langle Y, j\alpha + \beta\rangle = 0\}$ *there exists a subset* $W_j(\varepsilon_0)$ *such that for all* $y \in W_j(\varepsilon_0)$ *the Hamiltonian system with Hamiltonian* (11.7) *has for* $0 < \varepsilon < \varepsilon_0$ *hyperbolic n-dimensional tori which depend smoothly on* $\varepsilon \geq 0$. *Moreover, for each* $j \geq 0$ *there exists* $\varepsilon_0(j) > 0$ *such that the measure of* $W_j(\varepsilon_0(j))$ *is positive, and the measure of the set*

$$\pi_j \setminus \bigcup_{\varepsilon > 0} W_j(\varepsilon)$$

vanishes.

This result extends Theorem 2, §10 to non-autonomous systems. The proof uses the classical scheme of perturbation theory applied to the Hamilton system with Hamiltonian (11.7) (the details can be found in [110]), and also the generalized version of Theorem 1 from §11.3, concerning analytic Hamiltonians of the form

$$\sum_{r=0}^{k-1} \varepsilon^r H_r(y) + \varepsilon^k H_k(x, y, \phi) + o(\varepsilon^k) .$$

Theorem 3 is established in [110] for $n = 1$. More precisely, all resonant two-dimensional tori of the unperturbed problem, which correspond to non-zero $y \in \pi_j$, are destroyed by the perturbation. Moreover, the perturbed problem has an even number of non-degenerate periodic solutions. Half of them consist of hyperbolic solutions, and the other half of elliptic ones.

11.6 Now we apply the results of §5 to systems of pendulum type considered in §4, Chap. I.

a) Consider the forced oscillations of the mathematic pendulum which are described by the equation

$$\ddot{q} + \Omega^2(t)\sin q = 0\,, \quad \Omega^2 = \omega_0^2(1 - \varepsilon\cos\nu t)\,.$$

Here $\omega_0 = \text{const} > 0$, ε is a small parameter, and ν is the frequency of the external force. Introduce the canonical momentum $p = \dot{q}$ and replace this equation by the Hamilton system with Hamiltonian

$$H_\varepsilon = p^2/2 - \Omega^2\cos q\,.$$

For $\varepsilon = 0$ we have a completely integrable system—the mathematical pendulum. In this problem we can pass to the action–angle variables $I, \phi \bmod 2\pi$:

$$\begin{aligned} I &= \frac{1}{2\pi}\int_0^{2\pi}\sqrt{2(H_0 + \omega_0^2\cos q)}\,dq\,, \qquad H_0 = H_0(I), \\ \phi &= \omega\int_0^q \frac{dx}{\sqrt{H_0(I) + \omega_0^2\cos x}}\,, \qquad \omega(I) = \frac{dH_0}{dI}\,. \end{aligned} \tag{11.9}$$

Here, for definiteness, we consider the domain of rotational motions $\{H_0 > \omega_0^2\}$.

The perturbing function H_1 equals $\cos q\cos\nu t$. In order to expand H_1 into a multiple Fourier series in the variables $\phi, \nu t$, we replace x by $2u$ in the second relation of (11.9). Then

$$\phi = \frac{2\omega}{\sqrt{H_0 + \omega_0^2}}\int_0^{q/2}\frac{du}{\sqrt{1 - k^2\sin^2 u}}\,, \qquad k^2 = \frac{2\omega_0^2}{H_0 + \omega_0^2} < 1\,.$$

Consequently

$$\begin{aligned} &\sin\frac{q}{2} = \operatorname{sn}\left(\frac{\sqrt{a}}{2\omega}\phi\right), \quad \cos\frac{q}{2} = \operatorname{cn}\left(\frac{\sqrt{a}}{2\omega}\phi\right), \\ &\cos q = \operatorname{cn}^2\left(\frac{\sqrt{a}}{2\omega}\phi\right) - \operatorname{sn}^2\left(\frac{\sqrt{a}}{2\omega}\phi\right), \qquad a = H_0 + \omega_0^2\,. \end{aligned}$$

Since

$$\begin{aligned} \omega^{-1} = \frac{dI}{dH_0} &= \frac{1}{2\pi}\int_0^{2\pi}\frac{dq}{\sqrt{H_0 + \omega_0^2\cos q}} \\ &= \frac{1}{\pi\sqrt{a}}\int_0^{\pi}\frac{du}{\sqrt{1 - k^2\sin^2 u}} = \frac{2K}{\pi\sqrt{a}}\,, \end{aligned} \tag{11.10}$$

we get

$$\cos q = \operatorname{cn}^2\frac{K}{\pi}\phi - \operatorname{sn}^2\frac{K}{\pi}\phi\,. \tag{11.11}$$

Here K is the complete elliptic integral of the first kind with modulus k.

We use the well-known Jacobi formula

$$(kK)^2 \mathrm{sn}^2 \frac{Kx}{\pi} = K^2 - KE - 2\pi^2 \sum_{n=1}^{\infty} \frac{nq^n}{1-q^{2n}} \cos 2nx ,$$

where $q = \exp(-\pi K'/K)$. The constant K' (respectively E) is the complete elliptical integral of the first (second) kind with modulus $\sqrt{1-k^2}$ (k). Taking into account (11.11), we obtain the expansion of the function H_1 in Fourier series:

$$H_1 = \sum_{-\infty}^{\infty} h_{m,1}(I) e^{i(2m\phi+\nu t)} + \sum_{-\infty}^{\infty} h_{m,-1}(I) e^{i(2m\phi-\nu t)} .$$

The Fourier coefficients $h_{m,\pm 1}$ can be easily calculated with the help of the Jacobi formula; they do not vanish.

In this problem the Poincaré set $\mathbf{P}_*$ (see §1) consists of the values of the action variable which satisfy the resonance relations

$$\pm 2n\omega(I) + \nu = 0 , \qquad n \in \mathbb{Z} \setminus \{0\} .$$

According to (11.10), the frequency ω tends to zero as $H_0 \to \omega_0^2$. Note that for $H_0 = \omega_0^2$ we have the motion along the separatrices. Consequently, the set $\mathbf{P}_*$ consists of an infinite number of points $I \in \mathbb{R}$ which accumulate near the point

$$I_c = \frac{\omega_0}{2\pi} \int_0^{2\pi} \sqrt{2(1+\cos q)}\, dq = \frac{4\omega_0}{\pi} .$$

It is easy to verify that the values $I \in \mathbf{P}_*$ satisfy the conditions of the Poincaré theorem on the birth of pairs of non-degenerate periodic solutions. The existence of an infinite number of families of non-degenerate periodic solutions in the perturbed problem prevents its integrability.

b) The motion of a charged particle in the field of two equal electric waves, which move in opposite directions with equal velocities ω, is described by the equation

$$\ddot{x} = \varepsilon[\sin(x+\omega t) + \sin(x-\omega t)] . \tag{11.12}$$

Introduce the momentum $y = \dot{x}$ and represent Eq. (11.12) in the form of the canonical equations with Hamiltonian

$$H = y^2/2 - \varepsilon[\cos(x+\omega t) + \cos(x-\omega t)] .$$

The set Δ consists of four vectors: $(\pm 1, \pm 1)$. For the standard lexicographic order we have $\alpha = (1,1)$, $\beta = (1,-1)$. The vectors $j\alpha + \beta$ have the components $j+1$ and $j-1$. These integer numbers are relatively prime for all even $j \geq 0$. Consequently, according to the results of §11.4, the resonant tori

$$\omega(j+1) + y(j-1) = 0 , \qquad j = 0, 2, 4, \ldots$$

are destroyed for small values of $\varepsilon \neq 0$. For $j \to \infty$ these tori accumulate near the resonant torus $y = -\omega$ which is destroyed by the perturbation. Changing the lexicographic order, we can obtain a sequence of destroyed resonant tori which accumulate near the straight line $y = \omega$.

The relations $y = \pm\omega$ were called (by Landau) resonances in the problem of the interaction of particles with waves. The corresponding tori are destroyed on the first step of perturbation theory. The numerical analysis of Eq. (11.12) is discussed for $\varepsilon > 1$ in [241].

c) The restricted problem on the rotation of a heavy rigid body with a fixed point with zero area constant is described by the canonical Hamilton equations with Hamiltonian

$$H = \eta^2/2 + \sin\sqrt{2h}t \sin\xi \,.$$

Here h is the energy constant. We introduce the new time $\tau = \sqrt{2h}t$ and rewrite the equation of oscillations in form which is analogous to (11.12):

$$\xi'' + \frac{1}{2h}\sin\tau\cos\xi = 0 \,. \tag{11.13}$$

The role of a small parameter is played by h^{-1}. The results of §11.5 imply the existence in Eq. (11.13) of an infinite number of distinct families of non-degenerate periodic solutions if the energy h is sufficiently large. This effect prevents the integrability of Eq. (11.13).

V Splitting of Asymptotic Surfaces

Non-degenerate hyperbolic invariant toriof Hamiltonian systems have asymptotic invariant manifolds filled by trajectories which tend to quasi-periodic orbits in the hyperbolic tori as $t \to \pm\infty$. In integrable Hamiltonian systems such manifolds, as a rule, coincide. In the nonintegrable cases, the situation is different: asymptotic surfaces can have transverse intersections forming a complicated tangle. "One will be struck by the complexity of this figure which I do not even attempt to draw. Nothing more properly gives us an idea of the complication of the problem of three bodies and, in general, of all problems in Dynamics where there is no uniform integral..." (Poincaré [203]).

This chapter contains methods, going back to Poincaré, of proving nonintegrability by analysing of asymptotic surfaces of Hamiltonian systems which are close to completely integrable systems.

1 Asymptotic Surfaces and Splitting Conditions

1.1 Let M^{2n} be the phase space with symplectic structure Ω, and $H = H_0 + \varepsilon H_1 + O(\varepsilon^2)$ the Hamiltonian function. Suppose that for $\varepsilon = 0$ the Hamiltonian system has an m-dimensional hyperbolic invariant torus T_0^m (see Chap. IV, §9). Recall that in a neighborhood of such a torus it is possible to introduce symplectic coordinates $x \bmod 2\pi$, y, z^-, z^+, with the following properties:

a) $x = (x_1, \dots, x_m)$, $y = (y_1, \dots, y_m)$, $z^\pm = (z_1^\pm, \dots, z_{n-m}^\pm)$,

b) $\Omega = dy \wedge dx + dz^- \wedge dz^+$,

c) $T_0^m = \{(x, y, z^-, z^+) : y = 0,\ z^\pm = 0\}$,

d) $H_0 = (\nu, y) + (Ay, y)/2 + (z^-, Bz^+) + O_3(y, z)$.

Here ν is a constant vector satisfying the "strong non-resonance" condition:

$$|(\nu, k)| \geq \alpha |k|^{-\beta}, \qquad k \in \mathbb{Z}^m \setminus \{0\}, \qquad \alpha, \beta > 0 .$$

The square matrix A is constant and non-degenerate. The $(n-m) \times (n-m)$ matrix $B(x)$ satisfies the inequality

$$\operatorname{Re}(\zeta, B\bar{\zeta}) \geq \mu |\zeta|^2, \qquad \mu > 0, \tag{1.1}$$

for all $x \bmod 2\pi$ and $\zeta \in \mathbb{C}^{n-m}$. We call x, y, $z^\pm$ the canonical coordinates of the torus $\Gamma = T_0^m$.

In what follows we are most interested in the cases $m = 0$ and $m = 1$. If $m = 0$, then the hyperbolic torus is an unstable equilibrium point $z^- = z^+ = 0$, and z^-, z^+ are the conjugate symplectic coordinates. Then $H_0 = (z^-, Bz^+) + \cdots$, where $B = \text{const}$. In these coordinates, we have

$$\dot{z}^- = -B^T z^- + \cdots, \qquad \dot{z}^+ = Bz^+ + \cdots. \tag{1.2}$$

The assumption (1.1) implies that the eigenvalues of the matrix B $(-B^T)$ lie in the right (left) half-plane (Chap. IV, §4.9).

For $m = 1$, we have an unstable periodic solution. We restrict the Hamiltonian system to the $(2n-1)$-dimensional energy surface containing the trajectory of this solution. Then Eqs. (1.2) are the variational equations of the given periodic orbit. Thus, half of its multipliers lie inside the unit circle, and the other half outside.

We assume that the Hamiltonian function H is analytic in ε. Then, by the implicit function theorem, hyperbolic equilibria do not vanish after adding a small perturbation, but only change slightly. In the case of periodic solutions, the perturbed system has a periodic solution with the same period on a nearby energy level surface. It turns out that the same holds for multidimensional hyperbolic invariant tori: Graff's theorem [92] implies that for small ε the perturbed system has an invariant hyperbolic torus T^m_ε with the same frequencies of quasi-periodic motions $\nu = (\nu_1, \ldots, \nu_m) = \text{const}$. The perturbed tori T^m_ε depend analytically on ε.

It can be shown [92] that each hyperbolic torus T^m_ε lies in the intersection of two invariant Lagrangian surfaces Λ^+_ε and Λ^-_ε filled with trajectories of the Hamiltonian system which tend to the torus T^m_ε as $t \to +\infty$ and $t \to -\infty$ respectively. For example, if we drop the $O_3(y, z)$-terms in the expansion of the Hamiltonian, then near the torus T^m_ε the surface Λ^+ is given by the equations $y = 0$, $z^+ = 0$. The variables $x \bmod 2\pi$ are local coordinates on Λ^+. By (1.1), Eqs. (1.2) show that the variables z^- tend to zero exponentially fast. Hence, the solutions of the Hamiltonian system with trajectories in Λ^+ tend exponentially to quasi-periodic motions in the hyperbolic invariant torus. It is natural to call such solutions asymptotic. Therefore, the invariant manifolds $\Lambda^\pm_\varepsilon$ are called *asymptotic surfaces*. They are also called the stable and unstable invariant manifolds of the hyperbolic torus T^m_ε.

In the case of two degrees of freedom, we have two-dimensional asymptotic surfaces in the three-dimensional level surface of the energy integral. As a rule, for integrable systems such surfaces separate the regions with different topological type of phase trajectories (recall the phase portrait of the mathematical pendulum). For this reason asymptotic surfaces are often called *separatrices*.

We must keep in mind that the surfaces $\Lambda^\pm_\varepsilon$ are globally defined and depend analytically on ε (it is sufficient to extend the asymptotic trajectories to the whole time axis). However, in general they are immersed but not embedded submanifolds of the phase space. Recall (Chap. II, §2) that an n-dimensional surface $\Sigma \subset M^{2n}$ is called Lagrangian if the value of the 2-form Ω is zero for any pair of tangent vectors to Σ.

For hyperbolic equilibria, the existence of Lagrangian asymptotic surfaces was proved by Lyapunov, Kneser and Bohl with different generality. The case of hyperbolic periodic orbits was first studied by Poincaré [203, Chap. 7].

Suppose that for $\varepsilon = 0$ the Hamiltonian system is completely integrable: there exist n analytic integrals $F_1, \dots, F_n$ which are pairwise in involution and independent almost everywhere. Since the hyperbolic torus T_0^m is non-resonant and the manifolds $\Lambda_0^\pm$ consist of asymptotic trajectories, the functions F_j are constant on $\Lambda_0^\pm$. Thus, $\Lambda_0^\pm$ are contained in some closed subset

$$\{z \in M^{2n} : F_1(z) = c_1 \,, \ \dots \,, \ F_n(z) = c_n\}.$$

The results of Chap. II, §9 imply that the point $c = (c_1, \dots, c_n)$ is a critical value of the mapping $F : M^{2n} \to \mathbb{R}^n$.

For a typical integrable system, there are two possibilities:

1) $\Lambda_0^+ = \Lambda_0^-$,
2) the stable and unstable surfaces of different hyperbolic m-dimensional hyperbolic tori Γ_1 and Γ_2 coincide.

In both cases asymptotic solutions are doubly asymptotic. Following Poincaré, we divide doubly asymptotic solutionsinto two classes: *homoclinic* (when $\Gamma_1 = \Gamma_2$), and *heteroclinic* (when $\Gamma_1 \neq \Gamma_2$). In the heteroclinic case we also denote the merged asymptotic surfaces by Λ_0^+ and Λ_0^-. Note that in general the sets $\Lambda_0^\pm$, regarded as subsets of M^{2n}, are not manifolds, but stratified sets.

Poincaré [202] was the first to notice that in general for small values of $\varepsilon \neq 0$ the perturbed surfaces Λ_ε^+ and Λ_ε^-, regarded as subsets of M^{2n}, do not coincide. This phenomenon is called *splitting of asymptotic surfaces*. It is an obstruction to integrability of the perturbed Hamiltonian system (see §5.2).

Of course, it can happen that the hyperbolic tori $\Gamma_1(\varepsilon)$ and $\Gamma_2(\varepsilon)$ lie on different energy surfaces. Since the asymptotic surfaces are contained in the same energy levels as the corresponding hyperbolic tori, the problem of splitting of the surfaces Λ_ε^+ and Λ_ε^- is trivial in this case. Thus, we assume that the tori $\Gamma_1(\varepsilon)$ and $\Gamma_2(\varepsilon)$ lie on the same $(2n-1)$-dimensional level surface of the energy integral. Obviously, in the homoclinic case this assumption is always satisfied.

Poincaré obtained the conditions for splitting of asymptotic surfaces in the case $m = 1$. Later this result was generalized by many authors (see, for example, [34, 173, 100, 235]). The following analysis of the splitting problem was carried out by Treshchëv.

1.2 Let $t \to \gamma(t)$ be a doubly asymptotic solution of the unperturbed completely integrable Hamiltonian system. Denote [34]

$$I_i(\gamma) = \lim_{T\to\infty}\left[\int_{-T}^{T} \{F_i, H_1\}(\gamma(t))\,dt \right.$$
$$\left. + \{F_i, \chi\}(\gamma(-T)) - \{F_i, \chi\}(\gamma(T))\right],$$
$$J_{ij}(\gamma) = \lim_{T\to\infty}\left[\int_{-T}^{T} \{F_i\{F_j, H_1\}\}(\gamma(t))\,dt \right. \tag{1.3}$$
$$\left. + \{F_i, \{F_j, \chi\}\}(\gamma(-T)) - \{F_i, \{F_j, \chi\}\}(\gamma(T))\right].$$

In these expressions χ is an analytic function defined in the canonical coordinates of the hyperbolic torus by the equation

$$\left(\nu, \frac{\partial \chi}{\partial x}\right) + \eta(x) = \langle\eta\rangle\,, \tag{1.4}$$

where

$$\eta(x) = H_1(x, y, z^-, z^+)\big|_{y=0,\ z^{\pm}=0}\,, \qquad \langle\eta\rangle = \frac{1}{(2\pi)^m}\int_{\mathbb{T}^m} \eta(x)\, d^m x\,.$$

Due to the strong non-resonance condition on the frequencies $\nu_1, \ldots, \nu_m$, Eq. (1.4) has an analytic solution which is unique up to addition of a constant. The problem of extending the function χ to the whole phase space is irrelevant, since expressions (1.3) depend only on the values of the function χ in small neighborhoods of the hyperbolic tori.

Theorem 1. *The following statements hold*:
1) *the limits* (1.3) *exist*;
2) *if* $I_s \neq 0$ *for at least one* $s = 1, \ldots, n$, *then for small* $\varepsilon \neq 0$ *the surfaces* Λ_ε^+ *and* Λ_ε^- *do not coincide*;
3) *if* $I_1(\gamma) = \cdots = I_n(\gamma) = 0$ *and the rank of the matrix* $\|J_{ij}(\gamma)\|$ *is equal to* $n-1$, *then for small* ε *the* n*-dimensional surfaces* Λ_ε^+ *and* Λ_ε^- *intersect transversely along a doubly asymptotic trajectory*γ_ε *in the* $(2n-1)$*-dimensional energy level. Moreover,* $\gamma_\varepsilon \to \gamma$ *as* $\varepsilon \to 0$.

Remark. For $F_1 = H_0$, the condition rank $\|J_{ij}\| = n-1$ is equivalent to the condition that the $(n-1)\times(n-1)$ matrix $\|J_{ij}\|_{2\le i,j\le n}$ is non-degenerate.

In the proof of Theorem 1, we shall use the following result from symplectic topology. Let $\Lambda \subset M$ be a Lagrangian manifold. Then for every regular point of Λ there exists a neighborhood U with symplectic coordinates p, q, such that $\Omega = dp \wedge dq$ and the set $\Lambda \cap U$ is given by the equation $p = 0$ [15]. For example, suppose that Λ is given by the equation $y = \partial S/\partial x$ (see Chap. II, §2). Then the required canonical coordinates p, q are introduced by the canonical transformation

$$q = x\,, \quad p = y - \partial S/\partial x\,.$$

In the special canonical coordinates of the Lagrangian manifold $\Lambda_0^+ = \Lambda_0^-$ (see §1.1), the perturbed asymptotic surfaces $\Lambda_\varepsilon^\pm$ are defined by the equations $p = \varepsilon f^\pm(q, \varepsilon)$. Since $\Lambda_\varepsilon^\pm$ are Lagrangian, we have

$$df^\pm \wedge dq = \frac{1}{\varepsilon}\Omega|_{\Lambda_\varepsilon^\pm} = 0 .$$

Hence,

$$f^\pm = \frac{\partial S^\pm(q, \varepsilon)}{\partial q} .$$

It can be proved that the functions

$$S_0^\pm = S^\pm\big|_{\varepsilon=0} : \Lambda_0^\pm \to \mathbb{R}$$

are correctly defined (i.e., they do not depend on the choice of the coordinates p, q).

The assumptions of §1.1 imply that the asymptotic surfaces $\Lambda_\varepsilon^\pm$ lie on the same energy surface $H = H_0(p, q) + \varepsilon H_1(p, q) + o(\varepsilon) = h(\varepsilon)$. Hence, the functions $S^\pm(q, \varepsilon)$ satisfy the Hamilton–Jacobi equation

$$H\left(\varepsilon \frac{\partial S^\pm}{\partial q}, q, \varepsilon\right) = h(\varepsilon) .$$

The first order terms in ε yield

$$\frac{\partial H_0}{\partial p}(0, q)\,\frac{\partial S_0^\pm}{\partial q} + H_1(0, q) = h_1 , \tag{1.5}$$

where

$$h_1 = \frac{d}{d\varepsilon}\bigg|_{\varepsilon=0} h .$$

From (1.5) we obtain the equation

$$\frac{d}{dt} S_0^\pm(\gamma(t)) + H_1(\gamma(t)) = h_1 .$$

Hence,

$$S_0^\pm(\gamma(t_2)) - S_0^\pm(\gamma(t_1)) = \int_{t_1}^{t_2} (H_1 - h_1)|_{\gamma(t)}\, dt . \tag{1.6}$$

Lemma 1. $S_0^\pm - \chi = \mathrm{const} + O(y, z)$.

Corollary. *For any $i_1, \dots, i_k$, we have*

$$\{F_{i_1}, \{F_{i_2}, \dots, \{F_{i_k}, S_0 - \chi\} \cdots\}\} = O(y, z) . \tag{1.7}$$

Indeed, in a neighborhood of the hyperbolic torus every integral of the Hamiltonian equations has the form

$$\mathrm{const} + (Y_0, y) + O_2(y, z) , \qquad Y_0 = \mathrm{const}$$

(see Chap. IV, §9). Hence, the corollary follows from Lemma 1. The proof of Lemma 1 is contained in §1.3.

Suppose that the asymptotic surfaces Λ_ε^+ and Λ_ε^- coincide. Then

$$\frac{\partial}{\partial q}(S_0^+ - S_0^-) = 0 .$$

Extending the functions $S_0^\pm$ to a small neighborhood of $\Lambda_0^+ = \Lambda_0^-$, we obtain:

$$\{F_i, S_0^+ - S_0^-\}\big|_{\Lambda_0^\pm} = 0 , \qquad i = 1, \dots, n . \tag{1.8}$$

Then, using (1.6) and Lemma 1, we get

$$-\{F_i, S_0^+\}(\gamma(0)) = \int_0^T \{F_i, H_1\}(\gamma(t))\, dt - \{F_i, S_0^+\}(\gamma(T))$$
$$= \lim_{T\to\infty} \left[\int_0^T \{F_i, H_1\}(\gamma(t))\, dt - \{F_i, \chi\}(\gamma(T)) \right] .$$

There is a similar expression for the function S_0^-. Hence,

$$\{F_i, S_0^-\}(\gamma(0)) - \{F_i, S_0^+\}(\gamma(0)) = I_i(\gamma) . \tag{1.9}$$

Thus, the limits in the expressions (1.3) for I_i really exist and the non-splitting conditions (1.8) take the form $I_1 = \dots = I_n = 0$.

The proof of statement 3) of Theorem 1 is based on the implicit function theorem. We are looking for a solution of the equations

$$\frac{\partial S^-}{\partial q}(q_0) = \frac{\partial S^+}{\partial q}(q_0) . \tag{1.10}$$

If a solution exists, then the surfaces Λ_ε^+ and Λ_ε^- have non-empty intersection. Since the functions $S^\pm$ satisfy the same Hamilton–Jacobi equation, one of Eqs. (1.10) is dependent on the others.

By (1.10), we have

$$\{F_i, S^+ - S^-\} = 0 , \qquad i = 1, \dots, n . \tag{1.11}$$

The Poisson brackets are calculated at the point

$$q = q_0 , \quad p = \varepsilon \frac{\partial S^\pm}{\partial q}(q_0) .$$

From (1.9) it follows that for $\varepsilon = 0$ Eqs. (1.11) take the form $I_1(\gamma) = \dots = I_n(\gamma) = 0$, where γ is the doubly asymptotic trajectory of the unperturbed system passing through the point $(p, q) = (0, q_0)$. If these equations hold, then the existence of a solution of Eqs. (1.11) depends on the properties of the Jacobian matrix

$$J = \left\| \frac{\partial}{\partial q_i}\{F_j, S_0^+ - S_0^-\} \right\| ,$$

calculated at the point $(p, q) = (0, q_0)$. The elements of J are the limits J_{ij} in (1.3). Since the Eqs. (1.11) are dependent, the matrix J is degenerate. On the other hand, the condition $\operatorname{rank} J = n-1$ guarantees transverse intersection of the surfaces Λ_ε^+ and Λ_ε^- along a doubly asymptotic trajectory γ_ε, which is close to γ.

If $F_1 = H_0$, then $I_1(\gamma) = 0$ for every doubly asymptotic solution γ of the unperturbed system. Indeed, consider the integral

$$\int_{-T}^{T} \{H_0, H_1\}(\gamma(t))\, dt = \int_{-T}^{T} \frac{d}{dt} H_1(\gamma(t))\, dt = H_1(\gamma(T)) - H_1(\gamma(-T)) .$$

On the other hand, in the canonical coordinates of the torus Γ we have:

$$\{H_0, \chi\} = \left(\nu, \frac{\partial \chi}{\partial x}\right) + O(y, z) = \eta(x) - \langle \eta \rangle + O(y, z) .$$

Now it is sufficient to pass to the limit as $T \to \infty$.

1.3 Now we shall prove Lemma 1. In a neighborhood of the hyperbolic torus Γ, we can take $q = (x, z^\mp)$, $p = (y, z^\pm) + O_2(z^\mp)$ for the special canonical coordinates of the asymptotic surfaces $\Lambda^\pm$. Near Γ_ε we have

$$\Lambda_\varepsilon^\pm = \left\{ (y, x, z) : y = \varepsilon \frac{\partial S_0^\pm}{\partial x}\bigg|_\Gamma + O((z^\mp)^2, \varepsilon z^\mp, \varepsilon^2) , \right.$$
$$\left. z^\pm = \varepsilon \frac{\partial S_0^\pm}{\partial z^\mp}\bigg|_\Gamma + O((z^\mp)^2, \varepsilon z^\mp, \varepsilon^2) \right\} .$$

Since $\Gamma_\varepsilon \subset \Lambda_\varepsilon^+ \cap \Lambda_\varepsilon^-$,

$$\Gamma_\varepsilon = \left\{ (y, x, z) : z^\pm = \varepsilon \frac{\partial S_0^\pm}{\partial z^\mp}\bigg|_\Gamma + O(\varepsilon^2), \right.$$
$$\left. y = \frac{\partial S_0^+}{\partial x}\bigg|_\Gamma + O(\varepsilon^2) = \frac{\partial S_0^-}{\partial x}\bigg|_\Gamma + O(\varepsilon^2) \right\} .$$

Hence,

$$(S_0^+ - S_0^-)|_\Gamma = \text{const} .$$

In the relation $H|_{\Gamma_\varepsilon} = \text{const}$, the first order term in ε is as follows

$$\left(\nu, \frac{\partial}{\partial x} S_0^\pm|_\Gamma\right) + \eta(x) - \langle \eta \rangle .$$

Hence, the equations for the functions χ and $S_0^\pm|_\Gamma$ coincide. Thus, since the frequencies $\nu_1, \dots, \nu_m$ are non-resonance, these functions differ by a constant. The lemma is proved.

1.4 The case $m = 1$ is the most interesting for applications. Poincaré studied it in [203]. We consider briefly the bifurcation of asymptotic surfaces for non-autonomous time-periodic Hamiltonian systems. Let

$$H = H_0(x, y) + \varepsilon H_1(x, y, t) + o(\varepsilon) .$$

The perturbing function H_1 is periodic in t with period τ. Suppose that there exist two critical points x^-, y^- and x^+, y^+ of the function H_0 such that the eigenvalues of the linearized Hamiltonian system

$$\dot{y} = -\frac{\partial H_0}{\partial x}, \quad \dot{x} = \frac{\partial H_0}{\partial y}$$

are real and non-zero. Hence, the points $x^\pm$, $y^\pm$ are hyperbolic equilibrium positions. Suppose, for example, that $H_0 = T + V$ is the Hamiltonian function of a reversible system. Then non-degenerate local maxima of the potential energy correspond to hyperbolic equilibria with real eigenvalues.

In the extended phase space of variables x, y, $t \bmod 2\pi$, to the critical points $x^\pm, y^\pm$ there correspond τ-periodic hyperbolic solutions.

Let λ_0^+ (λ_0^-) be the stable (unstable) asymptotic manifold passing through the point x^+, y^+ (or x^-, y^-) in the phase space of the unperturbed system. In the extended phase space, the Cartesian products $\Lambda_0^\pm = \lambda_0^\pm \times \mathbb{T}_t^1$ are the asymptotic surfaces of the corresponding τ-periodic hyperbolic solution.

We assume that the unperturbed system is completely integrable: it admits n commuting integrals $F_1, \dots, F_n$, independent almost everywhere. One of them may be the Hamiltonian function H_0. Obviously, $dF_i = 0$ for $x = x^\pm$, $y = y^\pm$.

Suppose that $\lambda_0^+ = \lambda_0^-$. Then, obviously, $\Lambda_0^+ = \Lambda_0^-$. Let $x = x_a(t)$, $y = y_a(t)$ be a doubly asymptotic solution of the unperturbed system: $x_a(t) \to x^\pm$ and $y_a(t) \to y^\pm$ as $t \to \pm\infty$.

To solve the bifurcation problem for the perturbed asymptotic surfaces, Poincaré introduced the τ-periodic function

$$\begin{aligned} P(\alpha) &= \int_{-\infty}^{\infty} H_1(x_a(t+\alpha), y_a(t+\alpha), t)\, dt \\ &= \int_{-\infty}^{\infty} H_1(x_a(t), y_a(t), t-\alpha)\, dt , \end{aligned} \tag{1.12}$$

and showed that if $P(\alpha) \neq \text{const}$, then $\Lambda_\varepsilon^+ \neq \Lambda_\varepsilon^-$ for small $\varepsilon \neq 0$ ([203, Chap. 21]; see also [135]). This theorem of Poincaré can be easily proved by the method of §1.2. The integral (1.12) may be divergent. However, if $H_1(x^+, y^+, t) \equiv H_1(x^-, y^-, t)$, then, after subtracting from H_1 the function of time $H_1(x^\pm, y^\pm, t)$, the integral (1.12) can be assumed to be convergent. Clearly, this argument always works in the homoclinic case, when $x^+ = x^-$, $y^+ = y^-$.

To avoid the convergence problem, we differentiate the function P with respect to α:

$$I(\alpha) = \frac{dP}{d\alpha} = \int_{-\infty}^{\infty} \{H_0, H_1\}(x_a(t+\alpha), y_a(t+\alpha), t)\, dt .$$

We used the obvious identity

$$\frac{d}{d\alpha} f(t+\alpha) = \frac{d}{dt} f(t+\alpha) .$$

Therefore, the Poincaré condition can be rewritten in the form $I(\alpha) \not\equiv 0$. The splitting condition of the asymptotic surfaces seems to have first appeared in this form in Melnikov's paper [173].

For $n > 1$, it is natural to introduce the integrals along the doubly asymptotic trajectories

$$I_i = \int_{-\infty}^{\infty} \{F_i, H_1\}(x_a(t), y_a(t), t)\, dt . \qquad (1.13)$$

Since $dF_i = 0$ for $x = x^{\pm}$, $y = y^{\pm}$, the integrand tends to zero exponentially as $t \to \pm\infty$. In particular, the integrals (1.13) are always convergent.

The sufficient condition transverse intersection of the perturbed asymptotic surfaces was obtained by Bolotin (see [34]):

$$I_i = 0 , \quad \det \|J_{ij}\| \neq 0 ; \qquad 1 \leq i, j \leq n ,$$

where

$$J_{ij} = \int_{-\infty}^{\infty} \{F_i, \{F_j, H_1\}\}(x_a(t), y_a(t), t)\, dt .$$

Convenient expressions for solving the problem of splitting for asymptotic surfaces of hyperbolic periodic orbits in the autonomous case were pointed out in [134–135].

2 Theorems on Nonintegrability

2.1 Let M^3 be a three-dimensional analytic manifold and v an analytic vector field on M with no equilibrium points. As an example, we may mention a Hamiltonian system with two degrees of freedom restricted to a regular three-dimensional level surface of the energy integral.

Suppose that there exist two hyperbolic periodic trajectories γ_1 and γ_2 (we do not rule out the case when γ_1 and γ_2 coincide). Denote by Λ_1^+ (Λ_2^-) the stable (unstable) asymptotic surface of the trajectory γ_1 (γ_2). Recall that these surfaces are regular and analytic. However, they can be immersed in M in a very complicated way.

Theorem 1. *Suppose that Λ_1^+ and Λ_2^- intersect and do not coincide (as subsets of M). Then the system*

$$\dot{x} = v(x) , \qquad x \in M , \qquad (2.1)$$

has no non-constant analytic integrals in M, and every analytic symmetry field (2.1) is of the form $u = \lambda v$, $\lambda = $ const.

Proof. Since Λ_1^+ and Λ_2^- intersect, the system (2.1) has a doubly asymptotic trajectory $\gamma_a(t)$ such that $\gamma_a(t) \to \gamma_1$ as $t \to +\infty$ and $\gamma_a(t) \to \gamma_2$ as $t \to -\infty$. Hence, an arbitrary small neighborhood of the closed orbit γ_1 (γ_2) contains intersection points of Λ_1^+ and Λ_2^-.

Let Π be a two-dimensional regular surface in M transversely intersecting γ_1 at a point x_1. Then in a neighborhood of x_1 we obtain the Poincaré map $g : \Pi \to \Pi$ generated by the phase flow of the system (2.1). It is possible to choose the section surface Π in such a way that Λ_2^- intersects Π along a segment Δ of a regular curve, and the intersection $\Delta \cap \Lambda_1^+$ is non-empty (see Fig. 17).

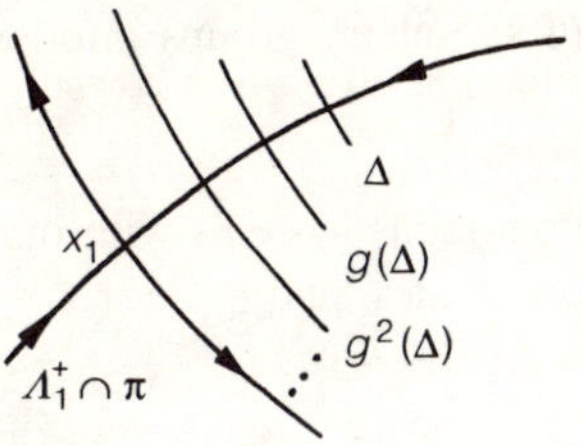

Fig. 17.

Consider the iterated images $g^n(\Delta)$; $n = 1, 2, \ldots$ of the segment Δ under the Poincaré map. As $n \to \infty$, the segment $g^n(\Delta)$ stretches along the separatrix $\Lambda_1^- \cap \Pi$ and approaches it unboundedly. This follows, for example, from the Grobman–Hartman theorem, which asserts that in a small neighborhood of the point x_1 the map g is topologically equivalent to a linear hyperbolic rotation. As a consequence, the union

$$\bigcup_{n=1}^{\infty} g^n(\Delta) \tag{2.2}$$

is a key set for the class of analytic functions on the surface Π.

Now suppose that the system (2.1) has an analytic integral F. Let f be the restriction of the function F to Π. Since Π is regular, f is analytic on Π. It is well known that F is constant on the trajectories γ_1 and γ_2, and also on the asymptotic surfaces Λ_1^+ and Λ_2^-. Hence, f is constant on the set (2.2). Since (2.2) is a key set, $f = \text{const}$ on the surface Π. Varying Π, we conclude that $F = \text{const}$ on the whole manifold M.

Now suppose that there exists an analytic symmetry field u. Let g_u be the phase flow of the system of differential equations generated by u. Since the periodic orbits γ_1 and γ_2 are non-degenerate, transformations from the group g_u take γ_1 and γ_2 into themselves. Hence, the transformations from g_u map doubly asymptotic trajectories into doubly asymptotic trajectories. The vector fields u and v are linearly dependent on these trajectories. Indeed, if this is not so, then Λ_1^+ and Λ_2^- intersect along two-dimensional analytic manifolds. Since Λ_1^+ and Λ_2^- are regular and analytic, they coincide. In particular, it follows that u and v are linearly dependent at all points of Λ_1^+ and Λ_2^-. We have already showed that under the assumptions of Theorem 1 the union $\Lambda_1^+ \cup \Lambda_2^-$ is a key set in M. Hence, the vectors $u(x)$ and $v(x)$ are dependent for all $x \in M$. Since $v \neq 0$, we have $u = \lambda v$, and the function λ is an integral of the system (2.1). The analyticity of λ on M implies that $\lambda = \text{const}$. This finishes the proof.

The nonexistence of analytic integrals under the assumption that the asymptotic surfaces intersect but do not coincide was first proved by Cushman [61] (where in fact non-autonomous Hamiltonian systems with one degree of freedom were studied). The nonexistence of nontrivial symmetry groups was established in [143]. Obviously, in the Hamiltonian case the nonexistence of symmetry groups implies the nonexistence of new integrals.

2.2 We apply the results of §2.1 to the case of non-autonomous systems with one degree of freedom. Let $z = (x, y)$ be the symplectic coordinates, and

$$H = H_0(z) + \varepsilon H_1(z, t) + \cdots$$

the Hamiltonian function which is periodic in t. We assume that for $\varepsilon = 0$ there exist two hyperbolic critical points of the function H_0. They are connected by a doubly asymptotic trajectory $z_0(t)$ (see §1.4). Theorem 1 and the results of §1 yield

Theorem 2. *Suppose that*

1) $\int_{-\infty}^{\infty} \{H_0, H_1\}(z_0(t), t)\, dt \neq 0$,
2) *for small ε the perturbed system has a doubly asymptotic solution $t \to z_\varepsilon(t)$ close to $t \to z_0(t)$.*

Then for small $\varepsilon \neq 0$ the perturbed Hamiltonian system is nonintegrable (it has no nontrivial analytic integrals and symmetry fields that are periodic in t).

In the homoclinic case, condition 2) of Theorem 2 can be omitted. As proved by Poincaré [203], for small ε the perturbed system always has homoclinic solutions (of course, if they exist for $\varepsilon = 0$). The proof is based on the conservation of area under the Poincaré map of a Hamiltonian system (see Fig. 18).

Let $W^\pm$ be the intersection curves of the asymptotic surfaces $\Lambda_\varepsilon^\pm$ with the plane $t = 0$. Suppose that for some sufficiently small $\varepsilon \neq 0$ these curves do not intersect: there is a small clearance between them. Let Δ be a small segment connecting close points in W^+ and W^- (see Fig. 18). The Poincaré map g shifts the segment in the direction shown by the arrow. Since $W^\pm$ are invariant with respect to g, the region D, bounded by pieces of the curves W^+, W^- and the segment Δ, is transformed into the subset of D obtained by deleting the small shaded curvilinear tetragon. However, this contradicts the identity $\operatorname{mes} D = \operatorname{mes} g(D)$, where mes is the standard area measure in the plane $\mathbb{R}^2 = \{x, y\}$.

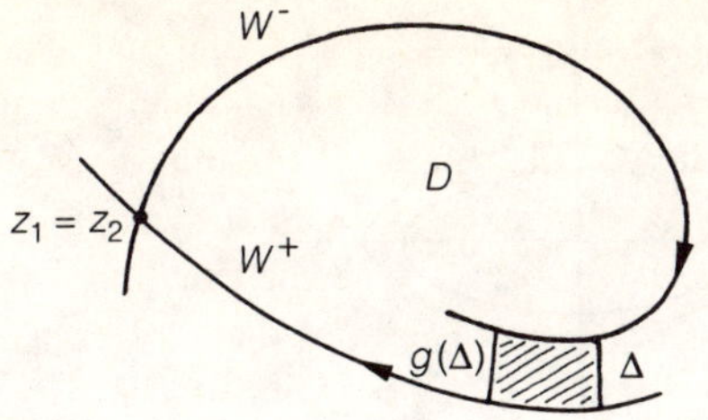

Fig. 18.

Now consider an analytic autonomous Hamiltonian system with two degrees of freedom. Let $H = H_0 + \varepsilon H_1 + o(\varepsilon)$ be the Hamiltonian function. Suppose that the unperturbed system has two hyperbolic periodic trajectories γ_1, γ_2 lying on the same level surface of the energy integral. Let F_0 be an integral of the unperturbed system such that $dF_0 = 0$ at the points of the trajectories γ_1 and γ_2. In this case Theorem 1 also holds, but condition 1) must be replaced by the following one:

$$\int_{-\infty}^{\infty} \{H_0, H_1\}\, dt \neq 0 \,. \tag{2.3}$$

Here the Poisson bracket is evaluated on the doubly asymptotic trajectory of the unperturbed system. If $dF \neq 0$ on the periodic trajectories γ_1 and γ_2, then the integral (2.3) must be replaced by the more complicated expressions (1.3) from §1.

It must be emphasized that in Theorem 2 nonintegrability is asserted for small but fixed values of the parameter ε.

We give an example showing that in the heteroclinic case splitting of asymptotic surfaces does not imply nonintegrability of the perturbed system. In other words, condition 2) in Theorem 2 cannot be omitted.

Let

$$H = \frac{y^2}{2} - \frac{\cos^2 x}{2} + \varepsilon \sin x \,.$$

Since the system is autonomous, for all ε the Hamiltonian H is a first integral of the equations of motion. Let us evaluate the integral (2.3) taken along an asymptotic trajectory $x_a(t)$ of the unperturbed system:

$$x_a(t) \to \pm\pi/2 \quad \text{as} \quad t \to \pm\infty \,.$$

We have $\{H_0, H_1\} = y \cos x$ and $y = \dot{x}$. Hence, the integral (2.3) is equal to

$$\int_{-\infty}^{\infty} \dot{x}_a \cos x_a \, dt = \int_{-\infty}^{\infty} d \sin x_a = 2 \neq 0 \,.$$

The picture of split separatrices is presented in Fig. 19.

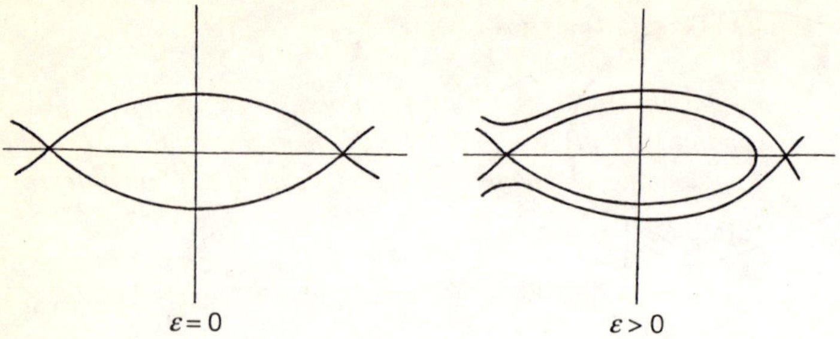

Fig. 19.

2.3 The conditions for the nonexistence of a complete set of involutive integrals for a multidimensional Hamiltonian system were pointed out by Bolotin [34]. Consider a non-autonomous Hamiltonian system with analytic time-periodic Hamiltonian $H = H_0(z)+\varepsilon H_1(z,t)+o(\varepsilon)$. Here $z = (x, y)$ is the set of $2n$ symplectic variables. We assume that the unperturbed system has two hyperbolic equilibrium positions $z^{\pm}$ joined by a doubly asymptotic solution $t \to z_0(t)$, $t \in \mathbb{R}$.

Theorem 3 [34]. *Suppose that*
1) $\int_{-\infty}^{\infty}\{H_0, \{H_0, H_1\}\}(z_0(t), t)\, dt \neq 0$,
2) *for small* ε *the perturbed system has a doubly asymptotic solution* $t \to z_\varepsilon(t)$ *close to* $z_0(t)$.

Then for small fixed values of $\varepsilon \neq 0$, *the perturbed Hamiltonian system does not have a complete set of independent integrals in involution in any neighborhood of the closure of the trajectory* $z_\varepsilon(t)$.

Remark. Condition 1) can be replaced by the following assumption: for some $m \geq 2$

$$\int_{-\infty}^{\infty} \{H_0, \ldots, \{H_0, H_1\} \cdots\}(z_0(t), t)\, dt \neq 0\,.$$

If 1) holds, then the asymptotic surfaces do not coincide.

Since the characteristic exponents are real, using Birkhoff's method of normal forms, we can find in small neighborhoods of the unstable periodic solutions $z^{\pm} + O(\varepsilon)$ formal canonical t-periodic changes of variables $z \to u$ transforming the Hamiltonian function $H(z, t, \varepsilon)$ into functions $H^{\pm}(u, \varepsilon)$, which are independent of t. Note that the coefficients of the normalizing transformations are smooth functions of ε, though in general the characteristic exponents are resonant for a dense set of ε. The reason is that we do not attempt to kill all resonant terms, but only those which are time-dependent. Of course, the Birkhoff transformation may diverge. However, for one degree of freedom ($n = 1$) the formal series of the transformation $z \to u$ converge everywhere and depend analytically on ε (Moser [178]).

Theorem 4. *Suppose that the Birkhoff transformation converges and depends analytically on ε. If condition* 1) *of Theorem* 1 *holds, then for small $\varepsilon \neq 0$ the Hamiltonian equations do not have a complete set of independent analytic integrals in involution.*

In particular, for $n = 1$ condition 1) is sufficient for nonintegrability (Ziglin [243]).

Of course, condition 1) can be replaced by the assumption that the periodic function $I(\alpha)$, introduced in §1.4, is non-constant. Note that in the example given in §2.2, we have $I(\alpha) \equiv \text{const}$.

Proof of Theorem 4 [34]. Define the functions $R^{\pm}$ on $\Lambda_0^{\pm}$ by the expressions

$$\begin{aligned} R^+(z) &= -\int_0^{\infty} \{H_0, \{H_0, H_1\}\}(z(t), t)\, dt\,, \\ R^-(z) &= \int_{-\infty}^{0} \{H_0, \{H_0, H_1\}\}(z(t), t)\, dt\,, \end{aligned} \tag{2.4}$$

where $t \to z(t)$ is the asymptotic motion of the unperturbed system with the initial condition $z(0) = z$.

Lemma 1. *The functions $R^{\pm}$ are defined by H_0, the family of surfaces $\Lambda_\varepsilon^{\pm}$, and the symplectic structure. Thus, they are symplectic invariants of the Poincaré map.*

According to the results of the previous section, the functions

$$\begin{aligned} S^+(z) &= -\varepsilon \int_0^{+\infty} (H_1(z(t), t) - H_1(z^+(t), t))\, dt\,, \\ S^-(z) &= \varepsilon \int_{-\infty}^{0} (H_1(z(t), t) - H_1(z^+(t), t)\, dt\,, \end{aligned}$$

are the generating functions of the Lagrangian surfaces $\Lambda_\varepsilon^{\pm}$ up to $O(\varepsilon^2)$. It can be shown that up to terms of order $O(\varepsilon^2)$ the generating functions are symplectic invariants (i.e., they do not depend on the choice of special canonical coordinates). Thus, the functions $\varepsilon R^{\pm} = \{H_0, \{H_0, S^{\pm}\}\}$ are symplectic invariants, as required.

By composing Birkhoff's transformations with powers of the Poincaré map at a period, we can extend the functions $H^{\pm}$ from the neighborhoods of the critical points $u^{\pm}(\varepsilon)$ to certain neighborhoods $W^{\pm}$ of the asymptotic surfaces $\Lambda_\varepsilon^{\pm}$. Since the possible splitting of the surfaces Λ_ε^{+} and Λ_ε^{-} is of order ε, for small ε the neighborhoods W^+ and W^- intersect.

Lemma 2. *In $W^+ \cap W^-$ we have $\{H^+, H^-\} \not\equiv 0$ for small $\varepsilon \neq 0$.*

Proof. Let $H^{\pm}(u,\varepsilon) = H_0^{\pm}(u) + \varepsilon H_1^{\pm}(u) + O(\varepsilon^2)$. Since $H_0^{\pm}(u) = H_0(u)$, we obtain

$$\{H^+, H^-\} = \varepsilon\{H_0, H_1^- - H_1^+\} + O(\varepsilon^2) .$$

Since Λ_0^- is an invariant asymptotic manifold of the unperturbed Hamiltonian system, by Lemma 1,

$$\{H_0, H_1^-\}(u) = \int_{-\infty}^{0} \{H_0, \{H_0, H_1^-\}\}(u_0(t))\, dt = R^-(u) , \qquad u \in \Lambda_0^- ,$$

where $u_0(t)$ is the solution of the unperturbed system with the initial condition $u_0(0) = u$. Here we used the fact that in the normal coordinates u the functions $R^{\pm}$ are given by the same expressions (2.4) with H_1 replaced by $H_1^{\pm}$ (Lemma 1). Similarly,

$$\{H_0, H_1^+\}(u) = -\int_{0}^{+\infty} \{H_0, \{H_0, H_1^+\}\}(u_0(t))\, dt = R^+(u) , \quad u \in \Lambda_0^+ .$$

Hence, (2.4) yields

$$\{H^+, H^-\}(z_0(0)) = \varepsilon \int_{-\infty}^{\infty} \{H_0, \{H_0, H_1\}\}(z_0(t), t)\, dt + O(\varepsilon^2) .$$

By condition 1), for small $\varepsilon \neq 0$ the Poisson bracket $\{H^+, H^-\} \not\equiv 0$.

In the new variables u, the integrals $F_1, \dots, F_n$ do not depend on t. Suppose that for $\varepsilon \neq 0$ the integrals $F_1, \dots, F_n$ are independent at some point of $W^+ \cap W^-$. Since $\{H^{\pm}, F_i\} \equiv 0$, the vector $v_{H^{\pm}}$ is a linear combination of the vectors v_{F_i}. Since $\{F_i, F_j\} \equiv 0$, at this point $\{H^+, H^-\} = 0$. To complete the proof it remains to note that the analytic function $\{H^+, H^-\}$ does not vanish on an everywhere dense set.

It is worthwhile mentioning that the proof of Theorem 4 essentially used the assumption that the multipliers of the perturbed periodic trajectories $z^{\pm} + O(\varepsilon^2)$ are real. If this is not so, then in general the Birkhoff normal form is not autonomous. In fact it is easier to prove nonintegrability in this case. Our assumption is always satisfied if the characteristic exponents of the unperturbed system linearized near the points $z^{\pm}$ are real and pairwise distinct (see the beginning of this section). Obviously, for $n = 1$ the assumption that the critical points $x^{\pm}$ are hyperbolic is sufficient.

The idea of the proof of Theorem 3 is similar to that of Theorem 4. However, the proof is more complicated, since it is possible that the Birkhoff transformation is divergent (convergence is proved for integrable systems only in the non-resonance case [103]). It is essential that the Birkhoff transformationis convergent on the asymptotic surfaces (see §11 of Chap. II). A complete proof of Theorem 3 is contained in [34]. The autonomous variant is also studied there. Suppose that the unperturbed Hamiltonian system with the Hamiltonian $H^{\varepsilon} = H_0 + \varepsilon H_1 + o(\varepsilon)$ has two hyperbolic periodic orbits γ_1^{ε} and γ_2^{ε} connected by a doubly asymptotic trajectory $\gamma_{\varepsilon}(t)$ smoothly depending on ε. In [34] it is proved that if the integral

J_{00} (obtained from (1.3) for $i = j = 0$) is not equal to zero, then for sufficiently small ε the system with Hamiltonian H^ε does not have a complete set of analytic integrals in involution on the level surface $H^\varepsilon = h$, where $h = H_\varepsilon(\gamma_\varepsilon)$. The proof is based on the reduction of the Hamiltonian system to a non-autonomous form with a time-periodic Hamiltonian using the integral F_0 and the corresponding symmetry group $\mathbb{T}^1$. It would be interesting to find out if the conditions of Theorem 3 are sufficient for the nonexistence of n analytic commuting symmetry fields of the perturbed system.

3 Some Applications

3.1 First we consider the simplest problem of oscillations of a pendulum with a vibrating suspension point. The Hamiltonian function H is $H_0 + \varepsilon H_1$, where

$$H_0 = p^2/2 - \omega^2 \cos x\,, \quad H_1 = -\omega^2 f(t) \cos x\,,$$

and $f(t)$ is a time-periodic function. When $\varepsilon = 0$, the upper position of the pendulum is an unstable equilibrium point. The unperturbed system has two families of homoclinic solutions:

$$\cos x_0 = \frac{2e^{\pm\omega(t-t_0)}}{e^{\pm 2\omega(t-t_0)} + 1}\,, \qquad x_0 \to \pm\pi \quad \text{as} \quad t \to \pm\infty\,. \tag{3.1}$$

Since $\{H_0, H_1\} = -\omega^2 f(t)\dot{x} \sin x$, the integral (2.4), up to multiplication by a constant, is equal to

$$\int_{-\infty}^{\infty} \dot{f}(t) \cos^2 \frac{x_0}{2}\, dt\,. \tag{3.2}$$

Let $f(t) = \sum f_n e^{int}$. Then the integral (3.2) can be expressed as a series

$$\sum in f_n J_n e^{int_0}\,, \qquad J_n = \int_{-\infty}^{\infty} \frac{e^{int}}{\cosh^2 \omega t}\, dt\,.$$

The integrals J_n are easily calculated by residues:

$$J_n = \frac{\pi n}{\omega^2 \sinh(\pi n/(2\omega))}\,.$$

Hence, if $f(t) \neq \text{const}$ (i.e., $f_n \neq 0$ for some $n \neq 0$), then the integral (2.4) is non-zero on at least one doubly asymptotic solution from the family (3.1). Thus, if $f(t) \neq \text{const}$, then by the results of §2 for sufficiently small (but fixed) $\varepsilon \neq 0$, the Hamiltonian system does not have an analytic integral $F(y, x, t)$ that is 2π-periodic in x and t.

3.2 Using this method Burov studied the integrability problem for the equation of planar oscillations of a satellite in an elliptic orbit:

$$(1 + e\cos\nu)\frac{d^2\delta}{d\nu^2} - 2e\sin\nu\frac{d\delta}{d\nu} + \mu\sin\delta = 4e\sin\nu\,. \tag{3.3}$$

Here e is the eccentricity of the orbit; the meaning of other parameters is described in Chap. I, §4. For $e = 0$, we again obtain an integrable problem governing oscillations of an ordinary pendulum. Let $\mu \neq 0$. Then, as shown in [47], one pair of separatrices of the unperturbed problem splits under the perturbation. Thus, for sufficiently small values of $e \neq 0$ (3.3) does not have an analytic integral that is 2π-periodic in δ and ν.

3.3 For the rapid rotation of an asymmetric rigid body the Hamiltonian function H is $H_0 + \varepsilon H_1$, where

$$H_0 = (Am, m)/2\,, \quad H_1 = r_1\gamma_1 + r_2\gamma_2 + r_3\gamma_3\,, \quad A = \operatorname{diag}(a_1, a_2, a_3)\,.$$

Here $m = I\omega$ is the angular momentum of the body; the numbers a_1, a_2, a_3 are the inverses of the principal moments of inertia. For $\varepsilon = 0$, we have the integrable Euler system. In this unperturbed problem, on all non-critical three-dimensional level surfaces

$$M_{h,c} = \{(m, \gamma) : H_0 = h,\ (m, \gamma) = c,\ (\gamma, \gamma) = 1\}$$

there exist two hyperbolic periodic solutions. If $a_1 < a_2 < a_3$, then they are given by the expressions:

$$\begin{cases} m_1 = m_3 = 0\,, \quad m_2 = m_2^0 = \pm\sqrt{2h/a_2}\,, \quad \gamma_2 = \gamma_2^0 = \pm c/m_2^0\,, \\ \gamma_1 = \alpha\cos(a_2 m_2^0 t)\,, \quad \gamma_3 = \alpha\sin(a_2 m_2^0 t)\,, \end{cases} \tag{3.4}$$

where $\alpha^2 = 1 - (c/m_2^0)^2$. These solutions have a clear mechanical interpretation: they are the unstable stationary rotations (in both directions) of the rigid body about the second principal inertia axis. The inequality $(m, \gamma)^2 \leq (m, m)(\gamma, \gamma)$ and the independence of the classical integrals on $M_{h,c}$ imply that $\alpha^2 > 0$.

The stable and unstable asymptotic surfaces of the periodic solutions (3.4) can be represented as the intersections of the manifolds $M_{h,c}$ with the hyperplanes

$$F = m_1\sqrt{a_2 - a_1} \pm m_2\sqrt{a_3 - a_2} = 0\,. \tag{3.5}$$

In the Euler problem, the asymptotic surfaces are "doubled": they are completely filled by doubly asymptotic trajectories, which tend unboundedly to the periodic trajectories (3.2) as $t \to \pm\infty$. The splitting of these surfaces was studied in [127] and [243]. It turned out that under perturbation the asymptotic surfaces always split except in the "Hess–Appelrot case":

$$r_2 = 0\,, \qquad r_1\sqrt{a_3 - a_2} \pm r_3\sqrt{a_2 - a_1} = 0\,. \tag{3.6}$$

In this case one pair of separatrices does not bifurcate, while the other does (see Fig. 20).

The reason for non-splitting is that under the condition (3.6) for all ε the perturbed system has a "particular integral" – the function F defined by (3.5) ($\dot F = 0$ when $F = 0$). It can be shown that for small $\varepsilon \neq 0$ the closed invariant surfaces

$$\{H = h\,,\ (m, \gamma) = c\,,\ (\gamma, \gamma) = 1\,,\ F = 0\}$$

are exactly the pair of separatrices of the perturbed problem (see [127]).

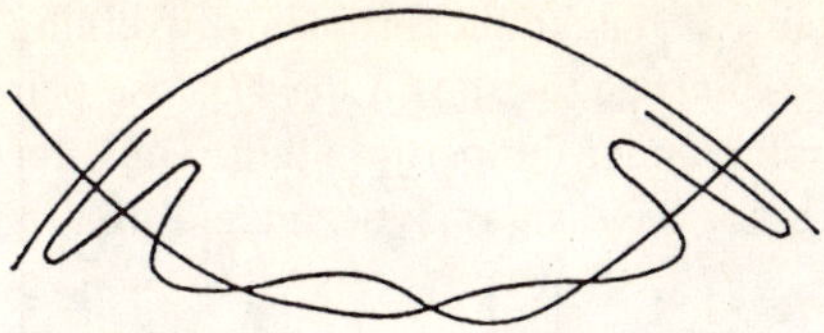

Fig. 20.

Apparently, in the problem of rapid rotation of a heavy non-symmetric top the split separatrices do not always intersect. However, Theorem 4 of §2 is applicable. With its help it can be established that there is no additional analytic integral of the perturbed problem for small fixed values of the parameter $\varepsilon \neq 0$ (see [127]).

Moreover, as proved by Dovbysh [68], in the non-symmetric case for small $\varepsilon \neq 0$ there always exist doubly asymptotic trajectories (not always heteroclinic, as in the unperturbed problem), and the corresponding intersecting asymptotic surfaces do not coincide. Thus, Theorem 2 of §2 implies nonintegrability in a much stronger sense: the equations of motion admit no nontrivial analytic symmetry fields. This result was obtained in [143].

The behavior of solutions of the perturbed problem has been studied numerically in [104]. In Fig. 21 the results of the calculations are presented for various values of ε.

Clearly, the picture of the invariant curves of the unperturbed problem starts disintegrating exactly in the neighborhoods of the separatrices.

The numerical calculations used the values 1, 2, 3 for the principal moments of inertia. The constant of the energy integral h was taken equal to 50, and the constant of the area integral equals zero. The special canonical coordinates L, G, l, g were used (see Chap. I, §2). The sections of the phase trajectories lying on the surface $H = 50$ produced by the plane $g = \text{const}$ are shown in Fig. 21. For $\varepsilon = 0$, we obtain the familiar phase portrait of the Euler problem (see Fig. 1).

Borisov investigated numerically the perturbed separatrices of the unstable periodic solutions for the same values of moments of inertia and energy integral. The separatrices are shown in Fig. 22.

In case a) the coordinates r_1, r_2, r_3 of the mass center of the rigid body are equal to 0, 1, 0, and in case b) to 10, 0, 0. It is fairly clear how the split separatrices start to oscillate when they approach the unstable periodic trajectory.

3.4 If the rigid body is dynamically symmetric, then the unperturbed Euler problem has no hyperbolic periodic orbits. Thus, the method of splitting the asymptotic surfaces cannot be applied directly. However, in this problem the small parameter can be introduced in a different way. Then it is possible to find homoclinic solutions.

Fix the values $I_1 = I_2$, r_3 and replace I_3, r_1 with μI_3, μr_1 respectively ($0 < \mu \leq 1$). Using dynamical symmetry, we can assume that the coordinate r_2 of the center of mass is zero. As μ tends to zero, we obtain the restricted problem of rigid

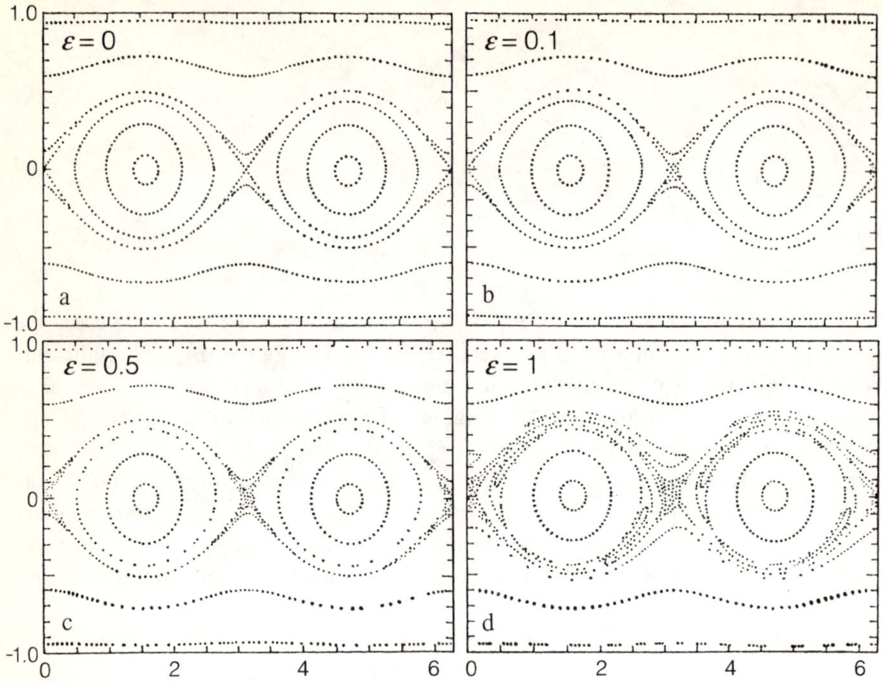

Fig. 21.

body rotation (see Chap. I, §4). Fixing the value of the area integral, we reduce the equations of motion to a Hamiltonian system with two degrees of freedom.

We shall show that if $r_1 \neq 0$ and $c \neq 0$, or $r_3 \neq 0$, then for sufficiently small $\mu > 0$ the reduced equations have no additional analytic integral and no nontrivial analytic symmetry group. This result was established in [154].

For simplicity, we consider the case $r_3 = 0$ (for $r_3 \neq 0$ the calculations are more complicated). In §4 of Chap. II we introduced the auxiliary variable ξ, which for $\varepsilon = 0$ satisfies the pendulum-type differential equation:

$$\ddot{\xi} = \frac{c}{\sqrt{2h}} \sin \xi - \sqrt{1 - \frac{c^2}{2h}} \sin\left(\sqrt{2h}t\right) \cos \xi \,. \tag{3.7}$$

Here h is the energy constant. Suppose that the equation of the reduced problem has an additional analytic integral that is analytic in the variables ω and γ ($|\gamma| = 1$). Taking into account the explicit expressions for ω and γ in terms of ξ and t (see Chap. II, §2), we obtain that the Eq. (3.7) has an integral which is analytic in ξ and t, 2π-periodic in ξ, and $2\pi/\sqrt{2h}$-periodic in t. However, (3.7) is nonintegrable if c^2 is close to $2h$.

We put $\varepsilon = \sqrt{1 - c^2/(2h)}$ and regard ε as a small parameter. Note that (3.7) makes sense for $\varepsilon = 0$, when the initial problem is degenerate. Denote $\eta = \dot{\xi}$ and

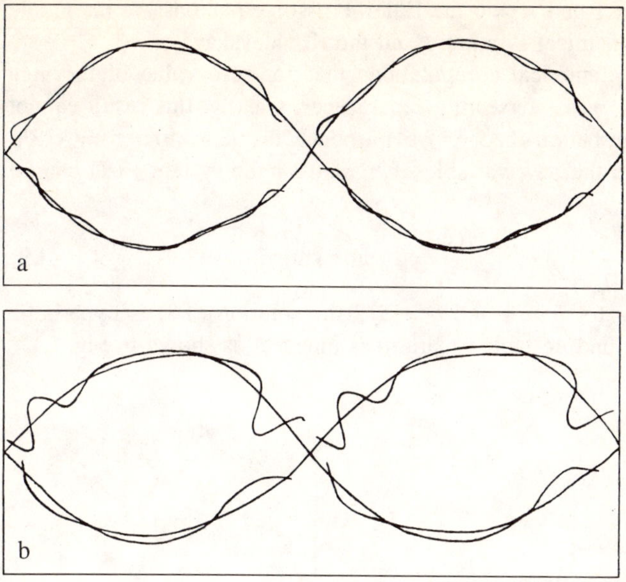

Fig. 22.

represent (3.7) in the Hamiltonian form:

$$\dot{\xi} = H'_\eta , \quad \dot{\eta} = -H'_\xi ; \qquad H = H_0 + \varepsilon H_1 + o(\varepsilon) ,$$
$$H_0 = \eta^2/2 + \cos x , \qquad H_1 = \sin\xi \sin\left(\sqrt{2h}t\right) .$$

For $\varepsilon = 0$ we have the unperturbed integrable problem – the mathematical pendulum. It has a family of homoclinic solutions:

$$\sin\xi/2 = \cosh^{-1}(t - t_0) , \quad \eta = 2\cosh^{-1}(t - t_0) , \qquad t_0 = \text{const} . \tag{3.8}$$

We shall show that for small $\varepsilon \neq 0$ the stable and unstable separatrices do not coincide. This implies nonintegrability of Eq. (3.7). The improper integral

$$\int_{-\infty}^{\infty} \{H_0, H_1\}\, dt ,$$

evaluated on the doubly asymptotic solutions (3.8), is not identically zero. For example, if $t_0 = \pi/(2\sqrt{2h})$, then it is equal to

$$-4\pi h \cosh^{-1}(\pi\sqrt{2h}/2) \neq 0$$

for $h \neq 0$.

Since the behavior of split separatrices is stable with respect to small variations of the parameters, for small $\mu \neq 0$ the Euler–Poisson equations also have a hyperbolic periodic trajectory with intersecting asymptotic surfaces. By Theorem 2 of §2, this contradicts the existence of an additional integral and a nontrivial symmetry group. Note that for $\mu = 1$ and $\mu = 0$ the Euler–Poisson equations are integrable (the case of complete dynamical symmetry and the Kovalevskaya case).

Dovbysh showed by numerical computations that for zero value of the area constant c, Eq. (3.7) also has intersecting separatrices. Clearly, this result cannot be obtained by the perturbation theory. We introduce the new time $\tau = \sqrt{2h}t$ and put $\xi = \pi/2 + x$. In the new variables $x, \tau \bmod 2\pi$ the system (3.7) can be rewritten as

$$\frac{dx}{d\tau} = y\,, \qquad \frac{dy}{d\tau} = 2h \sin\tau \sin x\,. \tag{3.9}$$

Dovbysh established that for $h = 1$ and $h = 5/2$ the solution (3.8) is hyperbolic and the pair of its stable and unstable separatrices intersect as shown in Fig. 23.

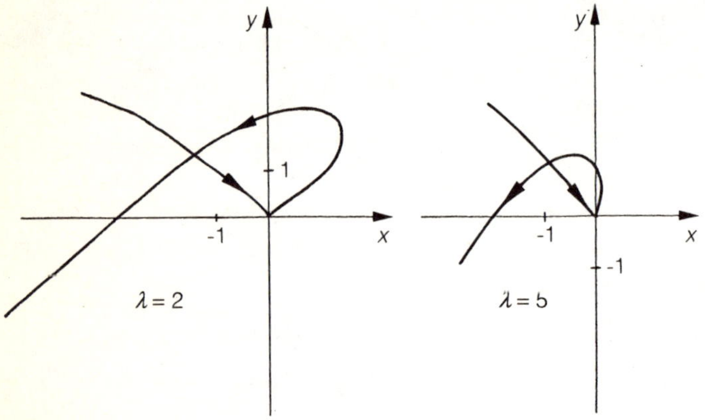

Fig. 23.

It seems that transverse intersection of the separatrices occurs for all values of h such that the solution (3.8) is hyperbolic.

3.5 Now consider inertial rotation about a fixed point of a rigid body with an asymmetric rotor, freely rotating about some axis rigidly attached to the body. Obviously, this mechanical problem has four degrees of freedom. The configuration space is the Cartesian product $SO(3) \times S^1$.

The Hamiltonian equations of motion have four integrals: the total energy H and three projections F_1, F_2, F_3 of the angular momentum onto the axes of a fixed orthogonal reference system. It is easy to show that

$$\{F_1, F_2\} = F_3\,, \qquad \{F_2, F_3\} = F_1\,, \qquad \{F_3, F_1\} = F_2\,.$$

Hence, the functions H, F_1, $F^2 = F_1^2 + F_2^2 + F_3^2$ are in involution. Thus, for complete integrability of the equations of motion, we lack one independent integral commuting with the functions H, F_1, F^2. If, for example, the rotor is symmetric with respect to the rotation axis, then the additional integral is the projection of the flywheel angular momentum onto its rotation axis (the Zhukovsky–Volterra problem).

In [104] Ivin used the method of separatrix splitting to show that in general this problem is nonintegrable. More precisely, he considered the rotation of a rigid body with a rotor of small mass. The unperturbed problem is the integrable Euler problem on rotation of a free rigid body. It has a double pair of asymptotic surfaces. Ivin showed that, with the addition of the rotor, the asymptotic surfaces split in a "nonintegrable way", and this implies the nonexistence of new analytic integrals and symmetry groups.

A similar result was obtained earlier by Marsden and Holmes [100] under additional simplifying assumptions of a technical kind: to simplify the calculations, some terms in the exact expression for the Hamiltonian function were omitted.

3.6 Using the method of splitting of asymptotic surfaces it is possible to establish nonintegrability of the four-vortex problem [242]. More precisely, consider this problem in a restricted form: a vortex of zero intensity (i.e., a particle in an ideal fluid) moves in the "field" of vortices with equal intensities. It turns out that the equations of motion of the zero vortex can be expressed in Hamiltonian form with a time-periodic Hamiltonian. These equations have hyperbolic periodic solutions with intersecting separatrices. Hence, the restricted four-vortex problem is not completely integrable, although, as in the unrestricted formulation, it has four independent non-commuting integrals.

3.7 The homogeneous two-component model of the classical Yang–Mills field (Chap. I, §8) is a Hamiltonian system with the Hamiltonian function

$$H = (y_1^2 + y_2^2)/2 + x_1^2 x_2^2/2 . \tag{3.10}$$

The Hamilton equations

$$\ddot{x}_1 + x_1 x_2^2 = 0 , \quad \ddot{x}_2 + x_1^2 x_2 = 0 \tag{3.11}$$

have unstable "cnoidal" periodic solutions

$$x_1 = x_2 = f , \quad x_1 = -x_2 = f ; \qquad f = \mathrm{cn}(t, 1/\sqrt{2}) . \tag{3.12}$$

Consider the two-dimensional section by the hyperplane $x_2 = 0$ of the energy integral level surface containing the solutions (3.11). The periodic trajectories (4.11) intersect this section at the fixed points of the Poincaré mapping. Since these points are of the hyperbolic type, we can investigate the mutual disposition of their stable and unstable separatrices. This problem was studied numerically in [191]. The result is illustrated in Fig. 24.

Since the Hamiltonian (3.10) is quasi-homogeneous, we obtain exactly the same picture of transverse separatrices on all energy surfaces with positive value of the total energy. Consequently, we see that the system (3.11) does not have an additional analytic integral. This result was first obtained by Ziglin [245] by analyzing branching of solutions of the system (3.10) in the complex time plane. In fact,

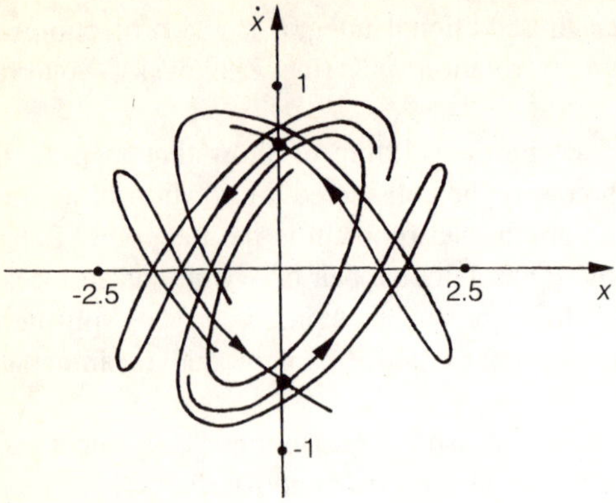

Fig. 24.

transverse intersection of separatrices implies a considerably stronger result – the nonexistence of nontrivial symmetry fields of the Hamiltonian system (3.11).

3.8 Following Chirikov, consider the standard symplectic map of the cylinder $x \bmod 2\pi$, y given by the equations

$$x' = x + y' \pmod{2\pi}\,, \qquad y' = y + \varepsilon \sin x\,. \tag{3.13}$$

For $\varepsilon = 0$, we have an integrable mapping: the coordinate y is an integral and all the points of the circle $y = \text{const}$ rotate under the mapping through the angle y. Thus, the unperturbed mapping (3.13) has no hyperbolic periodic points. However, for $\varepsilon > 0$ the fixed point $x = y = 0$ is hyperbolic. The eigenvalues (multipliers) of the linearized mapping are

$$1 + \frac{\varepsilon}{2} \pm \sqrt{\varepsilon + \left(\frac{\varepsilon}{2}\right)^2} = 1 \pm \sqrt{\varepsilon} + o(\sqrt{\varepsilon})\,. \tag{3.14}$$

The point $(x, y) = (\pi, 0)$ is also fixed, but it is of elliptic type for all sufficiently small $\varepsilon > 0$. This observation has a simple interpretation from the point of view of perturbation theory: for small values of the parameter $\varepsilon > 0$ the invariant circle $y = 0$ of the unperturbed problem disintegrates, and a pair of non-degenerate fixed points is born from the family of fixed points composing the circle. The coordinates of these points depend analytically on ε. One of them is stable (up to a first approximation), and the other unstable. The multipliers of the fixed points can be expressed as convergent series in powers of $\sqrt{\varepsilon}$ (compare with the results of Chap. IV, §8.7).

By (3.14), the stable and unstable separatrices Λ^+ and Λ^- of the hyperbolic fixed point intersect at this point at a small angle of order $\sqrt{\varepsilon}$. It turns out that there is another intersection close to the point $(x, y) = (\pi, 2\sqrt{\varepsilon})$. Lazutkin [160] obtained an asymptotic expansion for the intersection angle of the separatrices:

$$\phi = \frac{\pi\alpha}{\varepsilon} e^{-\frac{\pi^2}{\sqrt{\varepsilon}}}[1 + O(\varepsilon^{1/8-\delta})] , \tag{3.15}$$

where $\alpha = 1118.82770595\ldots$, δ is an arbitrary positive number, and the constant in the estimate for $O(\cdot)$ depends on δ. The exponential smallness of the angle ϕ can be deduced from Neishtadt's result [189], which is proved under very general assumptions.

From (3.15) it follows, in particular, that the separatrices Λ^+ and Λ^- intersect transversely and hence there is a "stochastic layer" close to $\Lambda^+ \cup \Lambda^-$. Earlier Chirikov [59] proved the existence of this layer by numerical computations, and established that its size increases with ε. With further increasing of ε, the layer merges with other stochastic layers of a similar origin. However, Lazutkin's main result is the derivation of the asymptotic expansion (3.14), which is still unique in problems of this kind. Formula (3.14) was obtained by continuation of the mapping (3.13) into the complex plane of variables x, y. Admittedly, as Lazutkin himself remarked, the proof of the expansion (3.15) given in [160] is not entirely strict and complete. The results of [160] were verified and developed in subsequent publications (see, for example, [86, 161]). Lately Lazutkin's technique was extended to the case of analytic Hamiltonian systems having no hyperbolic periodic solutions for the zero value of the perturbing parameter [63, 85, 101]. We discussed such systems in Chap. IV.

3.9 The method of splitting of asymptotic surfaces can be applied to the problem of plane flow of a homogeneous ideal fluid which was discussed in Chap. I, §8.2 [150]. We consider equations in the Hamiltonian form

$$\dot{x} = H'_y , \quad \dot{y} = -H'_x ; \qquad H = H_0 + \varepsilon H_1 . \tag{3.16}$$

Here H_0 is the stream function of the stationary flow (it satisfies the Laplace equation), and $H_1 = x\cos\lambda t$, $\lambda = \text{const}$. The system (3.16) describes the potential flow corresponding to a stationary flow perturbed by a small sinusoidal velocity field of constant direction.

Suppose that there exist two hyperbolic critical points (not necessarily distinct) of the function H_0 which are joined by a separatrix Λ_0. This curve is the trajectory of a one-parameter family of doubly asymptotic solutions of the unperturbed problem:

$$x = x_a(t-\mu) , \quad y = y_a(t-\mu) ,$$

where μ is a parameter. The functions $x_a(\cdot)$, $y_a(\cdot)$ are holomorphic on a certain strip of the complex plane containing the real axis.

For small values of ε, the conditions for splitting of the perturbed separatrices Λ_ε^+ and Λ_ε^- are reduced to analysis of the integral

$$I(\mu) = \int_{-\infty}^{\infty} \{H_0, H_1\}(x_a, y_a)\, dt = c_1(\lambda)\cos\lambda\mu + c_2(\lambda)\sin\lambda\mu ,$$

where

$$c_1 = \int_{-\infty}^{\infty} \dot{x}_a(t) \cos \lambda t \,, \qquad c_2 = \int_{-\infty}^{\infty} \dot{x}_a(t) \sin \lambda t \, dt \,.$$

Clearly,

$$c_1(\lambda) - ic_2(\lambda) = \int_{-\infty}^{\infty} \dot{x}_a(t) e^{-i\lambda t} \, dt$$

is the Fourier transform of the function $\dot{x}_a(t)$. Since $\dot{x}_a$ is analytic and tends to zero exponentially as $|t| \to \infty$, by the Paley–Wiener theorem $c_1 - ic_2$ also is an analytic function of ε.

If the zeros of the function $I(\mu)$ are simple, then for small $\varepsilon \neq 0$ the asymptotic surfaces $\Lambda_\varepsilon^\pm$ intersect transversely (see §1). In the present case the average of the $2\pi/\lambda$-periodic function I equals zero. Hence, the function I always vanishes somewhere. Obviously, the zeros are simple if $c_1^2 + c_2^2 \not\equiv 0$.

Let us show that if

$$\dot{x}_a(t) \not\equiv 0 \,, \tag{3.17}$$

then for almost all λ the sum $c_1^2 + c_2^2$ is non-zero. Indeed, suppose that $c_1^2 + c_2^2 = 0$. Then by the inversion theorem for the Fourier transform, we have

$$\dot{x}_a(t) = \frac{1}{2\pi} \int_{-\infty}^{\infty} (c_1(\lambda) - ic_2(\lambda)) e^{i\lambda t} \, d\lambda \equiv 0 \,.$$

Hence, condition (3.17) implies that $c_1^2 + c_2^2 \not\equiv 0$. Since c_1 and c_2 are analytic in λ, $c_1^2 + c_2^2 \neq 0$ almost everywhere.

Thus, if the unperturbed separatrix Λ_0 does not lie on a straight line, then almost any sinusoidal perturbation causes chaotic behavior of the flow near Λ_0. A similar result holds in the case of a rectilinear separatrix, with only additional assumption that the perturbing field is not parallel to the line containing Λ_0.

As an application, consider the problem of two vortices of equal intensity (see Chap. II, §8). In a certain rotating frame of reference, the motion of the fluid particles is described by the Hamiltonian equations with the Hamiltonian

$$H_0 = \frac{\Gamma}{2\pi} \left(\frac{y}{2a} + \ln \left[\frac{x^2 + (y-a)^2}{x^2 + (y+a)^2} \right]^{1/2} \right) .$$

The vortices with intensities $\Gamma_1 = -\Gamma_2 = \Gamma \neq 0$ are located at the points with coordinates $(0, \pm a)$.

Clearly, the unperturbed stationary velocity field has two hyperbolic critical points $(\pm\sqrt{3}a, 0)$ joined by three pairs of double separatrices (the phase portrait of the system is shown in Fig. 8). As indicated above, in the presence of the small perturbation $\varepsilon H_1 = \varepsilon x \cos \lambda t$ these pairs of separatrices split and intersect transversely for almost all values of λ. For small $\varepsilon \neq 0$, we obtain "islands" with chaotic behavior of fluid particles near the split separatrices.

In Fig. 25 this result is illustrated by numerical calculations for unit values of Γ, α and λ. The positions of certain fluid particles are marked at time moments $t =$

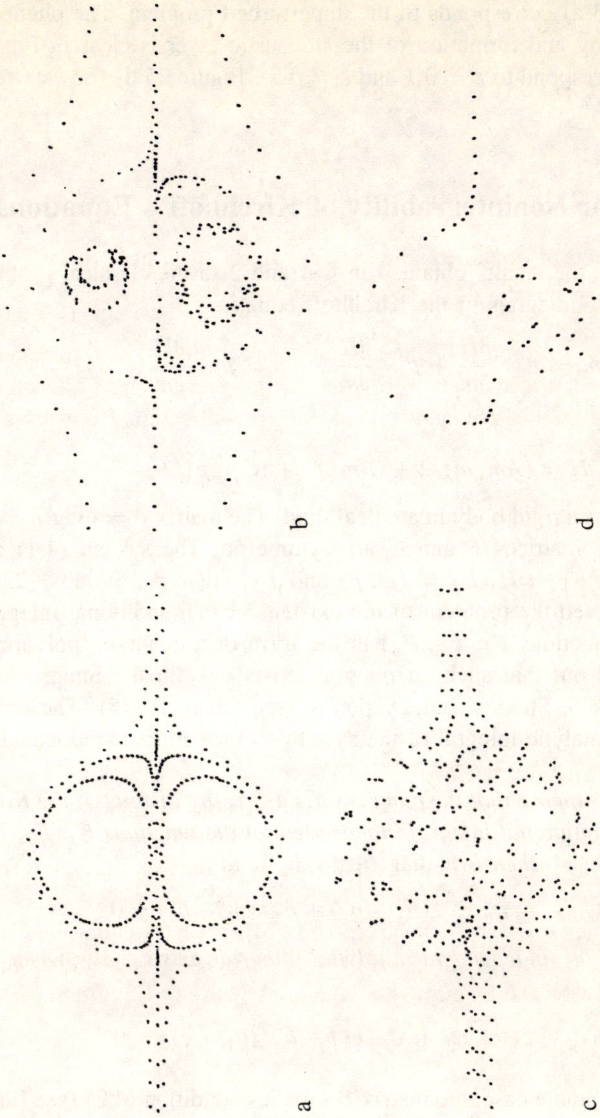

Fig. 25.

$2\pi n$, $n \in \mathbb{Z}$. Fig. 25 a) corresponds to the unperturbed problem. The phenomenon of separatrix splitting and formation of the stochastic layer is clear in Fig. 25 b) and 25 c) (they correspond to $\varepsilon = 0.1$ and $\varepsilon = 0.5$). Figure 25 d) shows a separate trajectory for $\varepsilon = 0.5$.

4 Conditions for Nonintegrability of Kirchhoff's Equations

4.1 We shall apply the results obtained in §§1 and 2 to the classical problem on the existence of new integrals of the Kirchhoff equations

$$\dot{m} = m \times \frac{\partial H}{\partial m} + p \times \frac{\partial H}{\partial m}, \quad \dot{p} = p \times \frac{\partial H}{\partial m}, \tag{4.1}$$

where

$$H = (Am, m)/2 + (Bm, p) + (Cp, p)/2, \tag{4.1'}$$

governing rotation of a rigid body in an ideal fluid. The matrix $A = \operatorname{diag}(a_1, a_2, a_3)$ is diagonal and the matrices B and C are symmetric. The system (4.1) always has three integrals: $F_1 = H$, $F_2 = (m, p)$ and $F_3 = (p, p)$. Steklov [221] and Lyapunov [166] solved the problem of the existence of an additional integral (independent of the functions F_1, F_2, F_3) in the form of a quadratic polynomial in m and p. It turned out that such an integral exists only in the integrable cases of Kirchhoff, Clebsch, Steklov and Lyapunov (see Chap. II, §5). The existence problem of a new analytic integral of the system (4.1) is more complicated.

Theorem 1 [153]. *Suppose that the numbers a_1, a_2, a_3 are distinct. If the Kirchhoff equations have an additional integral independent of the functions F_1, F_2, F_3 and analytic in $\mathbb{R}^6 = \{m, p\}$, then $B = \operatorname{diag}(b_1, b_2, b_3)$ and*

$$a_1^{-1}(b_2 - b_3) + a_2^{-1}(b_3 - b_1) + a_3^{-1}(b_1 - b_2) = 0. \tag{4.2}$$

If $B = 0$, then an independent analytic integral exists only when $C = \operatorname{diag}(c_1, c_2, c_3)$ and

$$a_1^{-1}(c_2 - c_3) + a_2^{-1}(c_3 - c_1) + a_3^{-1}(c_1 - c_2) = 0. \tag{4.3}$$

In Steklov's integrable case, the matrix B satisfies condition (4.2) (see Eq. (5.6) of Chap. II). Condition (3.6) gives the Clebsch integrable case (see Chap. II, §5). It is interesting to note that conditions (4.2) and (4.3) are of the same form.

Corollary. *In general, Kirchhoff's equations are nonintegrable.*

The proof of Theorem 1 is based on splitting of asymptotic surfaces. Replacing p by εp, we introduce into (4.1) a small parameter ε. On a fixed four-dimensional integral surface $M_{23} = \{(m, p) : F_2 = f_2,\ F_3 = f_3\}$, the Eqs. (4.1) are Hamiltonian with the Hamiltonian function $H_0 + \varepsilon H_1 + \varepsilon^2 H_2$, where H_0, H_1 and H_2 are the restrictions of the functions $(Am, m)/2$, (Bm, p) and $(Cp, p)/2$ on M_{23}. The case $\varepsilon \to 0$ is equivalent to the case when the energy constant f_1 is much larger than f_2 and f_3 (more precisely, the ratios f_2/f_1 and f_3/f_1 are small quantities of the same order). For $\varepsilon = 0$ we obtain the integrable Euler problem on inertial rotation of a free rigid body. The equations of motion have two unstable periodic solutions similar to (3.4):

$$\begin{aligned} &m_1 = m_3 = 0\,, \quad m_2 = m_2^0 = \pm\sqrt{2f_1/a_2}\,, \\ &p_2 = p_2^0 = \pm f_2/m_2^0\,, \\ &p_1 = \alpha\cos(a_2 m_2^0 t)\,, \quad p_3 = \alpha\sin(a_2 m_2^0 t)\,, \end{aligned} \tag{4.4}$$

where $\alpha^2 = f_3 - (f_2/m_2^0)^2$.

Let F_0 be an analytic integral of the Euler problem. Suppose that the improper integral

$$J = \int_{-\infty}^{\infty} \{F_0, H_1\}\, dt\,, \tag{4.5}$$

evaluated on the solutions of the unperturbed problem that are asymptotic to the periodic solutions (4.4), is non-constant on the asymptotic surfaces of the Euler problem. Then, according to the results of §2, for small $\varepsilon \neq 0$ the Kirchhoff equations have no non-constant analytic integrals. Moreover, they admit no nontrivial analytic symmetry groups.

Thus, the proof of Theorem 1 is reduced to verification that the integral (4.4) is non-constant. It is convenient to put $F_0 = (m, m)/2$. If $B = 0$, then, of course, we must replace in (4.5) H_1 by H_2. If $F_0 = (m, m)/2$, then the integral J exists only in the sense of the principal value. In this case we can put, for instance, $F_0 = [(m, m) - a_2^{-1}(Am, m)]/2$.

As an example, we shall obtain Steklov's condition (4.2) in the simplest case when $B = \mathrm{diag}(b_1, b_2, b_3)$. Since

$$\begin{aligned} \{F_0, H_1\} = (b_3 - b_2)(m_1 m_2 p_3 + m_1 m_3 p_2) \\ + (b_1 - b_3)(m_2 m_3 p_1 + m_1 m_2 p_3) \\ + (b_2 - b_1)(m_1 m_3 p_2 + m_2 m_3 p_1)\,, \end{aligned}$$

we obtain

$$\begin{aligned} J = (b_2 - b_3)(J_{123} - J_{132}) + (b_1 - b_3)(J_{231} - J_{123}) \\ + (b_2 - b_1)(J_{132} - J_{231})\,, \end{aligned}$$

where

$$J_{ijk} = \int_{-\infty}^{\infty} m_i m_j p_k\, dt\,.$$

The integrals J_{ijk} satisfy the following linear equations:

$$\begin{aligned} a_3 J_{132} - a_2 J_{123} + (a_3 - a_2) J_{231} = 0\,, \\ a_1 J_{123} - a_3 J_{231} + (a_1 - a_3) J_{132} = 0\,, \\ a_2 J_{231} - a_1 J_{132} + (a_2 - a_1) J_{213} = 0\,. \end{aligned} \tag{4.6}$$

Let us prove, for example, the first relation. From the Kirchhoff equations it follows that for $\varepsilon = 0$

$$\frac{d}{dt}(m_1 p_1) = a_3 m_1 m_3 p_2 - a_2 m_1 m_2 p_3 + (a_3 - a_2) m_2 m_3 p_1 .$$

Since $m_1 \to 0$ as $t \to \pm\infty$, we have

$$a_3 J_{132} - a_2 J_{123} + (a_3 - a_2) J_{231} = \int_{-\infty}^{\infty} \frac{d}{dt}(m_1 p_1)\, dt .$$

If $a_2 a_3 - a_1 a_2 - a_1 a_3 \neq 0$, then from the Eqs. (4.6) we obtain the equalities

$$J_{132} = \frac{a_1 a_3 - a_1 a_2 - a_2 a_3}{a_2 a_3 - a_1 a_2 - a_1 a_3} J_{231} , \qquad J_{123} = \frac{a_1 a_2 - a_1 a_3 - a_2 a_3}{a_2 a_3 - a_1 a_2 - a_1 a_3} J_{231} .$$

It is possible to calculate the integral J_{231} by means of residues and verify that it does not vanish. If condition (4.2) is not satisfied, then the evident equality

$$\frac{J(a_1 a_2 + a_1 a_3 - a_2 a_3)}{2 a_1 a_2 a_3 J_{231}} = a_1^{-1}(b_3 - b_2) + a_2^{-1}(b_1 - b_3) + a_3^{-1}(b_2 - b_1)$$

yields $J \neq 0$. Hence, the perturbed separatrices split. In the case when $a_2 a_3 - a_1 a_2 - a_1 a_3 = 0$, the integral J is proportional to J_{123} or J_{132}. From symmetry arguments and the existence of an invariant measure on M_{23} induced by the standard measure in $\mathbb{R}^6$ it follows that the perturbed separatrices intersect. Hence, the Kirchhoff equations are nonintegrable on the invariant manifolds M_{23} and, in particular, in the whole phase space $\mathbb{R}^6$.

Now suppose that $B = 0$ and the matrix C is diagonal. We are going to obtain the Clebsch condition (4.3). Put $2F_0 = (m, m) - a_2^{-1}(Am, m)$ and calculate the Poisson bracket

$$\{F_0, H_2\} = m_1 p_2 p_3 (c_2 - c_3)(1 - a_1/a_2) + m_3 p_1 p_2 (c_2 - c_1)(1 - a_3/a_2) .$$

Hence,

$$J = \int_{-\infty}^{\infty} \{F_0, H_2\}\, dt = J_1 (c_3 - c_2)(a_1^{-1} - a_2^{-1}) + J_3 (c_2 - c_1)(a_3^{-1} - a_2^{-1}) ,$$

where

$$J_1 = a_1 \int_{-\infty}^{\infty} m_1 p_2 p_3\, dt , \qquad J_3 = a_3 \int_{-\infty}^{\infty} m_3 p_1 p_2\, dt .$$

Let us show that $J_1 = J_3$. We multiply the Kirchhoff equation $\dot{p}_2 = a_1 m_1 p_3 - a_3 m_3 p_1$ by p_2 and integrate it from $t = -\infty$ to $t = \infty$. Then

$$J_1 - J_3 = \int_{-\infty}^{\infty} p_2 \dot{p}_2\, dt = \left. \frac{p_2^2}{2} \right|_{-\infty}^{\infty} = 0 .$$

Calculating the integral J_1 by means of residues, it is possible to make sure that it does not vanish. If the condition (4.3) is not satisfied, then an obvious equality

$$J/J_1 = (c_2 - c_3) a_1^{-1} + (c_1 - c_3) a_2^{-1} + (c_2 - c_1) a_3^{-1}$$

implies that $J \neq 0$. Hence, the perturbed asymptotic surfaces split. These surfaces always intersect. Indeed, the Hamiltonian system is invariant under the substitution $p \to -p$. Identifying the antipodal points of the sphere $p^2 = f_3 > 0$, we find that the unstable periodic trajectories (4.4) coincide. On the other hand, Poincaré established that in the homoclinic case there always exists a doubly asymptotic solution. To complete the proof of the nonexistence of an additional analytic integral, it is sufficient to apply the results of §2.

If the condition (4.2) (or (4.3) for $B = 0$) does not hold, then one pair of separatrices of the Euler problem always splits under the perturbation. It is interesting to note that for a suitable choice of the matrices B and C, one pair of separatrices remains doubled and the other splits. For example, suppose that $B = 0$, while the elements c_{ij} of the symmetric matrix C satisfy the following conditions:

$$\sqrt{a_2 - a_1}\, c_{13} \pm \sqrt{a_3 - a_2}\,(c_{22} - c_{11}) = 0\,,$$
$$\sqrt{a_2 - a_1}\,(c_{33} - c_{22}) \mp \sqrt{a_3 - a_2}\, c_{13} = 0\,.$$

Then for all ε the Kirchhoff equations have the "Hess–Appelrot particular integral" $F = m_1\sqrt{a_2 - a_1} \pm m_3\sqrt{a_3 - a_2} = 0$. For small values of ε, the separatrices of the Euler problem

$$\{F_k = f_k\,, \quad k = 1, 2, 3\,; \quad F = 0\}$$

are those of the perturbed periodic solutions (4.4).

4.2 Barkin and Borisov studied the Kirchhoff problem numerically for the case of diagonal matrices A, B and C. Equations (4.1) were represented in terms of the special canonical variables L, G, H, l, g, h. The coordinate H corresponds to the integral F_2, whose value was assumed to be zero. Following [84], we fix the three-dimensional level surface $F_1 = 50$ of the energy integral of the reduced system and take the section plane $g = \pi/2$. The variables L/G, $l \bmod 2\pi$ are coordinates on this plane. Since $|L| \leq G$, we have $|L/G| \leq 1$.

Some results of the calculations are presented in Fig. 26. The following values were taken for the diagonal elements of the matrices A and B: $a_1 = 1/3$, $a_2 = 1/2$, $a_3 = 1$; $b_1 = 3$, $b_2 = 2$, $b_3 = 1$. Clearly, condition (4.2) is satisfied. The trajectories in the integrable Steklov case, when $c_1 = 3$, $c_2 = 8$, $c_3 = 1$, are shown in Fig. 26 a). In the next figures the parameter c_1 starts increasing. Fig. 26 b) corresponds to $c_1 = 5$, and Fig. 26 c) to $c_1 = 10$. It is clear that the picture of the "integrable" behavior of the phase trajectories starts disintegrating exactly in the neighborhoods of separatrices. As we get further from the integrable problem, the "stochastic" layer near the split separatrices starts to spread. In Fig. 26 d) the picture of intersecting separatrices is shown for the following values of the parameters: $b_1 = 0.1$, $b_2 = b_3 = 0$; $c_1 = 3$, $c_2 = 8$, $c_3 = 1$. Obviously, the heteroclinic net of intersecting separatrices repeats itself with period π. This is a consequence of the invariance of the problem under the substitution $m \to -m$, $p \to -p$ (see §4.1). Calculations show that condition (4.2) is not sufficient for integrability of Kirchhoff's equations.

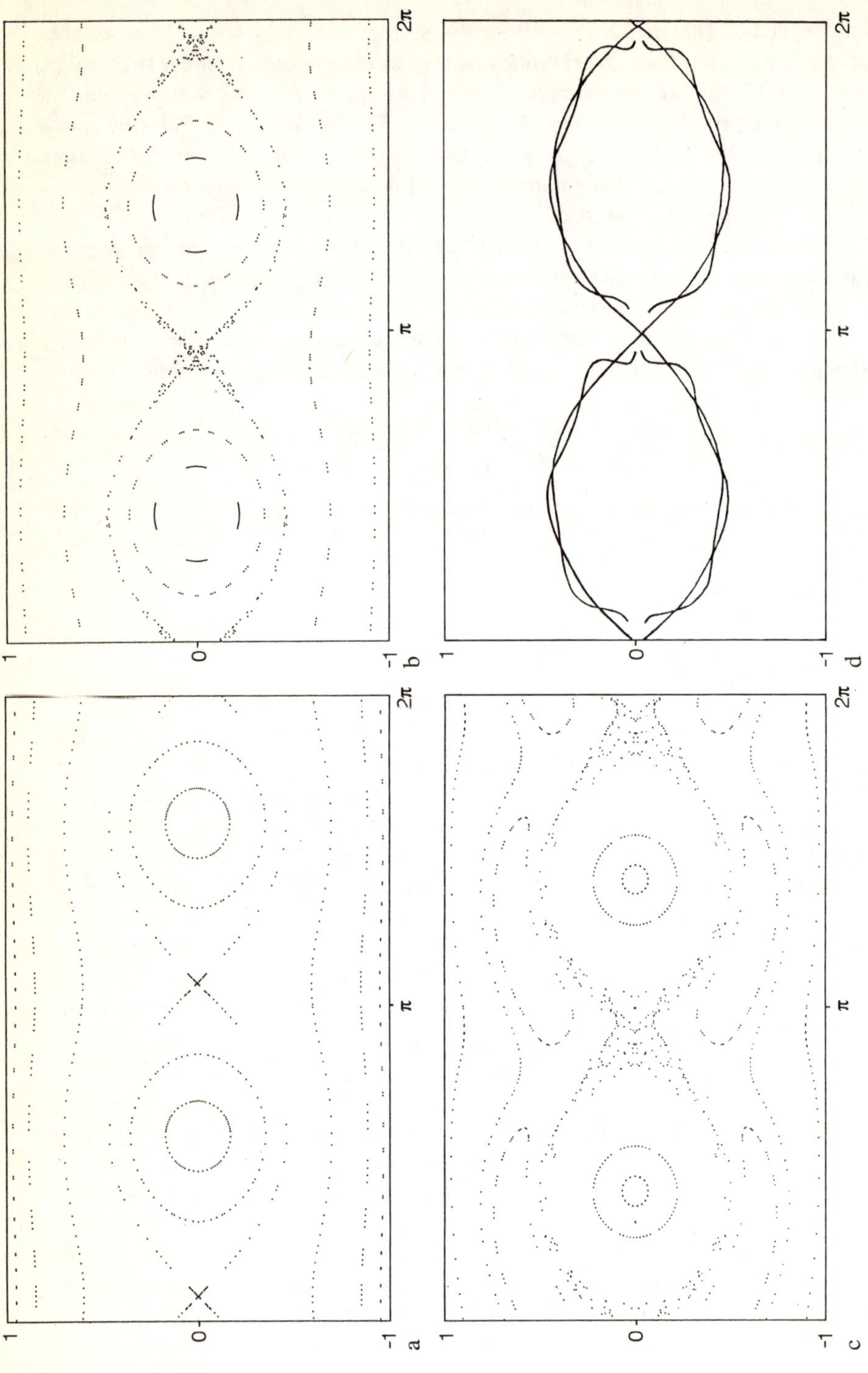

Fig. 26.

4.3 Barkin and Borisov [18] also investigated the existence of additional integrals for the system (4.1) under the assumption that all eigenvalues of the matrix A are distinct. As in §4.1, they introduced a small parameter ε by replacing p by εp. Then they studied the existence problem of additional integrals in the form of a power series

$$\Phi_0(m, p) + \varepsilon\Phi_1(m, p) + \varepsilon^2\Phi_2(m, p) + \cdots , \tag{4.7}$$

with coefficients which can be extended to single-valued holomorphic functions on the complexified phase space $\mathbb{C}^6 = \{m, p\}$. By using the method of [128] (see §1 of Chap. VII), Barkin and Borisov showed that a "single-valued" integral of type (4.7) exists only in the known integrable cases of Clebsch and Steklov. Since the considered integrals are defined in the complex space, the authors remark that this result is only formally related to dynamics. This is not exactly so. The right-hand terms of equations (4.1) are homogeneous. Hence, every homogeneous form of a real-analytic integral is an integral of the system (4.1). The substitution $p \to \varepsilon p$ transforms such forms into polynomials in ε. The coefficients are single-valued holomorphic functions in m and p, since they are polynomials in m and p.

Of course, this argument does not clarify any features of the behavior of real trajectories. However, it makes it possible to obtain the complete solution for the problem of analytic integrals of the Kirchhoff equations in the non-symmetric case.

4.4 For $a_1 = a_2$, the problem of integrability for the system (4.1) is much more involved. We already noted that the existence of an analytic integral of the system (4.1) implies the existence of an integral which is a homogeneous polynomial in m and p. This simple observation makes it possible to apply the Husson method, first used to solve the existence problem of an additional algebraic integral for the equations of a heavy rigid body with a fixed point (the history of the problem and the description of the Husson method can be found in [113]; see also [16]). This was accomplished by Sadetov [209] under the assumption that the matrices A, B and C are diagonal.

Theorem 2. *If $a_1 = a_2 \neq a_3$, then an additional analytic integral of the system* (4.1) *exists only in the Kirchhoff case* (*when $b_1 = b_2$ and $c_1 = c_2$*).

Note that for $a_1 = a_2 \neq a_3$ all known integrable problems are included in the Kirchhoff case (see Chap. II, §5).

In [209] a more complicated problem of particular integrals was also studied. By definition, a function F is a particular integral if $\dot{F} = 0$ on the surface $F_2 = 0$, and almost everywhere on this surface the functions F_1, F_2 and F are independent.

Theorem 3. *Let $B = 0$. Then an analytic particular integral of the system* (4.1) *exists only in the Kirchhoff and Chaplygin cases.*

Chaplygin's integrable case is particular (see Chap. II, §5). In [209] certain necessary conditions for integrability were obtained under the assumption $a_1 = a_2 = a_3$. However, apparently they are not sufficient.

4.5 The Husson method does not provide any information on the behavior of real phase trajectories. If $a_1 = a_2$, then in certain cases it is possible to indicate homoclinic structures clarifying the mechanism of nonintegrability of Kirchhoff's equations.

Let $B = 0$. Then the system (4.1) can be represented in the form of the Euler–Poisson equations:

$$I\dot{\omega} = I\omega \times \omega + e \times Je \,, \quad \dot{e} = e \times \omega \,. \tag{4.8}$$

Here I and J are the inverses of the matrices A and B; $\omega = Am$, $e = Cp$. We assume that $a_1 = a_2$ and hence, $I_1 = I_2$.

We can pass from the system (4.8) to the "restricted" problem by replacing I_3, J_1 and J_2 with μI_3, μJ_1 and μJ_2 respectively, where $0 < \mu \leq 1$. It turns out that if $J_1 \neq J_2$, then for small values of $\mu > 0$ the system (4.8) has no nontrivial symmetry fields with analytic components [154]. The proof follows the method of §3.4.

To get rid of ω_3, we divide both sides of Eqs. (4.8) by μ and then put $\mu = 0$. As a result, we obtain the restricted problem of a rigid body in an ideal fluid (Chap. I, §6.4). First we shall prove that it is nonintegrable. For $\mu = 0$, the integrals of the system (4.8) are transformed into the following integrals of the restricted problem:

$$\omega_1^2 + \omega_2^2 + \alpha e_3^2 = h \,, \quad \omega_1 e_1 + \omega_2 e_2 = c \,, \quad e_1^2 + e_2^2 + e_3^2 = e^2 \,.$$

Here $\alpha = J_3/I_1$ and $h, c, e = \text{const}$. Using the equations of motion, it is easy to obtain the following equation for the double rotation angle $\phi = 2 \arctan e_1/e_2$:

$$\ddot{\phi} + \Lambda(e^2 - u^2) \sin\phi = 2\frac{d}{dt}\frac{cu}{e^2 - u^2} \,, \tag{4.9}$$

where

$$\Lambda = (J_1 - J_2)/I_3 \,, \qquad \dot{u}^2 = (h - \alpha u^2)(e^2 - u^2) - c^2 \,.$$

Let $he^2 = c^2 + \varepsilon^2$. Then

$$u(t) = \varepsilon u_0(t) + o(\varepsilon) \,, \qquad u_0 = -\lambda \cos\left(\sqrt{h - \alpha e^2}(t - t_0)\right) \,,$$

where $\lambda, \alpha = \text{const}$. For small ε, Eq. (4.9) can be represented in the following form:

$$\ddot{\phi} + \Lambda e^2 \sin\phi = \frac{2\varepsilon c\lambda}{e^2\sqrt{h - \alpha e^2}} \sin\left(\sqrt{h - \alpha e^2}(t - t_0)\right) + o(\varepsilon) \,. \tag{4.10}$$

Using the methods of §1, it is easy to establish the existence of a transverse intersection point for the separatrices of (4.10) if $\varepsilon \neq 0$ is small (see also [176]). It follows, in particular, that Eq. (4.10) is nonintegrable. Since transverse intersection of separatrices is stable under small perturbations, the system (4.8) is also nonintegrable for small $\mu \neq 0$.

4.6 Another method for detecting homoclinic structures was suggested in [134]. Let $a_1 = a_2 \neq a_3$, $B = 0$, $C = \mathrm{diag}(c_1, c_2, c_3)$ and $c_1 = c_2 + \varepsilon$. For $\varepsilon = 0$

we have the integrable Kirchhoff case. In this unperturbed problem, there exist unstable periodic trajectories and homoclinic solutions. By using the results of §1, it is possible to establish that the separatrices split for small non-zero ε. A more general problem, when the matrix A has non-diagonal elements of order ε, was considered in [211].

4.7 The results of §4.1 make it possible to obtain conditions for nonintegrability of some problems of rigid body dynamics. Consider, for example, rotation of a ferromagnet in a magnetic field (Chap. I, §3.8). It is governed by Eqs. (3.18) of Chap. I:

$$I\dot{\omega} = I\omega \times \omega + \varepsilon(\omega \times \gamma)\,, \quad \dot{\gamma} = \gamma \times \omega\,, \qquad \varepsilon = \text{const} \neq 0\,. \tag{4.11}$$

Theorem 4 [138]. *The system* (4.11) *has an additional analytic integral if and only if the rigid body is dynamically symmetric.*

Indeed, the system (4.11) can be represented in the form of Kirchhoff's Eqs. (4.1), if we put (see §3 of Chap. I)

$$A = I^{-1}\,, \quad B = -\varepsilon I^{-1}\,, \quad C = \varepsilon^2 I^{-1}\,. \tag{4.12}$$

Let $I = \text{diag}(I_1, I_2, I_3)$. If $I_1 = I_2$, then we obtain the integrable Kirchhoff case (the function $I_3\omega_3 + \varepsilon\gamma_3$ is an additional integral). This integrable system was discovered and studied in [212]. Now consider the general case when all moments of inertia I_k are distinct. We apply Theorem 1 and write down the relation (4.2) under the assumption (4.12):

$$\varepsilon I_1 I_2 I_3 (I_3^{-1} - I_2^{-1})(I_1^{-1} - I_3^{-1})(I_1^{-1} - I_2^{-1}) = 0\,.$$

For a non-symmetric body, this equation is, obviously, never satisfied. This finishes the proof.

In [49] integrability was studied for the more general problem of rotation of a ferromagnetic body with a non-spherical magnetic tensor. The gyromagnetic effects were also taken into account.

Burov and Karapetyan [48] applied Theorem 1 and the results of §3.3 to the problem of complete integrability of the equations describing a heavy ellipsoid sliding on a horizontal plane. They assumed that the ellipsoid is close to a ball and its moments of inertia are distinct. This system has five degrees of freedom and four integrals: the total energy, the horizontal component of the momentum of the body, and the vertical component of the angular momentum. Thus, only one integral is lacking for complete integrability. In [48] the following necessary conditions for integrability were obtained:

1) the center of mass of the ellipsoid coincides with its geometric center,
2) the principal axes of the inertia ellipsoid and the surface ellipsoid coincide,
3) the moments of inertia and the half-axes of the surface ellipsoid satisfy the relation

$$I_1(a_2 - a_3) + I_2(a_3 - a_1) + I_3(a_1 - a_2) = 0\,.$$

5 Bifurcation of Separatrices

5.1 Let $(x, y) = z$ be the Cartesian coordinates in the plane $\mathbb{R}^2$. We furnish the plane with the standard symplectic structure $\Omega = dx \wedge dy$. Let

$$H = H_0(z) + \varepsilon H_1(z, t) + o(\varepsilon)$$

be the Hamiltonian function which is 2π-periodic in t and analytic in the domain $D \times \mathbb{T}^1\{t \bmod 2\pi\} \times (-\varepsilon_0, \varepsilon_0)$. Here D is a domain in the plane and $\varepsilon_0 > 0$.

For $\varepsilon = 0$, we have an integrable Hamiltonian system with one degree of freedom. Assume that the unperturbed system has in D three unstable non-degenerate equilibrium positions z_1, z_2 and z_3 connected by doubly asymptotic trajectories Γ_1 and Γ_2, as shown in Fig. 27.

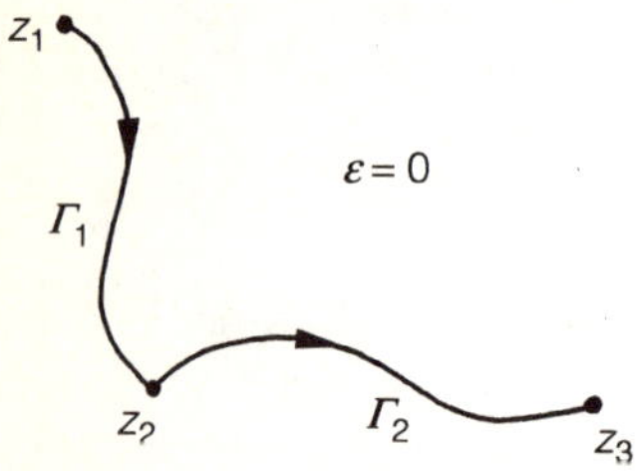

Fig. 27.

The points z_1 and z_3 may coincide, but we require that $z_1 \neq z_2$. The points z_1, z_2 and z_3 are fixed points of the Poincaré map g_0 over the period $t = 2\pi$ of the unperturbed system. The curves Γ_1 and Γ_2 are invariant curves of this mapping, filled by points which, under positive (negative) iterations of the mapping g_0, tend to the point z_2 (z_1) for the curve Γ_1, and to the point z_3 (z_2) for the curve Γ_2. For small values of the parameter $\varepsilon \neq 0$, the points z_1, z_2 and z_3 do not disappear, but transform into fixed points $z_1(\varepsilon)$, $z_2(\varepsilon)$ and $z_3(\varepsilon)$ of the perturbed Poincaré map g_ε. In general, as shown by Poincaré, the unstable separatrixΓ_1' of the point $z_1(\varepsilon)$ and the stable separatrix Γ_1'' of the point $z_2(\varepsilon)$ no longer coincide for $\varepsilon \neq 0$. The same holds for the separatrices Γ_2' and Γ_2'' of the points $z_2(\varepsilon)$ and $z_3(\varepsilon)$ (see Fig. 28).

The conditions for the separatrices Γ_s' and Γ_s'' to be distinct were discussed in §1. If they do not coincide as immersed curves, then they may or may not coincide as subsets of $\mathbb{R}^2$. If, for example, the separatrices Γ_1' and Γ_1'' intersect, then the trajectories of the heteroclinic doubly asymptotic solutions connecting $z_1(\varepsilon)$ and $z_2(\varepsilon)$ will pass through the points of intersection. In §1 we indicated sufficient conditions for intersection and non-intersection of the perturbed separatrices in the regions of $\mathbb{R}^2$ which do not contain the points z_1, z_2 and z_3. However, the question about the fate of the perturbed separatrices near the points $z_s(\varepsilon)$ remained

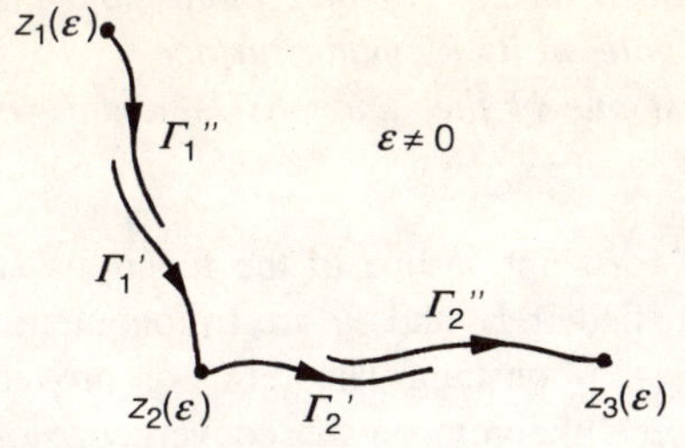

Fig. 28.

open. Below in §§5.1 and 5.2 we describe Dovbysh's results concerning the mutual disposition of the separatrices Γ_1'' and Γ_2'' in a neighborhood of the point $z_2(\varepsilon)$.

We introduce some notation. Let $t \to z_a^{(1)}(t)$ be a doubly asymptotic solution of the unperturbed problem such that

$$\lim_{t\to-\infty} z_a^{(1)}(t) = z_1 , \qquad \lim_{t\to\infty} z_a^{(1)}(t) = z_2 .$$

Denote by $z_a^{(2)}(\cdot)$ another doubly asymptotic solution joining the points z_2 and z_3. Let

$$I_s(\phi) = \int_{-\infty}^{\infty} \{H_0, H_1\}(z_a^{(s)}(t-\phi), t)\, dt , \qquad s = 1, 2 .$$

These functions played leading roles in our analysis of splitting of the separatrices Γ_s' and Γ_s'' in §1. Note that the functions I_s are analytic and 2π-periodic. In the case of homoclinic trajectories, their averages over the period are zero. However, this is usually not so in the present situation. A necessary condition for non-intersection of the perturbed separatrices Γ_s' and Γ_s'' is that the function I_s does not change its sign. We assume that this condition is satisfied. Moreover, we assume that $I_1 \geq 0$ and $I_2 \leq 0$. In this case for small positive ε, the picture of the disposition of the split separatrices is the same as shown in Fig. 28. For $\varepsilon = 0$, there exists a Birkhoff canonical transformation $x, y \to \xi, \eta$ in a neighborhood of the point z_2 such that in the new variables $H_0(\xi, \eta) = F_0(\zeta)$, where $\zeta = \xi\eta$, and

$$\frac{dF_0}{d\zeta}(0) = \Lambda > 0 . \tag{5.1}$$

The positive eigenvalue Λ of the unperturbed linearized system will appear in further discussion.

Theorem 1 [68]. *The perturbed separatrices Γ_1'' and Γ_2'' do not coincide for small positive values of the parameter $\varepsilon > 0$, if at least one of the following conditions is satisfied:*

1) *there exists ϕ such that $d \ln I_1(\phi)/d\phi \geq \Lambda$ or $d \ln(-I_2(\phi))/d\phi \leq -\Lambda$,*
2) *the value domains of the functions I_1 and $-I_2$ do not coincide,*

3) *one of the functions* I_1 *or* I_2 *is single-valued on the complex plane, and the other is not constant and has a zero or a pole on its Riemann surface,*
4) $d^2F_0(\zeta)/d\zeta^2 \neq 0$ *for* $\zeta = 0$ *and at least one of the functions* I_1 *and* I_2 *is non-constant.*

The first condition is certainly satisfied if for some ϕ one of the functions I_1 or I_2 vanishes. Similarly, condition 3) is satisfied if I_1 and I_2 are trigonometric polynomials in ϕ and at least one of them is not constant. Theorem 1 is proved with the help of the uniform version of Moser's theorem on the convergence of the normalizing Birkhoff transformation in a neighborhood of a hyperbolic point [178], and the technique of [243].

5.2 Dovbysh applied these general results to the well-known problem on rotation of a heavy non-symmetric rigid body with a fixed point in a weak homogeneous gravitational field. The small parameter is the product of the weight of the body and the distance from the mass center to the suspension point. By factorization with respect to the group of rotations about the vertical, the problem can be reduced to a Hamiltonian system with two degrees of freedom. Fixing a positive value of the energy integral and applying Whittaker's method of isoenergetic reduction, we reduce the equations of motion to a Hamiltonian system with one and a half degrees of freedom. The Hamiltonian is periodic in the new time and it is of the type considered above (all the details can be found in [131]). The diagram of the separatrices of the unperturbed Euler problem (in the non-symmetric case) is shown in Fig. 29 (the points z_1 and z_2 coincide, since the phase space of the system is not a plane but a cylinder).

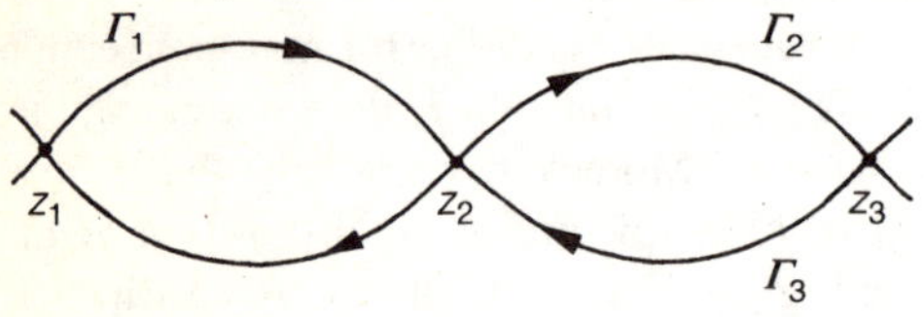

Fig. 29.

The distinctive feature of this problem is that the characteristic exponents of the hyperbolic equilibria z_1 and z_2 coincide. We isolate three separatrices Γ_1, Γ_2 and Γ_3 as shown in Fig. 29.

In §3.3 we have already discussed the splitting of separatrices in this problem for small values of $\varepsilon \neq 0$. Apparently, for certain values of the parameters the perturbed separatrices intersect, while for other values there is no intersection. Nevertheless, we have

Proposition 1. *For all values of the parameters not satisfying the Hess-Appelrot condition* (3.6) *there exist at least two doubly asymptotic* (*homoclinic*) *solutions to*

the perturbed unstable solutions $z_1(\varepsilon)$ and $z_2(\varepsilon)$. In the Hess–Appelrot case there are no such solutions.

This result was communicated to the author by S.L. Ziglin. Its proof is based on simple arguments involving Moser's theorem on invariant curves and the conservation of area under the Poincaré map. Arguments of this kind were first used by Poincaré for proving the existence of homoclinic solutions when a separatrix loop splits (see §2.2). Various arrangements of the perturbed separatrices are shown in Fig. 30 (case c) is impossible).

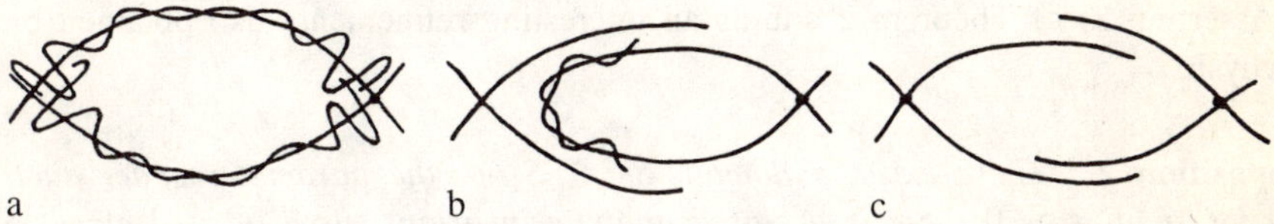

Fig. 30.

It remained unclear whether some of these homoclinic trajectories can lie on doubled separatrices. The answer is given by Theorem 1: the integrals $I_s(\phi)$, evaluated along the doubly asymptotic solutions corresponding to the separatrices Γ_1, Γ_2 and Γ_3, are non-constant trigonometric polynomials. Hence, condition 3) of Theorem 1 holds.

Theorem 2 [68]. *There exist domains S_1, S_2 and S_3 in the space of parameters (the constants of energy and momentum integrals are regarded as parameters) such that:*

1) *for parameter values in the domain $S_1 \cup S_2 \cap S_3$, the perturbed separatrices split, do not intersect, and are arranged as shown in Fig.* 31,

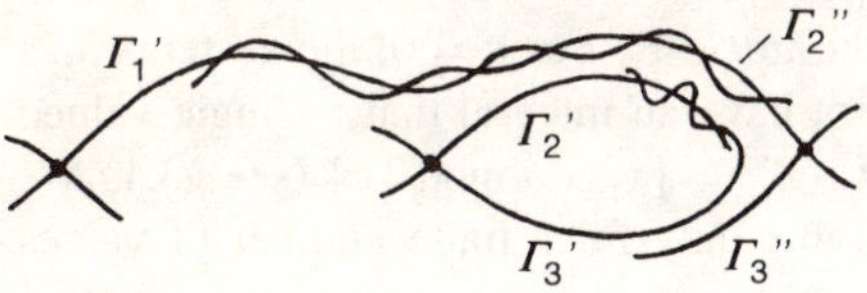

Fig. 31.

2) *for parameter values in the domain S_1 and all small $\varepsilon > 0$, the perturbed separatrices Γ_1' and Γ_3' do not intersect near the unperturbed separatrices Γ_1, Γ_2 and Γ_3,*

3) *for parameter values in the domain* S_2 *and all small* $\varepsilon > 0$, *the separatrices* Γ'_1 *and* Γ'_3 *intersect near the curves* Γ_1, Γ_2 *and* Γ_3,
4) *for parameter values in the domain* S_3 *there exist sequences of positive numbers* $\varepsilon_n^+ \to 0$ *and* $\varepsilon_n^- \to 0$ $(n \to \infty)$ *such that for* $\varepsilon = \varepsilon_n^+$ *the separatrices* Γ'_1 *and* Γ'_3 *intersect near the curves* Γ_1, Γ_2 *and* Γ_3, *while for* $\varepsilon = \varepsilon_n^-$ *they do not intersect.*

Thus, for points in the domain S_3 we obtain a nontrivial phenomenon: as the positive parameter ε tends to zero, there occur an infinite number of birth and death bifurcations of heteroclinic solutions passing close to the unperturbed separatrices Γ_1, Γ_2 and Γ_3.

Assertion 2) of Theorem 2 admits an interesting refinement, also obtained by Dovbysh.

Proposition 2. *There exists a domain* $S_0 \subset S_1$ *in the parameter space such that for small* $\varepsilon > 0$ *there is a closed analytic invariant curve located close to the unperturbed separatrices and separating the perturbed fixed points* $z_1(\varepsilon)$ *and* $z_2(\varepsilon)$.

In particular, in these cases there are no homoclinic motions: the separatricesof the hyperbolic points $z_1(\varepsilon)$ and $z_2(\varepsilon)$ remain on opposite sides of the closed invariant curve and thus they do not intersect. The proof of Theorem 2 is based on Moser's theorem on invariant curves and the technique developed in the proof of Theorem 2.

5.3 If a Hamiltonian system depends on a parameter, then in a typical situation integrable cases correspond to exceptional isolated values of the parameter. For a concrete problem, proving that the integrable cases are isolated can turn out to be very difficult. Following [135], we investigate this problem for the Hamiltonian equation

$$\ddot{x} + \omega^2[1 + \varepsilon f(t)] \sin x = 0 \,, \qquad \omega, \varepsilon = \text{const} \,, \tag{5.2}$$

describing oscillations of a mathematical pendulum. The analytic function $f(t)$ is assumed to be non-constant and 2π-periodic in $t \in \mathbb{R}$. For $\varepsilon = 0$, the system (5.2) is integrable, and for small $\varepsilon \neq 0$ it does not have an integral that is single-valued and analytic in the extended phase space $\mathbb{R} \times \mathbb{T}^2 = \{\dot{x};\ x, t \bmod 2\pi\}$ (see §3.1) We shall show that this equation can be integrable only for a finite number of values of ε in the interval $[-a, a]$, where $a = \left(\max_{\mathbb{R}} |f(t)|\right)^{-1}$.

For all $\varepsilon \in [-a, a]$, the periodic solution $x(t) \equiv \pi$ (or, equivalently, $x(t) \equiv -\pi$), corresponding to vertical oscillations of the inverted pendulum, is hyperbolic. Indeed, put $x = \pi + \xi$. Then the variational equation for the periodic solution $x(t) = \pi$ is

$$\ddot{\xi} - p(t)\xi = 0 \,, \qquad p = \omega^2(1 + \varepsilon f) \,.$$

Since $p(t) \geq 0$ and $p(t) \not\equiv 0$, the multiplicators of this solution are positive. One of them is larger and the other smaller than one (Lyapunov [167]). Thus, the solution $x(t) = \pi$ is hyperbolic. It has two-dimensional asymptotic surfaces Λ^+ and Λ^- completely filled by trajectories that tend unboundedly to the point $x = \pm\pi$ as $t \to \pm\infty$. Since the Hamiltonian H is analytic, Λ_ε^+ and Λ_ε^- are regular analytic surfaces in $\mathbb{R} \times \mathbb{T}^2$, and they depend analytically on ε.

It turns out that the surfaces Λ_ε^+ and Λ_ε^- intersect for all $\varepsilon \in (-a, a)$. This statement is, obviously, equivalent to the existence of a homoclinic solution $x(t)$ such that $x(t) \to \pm\pi$ as $t \to \pm\infty$. The proof can be derived from the following general result.

Theorem 3 [30]. *Let (M, T, V) be a natural mechanical system, where M is compact, the metric T does not depend on time, and the potential energy $V : M \times \mathbb{R}\{t\} \to \mathbb{R}$ is periodic in t. If $V(x, t) < V(x_0, t)$ for all $x \neq x_0$ and all $t \in \mathbb{R}$, then there exists a doubly asymptotic (homoclinic) solutionx(t) such that $x(t) \to x_0$ as $t \to \pm\infty$.*

In our case $M = S^1$, $T = \dot{x}^2/2$ and $U = -\omega^2(1+\varepsilon f)(1+\cos x)$. If $-a < \varepsilon < a$, then $V(x, t) < V(\pi, t)$ for all $x \in [0, 2\pi]$ and all t.

Since the surfaces Λ_ε^+ and Λ_ε^- do not coincide for small $\varepsilon > 0$, the values of $\varepsilon \in [-a - \delta, a + \delta]$ ($\delta > 0$), for which $\Lambda_\varepsilon^+ = \Lambda_\varepsilon^-$, are isolated. Since for $|\varepsilon| \leq a$ the surfaces Λ_ε^+ and Λ_ε^- intersect, (4.1) is integrable only for isolated values of ε.

It is worthwhile mentioning that the integrable cases are not always isolated. In Chap. I, §1 we gave an example of an analytic Hamiltonian system depending analytically on a parameter ε such that for an everywhere dense set of values of ε it is completely integrable, while for values of ε from another everywhere dense set the system does not even have continuous integrals.

6 Splitting of Separatrices and Birth of Isolated Periodic Solutions

6.1 We again consider the Hamiltonian equations

$$\dot{x} = H'_y, \quad \dot{y} = -H'_x; \qquad (x, y) = z \in \mathbb{R}^2,$$

with analytic Hamiltonian $H_0(z) + \varepsilon H_1(z, t) + o(\varepsilon)$ which is 2π-periodic in t. We assume that the unperturbed system with Hamiltonian H_0 satisfies the following two assumptions:

a) the point $z = 0$ is a non-degenerate critical point of the function H_0 with index 1 (saddle point);

b) $H_0(0) = 0$ and there is a bounded connected component σ of the set $\{z : H_0(z) = 0\} \setminus \{z : dH_0(z) = 0\}$, whose closure is $\sigma \cup \{0\}$.

Thus, the point $z = 0$ is an unstable equilibrium point of the unperturbed system, and the curve σ is its homoclinic trajectory (separatrix loop).

Let g_ε be a mapping over the period $t = 2\pi$ of the perturbed system. A point $\zeta \in \mathbb{R}^2$ is a periodic point of g_ε with period $m \in \mathbb{N}$ if $g_\varepsilon^m \zeta = \zeta$. The periodic points (and only they) are initial conditions for the periodic solutions of the Hamiltonian system. If m is the period of a point ζ, then $2\pi m$ is the period of the solution $t \to z(t, \zeta)$, $z(0, \zeta) = \zeta$. A periodic point ζ is called non-degenerate if the eigenvalues of the mapping $z \to g_\varepsilon^m z$, linearized at the point ζ, are distinct from 1. Obviously, regular bounded level curves of the function H_0 consist entirely of either degenerate periodic or non-periodic points of the mapping g_0.

Consider the function

$$I(\lambda) = \int_{-\infty}^{\infty} H_1(z_a(t + \lambda), t)\, dt \ ,$$

where $t \to z_a(t)$ is one of the homoclinic solutions corresponding to the separatrix σ. We may assume that $H_1(0, t) \equiv 0$, and so the integral is convergent. Since the average over the period of the function I is zero, I must vanish somewhere.

Theorem 1 [139]. *Suppose that the function I has a simple zero λ_0. Then there exists an infinite sequence of analytic functions $\zeta_n : (-\varepsilon_n, \varepsilon_n) \to \mathbb{R}^2$, $\varepsilon_n > 0$, such that:*

1) *the point $\zeta_n(\varepsilon)$ is a periodic point of the mapping g_ε for all $-\varepsilon_n < \varepsilon < \varepsilon$, and it is non-degenerate for $\varepsilon \neq 0$;*
2) *$H_0(\zeta_n(0)) < 0$ and the distance between the points $\zeta_n(0)$ and $z_a(\lambda_0)$ tends to zero as $n \to \infty$.*

In general, the sequence ε_n converges to zero. Hence, for sufficiently small $\varepsilon \neq 0$ the theorem implies the existence of a large but finite number of non-degenerate long-periodic solutionsof the perturbed Hamiltonian system.

It is useful to compare the assertion of Theorem 1 with the result of Chap. IV, §11 on the birth of an infinite number of distinct families of long-periodic solutions in the problem of forced oscillations of a pendulum.

The proof of Theorem 1 is based on the Poincaré small parameter method. We transfer from the coordinates x, y to the symplectic action–angle variables J, $\psi \bmod 2\pi$ of the unperturbed system in the domain of $\mathbb{R}^2$ given by the inequalities $-c < H_0(z) < 0$. Here c is a small positive constant. Recall that

$$J = \frac{1}{2\pi} \iint_{\{H_0 \leq h\}} dx\, dy \ .$$

The function $h \to J(h)$ is monotone. We denote the inverse function by $F_0(J)$. In the new variables,

$$H = F_0(J) + \varepsilon F_1(J, \psi, t) + o(\varepsilon) \ ,$$

where $H_0(z) = F_0(J)$ and $H_1(z, t) = F_1(J, \psi, t)$. Of course, the Hamiltonian is 2π-periodic in ψ and t. Let

$$\omega(J) = \frac{dF_0}{dJ}$$

be the frequency of oscillations of the unperturbed system. Obviously, $\omega(J)$ tends to zero when the level curve $H_0(z) = F_0(J)$ tends to the separatrix loop σ. Hence, there exist infinitely many different values J_n such that $\omega(J_n) = 1/n$. We claim that the birth of the non-degenerate periodic points $\zeta_n(\varepsilon)$ of the map g_ε which are mentioned in the theorem occurs for small $\varepsilon > 0$ exactly on the invariant curves $J = J_n$. In order to prove this, we must check the conditions of the non-autonomous version of the Poincaré theorem (see Chap. II, §11):
a) $d^2F_0/dJ^2 \neq 0$ at the point J_n,
b) the 2π-periodic function

$$f_n(\lambda) = \int_{-\pi n}^{\pi n} F_1(J_n, (t+\lambda)/n, t)\, dt$$

has a non-degenerate critical point λ_n.

We claim that if n is sufficiently large, then $d^2F_0/\,dJ^2 < 0$ for $J = J_n$. Consider the asymptotic expansion of the function $h \to J(h)$:

$$J = -\Lambda h \ln(-h) + \cdots\,, \tag{6.1}$$

where the positive constant Λ is given by (6.1), and the dots denote a power series which is convergent for small values of h.

Differentiate Eq. (6.1), regarding h as a function of J:

$$1 + \Lambda h'(\ln(-h) + \cdots) = 0\,. \tag{6.2}$$

It follows that for values of J close to $J(0)$ (where $J(0)$ is the area enclosed inside the separatrix σ), the frequency $\omega(J) = h' < 0$. If we again differentiate (6.2), we obtain

$$hh''(\ln(-h) + \cdots) + h'^2(1 + \cdots) = 0\,. \tag{6.3}$$

The dots in the second bracket denote a power series in h without the constant term. By (6.3), we obtain that for J close to $J(0)$ the second derivative h'' is negative. This proves the claim.

Now we analyze condition b). First note that critical points of the function f_n coincide with zeros of the function

$$f_n' = \int_{-\pi n}^{\pi n} \{H_0, H_1\}(z_n(t+\lambda), t)\, dt\,,$$

where $z_n(\,\cdot\,)$ is a $2\pi n$-periodic solution of the unperturbed system. If $z_n(0) \to z_a(0)$ as $n \to \infty$, then the sequence $f_n'(\lambda)$ is convergent to $I(\lambda)$ uniformly in λ. Using the obvious relation

$$\frac{d^n}{d\lambda^n} J = \int_{-\infty}^{\infty} \{H_0, \{H_0, \ldots, \{H_0, H_1\} \cdots\}\}\, dt\,,$$

we conclude that $f_n'' \to J'$ uniformly in λ. Since by the assumption of the theorem the function J has a simple zero, for sufficiently large n the functions f_n have the property b). Simultaneously, we obtain that the sequence of points $z_n(\lambda_n)$ is convergent to $z_a(\lambda_0)$, where λ_0 is a simple zero of the function I. The theorem is proved.

Note that it is possible that the perturbed system has no non-degenerate long-periodic solutions with period $2\pi/\omega = 2\pi n/m$, where $m \neq 1$. More precisely, in general the existence of such solutions does not follow from studying the first order terms of the perturbation. An example is provided by the well-known problem of plane oscillations of a satellite in an elliptic orbit (see Chap. I, §4). In this problem, the transverse intersection of the separatrices for small non-zero values of the eccentricity was established in [47].

6.2 If ε and $I'(\lambda_0)$ have different signs, then for small values of $\varepsilon \neq 0$ the points $\zeta_n(\varepsilon)$ mentioned in Theorem 1 correspond to periodic solutions of elliptic type. Hence, they are stable in the linear approximation. Using the results of [171], Dovbysh [69] proved that, under the additional assumption

$$[5(I'')^2 - 3I'I'''](\lambda_0) \neq 0 ,$$

for small ε these periodic solutions are stable.

The Poincaré periodic solutions from Theorem 1 depend on two parameters: continuous ε and discrete n. Suppose that n is fixed and ε small. Then under the assumptions of Theorem 1 the perturbed system has a $2\pi n$-periodic solution. Depending on the sign of the product $\varepsilon I'(\lambda_0)$, this solution can be elliptic or hyperbolic. The question arises about the fate of these solutions when we increase ε. This problem was studied in [69]. It turns out that there exists a positive constant c such that with increasing $|\varepsilon| < c/n$, the multipliers λ, λ^{-1} of the Poincaré periodic solution, starting from the point $\lambda = \lambda^{-1}$ for $\varepsilon = 0$, either move monotonically in opposite directions along the positive real half-axes (when $\varepsilon I'(\lambda_0) > 0$), or revolve along the unit circle in the complex plane, meet at the point $\lambda = \lambda^{-1} = -1$, and then move in opposite directions along the negative real half-axis (when $\varepsilon I'(\lambda_0) < 0$). For $|\varepsilon| \geq c/n$, the movement of the multipliers may cease to be monotone.

7 Asymptotic Surfaces of Unstable Equilibria

7.1 Following Devaney [65], we consider an autonomous analytic Hamiltonian system with two degrees of freedom. Let p be a critical point of the Hamiltonian H with eigenvalues $\pm(\alpha \pm i\beta)$; $\alpha, \beta \in \mathbb{R}$. If $\alpha \neq 0$, then p is a hyperbolic equilibrium position. Let Λ^+ and Λ^- be the stable and unstable asymptotic surfaces. Let γ be a homoclinic trajectory tending to p as $t \to \pm\infty$. Obviously, $\gamma \subset \Lambda^+ \cap \Lambda^-$. We assume that at every point of the trajectory γ the two-dimensional surfaces Λ^+ and Λ^- intersect transversely.

Theorem 1. *Let $\alpha\beta \neq 0$. Then the Hamiltonian system has no nontrivial analytic symmetry fields in any neighborhood of γ.*

Corollary. *Under the assumptions of Theorem* 1 *the Hamiltonian system has no analytic integrals independent of* H.

Strictly speaking, Theorem 1 is not contained in [65]. However, Devaney's theorem (the precise statement will be given in §8) implies the existence of an infinite number of long-periodic hyperbolic trajectories in a small neighborhood of γ. They form a key set for the class of analytic functions. Thus, the statement of Theorem 1 is a simple consequence of the results of Chap. IV, §8.

The condition $\alpha\beta \neq 0$ is necessary. This follows from the counterexample pointed out by Devaney [66]. Consider the Neumann problem of a particle moving on the n-dimensional sphere $S^n = \{x \in \mathbb{R}^{n+1} : |x| = 1\}$ in a force field with quadratic potential $V = (Ax, x)/2$. In order to obtain equilibria with homoclinic trajectories, we identify the opposite points of the sphere. Suppose that the eigenvalues of the operator A are distinct. Let p be the position of unstable equilibrium corresponding to the maximum point of the potential V on S^n. It turns out that

a) the equilibrium position p is hyperbolic,
b) it has $2n$ transverse homoclinic trajectoriesin the energy surface containing the point p,
c) the intersection of stable and unstable manifolds $\Lambda^+ \cap \Lambda^-$ contains an open neighborhood of each of the homoclinic trajectories in Λ^+ and Λ^-,
d) the system has n almost everywhere independent analytic integrals.

In this example, all eigenvalues of the equilibrium point p are real ($\beta = 0$).

Devaney's theorem holds also for heteroclinic trajectories. This has an interesting application to celestial mechanics. According to the Strömgren hypothesis, for certain rational values of the mass ratio the restricted three-body problem has several heteroclinic orbits connecting Lagrange's equilibrium solutions L_4 and L_5. If we manage to prove that they are transverse, then we obtain a new proof of non-integrability of the three-body problem. This idea was realized by Danby [62] for the so-called Copenhagen variant of the three-body problem (i.e., the plane circular restricted three-body problem with equal masses). A large number of transverse homoclinic and heteroclinic orbits was discovered.

7.2 Dovbysh [70] applied Theorem 1 to study perturbation of the integrable Lagrange problem in rigid body dynamics. By choosing appropriate units of mass, length and time, we can assume that the principal moments of inertia of the body are $I_1 = I_2 = 1$ and $I_3 \neq 1$, the coordinates of the center of mass in the principal axes of inertia are 0, 0, 1, and the weight of the body equals 1. Let c be the constant of the area integral.

We consider a permanent rotation

$$\omega_1 = \omega_2 = 0\,, \quad \omega_3 = r_0\,; \qquad \gamma_1 = \gamma_2 = 0\,, \quad \gamma_3 = 1\,. \tag{7.1}$$

It corresponds to an equilibrium position of the Euler–Poisson equations. If the well-known Majevsky condition $|c| < 2$ is satisfied, then the equilibrium position (7.1) is unstable. Obviously, the variational equations of the solution (7.1) have two zero characteristic exponents (since there exist the area integral and the geometric integral $\gamma^2 = 1$). We write down expressions for the doubly asymptotic solutions to the rotation (7.1):

$$\begin{aligned}
&\omega_1 = 2\alpha \cosh^{-1} \alpha t \cos(\beta t + s)\,, \quad \omega_2 = 2\alpha \cosh^{-1} \alpha t \sin(\beta t + s)\,, \\
&\gamma_1 = c\alpha \cosh^{-1} \alpha t \sin(\beta t + s) + 2\alpha^2 \cosh^{-2} \alpha t \sinh \alpha t \sin(\beta t + s)\,, \\
&\gamma_2 = c\alpha \cosh^{-1} \alpha t \sin(\beta t + s) - 2\alpha^2 \cosh^{-2} \alpha t \sinh \alpha t \cos(\beta t + s)\,, \\
&\gamma_3 = 1 - 2\alpha^2 \cosh^{-2} \alpha t\,, \quad \omega_3 = r_0\,.
\end{aligned} \tag{7.2}$$

Here $\alpha^2 = 4 - c^2/2$, $\beta = (I_3 - 2)r_0/2$, and s is a real parameter enumerating the homoclinic trajectories.

Non-zero characteristic exponents of the solution (7.1) are equal to $\pm\alpha\pm i\beta$. The triangle inequality for the moments of inertia yields $I_3 < 2$. Therefore, if $r_0 \neq 0$ and Majevsky's condition is satisfied, then $\alpha\beta \neq 0$. In this case the equilibrium position (7.1) of the reduced system (on the fixed four-dimensional level surface of the geometric and area integrals) is a critical point of the "saddle-focus" type.

In the unperturbed system, the asymptotic surfaces (7.2) are doubled, and all homoclinic trajectories are non-transversal. Apparently, for a small perturbation of Lagrange's problem equilibrium (7.1) does not disappear and still is a "saddle-focus", but there exist transverse homoclinic trajectories. Thus, Theorem 1 can be applied to the perturbed Lagrange problem.

To elucidate the question of splitting of the doubled asymptotic surfaces (7.2), we consider the 2π periodic function of s:

$$I(s) = \int_{-\infty}^{\infty} \{F, H_1\}\, dt\,. \tag{7.3}$$

Here $F = I_3\omega_3$ is the additional integral of Lagrange's problem, and H_1 is the perturbing function. The Poisson bracket is evaluated on the trajectories from the family (7.2).

A general perturbation of Lagrange's top can be reduced to a perturbation of the inertia tensor and a shift of the center of mass along the dynamical symmetry axis. The latter perturbation is of no significance. Hence,

$$H_1 = \sum \varepsilon_{ij}\omega_i\omega_j/2\,,$$

where ε_{ij} are small quantities. By symmetry, we can assume that $\varepsilon_{12} = \varepsilon_{22} = 0$. It is convenient to put $\varepsilon = (\varepsilon_{11}^2 + \varepsilon_{13}^2 + \varepsilon_{23}^2)^{1/2}$. The integral (7.3) can be evaluated by residues:

$$\begin{aligned}
I = 2\pi r_0 \Bigg[&\sinh^{-1}\left(\frac{\pi}{\lambda}\right)\frac{\varepsilon_{11}}{\varepsilon}(2 - I_3)\sin 2s \\
&+ \cosh^{-1}\left(\frac{\pi}{2\lambda}\right)\left(\frac{\varepsilon_{23}}{\varepsilon}\cos s - \frac{\varepsilon_{13}}{\varepsilon}\sin s\right)\Bigg]\,,
\end{aligned}$$

where $\lambda = \alpha/\beta$. The 2π-periodic function $I(s)$ always has simple zeros. Thus, by Theorem 1 of §1, for small non-zero ε the reduced perturbed Lagrange problem has a transverse homoclinic trajectory. By Theorem 1, the perturbed system admits no nontrivial analytic symmetry fields (consequently, it has no new analytic integrals).

7.3 Bolotin obtained sufficient conditions for nonintegrability of Hamiltonian systems in the case of real characteristic exponents. They are expressed in terms of Birkhoff's normal form [35].

Let p be an equilibrium position of an analytic Hamiltonian system with different real characteristic exponents $\lambda_1 > 0, \ldots, \lambda_n > 0, -\lambda_1, \ldots, -\lambda_n$. These may satisfy the resonance relations:

$$(k, \lambda) = 0\,, \qquad k \in \mathbb{Z}^n \setminus 0\,. \tag{7.4}$$

We assume that among (7.4) there are no combinational resonances:

$$\lambda_i = (l, \lambda)\,, \tag{7.5}$$

$$\lambda_i + \lambda_j = (l, \lambda)\,, \tag{7.6}$$

where $l = (l_1, \ldots, l_n)$, $l_k \geq 0$, $|l| = \sum l_k > 1$. Note that the set of vectors $\lambda \in \mathbb{R}^n_+$ satisfying resonance relations of type (7.4) is everywhere dense in $\mathbb{R}^n_+$, while the set of vectors satisfying relations (7.5) and (7.6) is discrete.

If there are no resonances of type (7.5) and (7.6), then there are no resonances (7.4) of third and fourth order (with $|k| = \sum |k_i| = 3$ or 4). Therefore, according to the classical Birkhoff result, in a neighborhood of the point p the Hamiltonian function can be reduced to the normal form:

$$H = \sum_{i=1}^{n} \lambda_i \sigma_i + \sum_{i,j=1}^{n} a_{ij} \sigma_i \sigma_j + O(|x|^5 + |y|^5)\,,$$

where x, y are analytic canonical coordinates in the neighborhood of the point p, and $\sigma_i = x_i y_i$. We call the equilibrium position weakly non-resonant if among the resonances (7.4) there are no combinational resonances (7.5) and (7.6), and the matrix $A = \|a_{ij}\|$ satisfies the non-degeneracy condition

$$Ak \neq 0 \quad \text{for} \quad k \in \mathbb{Z}^n \setminus \{0\}\,, \quad (k, \lambda) = 0\,. \tag{7.7}$$

If the equilibrium position is non-resonant, or the matrix A is non-degenerate, then condition (7.7) always holds.

Let Λ be one of the asymptotic surfaces of the point p. Then Λ is the union of trajectories of the Hamiltonian system tending to the point p as $t \to +\infty$ (or $t \to -\infty$). Since there are no resonances of type (7.5), there exist analytic coordinates $x_1, \ldots, x_n$ in a neighborhood of p in Λ such that the Hamiltonian system on Λ takes the form

$$\dot{x}_k = \mp \lambda_k x_k\,.$$

Let γ be one of the trajectories asymptotic to p. We call γ a principal trajectory if γ is contained in one of the planes $x_k = 0$ as $t \to +\infty$ (or $t \to -\infty$).

Theorem 2 [35]. *Let $p_\pm$ be weakly non-resonant equilibrium positions with real characteristic exponents. If there exists a transverse doubly asymptotic trajectory to $p_\pm$ that is not principal as $t \to \pm\infty$, then the Hamiltonian system has no analytic integrals independent of the energy integral in any neighborhood of the set $\gamma \cup p_+ \cup p_-$.*

The condition that γ is a non-principal trajectory is essential, since in a typical case the Neumann problem (see §7.1) satisfies all other assumptions of the theorem. The proof of Theorem 2 is similar to the proof of Theorem 3 in §2. It would be interesting to find out if the conditions of Theorem 2 imply the nonexistence of nontrivial symmetry fields.

Bolotin indicated an application of Theorem 2 to rigid body dynamics. Consider the perturbation of the reduced Lagrange problem which was studied in §7.2. If the area constant is zero, then the characteristic exponents of unstable equilibrium are real. Therefore, Devaney's theorem does not hold. In [35] it is proved that if the inertial ellipsoid is not a sphere and the center of mass is slightly displaced from the axis of dynamical symmetry (and its z coordinate does not vanish), then the perturbed Lagrange problem has a non-principal homoclinic trajectory to a weakly non-resonant equilibrium position. Construction of this trajectory is based on the methods of perturbation theory (see §1). This problem is discussed also in [70].

8 Symbolic Dynamics

8.1 We start with a model example. Let S be a map of the square $B = \{(x, y) \in \mathbb{R}^2 : 0 \le x, y \le 1\}$ into itself given by the rule:

$$\begin{aligned} x \to 3x\,, \quad & y \to y/3\,, \quad \text{if} \quad 0 \le x \le 1/3\,, \\ x \to 3x - 2\,, \quad & y \to y/3 + 2/3\,, \quad \text{if} \quad 2/3 \le x \le 1\,. \end{aligned} \tag{8.1}$$

In the strip $1/3 < x < 2/3$, $0 \le x \le 1$, the map S is not defined. The geometrical meaning of the transformation $S : B \to B$ is clear from Fig. 32.

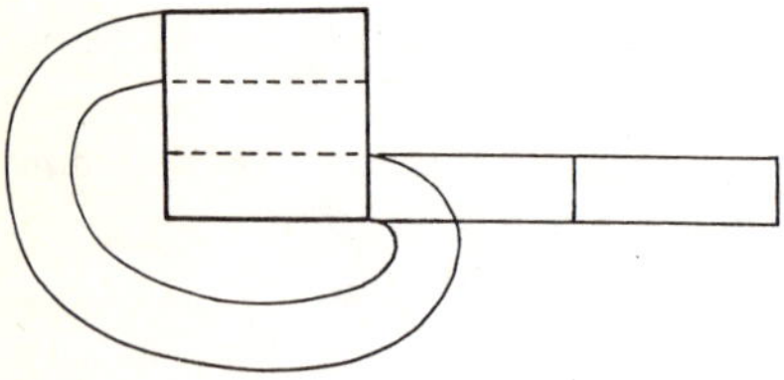

Fig. 32.

The map S can be interpreted as the Poincaré map of some dynamical system. We are interested in trajectories that are completely contained in the square B.

We clarify the structure of the sets $S^n B$ for $n > 1$. The set $S^1 B$ is obtained from the square B by removing the horizontal strip $[0, 1] \times (1/3, 2/3)$. If we delete from the remaining two strips the thinner strips $[0, 1] \times (1/9, 2/9)$ and $[0, 1] \times (7/9, 8/9)$, then we get the set $S^2 B$, and so on. Continuing this process indefinitely, we obtain the set $[0, 1] \times K \subset B$ (here K is the Cantor set in the segment $[0, 1]$). This set consists of points of the square that remain in B under all positive iterations by S. In a similar way we conclude that all negative iterates of the mapping S are defined on the set $K \times [0, 1]$. Therefore, all iterates of S are defined on the "punctured" set $\Lambda = K \times K \subset B$: the trajectory of every point $z \in \Lambda$ (i.e., the set of points $S^n(z)$, $n \in \mathbb{Z}$) lies entirely in $\Lambda \subset B$.

What is the structure of the map $S : \Lambda \to \Lambda$? To answer this question we introduce the space of doubly infinite sequences $\omega = \{\omega_n\}$, $n \in \mathbb{Z}$, of zeros and ones. We endow Ω with a topology $\mathcal{T}$, defining convergence in the following way: a sequence $\omega^{(k)}$ is convergent to $\omega \in \Omega$ if

$$\lim_{k \to \infty} \omega_n^{(k)} = \omega_n$$

for all n.

Lemma 1. *The topological space* $(\Omega, \mathcal{T})$ *is homeomorphic to* Λ.

To prove this, for given $\omega = \{\omega_n\}$ we define two numbers

$$x = 2 \sum_{s \geq 0} \omega_s / s^{s+1}, \quad y = 2 \sum_{s \geq 0} \omega_{-s} / 3^s, \tag{8.2}$$

which, evidently, belong to $K \subset [0, 1]$. It is easy to see that the map $\omega \to (x, y)$ is a homeomorphism.

Let T be the map of Ω into itself which sends $\omega = \{\omega_n\}$ into $\omega' = \{\omega_{n+1}\}$ (it is the left shift of all symbols by one).

Theorem 1. *There exists a homeomorphism* $f : \Lambda \to \Omega$ *such that the following diagram*

$$\begin{array}{ccc} & S & \\ \Lambda & \to & \Lambda \\ f \downarrow & & \downarrow f \\ \Omega & \to & \Omega \\ & T & \end{array}$$

is commutative (i.e., $S \circ f = f \circ T$*).*

The required homeomorphism $\omega \to (x, y)$ is given by (8.2). The proof is based on comparison of expressions (8.1) and (8.2).

Therefore, to each trajectory $S^n(z)$, $z \in B$, $n \in \mathbb{Z}$, lying entirely in the square B, we assigned a sequence of symbols $\omega = \{\omega_n\}$ in such a way that the action of the map S becomes the left shift of all the symbols of the sequence by one. This method of coding trajectories goes back to the investigations of Hadamard, Birkhoff, Morse and Hedlund of geodesics on closed surfaces of negative curvature. It constitutes the basis of "symbolic dynamics". For a more detailed presentation we refer the reader to [7, 184].

The symbolic coding of the trajectories makes it possible to establish several important properties of the transformation S. Lemma 1 and Theorem 1 imply the following result.

Proposition 1. *The mapping $S : \Lambda \to \Lambda$ has the following properties*:
1) *the periodic points are everywhere dense in Λ,*
2) *every pair of periodic points can be joined by a doubly asymptotic trajectory,*
3) *any periodic point has a homoclinic trajectory,*
4) *there exist trajectories that are everywhere dense in Λ.*

Proof. To a periodic trajectory of the map S there corresponds a point $(a) = (\dots, a, a, a, \dots) \in \Omega$, where a is a finite string of zeros and ones. For every element $\omega \in \Omega$ we can construct a sequence of periodic trajectories $\omega^{(n)} = (a_n)$, where $a_n = \{\omega_{-n}, \dots, \omega_n\}$. Evidently, $\omega^{(n)} \to \omega$ as $n \to \infty$. Further, suppose that to the points $(a), (b) \in \Omega$ there correspond two periodic trajectories. Then the sequence $(\dots, a, a, b, b, \dots)$ provides the required doubly asymptotic trajectory. Homoclinic trajectories correspond to sequences of type $(\dots, a, a, c, a, a, \dots)$. Therefore, properties 2) and 3) are proven. Finally, consider a point $\omega^* \subset \Omega$ such that in the sequence ω_n^*, starting from some position, all possible finite strings of zeros and ones are written in succession. It is easy to verify that the closure of the trajectory $\cup\{T^n \omega^*\}$, $n \in \mathbb{Z}$, is equal to Ω. This finishes the proof.

Now we discuss the integrability problem of the discrete dynamical system (B, S). It is natural to call this system integrable, if there exists a locally non-constant function F (integral) that is invariant under the transformation S: $F(S(z)) = F(z)$ for all $z \in B$.

Proposition 2. *The map $S : B \to B$ has no non-constant analytic integrals.*

Proof (following Alekseev [5]*).* According to statement 4) of Proposition 1, the integral F takes the same value at all points of the subset $\Lambda \subset B$. From the construction of the perfect set Λ it follows that for every point $z \in \Lambda$ there exist two sequences of points of Λ converging to z from two different directions (say, horizontal and vertical). Hence, all the derivatives of f with respect to x or y vanish at the point z. To complete the proof it remains to use the analyticity of F.

Since Λ is nowhere dense, this argument does not imply the nonexistence of smooth integrals. One can show that all periodic points in Λ are hyperbolic and, consequently, non-degenerate. On the other hand, they are everywhere dense in Λ, and Λ is a key set for the class of analytic functions. Thus, the nonexistence of analytic integrals can be established by using the results of Chap. IV, §8.1. Another way to prove nonintegrability is to apply assertion 3) of Proposition 1. It is easy to show that the stable and unstable separatrices of the periodic points intersect and do not coincide. This makes it possible to apply Theorem 2 of §2.

If a dynamical system has trajectories that are everywhere dense in the phase space, then it is called *transitive*. Since our dynamical system is transitive on the subset $\Lambda \subset B$, it is a system of "transitive type" (Birkhoff). "Every nonintegrable system of transitive type can be considered as "solved", if there exists a special algorithm which is powerful enough to solve all problems on type and disposition of motions" (Birkhoff [23, Chap. 8]). In the considered case, such an algorithm is provided by the symbolic representation of trajectories inside the square B which is described in Theorem 1.

8.2 The transformation $T : \Omega \to \Omega$ which appears in Theorem 1 is called the *Bernoulli shift*. Clearly, this term originates in probability theory. We take a "chance" point $\omega \in \Omega$ and consider the iterations $T^n\omega$, $\omega \in \mathbb{Z}$. The "components" of $T^n\omega$ with zero number form a sequence of zeros and ones $\dots, \omega_{-n}, \dots, \omega_0, \dots, \omega_m, \dots$. The transition from ω_n to ω_{n+1} can be interpreted as a sequence of Bernoulli trials, i.e., of independent trials with two outcomes, where the probability of each outcome is equal to $1/2$. Therefore, the dynamical system (Λ, S) is often called quasi-random.

We consider a positive half-orbit $\{T^n\omega\}$ $(n > 0)$ of a point ω. The sequence $\omega_0, \omega_1, \dots$ defines a point x of the segment $[0, 1]$ according to the rule

$$x = \omega_0/2 + \omega_1/2^2 + \cdots .$$

Thus, the Bernoulli shift can be represented as a map f of the unit segment into itself given by the formula $f : x \to 2x \bmod 1$. It is easy to show that the map f has no integrals that are non-constant integrable functions. However, we can construct "quasi-integrals", i.e., functions that are nearly constant on long but finite segments of the trajectory of every point x. Indeed, consider the orthogonal system of Rademeyer functions on the segment $[0, 1]$:

$$\phi_m(x) = (-1)^{[2^m x]}, \qquad m \geq 0 .$$

Clearly, $\phi_m(f(x)) = \phi_{m+1}(x)$ for all $x \in [0, 1]$. We put

$$F_k = (\phi_1 + \cdots + \phi_{k-1})/\sqrt{k} .$$

It is easy to verify that

$$\int_0^1 F_k \, dx = 0 , \qquad \int_0^1 F_k^2 \, dx = 1 .$$

In particular, the variation of every function F_k on the segment $[0, 1]$ is not less than one.

We have an obvious estimate

$$|F_k(f(x)) - F_k(x)| = |\phi_k - \phi_1|/\sqrt{k} \leq 2/\sqrt{k} .$$

Hence,

$$|F_k(f^m(x)) - F_k(x)| \leq 2m/\sqrt{k} \,.$$

We fix small $\varepsilon > 0$ and large m. Let $k > 4m^2/\varepsilon^2$. Then for all $x \in [0, 1]$ the values of the function F_k at the points $x, f(x), \dots, f^m(x)$ differ by less than ε. For large k, it is natural to call the function F_k a quasi-integral of the mapping f.

By de Moivre's theorem,

$$\lim_{k\to\infty} \operatorname{mes}\{0 \leq x \leq 1 : a < F_k(x) < b\} = \frac{1}{2\pi}\int_a^b e^{-z^2/2}\,dz \,.$$

Therefore, as $k \to \infty$, the values of the quasi-integrals are distributed according to the normal law. This is another manifestation of the quasi-random behavior of the trajectories of the dynamical system (Λ, S).

8.3 The methods of symbolic dynamics can be applied to describe behavior of a system in a neighborhood of a transverse homoclinic trajectory. Let S be a map of an arbitrary manifold M into itself, and p hyperbolic fixed point of S. The manifold M can be regarded as the phase space of a non-autonomous periodic Hamiltonian system, and S as the map over the period (see §1.4). Let $\Lambda^{\pm}$, as usual, be the asymptotic invariant surfaces of the point p. We assume that Λ^+ and Λ^- intersect transversely. It is natural to call any point $q \in \Lambda^+ \cap \Lambda^-$ a transverse homoclinic point:

$$\lim_{n\to\pm\infty} S^n(q) = p \,.$$

The set in M consisting of the point p and the trajectory of q is compact. We cover it by a finite number of sufficiently small neighborhoods in the following way. Let U_0 be a neighborhood of the point p. Then U_0 covers all but a finite number of points of the trajectory $\{S^n(q)\}$. We cover the remaining points by non-intersecting neighborhoods $U_1, \dots, U_{N-1}$. Obviously, the intersections $S(U_0) \cap U_0$, $S(U_0) \cap U_1, \dots, S(U_{N-1}) \cap U_0$ are non-empty (the first set contains the point p, and every other set contains at least one point from the trajectory of q). This situation can be displayed in the form of an "intersection graph" (Fig. 33). The vertices enumerated by the symbols $0, 1, \dots, N-1$ are joined by a certain number of branches: a branch connecting a vertex i with a vertex j corresponds to the non-empty intersection $S(U_i) \cap U_j$. For arbitrary neighborhoods $U_0, \dots, U_{N-1}$, the intersection graph in Fig. 33 can have additional branches.

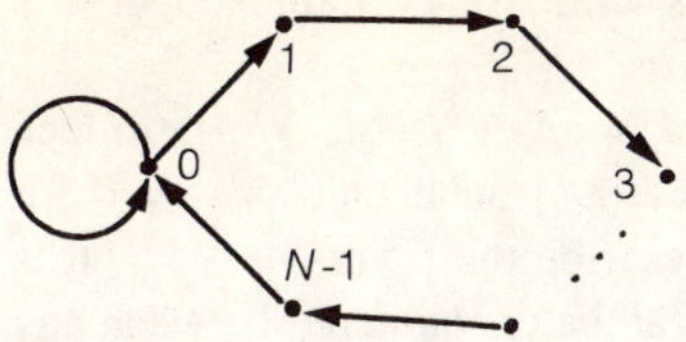

Fig. 33.

Lemma 2 [6]. *For any open set $V \subset M$ containing the closure of the trajectory of the point q it is possible to choose the neighborhoods $U_0, \ldots, U_{N-1} \subset V$ in such a way that the intersection graph coincides with the graph in Fig.* 33.

Obviously, the number N depends on the neighborhood V and grows as the size of the neighborhood decreases.

We introduce the space Ω^N consisting of infinite sequences of N symbols $0, 1, \ldots, N-1$. Let $\Omega(N)$ be the subset of Ω^N consisting of "admissible" sequences $\{\omega_n\}$ such that a pair of symbols may appear as adjacent entries in the symbol sequence only if they are connected by a branch of the intersection graph (Fig. 33). We endow $\Omega(N)$ with the natural product topology (see §8.1). Let T be the homeomorphism of $\Omega(N)$ which is the left shift of $\omega = \{\omega_n\} \in \Omega(N)$ by one.

Theorem 2 (Alekseev [6]). *Let $U_0, \ldots, U_{N-1}$ be the neighborhoods defined in Lemma* 1. *Denote $U = U_0 \cup \cdots \cup U_{N-1}$, and let*

$$\Lambda = \bigcap_{n=-\infty}^{\infty} S^n(U)$$

be the maximal invariant set contained in U. Then there exists a homeomorphism $f : \Lambda \to \Omega(N)$ such that the diagram

$$\begin{array}{ccc} & S & \\ \Lambda & \rightarrow & \Lambda \\ f \downarrow & & \downarrow f \\ \Omega(N) & \rightarrow & \Omega(N) \\ & T & \end{array}$$

is commutative.

We discuss some consequences of Theorem 2. As in §8.1, we can prove that S is transitive on Λ (there exists a dense trajectory), and that periodic trajectories are everywhere dense in Λ.

Every trajectory $\{S^n(x)\}$, $x \in \Lambda$, is coded by a sequence $\omega = \{\omega_n\} \in \Omega(N)$. What is the structure of such sequences? Obviously, it contains zeros, and each zero is followed by zero or one, while all the ones are initial points of the string

$\sigma = \{1, 2, \dots, N-1, 0\}$. Thus, ω consists of zeros with strings σ imbedded. It is possible that the strings σ are not separated by zeros.

Let s be an integer, $2 \leq s \leq N+1$, and let $k = N + s - 2$. We introduce the subset $\Omega_s \subset \Omega(N)$ consisting of all sequences $\{\omega_n\}$ such that $\omega_{mk} = 0$ for all $m \in \mathbb{Z}$. Then Ω_s is T^k-invariant. In order to describe the points $\omega \in \Omega_s$, it is sufficient to find out what strings of $k-1$ symbols can be encountered between any two zeros. Since $s < N+2$, Fig. 33 implies that they are the following s strings:

$$A_0 = \{0, \dots, 0\}\,,$$
$$A_1 = \{1, 2, \dots, N-1, 0, \dots, 0\}\,,$$
$$A_2 = \{0, 1, \dots, N-1, 0, \dots, 0\}\,,$$
$$\dots \quad \dots \quad \dots \quad \dots \quad \dots \quad \dots$$
$$A_{s-1} = \{0, \dots, 0, 1, \dots, N-1\}\,.$$

Therefore, any sequence $\{\omega_n\} \in \Omega_s$ has the following form:

$$\dots\,,\ A_{i_{-1}}\,,\ 0\,,\ A_{i_0}\,,\ 0\,,\ A_{i_1}\,,\ 0\,,\ A_{i_2}\,,\ \dots\,, \tag{8.3}$$

where the subscripts i_m can independently take any value $1, \dots, s-1$. Obviously, the set Ω_s is in one-to-one correspondence with the set Ω^s of all bi-infinite sequences of s symbols. The transformation T^k shifts the sequence (8.3) to the left by k symbols. This means that the string A_{i_m} is replaced by the string $A_{i_{m+1}}$. Apparently, this is exactly the shift T of the set Ω^s. Hence, the homeomorphism $T : \Omega^s \to \Omega^s$ is conjugate to the homeomorphism $T^k : \Omega_s \to \Omega_s$, which is conjugate to the map $S^k : \Lambda \to \Lambda$ (Theorem 2). Thus, we obtain

Corollary 1 (Smale [219]). *For any neighborhood U and any natural s there exist $k \in \mathbb{N}$ and a subset $\Lambda \subset U$ such that the map $S^k : \Lambda \to \Lambda$ is topologically conjugate to the shift homeomorphism of the sequences space with s symbols.*

This shift homeomorphism is also called Bernoulli's shift. In the present case we get a more complicated probability problem: independent trials with s outcomes of equal probability.

Now we give the exact statement of Devaney's theorem mentioned in §7.1. Let p be an equilibrium point of a Hamiltonian system with two degrees of freedom, $\pm\alpha \pm i\omega$ its characteristic exponents, and γ a transverse homoclinic trajectory of the point p.

Theorem 3 [65]. *If $\alpha\omega \neq 0$, then for any local transverse section Σ of the trajectory γ and every natural s there exists a compact invariant hyperbolic set $\Lambda \subset \Sigma$ such that the restriction of the Poincaré map on Λ is topologically conjugate to the Bernoulli shiftin the space of infinite sequences of s symbols.*

Therefore, in a neighborhood of a transverse homoclinic trajectory to an equilibrium point of saddle-focus type, the Hamiltonian system has quasi-random behavior.

8.4 Alekseev applied the methods of symbolic dynamics to the problem of a particle in the gravitational field of a double star (see Chap. II, §5). It turns out that if the eccentricity e of the orbits of the massive bodies does not vanish, then the trajectories of the particle are very complicated. This makes it possible to prove nonintegrability of the equations of motion [8]. More precisely, the quasi-random nature of the particle trajectories is established for small values of $e \neq 0$. By using Poincaré's method (Chap. IV, §1), it is possible to prove the nonexistence of non-constant integrals and nontrivial symmetry groups in the form of formal power series in e. Llibre and Simo [164] applied Alekseev's method to the restricted circular three-body problem under the assumption that the mass of Jupiter is much less than the mass of the Sun.

VI Nonintegrability in the Vicinity of an Equilibrium Position

There is another method of proving nonintegrability. It is based on estimates from below for coefficients of the power series representing formal integrals which exist by Birkhoff's theorem. It turns out that the reason for divergence of such series is the existence of anomalously small denominators, i.e., almost resonance relations between the frequencies of small oscillations near equilibria.

1 Siegel's Method

1.1 Consider a canonical system of differential equations

$$\dot{x}_k = \frac{\partial H}{\partial y_k}, \quad \dot{y}_k = -\frac{\partial H}{\partial x_k}, \qquad 1 \le k \le n, \tag{1.1}$$

and assume that H is an analytic function in a neighborhood of the point $x = y = 0$, where $H(0) = 0$ and $dH(0) = 0$. Let $H = \sum_{s \ge 2} H_s$, where H_s is a homogeneous polynomial in x and y of degree s.

Let $\lambda_1, \ldots, \lambda_n$ be the eigenvalues of the linearized canonical system with Hamiltonian H_2. We may assume that $\lambda_{n+k} = -\lambda_k$ $(1 \le k \le n)$. We consider the case when the numbers $\lambda_1, \ldots, \lambda_n$ are purely imaginary and independent over the field of rational numbers: the sum $m_1\lambda_1 + \cdots + m_n\lambda_n$ with integer m_i equals zero only if all $m_i = 0$. Under this assumption, Birkhoff found a formal canonical transformation reducing the system (1.1) to the normal form. In particular, the Hamiltonian equations (1.1) have n integrals in involution in the form of formal series in x and y (Chap. II, §11).

In this section we investigate obstructions to the complete integrability of the system (1.1) in the vicinity of the equilibrium position $x = y = 0$ and the convergence of Birkhoff's normalizing transformations. Let $\mathcal{H}$ be the set of all power series

$$H = \sum h_{ks} x^k y^s, \qquad k = (k_1, \ldots, k_n), \quad s = (s_1, \ldots, s_n),$$

converging in some neighborhood of the point $x = y = 0$. We endow $\mathcal{H}$ with the following topology $\mathcal{T}$: a neighborhood of a power series H^* with coefficients h^*_{ks} is the set of all power series H whose coefficients h_{ks} satisfy the inequalities $|h_{ks} - h^*_{ks}| < \varepsilon_{ks}$, where ε_{ks} is an arbitrary sequence of positive numbers.

Theorem 1 (Siegel [215]). *In every neighborhood of an arbitrary point $H^* \in \mathcal{H}$ there exists a Hamiltonian H such that the corresponding canonical system* (1.1) *does not have integrals independent of H and analytic in a neighborhood of the point $x = y = 0$.*

The proof is contained in §1.2. The theorem implies that nonintegrable systems are everywhere dense in $\mathcal{H}$. In particular, the set of Hamiltonian systems for which the Birkhoff transformation diverges is everywhere dense. Concerning the divergence of Birkhoff's transformations, the following stronger result holds:

Theorem 2 (Siegel [216]). *The Hamiltonian functions $H \in \mathcal{H}$ with converging Birkhoff transformations form a subset of first Baire category in the topology $\mathcal{T}$.*

Recall that a subset of a topological space is a set of first Baire category if it is a countable sum of nowhere dense sets. A set is nowhere dense if every neighborhood of any point of the topological space contains a non-empty open domain with no points from this set.

More precisely, Siegel proved the existence of a countably infinite set of independent analytic power series $\Phi_1, \Phi_2, \ldots$ in an infinite number of variables h_{ks} which are absolutely convergent for $|h_{ks}| < \varepsilon$ (for all k, s), and such that if $H \in \mathcal{H}$ is reduced to the normal form by a convergent transformation, then almost all Φ_s (except possibly a finite number) vanish at the point H. Since the functions Φ_s are analytic, their zeros are nowhere dense in $\mathcal{H}$. Consequently, the set of points of $\mathcal{H}$ satisfying at least one equation $\Phi_s = 0$ is of first Baire category. If we attempt to investigate the convergence of the Birkhoff transformation for any concrete Hamiltonian system, then we must check infinitely many conditions. No finite method for doing this is known, although all the coefficients of the series Φ_s can be calculated explicitly.

For example, it is still unknown whether the Birkhoff transformation is convergent near the Lagrangian positions of equilibrium in the restricted three-body problem with a fixed mass ratio. Concerning this problem, Siegel remarked that "...probably it is beyond the possibilities for known methods of calculus" [216].

1.2 According to the results of §11, Chap. II, in certain canonical coordinates the Hamiltonian H takes the form:

$$H = \frac{1}{2}\sum \omega_j(x_j^2 + y_j^2) + \cdots .$$

The eigenvalues are precisely $\pm i\omega_1, \ldots, \pm i\omega_n$. Perform the following linear canonical transformation $x, y \to u, v$ with complex coefficients:

$$y = \frac{iu + v}{\sqrt{2}}, \qquad x = \frac{u + iv}{\sqrt{2}} .$$

In the new coordinates

$$H = i\sum \omega_j u_j v_j + \cdots .$$

For the sake of simplicity, we restrict ourselves to the case of two degrees of freedom. Let $\omega_1 = 1$ and $\omega_2 = \omega$ be irrational.

Thus, we consider the canonical equations with Hamiltonian function of the following form:

$$H = i(u_1 v_1 + \omega u_2 v_2) + \sum_{p+q\geq 3} h_{p_1 p_2 q_1 q_2} u_1^{p_1} u_2^{p_2} v_1^{q_1} v_2^{q_2} . \tag{1.2}$$

The coefficients $h_{p,q}$ may be complex.

Since ω is irrational, by Birkhoff's theorem the Hamilton equations with the Hamiltonian (1.2) admit two formal integrals

$$S_l = u_l v_l + \cdots , \qquad l = 1, 2 .$$

The quadratic terms of these expansions do not uniquely determine the integrals S_1 and S_2. The terms of the special form

$$c(u_1 v_1)^{\alpha_1} (u_2 v_2)^{\alpha_2} , \qquad \alpha_1 + \alpha_2 = g \geq 2 , \tag{1.3}$$

are responsible for this non-uniqueness. We call g the degree of the monomial (1.3).

Lemma 1. *There exists an integral*

$$s = u_1 v_1 + \cdots$$

without the terms (1.3), *and every integral of the system with the Hamiltonian* (1.2) *is a power series in H and s.*

Proof. Suppose that the integral S_l contains a term of the special form (1.3) with the least possible degree g. Then the integral

$$S_l - c S_1^{\alpha_1} S_2^{\alpha_2}$$

does not contain terms of type (1.3) with degree less than or equal to g. This operation does not change the quadratic term. Applying it successively, we obtain two formal integrals

$$s_l = u_l v_l + \cdots , \qquad l = 1, 2 ,$$

which have no special terms. Every formal integral of the Hamilton system with Hamiltonian (1.2) is a power series in s_1 and s_2 (see §11, Chap. II). In particular, this holds for the integral H. By (1.2), the expansion of H in powers of s_1, s_2 has the form

$$H = i s_1 + i\omega s_2 + \cdots .$$

Hence, s_2 can be represented as a power series in H and s_1. The lemma is proved.

Let F be a power series in u, v and F_l ($l = 0, 1, \ldots$) the sum of its terms of order l. Denote by $\|F_l\|$ the maximal absolute value of the coefficients of the homogeneous polynomial F_l. The next result is technical.

Lemma 2. *Let ξ, η be complex numbers and G a homogeneous polynomial in u_1, u_2, v_1, v_2 of degree r. Then*

$$(|\xi| + |\eta|)\|G\| \leq (2r+2)\|(\xi u_1 v_1 + \eta u_2 v_2)G\| . \tag{1.4}$$

Proof. We have

$$G = \sum_{l=0}^{r} G_l ,$$

where G_l is a homogeneous form in u_1, u_2 of degree l. Since $\|G\|$ is the maximum of $\|G_l\|$ $(l = 0, \ldots, r)$ and $\|(\xi u_1 v_1 + \eta u_2 v_2)G\|$ is the maximum of $\|(\xi u_1 v_1 + \eta u_2 v_2)G_l\|$, inequality (1.4) is satisfied if it holds for G_l instead of G with $l = 0, \ldots, r$. Hence we can assume that

$$G = \sum_{k=0}^{l} \phi_k u_1^k u_2^{l-k} ,$$

where ϕ_k are homogeneous polynomials in v_1, v_2 of degree $r - l$. If $|\xi| > |\eta|$, then we interchange ξ, u_1, v_1 and η, u_2, v_2. Therefore, we may assume that

$$|\xi| \leq |\eta| . \tag{1.5}$$

Let $\phi_{-1} = 0$, $\phi_{l+1} = 0$, and

$$\Phi_k = \xi v_1 \phi_{k-1} + \eta v_2 \phi_k , \qquad k = 0, \ldots, l+1 .$$

Then we obtain

$$(\xi u_1 v_1 + \eta u_2 v_2)G = \sum_{k=0}^{l+1} \Phi_k u_1^k u_2^{l+1-k} ,$$

$$(\eta v_2)^{k+1} \phi_k = \sum_{p=0}^{k} (-\xi v_1)^{k-p} (\eta v_2)^p \Phi_p , \qquad k = 1, \ldots, l .$$

By (1.5), it follows that

$$|\eta| \|\phi_k\| \leq \sum_{p=0}^{k} \|\Phi_p\| .$$

Since $\|G\|$ is the maximum of the numbers $\|\phi_k\|$ $(0 \leq k \leq l)$, and $\|(\xi u_1 v_1 + \eta u_2 v_2)G\|$ is the maximum of the numbers $\|\Phi_k\|$ $(0 \leq k \leq l+1)$, we obtain the inequality

$$|\eta| \|G\| \leq (l-1) \|(\xi u_1 v_1 + \eta u_2 v_2)G\| .$$

Since $|\xi| + |\eta| \leq 2|\eta|$ and $l + 1 \leq r + 1$, the estimate (1.4) follows immediately. The lemma is proved.

The following result is the key for Siegel's method.

Lemma 3. *Let*

$$s = \sum_{k=2}^{\infty} s_k \,, \qquad s_2 = u_1 v_1 \,,$$

be the formal integral defined in Lemma 1. If the Hamiltonian equations have a converging integral independent of H, then the sequence

$$\ln \|s_k\| / (k \ln k) \,, \qquad k = 2, 3, \ldots \,,$$

is bounded.

Proof. Lemma 1 implies that every integral $P(u, v)$ of the Hamiltonian system can be represented in the form of a power series in s and H:

$$P = \sum_{\alpha=\beta=0} c_{\alpha\beta} s^{\alpha} H^{\beta} \,.$$

If P and H are independent, then at least one coefficient $c_{\alpha\beta} \neq 0$ ($\alpha > 0$). We take the coefficient with the least possible value of $\alpha + \beta = g$. Let P be a converging integral. Then the same holds for

$$p(u, v) = P - \sum_{\beta=0}^{g-1} c_{0\beta} H^{\beta} \,.$$

Expand this integral into a power series in homogeneous forms:

$$p = \sum_{k=2g}^{\infty} p_k \,,$$

where

$$p_{2g} = \sum_{\alpha+\beta=g} c_{\alpha\beta} s_2^{\alpha} H_2^{\beta} \,, \qquad s_2 = u_1 v_1 \,, \qquad H_2 = \lambda_1 u_1 v_1 + \lambda_2 u_2 v_2 \,.$$

Here for the sake of symmetry we used the notation $\lambda_1 = i$, $\lambda_2 = i\omega$.

The polynomial

$$\Delta = \frac{\partial p_{2g}}{\partial s_2} = \sum_{\alpha+\beta=g} \alpha c_{\alpha\beta} s_2^{\alpha-1} H^{\beta} \tag{1.6}$$

does not vanish completely. Since it is homogeneous in $u_1 v_1$ and $u_2 v_2$, we can rewrite Δ as follows:

$$\Delta = c \prod_{q=1}^{q-1} (\xi_q u_1 v_1 + \eta_q u_2 v_2) \,, \tag{1.7}$$

where $c \neq 0$ and not all constants ξ_q, η_q ($q = 1, \ldots, g-1$) are equal to zero.

Denote by x, y, z any three variables among u_1, u_2, v_1, v_2. Since p is a power series in s and H, the Jacobian

$$\frac{\partial(H, p, s)}{\partial(x, y, z)}$$

is identically zero. Calculating the terms of order $q-3$, we obtain

$$\sum_{\alpha+\beta+\gamma=q} \frac{\partial(H_\alpha, p_\beta, s_\gamma)}{\partial(x, y, z)} = 0 , \qquad q \geq 2g+4 .$$

We shall use this relation for two choices of the variables x, y, z: $(x, y, z) = (u_1, u_2, v_1)$ and $(x, y, z) = (u_1, u_2, v_2)$. Denoting the expressions

$$\sum_{\alpha+\beta=l} \frac{\partial(H_\alpha, p_\beta,)}{\partial(y, z)} , \qquad \sum_{\alpha+\beta=l} \frac{\partial(H_\alpha, p_\beta,)}{\partial(z, x)} , \qquad \sum_{\alpha+\beta=l} \frac{\partial(H_\alpha, p_\beta,)}{\partial(x, y)} ,$$

by A_{1l}, A_{2l}, A_{3l} and B_{1l}, B_{2l}, B_{3l} respectively ($l \geq 2g+2$), we get

$$\sum_{l+\gamma=q} \left(A_{1l}\frac{\partial s_\gamma}{\partial u_1} + A_{2l}\frac{\partial s_\gamma}{\partial u_2} + A_{3l}\frac{\partial s_\gamma}{\partial v_1} \right) = 0 , \tag{1.8}$$

$$\sum_{l+\gamma=q} \left(B_{1l}\frac{\partial s_\gamma}{\partial u_1} + B_{2l}\frac{\partial s_\gamma}{\partial u_2} + B_{3l}\frac{\partial s_\gamma}{\partial v_v} \right) = 0 . \tag{1.9}$$

In what follows, we denote by $\mu_1, \ldots, \mu_4$ certain positive constants independent of the subscript k in the subsequent formulas. Since the power series p and H converge,

$$\|H_k\| < \mu_1^k , \quad \|p_k\| < \mu_1^k , \qquad k \geq 2g .$$

Hence,

$$\|\partial H_k/\partial x\| < k\mu_1^k , \qquad \|\partial p_k/\partial x\| < k\mu_1^k .$$

Denote by ψ_k any of the six polynomials $A_{1k}, \ldots, B_{3k}$. Then

$$\|\psi_k\| < \mu_2^k , \qquad k \geq 2g+2 ,$$

which implies the following estimate:

$$\left\| \psi_l \frac{\partial s_\gamma}{\partial x} \right\| \leq \gamma C_{\gamma+2}^3 \mu_2^l \|s_\gamma\| . \tag{1.10}$$

By (1.6), there is an identity

$$\frac{\partial(H_2, p_{2g})}{\partial(x, y)} = \Delta \frac{\partial(H_2, s_2)}{\partial(x, y)} .$$

Thus, with $l = 2g+2$ we obtain for $A_{1l}, \dots, B_{3l}$ the following values: $\lambda_2 u_1 v_2 \Delta$, 0, $-\lambda_2 v_1 v_2 \Delta$, 0, $\lambda_2 v_1 u_2 \Delta$, $-\lambda_2 v_1 v_2 \Delta$. From (1.8–1.10) it follows that

$$\left\| \lambda_2 v_2 \Delta \left(u_1 \frac{\partial s_k}{\partial u_1} - v_1 \frac{\partial s_k}{\partial v_1} \right) \right\| + \left\| \lambda_2 v_1 \Delta \left(u_2 \frac{\partial s_k}{\partial u_2} - v_2 \frac{\partial s_k}{\partial v_2} \right) \right\| < \sum_{\gamma=2}^{k-1} (\gamma+3)^4 \|s_\gamma\| \mu_2^{k+2g+2-\gamma} .$$

In addition, by (1.7) and Lemma 2, we have

$$\left\| u_1 \frac{\partial s_k}{\partial u_1} - v_1 \frac{\partial s_k}{\partial v_1} \right\| + \left\| u_2 \frac{\partial s_k}{\partial u_2} - v_2 \frac{\partial s_k}{\partial v_2} \right\| < k^{g+3} \sum_{\gamma=2}^{k-1} \|s_\gamma\| \mu_3^{k-\gamma} , \qquad k \geq 3 . \tag{1.11}$$

Recall that the integral s contains no terms of the special type. If $w = a u_1^{\alpha_1} v_1^{\beta_1} u_2^{\alpha_2} v_2^{\beta_2}$ is some term of s_k, then

$$\alpha_1 - \beta_1| + |\alpha_2 - \beta_2| \geq 1 ,$$

$$u_1 \frac{\partial w}{\partial u_1} - v_1 \frac{\partial w}{\partial v_1} = (\alpha_1 - \beta_1) w , \qquad u_2 \frac{\partial w}{\partial u_2} - v_2 \frac{\partial w}{\partial v_2} = (\alpha_2 - \beta_2) w .$$

Therefore, (1.11) implies the inequality

$$\|s_k\| < k^{g+3} \sum_{\gamma=2}^{k-1} \|s_\gamma\| \mu_3^{k-\gamma} , \qquad k = 3, 4, \dots . \tag{1.12}$$

Now we prove the estimate

$$\|s_l\| \leq \left(2 l^{g+3} \mu_3\right)^{l-2} \tag{1.13}$$

using induction. Obviously, inequality (1.13) holds for $l = 2$. If it is proven for $l = 2, 3, \dots, k-1$, then (1.12) implies the inequality

$$\|s_k\| < k^{g+3} \sum_{\gamma=2}^{k-1} (2\gamma^{g+3} \mu_3)^{\gamma-2} \mu_3^{k-\gamma}$$
$$\leq (k^{g+3} \mu_3)^{k-2} \sum_{\gamma=2}^{k-1} 2^{\gamma-2} < (2k^{g+3} \mu_3)^{k-2} .$$

Hence, (1.13) holds also for $l = k$.

Therefore,

$$\|s_k\| < k^{\mu_4 k} , \qquad \ln \|s_k\| / (k \ln k) < \mu_4 .$$

The proof of Lemma 3 is completed.

Let $\varepsilon_{pq} < 1$ be an arbitrary sequence of positive numbers. Assume that the irrational number ω in (1.2) can be approximated sufficiently well by rationals: the inequality

$$0 < |a - \omega b| < \varepsilon_{pq}/b^{b^2} , \qquad p = (a, 0) , \quad q = (0, b) , \tag{1.14}$$

has infinitely many solutions given by natural numbers a, b ($b > 0$). From the theory of Diophantine approximations it is known that the measure of the set of such numbers is zero, but they are everywhere dense in $\mathbb{R}$ (see, for example, [52]).

Since ω is irrational, by Lemma 1 the Hamilton equations with the Hamiltonian (1.2) have a formal integral

$$s(u, v) = u_1 v_1 + \sum_{p+q \geq 3} s_{p_1 p_2 q_1 q_2} u_1^{p_1} u_2^{p_2} v_1^{q_1} v_2^{q_2} ,$$

which contains no terms of the special type (with $p_1 = q_1$ and $p_2 = q_2$).

Lemma 4. *In an ε_{pq}-neighborhood of each function* (1.2) *there is a point $H \in \mathcal{H}$ such that for integer a, b satisfying the inequality* (1.14) *the coefficients s_{a00b} admit the estimate*

$$|s_{a00b}| \geq b^{b^2} . \tag{1.15}$$

Corollary. *The set of points $H \in \mathcal{H}$ for which the Birkhoff transformation diverges is everywhere dense.*

Proof of Lemma 4. Let $s = u_1 v_1 + \sum s_l$, where s_l is a homogeneous polynomial of degree $l \geq 3$. The series s is a formal solution of the equation

$$\sum_{k=1}^{2} \left(\frac{\partial s}{\partial u_k} \frac{\partial H}{\partial v_k} - \frac{\partial s}{\partial v_k} \frac{\partial H}{\partial u_k} \right) \equiv 0 .$$

Equating the terms of l-th order to zero we arrive at the equation for the coefficient s_l:

$$u_1 \frac{\partial s_l}{\partial u_1} - v_1 \frac{\partial s_l}{\partial v_1} + \omega \left(u_2 \frac{\partial s_l}{\partial u_2} - v_2 \frac{\partial s_l}{\partial v_2} \right) + i \sum_{p+q=l} (p_1 - q_2) h_{pq} u^p v^q = \cdots ,$$

where the right-hand side is some polynomial of degree l whose coefficients can be expressed in terms of the coefficients of the polynomials $s_2, \ldots, s_{l-1}$ and h_{pq} with $p + q < l$. For the terms $s_{a00b} u_1^a v_2^b$ of the function s_l, we obtain the equation

$$s_{a00b}(a - \omega b) + i a h_{a00b} = g_{a00b} . \tag{1.16}$$

Finally, g_{a00b} can be expressed in terms of the coefficients h_{pq} with $p + q < l$. Now let a and b be natural numbers satisfying (1.14). It is possible to change the coefficient h_{a00b} by not more than ε_{a00b} in such a way that the inequality $|i a h_{a00b} - g_{a00b}| \geq \varepsilon_{a00b}$ holds. Then, by (1.14) and (1.15), we have the required estimate (1.15). This completes the proof.

It is important to note that, while constructing the "perturbed" Hamiltonian function H, we changed only the coefficients of type h_{a00b}.

According to (1.15), if $k = a+b$, then $\ln \|s_k\| \geq b^2 \ln b$. From (1.14) for $\varepsilon < 1$ we have an estimate for a: $a \leq \omega b + 1$. Hence, the sequence

$$\frac{\ln \|s_{a+b}\|}{(a+b)\ln(a+b)} \geq \frac{b^2 \ln b}{[(\omega+1)b+1]\ln[(\omega+1)b+1]}$$

is unbounded as $b \to \infty$. Using Lemma 3, we obtain that the Hamiltonian equations with the perturbed Hamiltonian have no additional integral in the form of a convergent power series. This completes the proof of Theorem 1.

1.3 We endow the set $\mathcal{H}$ of power series with a new topology $\mathcal{T}'$ in which the neighborhoods of a power series with coefficients h^*_{ks} consist of power series whose coefficients h_{ks} satisfy the inequalities $|h_{ks} - h^*_{ks}| < \varepsilon$, $|k| + |s| \leq N$, for some $\varepsilon > 0$ and $N \geq 3$. It is easy to show that in the topology $\mathcal{T}'$ the set of Hamiltonians for which the Birkhoff transformation converges is dense in $\mathcal{H}$. Indeed, if we drop the terms of order greater than N in the formal series representing the Birkhoff transformation, and then modify the coefficients of the higher order terms in the expansion of the given Hamiltonian, we obtain a converging canonical transformation reducing the modified Hamiltonian to the normal form. Of course, the topology $\mathcal{T}'$ is much weaker than $\mathcal{T}$.

It is still unknown whether there is a point in the topological space $(\mathcal{H}, \mathcal{T})$ having a neighborhood consisting of Hamiltonians with divergent Birkhoff transformations. Here is another unsolved problem: is it true that Hamiltonian systems admitting an additional analytic integral form a subset of first Baire category in $\mathcal{H}$? Probably, the answer to this question is positive.

1.4 In [179] Moser studied a related problem on nonintegrability of Hamiltonian systems with one degree of freedom and time-periodic Hamiltonians in a neighborhood of an equilibrium position of elliptic type. More precisely, consider the Hamiltonian system

$$\dot{x} = H'_y, \quad \dot{y} = -H'_x; \qquad x, y \in \mathbb{R}, \tag{1.17}$$

where

$$H(x, y, t) = -\frac{\alpha}{2}(x^2 + y^2) + \sum_{k+l \geq 3} H_{kl}(t) x^k y^l .$$

We assume that this power series with 2π-periodic coefficients H_{kl} is convergent in some neighborhood of the origin for all t. The point $x = y = 0$ is an equilibrium position. This equilibrium point can be interpreted as a 2π-periodic solution of the system (1.17). If α is a real non-zero number, then $x = y = 0$ is an elliptic periodic solution.

For the system (1.17), Moser investigated the existence of an integral

$$G = x^2 + y^2 + F(x, y, t) \tag{1.18}$$

with the following properties:

1) $F(x, y, t + 2\pi) = F(x, y, t)$,
2) the function F is defined for small $|x| + |y|$ and has continuous derivatives in x and y,
3) $\lim\limits_{r\to 0}(xF'_x + yF'_y)/r^2 = 0$, $r^2 = x^2 + y^2$.

It is clear that an analytic function in a neighborhood of the origin satisfies conditions 2) and 3).

Theorem 3 [179]. *For every sequence of positive numbers* ε_2, ε_{kl} $(k, l \geq 0$, $k + l \geq 3)$ *there exists a Hamiltonian*

$$H(x, y, t) = -\frac{\alpha^*}{2}(x^2 + y^2) + \sum_{k+l\geq 3} H^*_{kl}(t)x^k y^l ,$$

$$|\alpha - \alpha^*| < \varepsilon_2 , \qquad |H_{kl}(t) - H^*_{kl}(t)| < \varepsilon_{kl} ,$$

such that the system

$$\dot{x} = (H^*)'_y , \qquad \dot{y} = -(H^*)'_x \tag{1.19}$$

has no integrals of type (1.18) *with properties* 1)–3).

It turns out that an appropriately perturbed Hamiltonian system (1.19) has an isolated periodic orbit in any punctured neighborhood of the point $x = y = 0$. This property of the system (1.19) is incompatible with the existence of an integral of type (1.18).

The Maclaurin series of the integral (1.18) begins with a non-degenerate quadratic form. Of course, it is possible that the Hamiltonian equations admit a "degenerate" integral. It is likely that Theorem 3 holds in the case when, instead of continuously differentiable integrals of type (1.18), we consider integrals that are 2π-periodic in t and can be represented by convergent power series in a neighborhood of the point $x = y = 0$. Probably, this result can be proved by the method of [215]. It is necessary to make sure that the isolated periodic points of the period mapping of the perturbed system (1.19) form a key set for the class of analytic functions in a neighborhood of the origin.

2 Nonintegrability of Reversible Systems

2.1 We have already noted that it is impossible to prove nonintegrability of a concrete Hamiltonian system by Siegel's method. However, this method can be used to prove that nonintegrable systems are dense in certain subspaces of $\mathcal{H}$.

As an example, consider a reversible system

$$\ddot{x} = -\frac{\partial V}{\partial x} , \qquad x \in \mathbb{R}^n , \tag{2.1}$$

which governs the motion of a material particle in a force field with the potential $V(x)$. Of course, this equation can be rewritten in the Hamiltonian form:

$$\dot{x} = H'_y , \quad \dot{y} = -H'_x ; \qquad H = y^2/2 + V(x) . \tag{2.2}$$

Suppose that $V(0) = 0$ and $dV(0) = 0$. Then the point $x = 0$ is an equilibrium position. Put $V = \sum V_k$ and assume that the quadratic form V_2 is positive definite. In particular, the equilibrium point $x = 0$ is stable. By an appropriate orthogonal transformation conserving the form of equation (2.1), we can reduce V_2 to the quadratic form

$$\sum \omega_k^2 x_k^2 / 2 .$$

The numbers $\omega_1, \ldots, \omega_k$ are the frequencies of small oscillations near the equilibrium point.

Now introduce the space $\mathcal{V}$ of power series

$$\sum_{|k| \geq 2} v_k x^k , \qquad k = (k_1, \ldots, k_n)$$

converging in some neighborhood of the point $x = 0$. We equip $\mathcal{V}$ with the topology $\mathcal{T}$ which was introduced in §1 for the space $\mathcal{H}$.

Theorem 1 [135]. *In the space $\mathcal{V}$ with the topology $\mathcal{T}$, the points for which the system* (2.1) *has no integral $F(x, \dot{x})$ analytic in a neighborhood of the point $x = \dot{x} = 0$ and independent of the energy integral $E = \dot{x}^2/2$ are everywhere dense.*

Apparently, the points $V \in \mathcal{V}$ for which Birkhoff's normalizing transformation converges, form a subset of first category in $\mathcal{V}$.

2.2 Theorem 1 can be proved by Siegel's method (§1.2). For simplicity, we restrict ourselves to two degrees of freedom and put $\omega_1 = 1$, $\omega_2 = \omega$. Expand the Hamiltonian (2.2) in a power series in homogeneous forms:

$$H = H_2 + H_3 + \cdots , \qquad H_2 = (y_1^2 + x_1^2)/2 + \omega(y_2^2 + x_2^2)/2 .$$

Now carry out a linear canonical transformation of variables $x, y \to \xi, \eta$ with complex coefficients:

$$y_1 = \frac{\xi_1 + i\eta_1}{\sqrt{2}} , \qquad x_1 = \frac{i\xi_1 + \eta_1}{\sqrt{2}} ,$$
$$y_2 = \sqrt{\omega}\frac{\xi_2 + i\eta_2}{\sqrt{2}} , \qquad x_2 = \frac{i\xi_2 + \eta_2}{\sqrt{2\omega}} .$$

In the new variables, the Hamiltonian is of type (1.2):

$$H = i(\xi_1\eta_1 + \omega\xi_2\eta_2) + \sigma v_{k_1 k_2} \left(\frac{1}{\omega}\right)^{k_2/2} \left(\frac{i\xi_1 + \eta_1}{\sqrt{2}}\right)^{k_1} \left(\frac{i\xi_2 + \eta_2}{\sqrt{2}}\right)^{k_2} .$$

The coefficients h_{pq} of the terms $\xi^p \eta^q$ are linear in v_k, and

$$h_{a00b} = \frac{i^a v_{ab}}{(\sqrt{\omega})^b (\sqrt{2})^{a+b}} . \tag{2.3}$$

In §1.2 we have shown that by arbitrary small modifications of the coefficients of type h_{a00b} it is possible to construct Hamiltonians of nonintegrable systems. From (2.3) it follows that by varying the coefficients v_{ab} in the expansion of the potential energy we vary the required coefficients h_{a00b}. The theorem is proved.

3 Nonintegrability of Systems Depending on Parameters

3.1 Let $x = y = 0$ be an equilibrium position of an analytic Hamiltonian system with the Hamiltonian function

$$H(x, y, \varepsilon) = H_2 + H_3 + \cdots , \qquad (x, y) \in \mathbb{R}^{2n} .$$

Here ε is a parameter taking values in some connected domain $D \subset \mathbb{R}^2$; the Hamiltonian function is analytic in ε. Assume that for all $\varepsilon \in D$ the frequencies of linear oscillations $\omega(\varepsilon) = (\omega_1(\varepsilon), \ldots, \omega_n(\varepsilon))$ do not satisfy any resonance relation

$$(m, \omega) = m_1\omega_1 + \cdots + m_n\omega_n = 0$$

of order $|m_1| + \cdots + |m_n| \le n-1$. Using Birkhoff's method we can find a canonical transformation $x, y \to p, q$ such that in the new coordinates

$$H_2 = \frac{1}{2}\sum_{i=1}^{n} \omega_i \rho_i , \qquad H_k = H_k(\rho_1, \ldots, \rho_n) , \qquad k \le m-1 ,$$

where $\rho_i = p_i^2 + q_i^2$. The transformation is analytic in ε. Now introduce the canonical action–angle variables I, ϕ by the formulas

$$I_i = \rho_i/2 , \quad \phi_i = \arctan(p_i/q_i) , \qquad 1 \le i \le n .$$

In the variables I, ϕ we have

$$H = H_2(I, \varepsilon) + \cdots + H_{m-1}(I, \varepsilon) + H_m(I, \phi, \varepsilon) + \cdots .$$

We represent the trigonometric polynomial H_m as a finite Fourier series:

$$H_m = \sum h_k^{(m)}(I, \varepsilon) e^{i(k, \phi)} .$$

Theorem 1 [126]. *Let $(k, \omega(\varepsilon)) \not\equiv 0$ for all $k \in \mathbb{Z}^n$, $k \neq 0$. Suppose that for some $\varepsilon_0 \in D$ the resonance relation $(k_0, \omega(\varepsilon_0)) = 0$, $|k_0| = m$, is satisfied and $h_{k_0}^{(m)} \not\equiv 0$. Then the canonical equations with Hamiltonian $H = \sum H_s$ do not have a complete set of (formal) integrals $F_j = \sum F_s^{(j)}$ whose quadratic terms $F_2^{(j)}(x, y, \varepsilon)$ are independent for all $\varepsilon \in D$.*

Note that under the assumptions of the theorem there may exist independent integrals with (for certain values of ε) dependent quadratic parts of their Maclaurin expansions. Here is a simple example: the canonical equations with Hamiltonian

$$H = (x_1^2 + y_1^2)/2 + \varepsilon(x_2^2 + y_2^2)/2 + 2x_1y_1y_2 - x_2y_1^2 + x_1^2x_2$$

have an integral $F = x_1^2 + x_2^2 + 2(x_2^2 + y_2^2)$. For $\varepsilon = 2$, it is dependent on the quadratic form H_2. However, all conditions of the theorem are satisfied.

Lemma. *Let $\Phi(I, \phi, \varepsilon)$ be an analytic function in all variables I, ϕ, ε and 2π-periodic in ϕ. If $\{H_2, \Phi\} \equiv 0$, then Φ does not depend on ϕ.*

Proof. Let

$$\Phi = \sum \Phi_k(I, \phi)e^{i(k,\phi)} .$$

Since

$$\{H_2, \Phi\} = \sum i(k, \omega(\varepsilon))\Phi_k(I, \varepsilon)e^{i(k,\phi)} \equiv 0$$

and $(k, \omega(\varepsilon)) \not\equiv 0$ for $k \neq 0$, we conclude that $\Phi_k(I, \varepsilon) \neq 0$ only when $k = 0$, as required.

Let $F(x, y, \varepsilon) = \sum F_s(I, \phi, \varepsilon)$ be a formal analytic integral of the canonical equations with the Hamiltonian H. From the condition $\{H, F\} = 0$ we obtain the sequence of equations

$$\{H_2, F_2\} = 0 , \quad \{H_2, F_3\} + \{H_3, F_2\} = 0 , \ \dots ,$$
$$\{H_2, F_m\} + \dots + \{H_m, F_2\} = 0 , \ \dots .$$

We claim that the functions $F_2, \dots, F_{m-1}$ do not depend on ϕ. For F_2, this has already been proved in the lemma. Since H_3 does not depend on ϕ, we have $\{H_3, F_2\} = 0$. Hence, $\{H_2, F_3\} = 0$. According to the lemma, F_3 also does not depend on ϕ, and so on. Taking into account this remark, we can write down the equation for F_m in the following form:

$$\{H_2, F_m\} + \{H_m, F_2\} = 0 .$$

If

$$F_m = \sum f_k^{(m)}(I, \phi)e^{i(k,\phi)} ,$$

then

$$(\omega(\varepsilon), k)f_k^{(m)} = \left(\frac{\partial F_2}{\partial I}, k\right)h_k^{(m)} , \qquad k \in \mathbb{Z}^n .$$

Put $k = k_0$ and $\varepsilon = \varepsilon_0$. Then $(\omega, k) = 0$ and $h_k^{(m)} \neq 0$. Hence,

$$\left(\left.\frac{\partial F_2}{\partial I}\right|_{\varepsilon_0}, k_0\right) = 0 .$$

If our equations have n integrals $F_1, \dots, F_n$, then, for $\varepsilon = \varepsilon_0$, we obtain n linear equations

$$\left(\frac{\partial F_2^{(1)}}{\partial I}, k_0\right) = \dots = \left(\frac{\partial F_2^{(n)}}{\partial I}, k_0\right) = 0 .$$

Since $k_0 \neq 0$, the quadratic forms $F_2^{(1)}, \dots, F_2^{(n)}$ are dependent for $\varepsilon = \varepsilon_0$, as required.

Although the proof of the theorem is simple, its application to a concrete problem encounters rather difficult computations associated with normalization of the Hamiltonian.

3.2 As an example consider rotation of a dynamically symmetric rigid body ($I_1 = I_2$) with a fixed point. Assume that the mass center lies in the equatorial plane of the inertia ellipsoid. In the majority of integrable cases this condition is satisfied. It is possible to choose the units of measure for mass and length in such a way that $I_1 = I_2 = \varepsilon = 1$, where ε is the product of the weight of the body by the distance from the mass center to the suspension point. The only free parameter in this problem is the moment of inertia I_3.

In all integral manifolds $\{(I\omega, \gamma) = c, \ (\gamma, \gamma) = 1\}$ the reduced Hamiltonian system has two equilibrium positions. They correspond to permanent rotations of the body about the vertical passing through the center of mass which is exactly above (below) the suspension point. The angular velocity of such a rotation is connected with the area constant c by the formula $c = \pm I_3 \omega$.

Consider the case when the center of mass is below the suspension point. In a neighborhood of the equilibrium point, the Hamiltonian function H of the reduced system with two degrees of freedom has the form $H_2 + H_4 + \cdots$ (the terms of degree three are missing). The coefficients depend on two parameters $x = c^2$ and $y = I_3^2$. It can be shown that if $y > x/(x+1)$, then the characteristic roots of the secular equation are purely imaginary. Denote by Σ the subset of $\mathbb{R}^2 = \{x, y\}$ in which this inequality is satisfied. The ratio of the frequencies is 3 if the parameters x and y satisfy the relation

$$9x^2 - 82xy + 9y^2 + 118x - 82y + 9 = 0 . \tag{3.1}$$

This is the equation of a hyperbola. For $x > 0$, $y > 0$, its branches lie entirely in Σ.

From the triangle inequality for the moments of inertia ($I_1 + I_2 \geq I_3$) it follows that $y \geq 1/2$. For any fixed $y_0 \geq 1/2$ there exists x_0 such that the point (x_0, y_0) satisfies Eq. (3.1). The condition for vanishing of the coefficient $h_{1,-3}^{(4)}$ in the expansion of the function H_4 can be reduced to the following form:

$$\begin{aligned} 9x^4 - 10x^3y + x^2y^2 - 17x^3 + 58x^2y - 7xy^2 \\ -375x^2 - 86xy - 170y^2 + 541x + 1700y - 1530 = 0 . \end{aligned} \tag{3.2}$$

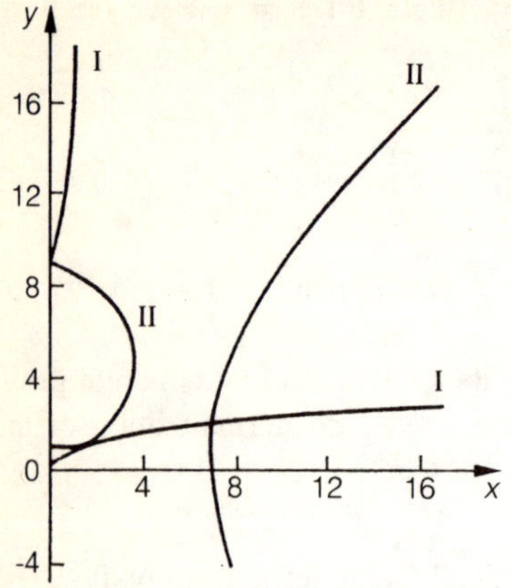

Fig. 34.

The algebraic curves (3.1) and (3.2) intersect at the points (4/3, 1) and (7, 2), which correspond to the integrable cases of Lagrange ($I_1 = I_3$) and Kovalevskaya ($I_1 = 2I_3$) (see Fig. 34).

3.3 Now consider the planar circular restricted three-body problem. The equations of motion of the asteroid with respect to the coordinate frame rotating with the Sun and Jupiter can be represented in the Hamiltonian form:

$$\dot{x}_s = \frac{\partial H}{\partial y_s}, \quad \dot{y}_s = -\frac{\partial H}{\partial x_s}, \qquad s = 1, 2,$$

where

$$H = \frac{1}{2}(y_1^2 + y_2^2) + x_2 y_1 - x_1 y_2 - F(x_1, x_2, \mu),$$

$$F = \frac{1-\mu}{\sqrt{(x_1+\mu)^2 + x_2^2}} + \frac{\mu}{\sqrt{(x_1+\mu-1)^2 + x_2^2}}.$$

This Hamiltonian system has equilibrium positions at the points $x_1 = 1/2 - \mu$, $x_2 = \pm\sqrt{3}/2$, $y_1 = y_2 = 0$. They are called Lagrange's solutions, or the triangular libration points (see Chap. I). If $0 < 27\mu(1-\mu) < 1$, then the eigenvalues of the linearized system are purely imaginary and distinct, and their ratio is a non-constant function of μ. For commensurability of third or fourth degree, the coefficients $h_{1,2}^{(3)}$ and $h_{1,3}^{(4)}$ were calculated by Markeev [169] who investigated the stability of the triangular libration points. These coefficients are non-zero. Apparently, the same is true for all (or almost all) resonance relations. From Theorem 1 it follows, in particular, that in a neighborhood of a libration point there does not exist even a formal Birkhoff normalizing transformation that is analytic in μ.

3.4 Onishchenko carried out similar calculations for the Kirchhoff problem on the motion of a rigid body in a fluid. More precisely, he considered the system of equations

$$I\dot{\omega} + \omega \times I\omega = e \times Je\,, \quad \dot{e} + \omega \times e = 0\,, \tag{3.3}$$

where $I = \operatorname{diag}(A, B, C)$, $J = \operatorname{diag}(L, M, N)$. As shown in §4 of Chap. V, the criteria for nonintegrability of the system (3.3) in the non-symmetric case (when none of the numbers A, B, C coincide) is given by the Clebsch condition. For $A = B \neq C$, the Clebsch condition gives $L = M$.

Thus, consider the symmetric case $A = B$. Let $\alpha = A/C$. We fix the positive constant of the integral (e, e) and vary the constant of the "area integral" $(I\omega, e) = c$. If the system (3.3) has an additional analytic integral independent of the classical integrals, then the Hamiltonian equations of the reduced system admit an integral which is independent of the energy integral and analytic in the parameter c.

The reduced system has equilibrium positions at which the vectors ω and e are collinear. In a certain domain of the space of physical parameters, the frequencies of the linearized system are purely imaginary. It turns out that if $\alpha \neq 1$ and $L \neq M$, then the reduced equations have no integral that is analytic in c and whose first nontrivial form is independent of the quadratic form H_2. It is clear that the value $\alpha = 1$ corresponds to an integrable case. For $\alpha = 2$ and zero value of the "area" constant c there is a particular integrable case discovered by Chaplygin. However, Chaplygin's case cannot be extended to the domain of non-zero values of c.

Onishchenko's result supplements Theorem 2 of §4, Chap. V. However, this result establishes a weaker type of nonintegrability of the motion equations.

4 Symmetry Fields in the Vicinity of an Equilibrium Position

4.1 Again, let $x = y = 0$ be an equilibrium position of a Hamiltonian system with Hamiltonian

$$H = \sum_{k \geq 2} H_k\,.$$

Assume that the eigenvalues $\lambda_1, \ldots, \lambda_n$ of the linearized system are independent over the field of rational numbers. In particular, in appropriate canonical coordinates H_2 takes the form

$$\sum_{j=1}^{n} \lambda_j x_j y_j\,. \tag{4.1}$$

If λ_j are purely imaginary, then the reduction of H_2 to the form (4.1) uses a linear canonical transformation with complex coefficients (see §1).

The Hamiltonian H defines the operator of differentiation with respect to the Hamiltonian field v_H:

$$L_v = \sum x_j(\lambda_j + \cdots)\frac{\partial}{\partial x_j} - \sum y_j(\lambda_j + \cdots)\frac{\partial}{\partial y_j} \,. \tag{4.2}$$

Dots denote terms of order ≥ 1. Let u be a vector field with analytic (or formally analytic) components which is a symmetry field of the Hamiltonian system with Hamiltonian H. Then the linear differential operator

$$L_u = \sum_s \left(\sum_r X_r^s \frac{\partial}{\partial x_r}\right) - \sum_s \left(\sum_r Y_r^s \frac{\partial}{\partial y_r}\right) \tag{4.3}$$

commutes with the operator (4.2). Here X_r^s, Y_r^s are homogeneous forms of degree s in the coordinates x, y. Since all $\lambda_j \neq 0$, the equilibrium position of the Hamiltonian system is isolated. It follows that $u = 0$ for $x = y = 0$. Hence, in (4.3) we have $s \geq 1$.

If at least one of the polynomials X_r^s and Y_r^s has terms of the form

$$a x_r (x_1 y_1)^{\alpha_1} \cdots (x_n y_n)^{\alpha_n} \,,$$

then we call the vector field u special. In what follows we are interested only in non-special symmetry fields.

Under the assumption that the eigenvalues $\lambda_1, \ldots, \lambda_n$ are rationally independent, the Hamiltonian system has n non-special linearly independent Hamiltonian vector fields. According to Lemma 1 of §1, the Hamiltonian equations admit n independent formal integrals

$$F_j = x_j y_j + \cdots$$

containing no special terms of the form $c x^\alpha y^\alpha$. Obviously, the Hamiltonian fields v_{F_j} are non-special symmetry fields.

4.2 The characteristic property of non-special fields is established by the following proposition.

Proposition 1. *If u is a non-special symmetry field, then the forms X_r^s, Y_r^s $(s \geq 2)$ are uniquely determined by the linear forms $X_1^1, \ldots, X_n^1$, $Y_1^1, \ldots, Y_n^1$.*

Proof. The condition for the operators (4.2) and (4.3) to commute implies the relations

$$\begin{aligned} \sum \lambda_j x_j \frac{\partial X_r^s}{\partial x_j} - \sum \lambda_j y_j \frac{\partial X_r^s}{\partial y_j} - \lambda_r X_r^s &= \Phi_r^s \,, \\ \sum \lambda_j x_j \frac{\partial Y_r^s}{\partial x_j} - \sum \lambda_j y_j \frac{\partial Y_r^s}{\partial y_j} - \lambda_r Y_r^s &= \Psi_r^s \,, \end{aligned} \tag{4.4}$$

where Φ_r^s, Ψ_r^s are homogeneous forms in x, y of degree s with coefficients which depend on the coefficients of the forms X_p^l, Y_q^l $(1 \leq l \leq s-1)$. Hence, it is possible to solve the sequence of equations (4.4) successively.

Let

$$X_r^s = \sum U_{k,m} x_1^{k_1} \cdots x_n^{k_n} y_1^{m_1} \cdots y_n^{m_n}, \qquad |k| + |m| = s \geq 2. \tag{4.5}$$

By the first equation in (4.4), we obtain the relation

$$\left[\lambda_r - \sum_{j=1}^{n} (k_j - m_j)\lambda_j\right] U_{k,m} = V_{k,m}, \tag{4.6}$$

where $V_{k,m}$ are certain known numbers. Since $\lambda_1, \ldots, \lambda_n$ are rationally independent, the coefficients in the square brackets vanish only when $k_j = m_j$ $(j \neq r)$ and $k_r = m_r + 1$. These relations correspond to the special terms in the sum (4.5) which are nonexistent by the assumption. By (4.6), the remaining terms in (4.5) are uniquely determined.

The second equation in (4.4) is solved by the same method. The proposition is proved.

4.3 Let us evaluate the linear forms X_r^1 and Y_r^1 in (4.3). These forms satisfy equations (4.4) with zero right-hand sides. Let

$$X_r^1 = \sum \alpha_j x_j + \sum \beta_j y_j, \qquad \alpha_j, \beta_j = \text{const}.$$

Then by the first equation (4.4),

$$\sum (\lambda_j - \lambda_r)\alpha_j x_j - \sum (\lambda_j + \lambda_r)\beta_j y_j \equiv 0.$$

Hence,

$$(\lambda_j - \lambda_r)\alpha_j = 0, \quad (\lambda_j + \lambda_r)\beta_j = 0.$$

Since $(\lambda_j - \lambda_r)\alpha_j = 0$ only for $j = r$, and $\lambda_j + \lambda_r \neq 0$, it follows that $\alpha_j = 0$ for $j \neq r$ and $\beta_j = 0$ for all j. Therefore, $X_r^1 = \alpha_r x_r$. In a similar way we obtain $Y_r^1 = \beta_r y_r$.

Thus, the linearized operator (4.3) has the following form:

$$\sum \alpha_r x_r \frac{\partial}{\partial x_r} - \sum \beta_r y_r \frac{\partial}{\partial y_r}. \tag{4.7}$$

This is a differential operator of a Hamiltonian vector field only under the condition $\alpha_r + \beta_r = 0$ $(1 \leq r \leq n)$.

Suppose that $H \equiv H_2$. Then for all α_r, β_r the operators (4.2) and (4.7) commute. Hence, a Hamiltonian system can have non-Hamiltonian symmetry fields in a neighborhood of an equilibrium point. Note that the linear fields (4.7) are obviously non-special. It turns out that, in general, non-special symmetry fields are Hamiltonian.

By Birkhoff's theorem the Hamiltonian H can be reduced to the normal form

$$H = K_2 + K_4 + K_6 + \cdots,$$

where K_{2m} is a homogeneous form of degree m in the products $\omega_j = x_j y_j$. In particular,

$$K_4 = \sum a_{kj}\omega_k\omega_j \, .$$

Proposition 2. *If* $\det \|a_{kj}\| \neq 0$, *then in* (4.7) *we have* $\alpha_r = -\beta_r$ *for all* $r = 1, \ldots, n$.

Since every non-special symmetry field is uniquely determined by its linear part (Proposition 1), if the form K_4 is non-degenerate, then the symmetry field is Hamiltonian. In particular, the existence problem for such fields is reduced to that of additional integrals that are analytic in a neighborhood of the equilibrium position.

The proof of Proposition 2 is based on analysis of the commutation condition for the operators (4.2) and (4.3) involving the forms of fourth order. We omit straightforward but cumbersome calculations.

VII Branching of Solutions and Nonexistence of Single-Valued integrals

Let $M_{\mathbb{C}}^{2n}$ be a symplectic complex analytic manifold (M is covered by coordinate neighborhoods of a set of complex charts with coordinates $(p, q) \in \mathbb{C}^{2n}$, and the transition maps from chart to chart are invertible holomorphic canonical transformations). Every complex analytic function $H(p, q, t) : M^{2n} \times \mathbb{C} \to \mathbb{C}$ defines the *complex Hamiltonian system*

$$\frac{dp}{dt} = -\frac{\partial H}{\partial q}, \quad \frac{dq}{dt} = \frac{\partial H}{\partial p}.$$

For this system, it is natural to investigate the existence of additional analytic (or, more generally, meromorphic) first integrals. In the majority of integrable problems of Hamiltonian mechanics, the known first integrals can be extended to holomorphic or meromorphic functions in the complex domain of variation of the canonical variables. In this chapter we show that, in general, branching of solutions of Hamiltonian systems in the complex time plane is an obstruction to the existence of new single-valued first integrals.

Problems of this kind originated in investigations by Kovalevskaya, Lyapunov, Husson and others in rigid body dynamics. Apparently, the general solution of the equations of motion can be represented by single-valued meromorphic functions of time only in the classical cases of Euler, Lagrange and Kovalevskaya, when there exists an additional single-valued polynomial integral. For a long time it was not known whether this is pure coincidence, or has some deep cause.

Painlevé formulated a general problem on the connection between the condition for the general solution to be meromorphic and the existence of nontrivial polynomial (or, more generally, algebraic) integrals. Apparently, there is no direct connection. We give examples for Hamiltonian systems with two degrees of freedom. Let

$$H = p_1^2 + q_1^2 + p_2^2 + f(q_2),$$

where f is a polynomial of degree not less than five. The corresponding Hamiltonian equations have two independent polynomial integrals

$$p_1^2 + q_1^2, \quad p_2^2 + f(q_2).$$

However, almost all solutions are multivalued in the complex time plane.

On the other hand, let

$$H = p_1 + p_2^2/2 - q_1 q_2 - 2q_2^3.$$

It can be shown that all solutions of the Hamilton equations with this Hamiltonian are meromorphic functions, but there is no integral independent of H that is polynomial in p, q (see [131, Chap 5]).

The importance of the Painlevé problem was emphasized by Golubev in his well known books [88–89]. He suggested a generalization of this problem: to investigate the connection between *branching of solutions* as functions of complex time and the existence of *single-valued integrals* . We will show that under certain general assumptions the Painlevé–Golubev problem can be solved.

1 The Poincaré Small Parameter Method

1.1 First recall certain simple results of the theory of analytic differential equations. Consider a system of differential equations

$$\frac{dz}{dt} = v(z) , \tag{1.1}$$

where $z = (z_1, \dots, z_n)$ are n complex coordinates, and t is a complex variable which plays the role of time. For the sake of simplicity, we assume that the components of the vector field v are holomorphic functions in $\mathbb{C}^n$. The following results can be extended (with small modifications) to the case when z are local coordinates on an n-dimensional complex manifold.

According to Cauchy's theorem, the system (1.1) has a unique solution $z(t)$ taking the prescribed value $z_0 \in \mathbb{C}^n$ for given $t_0 \in \mathbb{C}$: $z(t_0) = z_0$. It is holomorphic in a small neighborhood of t_0. The solution $z(t)$ can be represented in the form of a power series in $t - t_0$; the coefficients are uniquely determined by the initial condition, and the series converges for small $t - t_0$.

The problem is to determine the value of the function $z(t)$ at a point t_1 which is comparatively far from t_0. The solution is connected with the problem of analytic continuation of a locally holomorphic function. Let l be a continuous path connecting the points t_0 and t_1 in the complex plane $\mathbb{C} = \{t\}$. We say that the holomorphic function $z(\cdot)$ admits analytic continuation along the path l, if it is possible to cover the path by a finite number of open disks centered at the points of l and define holomorphic functions in these disks in such a way that:
1) in the intersection of adjacent disks the corresponding functions coincide,
2) the holomorphic function defined on the first disk coincides with the function $z(t)$ in a small neighborhood of the point t_0.

If we know the holomorphic function defined in the last disk containing the point t_1, then we can determine the value of $z(t)$ for $t = t_1$. However, this value essentially depends on the curve l connecting the points t_0 and t_1.

Starting with the local holomorphic solution $t \to z(t)$ defined for small $t - t_0$, we can construct an analytic function by analytic continuation of $z(\cdot)$ along all curves emanating from the point t_0 such that continuation is possible. Evidently, this function satisfies Eq. (1.1).

It may happen that the analytic function is multivalued: analytic continuation along different curves may yield several (even infinite) values at a point of $\mathbb{C} = \{t\}$. Since dealing with multivalued functions is tiresome, it is convenient to introduce a multisheeted surface which can be thought of as lying above the complex plane and having as many sheets above a point of $\mathbb{C}$ as the number of values of the analytic function at this point. On such Riemann surfaces analytic functions are single-valued holomorphic functions. A detailed exposition of these problems can be found, for example, in [82].

Now consider an analytic system of differential equations depending on a small parameter:

$$\dot{z} = v(z, \varepsilon) , \qquad z \in \mathbb{C}^n . \tag{1.2}$$

It turns out that solutions of this system can be expanded not only in powers of $t - t_0$, but also in powers of ε.

Theorem (Poincaré). *Suppose that the following conditions are satisfied*:

1) *for* $\varepsilon = 0$ *the system* (1.2) *has a solution* $z_0(t)$ *which is analytic along some continuous curve* l *joining the points* t_0 *and* t_1,
2) *the components of the vector field* $v_k(z, \varepsilon)$ $(1 \le k \le n)$ *are holomorphic functions in the direct product* $E \times \{\varepsilon : |\varepsilon| < \varepsilon_0\}$, *where* E *is a neighborhood of the set* $\{z \in \mathbb{C}^n : z = z_0(t),\ t \in l\}$.

Then there exists an analytic solution of the system (1.2):

$$z(t, \varepsilon) = z_0(t) + \varepsilon z_1(t) + \cdots \tag{1.3}$$

with the initial condition $z(t_0, \varepsilon) = z_0(t_0)$ *such that the series* (1.3) *converges for sufficiently small* ε *and all* $t \in l$.

The proof of the theorem can be found in [203, Chap. 2], or in [88].

If $v = v_0(z) + \varepsilon v_1(z) + \cdots$, then the function $z_1(t)$ satisfies the equation

$$\dot{z}_1 = v_0' z_1 + v_1(z_0(t)) , \qquad v_0' = \frac{\partial v_0}{\partial z}(z_0(t)) .$$

The initial condition for this linear system is $z_1 = 0$ for $t = t_0$. In particular, if $v_0 = \text{const}$, then

$$z_1 = \int_{t_0}^{t} v_1(z_0(t))\, dt , \tag{1.4}$$

where the integral is evaluated along the curve l.

Now let l be a loop (a closed curve) in the complex time plane. We say that an analytic vector-function is multivalued along l if its value changes after tracing the curve l. We assume that all solutions of the "unperturbed" system (1.2) are single-valued in the plane $\mathbb{C} = \{t\}$. Then the Poincaré theorem makes it possible to investigate branching of solutions of the system (1.2) for small non-zero values of ε. Everything is reduced to calculating integrals of type (1.4) along closed curves.

In applications, the integrands are usually meromorphic. Therefore, by Cauchy's theorem, the problem of branching of solutions is reduced to the existence problem of poles with non-zero residues.

1.2 Next consider Poincaré's "basic problem of dynamics" in the complex domain. Let

$$D_{\mathbb{C},\delta} = \{y \in \mathbb{C}^n : \operatorname{Re} y \in D \subset \mathbb{R}^n,\ |\operatorname{Im} y| < \delta\}\,.$$

Denote by $\mathbb{T}^n_{\mathbb{C}} = \mathbb{C}^n/2\pi\mathbb{Z}^n$ the complex torus (over $\mathbb{R}$ it is a "cylinder", i.e., the Cartesian product $\mathbb{T}^n \times \mathbb{R}^n$) with complex angular coordinates $x_1, \dots, x_n \bmod 2\pi$. Let E be a neighborhood of zero in $\mathbb{C}$. Let $H(y, x, \varepsilon) : D_{\mathbb{C},\delta} \times \mathbb{T}^n_{\mathbb{C}} \times E \to \mathbb{C}$ be a holomorphic function taking real values for real values of y, x, ε and such that $H(y, x, 0) = H_0(y)$.

The Cartesian product $D_{\mathbb{C},\delta} \times \mathbb{T}^n_{\mathbb{C}}$ is endowed with the obvious symplectic structure. The Hamilton equations with Hamiltonian H have the canonical form:

$$\frac{dy}{dt} = -\frac{\partial H}{\partial x}\,, \quad \frac{dx}{dt} = \frac{\partial H}{\partial y}\,, \qquad H = H_0 + \varepsilon H_1 + \cdots\,. \tag{1.5}$$

All solutions of the system with Hamiltonian H_0 are single-valued in the complex time plane $t \in \mathbb{C}$: they are given by

$$y = y_0\,, \quad x = x_0 + \omega(y_0)t\,.$$

For $\varepsilon \neq 0$, the solutions of the "perturbed" Eqs. (1.5) are, in general, no longer single-valued. Let γ be a closed loop in the complex time plane. According to Poincaré's theorem, solutions of the system (1.5) can be expanded in power series

$$\begin{gathered} y = y^0 + \varepsilon y^1(t) + \cdots\,, \qquad x = x^0 + \omega t + \varepsilon x^1(t) + \cdots\,, \\ y^1(0) = \cdots = x^1(0) = \cdots = 0\,, \end{gathered} \tag{1.6}$$

which converge for sufficiently small values of the parameter ε and all $t \in \gamma$.

We shall study branching of the action variables y for small $\varepsilon \neq 0$. By (1.6), if the vector-function $y^1(t)$ undergoes a jump after tracing the loop γ, then the same holds for the function $y(t, \varepsilon)$ for small $\varepsilon \neq 0$. By (1.4), the jump of the function $y(\cdot)$ is equal to

$$\xi = \int_\gamma \Phi(t)\, dt\,, \qquad \Phi = -\left.\frac{\partial H_1}{\partial x}\right|_{y=y^0,\ x=x^0+\omega(y^0)t}\,. \tag{1.7}$$

If for fixed y the function $H_1(y, x)$ is holomorphic in $\mathbb{C}^n$, then, of course, $\xi = 0$. However, in many important cases this function has singularities (for example, poles). Therefore, we assume only that the function H is holomorphic in the domain $D_{\mathbb{C},\delta} \times \Omega \times E$, where Ω is a connected domain in $\mathbb{T}^n_{\mathbb{C}}$ containing the real torus $\mathbb{T}^n_{\mathbb{R}}$ and the closed curve Γ, which is the image of the loop γ under the map $x = x^0 + \omega(y^0)t$, $t \in \gamma$.

Fix the initial values y^0, x^0 and deform γ continuously in such a way that Γ does not intersect any singular point of H. Then, by Cauchy's theorem, going around the deformed loop changes the function $y^1(t)$ by the same value $\xi \neq 0$. On the other hand, since the solutions (1.6) are continuous with respect to the initial conditions, the function $y^1(t, y^0, x^0)$ is multivalued along the curve γ for all initial points close to y^0, x^0.

Theorem 1 [128]. *Suppose that the following conditions are satisfied:*
1) $\det \|\partial^2 H_0/\partial y^2\| \not\equiv 0$ *in* $D_{\mathbb{C},\delta}$,
2) *for some initial point* y^0, x^0 *the function* y^1 *is not single-valued along a closed curve* $\gamma \subset \mathbb{C}$.

Then the system (1.5) *does not have a complete set of independent formal integrals*

$$F_s = \sum_{i=0}^{\infty} F_i^s(y, x)\varepsilon^i , \qquad 1 \le s \le n ,$$

whose coefficients are single-valued holomorphic functions in the Cartesian product $V \times \Omega \subset D_{\mathbb{C},\delta} \times \mathbb{T}^n_{\mathbb{C}}$, *where* V *is a neighborhood of the point* y^0 *in* $D_{\mathbb{C},\delta}$.

Again we call a formal series $F = \sum F_i \varepsilon^i$ an integral of the canonical Eqs. (1.5) if, formally, $\{H, F\} \equiv 0$. It is easy to see that in this case the composition of the formal series (1.6) and $\sum F_i \varepsilon^i$ is a formal series with constant coefficients.

We sketch the main steps of the proof of Theorem 1. As usual, first we show that the functions $F_0^s(y, x)$ do not depend on x. Let $(y, x) \in D \times \mathbb{T}^n_{\mathbb{R}}$ and $F_0^s = \Phi_0^s + i\Psi_0^s$. Then Φ_0^s and Ψ_0^s are first integrals of the non-degenerate unperturbed system. According to Poincaré's lemma (§1, Chap. II), they do not depend on $x \in \mathbb{T}^n_{\mathbb{R}}$. Since the domain Ω is connected, uniqueness of the analytic continuation implies that F_0^s is constant for $x \in \Omega \subset \mathbb{T}^n_{\mathbb{C}}$.

Next we prove that the functions $F_0^1, \dots, F_0^n$ are dependent in the domain $V \subset D_{\mathbb{C},\delta}$. Since $F_s(y, x, \varepsilon)$ is an integral of the canonical system (1.5), it is constant on the solutions (1.6). Therefore, its values at the time $\tau \in \gamma$ and after tracing the loop γ coincide:

$$\begin{aligned}
&F_0^s(y^0 + \varepsilon y^1(\tau) + \cdots) \\
&\qquad + \varepsilon F_1^s(y^0 + \varepsilon y^1(\tau) + \cdots, x^0 + \omega\tau + \varepsilon x^1(\tau) + \cdots) + \cdots \\
&\quad \equiv F_0^s(y^0 + \varepsilon(y^1(\tau) + \xi(y^0)) + \cdots) \\
&\qquad + \varepsilon F_1^s(y^0 + \cdots, x^0 + \omega\tau + \cdots) + \cdots .
\end{aligned}$$

Expanding both sides of this identity in power series in ε and equating coefficients of order ε, we obtain

$$\left(\frac{\partial F_0^s}{\partial y}, \xi\right) = 0 , \qquad 1 \le s \le n .$$

Since the jump ξ is non-zero in a neighborhood of the point y^0, the Jacobian

$$\frac{\partial(F_0^1, \dots, F_0^n)}{\partial(y_1, \dots, y_n)}$$

vanishes everywhere in the domain V containing the point y^0.

On the other hand, applying Poincaré's method of Chap. IV one can prove the nonexistence of independent integrals

$$\Phi_s(y, x, \varepsilon) = \sum \Phi_i^s(y, x)\varepsilon^i$$

such that the coefficients Φ_i^s are holomorphic in the domain $W \times \Omega$ (W being a small subdomain of V) and the functions Φ_0^s $(1 \leq s \leq n)$ are independent.

1.3 As an example, consider a system with two degrees of freedom with a pe turbing function of the following form:

$$H_1 = f \sin x_1 + g \cos x_1 + h\ , \tag{1.8}$$

where f, g and h are meromorphic functions with simple poles. The first component of the integrand in (1.7) equals

$$\Phi(t) = -\cos(\omega_1 t + c) f(\omega_2 t) + \sin(\omega_1 t + c) g(\omega_2 t)\ .$$

Here ω_1, ω_2 are the frequencies, and c an arbitrary constant. Suppose that for $y = y^0$ the frequency ω_2 is non-zero and at least one of the functions f and g has a simple pole. Then for an appropriate choice of the constant c the function Φ also has a simple pole. Hence, the function $y^1(t, \varepsilon)$ branches for small $\varepsilon \neq 0$. If, moreover, the unperturbed problem is non-degenerate, then, by Theorem 1, the Hamiltonian system (1.5) does not admit an additional analytic integral.

The perturbing function for the reduced problem of rotation of a heavy rigid body with a fixed point has the form (1.8). If the body is dynamically symmetric, then f, g and h are entire functions. Hence, Theorem 1 is not applicable. However, if none of the principal moments of inertia coincide, then f, g, and h are elliptic functions with simple poles. Consequently, in the case of a non-symmetric body, branching of solutions in the complex time plane for small values of Poincaré's parameter implies the nonexistence of additional single-valued integrals. This result, first obtained in [128], provides a positive answer to the Painlevé–Golubev problem.

1.4 Using the phenomenon of branching of solutions , it is possible to establish the nonexistence of single-valued analytic integrals for small but fixed values of $\varepsilon \neq 0$. We quote a relevant result which is due to Ziglin [244].

Let $M^3 = \mathbb{C}^2 \times \mathbb{T}^1_{\mathbb{C}}$, and let $H(z, t, \varepsilon) : M^3 \times E \to \mathbb{C}$ be a holomorphic function taking real values for real z, t, ε such that $H(z, t, 0) = H_0(z)$. Consider the Hamiltonian system

$$\dot{z} = JH'_z\ , \qquad H = H_0(z) + \varepsilon H_1(z, t) + \cdots\ . \tag{1.9}$$

Let $z = z_0 \in \mathbb{C}^2$, $\operatorname{Im} z_0 = 0$, be a hyperbolic fixed point of the unperturbed system

$$\dot{z} = JH'_0(z)\ , \qquad dH_0(z_0) = 0\ .$$

The eigenvalues $\pm\lambda$ of the linearized system have non-zero real parts ($\operatorname{Re}\lambda > 0$). We can regard z_0 as a periodic solution with the period 2π. According to

Poincaré, for sufficiently small $|\varepsilon|$ the system (1.3) has a 2π-periodic solution $z = p(t, \varepsilon)$, $p(t, 0) = z_0$. Continuing the solutions of the system (1.9) asymptotic to the trajectory $p(t, \varepsilon)$ as $t \to -\infty$ to the maximal (non-extendible) analytic functions of $t \in \mathbb{C}$ (possibly multivalued), we obtain a two-dimensional complex surface Λ_ε^-. We call it the unstable complex asymptotic surface of the hyperbolic periodic solution $p(t, \varepsilon)$.

We have seen in Chap. V that the stable and unstable asymptotic surfaces Λ_ε^+ and Λ_ε^- may intersect transversely in the real domain, and this leads to the nonexistence of real analytic integrals in $\mathbb{R}^2 \times \mathbb{T}^1_{\mathbb{R}}$. Hence, there are no complex analytic integrals in $\mathbb{C}^2 \times \mathbb{T}^1_{\mathbb{C}}$. In contrast to the real case, the complex asymptotic surfaces Λ_ε^- and Λ_ε^+ may have transverse self-intersections. This also prevents the existence of holomorphic integrals of the system (1.9).

We give a sufficient condition for the existence of self-intersections. Assume that the asymptotic solution $z = z_a(t)$ of the unperturbed system ($\lim_{t\to-\infty} z_a(t) = z_0$) has a single-valued analytic continuation along a closed continuous path $\gamma : [0, 1] \to \mathbb{C}$, $\gamma(0) = \gamma(1) \in \mathbb{R} \subset \mathbb{C}$. Then for sufficiently small $|\varepsilon|$ the solution $z(t, t_0, \varepsilon)$ of the system (1.9) with the initial condition $z(\gamma(0)+t_0, t_0, \varepsilon) = z_a(\gamma(0))$ also has an analytic (but, in general, not single-valued) continuation along the "shifted" path $\gamma + t_0$. Let

$$h(t_0, \varepsilon) = H_0(z(\gamma(1) + t_0, t_0, \varepsilon)) - H_0(z_a(\gamma(0))) = \varepsilon h_1(t_0) + o(\varepsilon)$$

be the jump of the function $H_0(z(t, t_0, \varepsilon))$ while t is tracing the loop $\gamma + t_0$.

Theorem 2 [244]. *If the function h_1 has a simple zero, then for sufficiently small $|\varepsilon| \neq 0$ the complex surface Λ_ε^- has a transverse self-intersection point, and the system* (1.9) *has no single-valued analytic first integrals.*

Note that $h_1(t_0)$ can be calculated by the formula

$$h_1(t_0) = \int_\gamma \frac{\partial H_1}{\partial t}(z_a(t), t + t_0)\, dt = \int_\gamma \{H_0, H_1\}(z_a(t), t + t_0)\, dt\,.$$

As an example, we consider the problem of pendulum oscillations. Suppose that the suspension point of the pendulum performs small harmonic oscillations. The corresponding Hamiltonian is

$$H = H_0 + \varepsilon H_1\,, \qquad H_0 = y^2/2 + \cos x\,, \quad H_1 = \cos t \cos x\,.$$

Here $z = (x, y)$ are symplectic coordinates and ε is a small parameter. The non-perturbed problem has a hyperbolic fixed point $x = y = 0$. The asymptotic solution

$$y = 2/\cosh t\,, \quad \sin x = -2\sinh t/\cosh^2 t\,, \quad \cos x = 1 - 2/\cosh^2 t\,,$$

is single-valued, meromorphic, and has poles at the points $a_k = i(\pi/2 + \pi k)$, $k \in \mathbb{Z}$.

Let γ be a closed path circling around the pole a_0 in the positive direction. Using the residue formula, it is easy to obtain

$$h_1(t_0) = -4\pi \cos(\pi i/2 + t_0)\,.$$

This function has simple zeros and, thus, Theorem 2 can be applied.

We mention also paper [47], where the same method was used to prove the nonexistence of holomorphic integrals in the problem of planar oscillations of a satellite in an elliptic orbit.

Ziglin generalized Theorem 2 to the non-Hamiltonian case.

2 Branching of Solutions and Polynomial Integrals of Reversible Systems on a Torus

2.1 Let $\mathbb{T}^n = \{x_1, \dots, x_n \bmod 2\pi\}$ be the configuration space of a reversible Hamiltonian system with n degrees of freedom,

$$T = \frac{1}{2}\sum a_{jk}\dot{x}_j\dot{x}_k\,, \qquad a_{jk} = \text{const}\,,$$

the kinetic energy, and $F = (F_1, \dots, F_n)$ a force field defined on $\mathbb{T}^n$. The equations of motion of this reversible system are as follows:

$$\sum a_{kj}\ddot{x}_j = F_k\,, \qquad 1 \le k \le n\,. \tag{2.1}$$

Assume that the components of the force field are analytic on $\mathbb{T}^n$ and can be extended to meromorphic functions defined on the affine space of complex variables $x_1, \dots, x_n$. Then the system (2.1) can be regarded as a system of differential equations in $\mathbb{C}^n$ with complex time $t \in \mathbb{C}$. Following [144], we are concerned with conditions for the general solution of the system (2.1) to be single-valued, and for the existence of $k \le n$ independent integrals that are polynomial in the velocity.

We call a polynomial integral (in the velocity) single-valued if its coefficients are

1) periodic in $x_1, \dots, x_n$ with real period 2π;
2) holomorphic in the domain $\mathbb{C}^n \setminus \mathcal{P}$, where $\mathcal{P}$ is the union of the sets of poles of the meromorphic functions $F_1, \dots, F_n$.

Let

$$x = az + b\,, \qquad a, b \in \mathbb{C}^n\,, \quad z \in \mathbb{C}\,, \tag{2.2}$$

be a straight line in the complex space $\mathbb{C}^n$. Suppose that the restrictions of the meromorphic functions $F_1, \dots, F_n$ to this line are meromorphic functions on the complex plane $z \in \mathbb{C}$. Denote them by $f_1, \dots, f_n$. When the line (2.2) intersects the pole set transversely at the points that are not uncertainty points of the functions F_s, the functions $f_1, \dots, f_n$ are always meromorphic. Since $\mathcal{P}$ is a complex hypersurface in $\mathbb{C}^n$, and the set of uncertainty points has complex codimension two, this property holds for almost all a and b. The meromorphic functions $f_1, \dots, f_n$ determine the meromorphic vector-function f. Therefore, we can speak about the residues of f at its poles. The residues are vectors in $\mathbb{C}^n$ depending on a and b.

Theorem 1. *Suppose that for some $a, b \in \mathbb{C}^n$ the function $z \to f(z)$ has a pole with a non-zero residue. Then the general solution of the system* (2.1) *is not a single-valued function of complex time.*

Theorem 2. *Suppose that*
1) *for some $a, b \in \mathbb{C}^n$ the function f has m poles with residues which are linearly independent over $\mathbb{C}$;*
2) *the system* (2.1) *has k single-valued polynomial integrals with almost everywhere independent principal terms.*

Then $m + k \leq n$.

Consider a simple example. Let $n = 1$ and $F = \operatorname{sn}(2Kx/\pi, \kappa)$, where K is the elliptic integral with modulus $\kappa > 0$. Since f has simple poles, we can apply Theorems 1 and 2. Therefore, the general solution is multivalued and the equations of motion have no single-valued polynomial integral. We note that in the real domain there exists a single-valued polynomial integral: namely, the energy integral. However, in the complex domain this function has logarithmic singular points. The question of the nonexistence of polynomial integrals of the system (2.1) for real x is far more complex. For potential fields with trigonometric potentials, this problem is solved in §5 of Chap. IV.

Suppose that the force field is potential ($F_s = -\partial V/\partial x_s$) and the potential V is a periodic meromorphic function. Then system (2.1) admits an energy integral which is a single-valued meromorphic function. Hence, $m \leq n - 1$. It is easy to give examples of potential fields such that $m = n - 1$ (see §2.3).

2.2 The proof of Theorems 2 and 3 is based on the introduction of an artificial small parameter and the ideas of §2.

Proposition 1. *Suppose that for some a, b the function $z \to f_j(z)$ has at a point z_0 a pole with the residue ζ_j. Then for sufficiently large $|\alpha|$, $\alpha \in \mathbb{C}$, the solution of the system* (2.1) *with the initial condition $x(0) = b$, $\dot{x}(0) = \alpha a$, can be continued onto a circular domain centered at the point $t_0 = z_0/a$. As t goes around the point t_0, the velocity $\dot{x}_s$ undergoes a jump $2\pi i \alpha^{-1} a^{sj} \zeta_j + o(\alpha^{-1})$, where $\|a^{js}\|$ is the inverse of the matrix $\|a_{lj}\|$.*

Theorem 1 is an evident consequence of this result. To prove Proposition 1, we rewrite the system (2.1) as the following system of first order equations

$$\dot{x}_s = v_s\,, \quad \dot{v}_s = \sum a^{sj} F_j\,, \qquad 1 \leq s \leq n\,, \tag{2.3}$$

and carry out the transformation $v_s = \alpha u_s$, $t = \tau/\alpha$. By (2.3), we obtain the following equations:

$$x_s' = u_s\,, \quad u_s' = \varepsilon \sum a^{sj} F_j\,, \qquad \varepsilon = \alpha^{-2}\,, \tag{2.4}$$

where a prime denotes differentiation by τ. We regard ε as a small parameter. Then the straight line $x = a\tau + b$ is a solution of the unperturbed system. To complete the proof, it is sufficient to use Poincaré's theorem on expansion of solutions of (2.4) in converging power series in ε, and Cauchy's theorem on residues.

To prove Theorem 2, we shall use the following

Lemma 1. *The principal homogeneous form of a single-valued integral of the system* (2.1) *does not depend on* x.

The principal homogeneous form of a polynomial integral is an integral of the equations of inertial motion on the n-dimensional torus $\mathbb{T}^n = \{x \bmod 2\pi\}$. Let x_s be real angular coordinates. Then both real and imaginary parts of the homogeneous form are also integrals of the equations $\ddot{x}_s = 0$. Since $\dot{x}_s = \text{const}$ and almost all trajectories of this system are dense in $\mathbb{T}^n$, any real periodic integral depends only on the velocities $\dot{x}_s$. This completes the proof.

Lemma 2. *Suppose that the conditions of Proposition* 1 *are satisfied and let* $\Phi(v_1, \dots, v_n)$, $v_s = \dot{x}_s$, *be the principal homogeneous form of a single-valued integral. Then*

$$\sum a^{sj} \frac{\partial \Phi}{\partial v_s} \zeta_j \equiv 0 \,. \tag{2.5}$$

The proof is based on the fact that the integral is constant on the branching solutions defined in Proposition 1. The substitution $v_s = \alpha u_s$ transforms the polynomial integral of the system (2.3) into an integral of the system (2.4) which is analytic in α^{-1}:

$$\Phi(u) + \alpha^{-1}\Psi(u, x) + \cdots \,.$$

This function is invariant under the transformation

$$u_s \to u_s' = u_s + 2\pi i \alpha^{-2} \sum a^{sj} \zeta_j + o(\alpha^{-2}) \,,$$

$$x_s \to x_s' = x_s + O(\alpha^{-2}) \,.$$

Differentiating this inequality with respect to α, multiplying by α^3, and taking the limit as $\alpha \to \infty$, we obtain (2.5).

Condition (2.5) means that the gradient Φ_v' and the residue ζ are orthogonal. If the system has m independent residues and k integrals with independent gradients of the principal forms, then, evidently, $m + k \leq n$. This proves Theorem 2.

2.3 We give an example of a potential force field such that $m = n - 1$. Let

$$V = -\sum_{j=1}^{n-1} \frac{\sin x_j}{\sin(x_n - a_j)} \,,$$

where the real numbers $a_1, \dots, a_n$ are distinct modulo 2π. Consider the straight line in $\mathbb{C}^n$: $x_1 = \cdots = x_{n-1} = 0$, $x_n = z$. Clearly,

$$f_1 = \sin^{-1}(z - a_1) \,, \quad \dots \,, \quad f_{n-1} = \sin^{-1}(z - a_{n-1}) \,.$$

These meromorphic functions have distinct poles of first order at the points $a_1, \dots, a_{n-1}$. The residues of the vector-function f at these points are, obviously, linearly independent. Thus, $m = n - 1$.

3 Integrals and Symmetry Groups of Quasi-Homogeneous Systems of Differential Equations

3.1 Recall (see §9, Chap. II) that a system of differential equations

$$\dot{z}_i = v_i(z_1, \dots, z_n), \qquad 1 \leq i \leq n, \tag{3.1}$$

is called quasi-homogeneous (or weight-homogeneous) if

$$v_i(\alpha^{g_1} z_1, \dots, \alpha^{g_n} z_n) = \alpha^{g_i+1} v_i(z_1, \dots, z_n), \qquad g_j \neq 0, \tag{3.2}$$

for all z and $t > 0$. The numbers $g_1, \dots, g_n$ are called the weights. The equations governing many important dynamical problems turn out to be quasi-homogeneous. Examples include the problem of many gravitating particles, the rotation of a heavy rigid body with a fixed point, and Kirchhoff's problem on the motion of a rigid body in an ideal fluid.

Differentiating equality (3.2) with respect to α and putting $\alpha = 1$, we obtain Euler's formula

$$\sum_{j=1}^{n} g_j z_j \frac{\partial v_i}{\partial z_j} = (g_i + 1) v_i . \tag{3.3}$$

The system (3.3) may have particular solutions of the form

$$z_j = c_j t^{-g_i}, \qquad 1 \leq i \leq n . \tag{3.4}$$

The constant coefficients c_i satisfy the system of algebraic equations

$$v_i(c_1, \dots, c_n) = -g_i c_i, \qquad 1 \leq i \leq n,$$

which generally has nontrivial solutions.

Now write down the variational equations for the particular solution (3.4):

$$\dot{\xi}_i = \sum_{j=1}^{n} \frac{\partial v_i}{\partial z_j}(c_1 t^{-g_1}, \dots, c_n t^{-g_n}) \xi_j . \tag{3.5}$$

Differentiating equality (3.2) with respect to z_j, we obtain the equality

$$\frac{\partial v_i}{\partial z_j}(\alpha^{g_1}, \dots, \alpha^{g_n} z_n) = \alpha^{g_i - g_j + 1} \frac{\partial v_i}{\partial z_j}(z_1, \dots, z_n) . \tag{3.6}$$

Substituting $\alpha = 1/t$ in (3.2), rewrite (3.5) as follows:

$$\dot{\xi}_i = \sum_{j=1}^{n} \frac{\partial v_i}{\partial z_j}(c) t^{g_j - g_i + 1} \xi_j . \tag{3.7}$$

This linear system has particular solutions

$$\xi_1 = \phi_1 t^{\rho - g_1} \ , \ \ldots \ , \ \xi_n = \phi_n t^{\rho - g_n} \ ,$$

where ρ is an eigenvalue and ϕ an eigenvector of the matrix $K = \|K_{ij}\|$,

$$K_{ij} = \frac{\partial v_i}{\partial z_j}(c) + \delta_{ij} g_i \ .$$

The matrix K and its eigenvalues are called the Kovalevskaya matrix and exponents respectively. If the general solution of system (3.1) is represented by single-valued (meromorphic) functions of complex time, then the Kovalevskaya exponents are integers (respectively, non-negative integers) (see §9, Chap. II).

3.2 The existence problem for quasi-homogeneous integrals of system (3.1) was studied by Yoshida [237]. Recall that a function $f(z)$ is called quasi-homogeneous of degree m with the weights $g_1, \ldots, g_n$ if

$$f(\alpha^{g_1} z_1, \ldots, \alpha^{g_n} z_n) = \alpha^m f(z_1, \ldots, z_m) \ . \tag{3.8}$$

For example, the functions v_i/z_i are quasi-homogeneous of degree 1.

Theorem 1 [237]. *Let f be a quasi-homogeneous integral of the system* (3.1) *and suppose that $df(c_1, \ldots, c_n) \neq 0$. Then $\rho = m$ is a Kovalevskaya exponent.*

This result establishes a remarkable connection between the fact that the general solution is meromorphic and the existence of non-constant integrals. We note that if the system (3.1) has one more integral g of the same degree m, and the differentials df and dg are linearly independent at the point $z = c$, then $\rho = m$ is a Kovalevskaya exponent of multiplicity ≥ 2.

Proof of Theorem 1. Since f is an integral of the system (3.1), the function

$$\sum \frac{\partial f}{\partial z_i}(ct^{-g})\xi_i$$

is an integral of the variational equations (3.7) (see §8, Chap. IV). Substituting $\alpha = t^{-1}$ in (3.8) and differentiating with respect to z_i, we obtain the equalities

$$\frac{\partial f}{\partial z_i}(ct^{-g}) = t^{g_i - m} \frac{\partial f}{\partial z_i}(c) \ .$$

Hence,

$$\frac{\partial f}{\partial z_i}(c)\xi_i t^{g_i - m} = \text{const} \ .$$

Differentiating this equality with respect to t and using equations (3.7), we obtain the relations

$$\sum_j \frac{\partial f}{\partial z_i}(c)\frac{(g_j - m)\xi_j}{t^{m-g_j+1}} + \sum_{i,j} \frac{\partial f}{\partial z_i}(c)\frac{\partial v_j}{\partial z_i}(c)\frac{\xi_i}{t^{m-g_i+1}} = 0 \ .$$

Let $\xi_i = \phi_i t^{\rho - g_i}$ be a solution of the variational equations. Then

$$\left((K - mE)^T \frac{\partial f}{\partial z}(c), \phi\right) = 0 .$$

This equality holds for all eigenvectors ϕ of the Kovalevskaya matrix K. Since there are n linearly independent vectors, it follows that

$$(K - mE)^T \frac{\partial f}{\partial z}(c) = 0 .$$

The theorem is proven.

3.3 Consider another system of autonomous differential equations

$$\frac{dz_i}{d\tau} = u_i(z_1, \dots, z_n) , \qquad 1 \le i \le n . \tag{3.9}$$

Assume that the right-hand side satisfies the equalities

$$u_j(\alpha^{g_1} z_1, \dots, \alpha^{g_n} z_n) = \alpha^{g_j + m} u_j(z_1, \dots, z_n) .$$

In other words, $u_j(z)/z_j$ are quasi-homogeneous functions of degree m with the same weights $g_1, \dots, g_n$. Clearly, the system (3.9) is invariant under the substitutions

$$z_i \to \alpha^{g_i} z_i , \qquad \tau \to \alpha^m \tau .$$

The functions u_i satisfy equalities similar to (3.3):

$$\sum_{j=1}^{n} g_j z_j \frac{\partial u_i}{\partial z_j} = (g_i + m) u_j . \tag{3.10}$$

The vector field u is a symmetry field of the system (3.1) if the linear differential operators

$$L_v = \sum v_i \frac{\partial}{\partial z_i}, \quad L_u = \sum u_j \frac{\partial}{\partial z_j} \tag{3.11}$$

commute. Clearly, the system (3.1) always admits the trivial symmetry field βv, $\beta = \text{const}$.

Theorem 2. *Suppose that the system* (3.1) *admits a quasi-homogeneous symmetry field u of degree m such that $u(c_0) \neq 0$. Then $\rho = -m$ is a Kovalevskaya exponent.*

Proof. The conditions for the operators (3.11) to commute is equivalent to the relations

$$\sum v_j \frac{\partial u_i}{\partial z_j} = \sum u_j \frac{\partial v_i}{\partial z_j} . \tag{3.12}$$

Substituting the solution (3.3) for z in both sides of these equations, we obtain

$$\sum_j v_j \frac{\partial u_i}{\partial z_j}\bigg|_{z=ct^{-g}} = -\frac{1}{t}\sum g_j \frac{c_j}{t^{g_j}} \frac{\partial u_i}{\partial z_j} ct^{-g} = -\frac{1}{t}\sum g_j z_j \frac{\partial u_i}{\partial z_j}\bigg|_{z=ct^{-g}}$$

$$= -\frac{1}{t}(g_i + m)u_i(ct^{-g}) = -\frac{g_i + m}{t^{g_i+m+1}} u_i(c) \, .$$

Using (3.6), we rearrange the right-hand side of (3.12):

$$\sum u_j(ct^{-g}) \frac{\partial v_i}{\partial z_j}\bigg|_{ct^{-g}} = \sum \frac{u_j(c)}{t^{g_j+m} t^{g_i - g_j + 1}} \frac{\partial v_i}{\partial z_j}(c) \, .$$

Hence,

$$\sum_j \frac{\partial v_i}{\partial z_j}(c) u_j(c) + (g_i + m) u_i(c) = 0 \, ,$$

or, equivalently,

$$(K + mE)u(c) = 0 \, .$$

Therefore, $u(c) \neq 0$ is an eigenvector of the matrix K with the eigenvalue $\rho = -m$.

Corollary 1. *If $c \neq 0$, then $\rho = -1$ is a Kovalevskaya exponent.*

Indeed, the symmetry field $u = v$ is quasi-homogeneous with degree $m = 1$. Thus, it is sufficient to note that $v_i(c) = -g_i c_i$ and $g_i \neq 0$.

Here is another proof of this result. Since the system (3.1) is homogeneous, it has the family of solutions

$$z_i(a) = c_i(t + a)^{-g_i} \, , \qquad 1 \leq i \leq n \, ,$$

where a is a real parameter. The derivatives

$$\frac{dz_i(a)}{da}\bigg|_{a=0} = -\frac{c_i g_i}{t^{g_i+1}}$$

satisfy the variational Eqs. (3.7). Therefore, $\rho = -1$ is an eigenvalue of the matrix K with the eigenvector $(-c_1 g_1, \ldots, -c_n g_n)^T \neq 0$.

Corollary 2. *Let u be a quasi-homogeneous symmetry field of degree 1. Suppose that the vectors v and u are linearly independent at the point $z = c$. Then $\rho = -1$ is a Kovalevskaya exponent with multiplicity ≥ 2.*

Theorem 2 establishes an interesting connection between single-valuedness of the general solution and the existence of nontrivial symmetry fields.

We consider more closely the case when v_i are polynomial in $z_1, \ldots, z_n$, and the system (3.1) is quasi-homogeneous with natural weights g_i. Obviously, these conditions are satisfied if v_i are polynomials of second degree in z; in this case $g_i =$

1. Let u be an analytic symmetry field of the system (3.1). Under the substitution $z_i \to \alpha^{g_i} z_i$ system (3.1) transforms to the system $\dot{z} = \alpha v(z)$, and the field u becomes the field

$$\sum_{m \geq 1} \alpha^m u_m(z) ,$$

where u_m is a quasi-homogeneous field of degree m. If the fields u and v commute, then, obviously, all the fields u_m ($m \geq 1$) are symmetry fields of the system (3.1).

Corollary 3. *Suppose that $\rho = -1$ is the only negative integer among the Kovalevskaya exponents, and it is a root of the equation* $\det \|K - \rho E\| = 0$ *with multiplicity one. Then the system* (3.1) *admits no analytic symmetry field u such that the vectors $u(c)$ and $v(c)$ are linearly independent.*

3.4 Now we consider the Hamiltonian equations

$$\dot{x}_k = \frac{\partial H}{\partial y_k} , \quad \dot{y}_k = -\frac{\partial H}{\partial x_k} , \qquad 1 \leq k \leq n , \tag{3.13}$$

with a quasi-homogeneous Hamiltonian of degree h:

$$H(\alpha^{g_1} x_1, \alpha^{f_1} y_1, \ldots, \alpha^{g_n} x_n, \alpha^{f_n} y_n) = \alpha^h H(x, y) . \tag{3.14}$$

Here g_k and f_k are the quasi-homogeneous weights. It is easy to show that the equations (3.13) are quasi-homogeneous in the sense of §3.1, if

$$f_k + g_k = h - 1$$

for all $k = 1, \ldots, n$. In this case the numbers $\rho_1 = h$ and $\rho_2 = -1$ are Kovalevskaya exponents.

Suppose that the system (3.13) admits the particular solution

$$x_k = u_k t^{-g_k} , \quad y_k = v_k t^{-f_k} , \qquad 1 \leq k \leq n , \tag{3.15}$$

where $\sum(|u_k| + |v_k|) \neq 0$.

Theorem 3. *Let $\Phi(x, y)$ be a quasi-homogeneous integral of degree m of the system* (3.1). *Suppose that Φ is independent of the Hamiltonian* (3.14) *at the point $(x, y) = (u, v)$: the rank of the Jacobian matrix of the functions H and Φ equals two. Then $\rho_1 = m$ and $\rho_2 = h - m - 1$ are Kovalevskaya exponents.*

Proof. Since Φ is a quasi-homogeneous integral of the system (3.13), by Yoshida's theorem $\rho = m$ is a Kovalevskaya exponent. The system of Hamiltonian equations

$$x_k' = \frac{\partial \Phi}{\partial y_k} , \quad y_k' = -\frac{\partial \Phi}{\partial x_k} , \qquad 1 \leq k \leq n , \tag{3.16}$$

is quasi-homogeneous of degree $m + 1 - f_k - g_k = m + 1 - h$. Since the functions H and Φ are in involution, the Hamiltonian field (3.16) is a symmetry field of the system (3.13). Since the functions H and Φ are independent at the point (u, v), it remains to apply Theorem 2.

Remark. If $m \neq h$, then the assumption that the functions H and Φ are independent can be replaced by a weaker assumption: $d\Phi \neq 0$ for $x = u$, $y = v$.

Note that the sum $\rho_1 + \rho_2 = h - 1$ is independent of the integral's degree. Apparently, this is not a coincidence: the Kovalevskaya exponents of a quasi-homogeneous Hamiltonian system break into pairs with the sum of each pair equal to $h - 1$. This statement is similar to the Poincaré–Lyapunov theorem on the multiplicators of periodic solutions of Hamiltonian systems. The proof is based on the obvious fact that the variational equations (3.5) are Hamiltonian:

$$\begin{aligned} \dot{\xi}_i &= \sum \frac{\partial^2 H}{\partial y_i \partial x_k} \xi_k + \sum \frac{\partial^2 H}{\partial y_i \partial y_k} \eta_k \,, \\ \dot{\eta}_i &= -\sum \frac{\partial^2 H}{\partial x_i \partial x_k} \xi_k - \sum \frac{\partial^2 H}{\partial x_i \partial y_k} \eta_k \,, \qquad i = 1, \ldots, n \,. \end{aligned} \tag{3.17}$$

Here the solution (3.15) must be substituted into the expressions for the second partial derivatives. Let (ξ, η) and (ξ^*, η^*) be two solutions of the equations (3.17). It is easy to verify that the sum

$$\sum (\xi_k \eta_k^* - \xi_k^* \eta_k) \tag{3.18}$$

is constant. We put

$$\xi_k = \phi_k t^{g_k - \rho_1} \,, \quad \eta_k = \psi_k t^{f_k - \rho_1} \,, \quad \xi_k^* = \phi_k^* t^{g_k - \rho_2} \,, \quad \eta_k^* = \psi_k^* t^{f_k - \rho_2} \,.$$

Then (3.18) takes the form

$$t^{f_k - g_k - \rho_1 - \rho_2} \sum (\phi_k \psi_k^* - \phi_k^* \psi_k) \,.$$

This sum is independent of time in two cases:
1) $\rho_1 + \rho_2 = f_k + g_k = h - 1$,
2) the sum

$$\sum (\phi_k \psi_k^* - \phi_k^* \psi_k) \tag{3.19}$$

is zero.

In the second case, the vectors

$$\mu = (\phi_1, \ldots, \phi_n, \psi_1, \ldots, \psi_n)^T, \quad \mu^* = (\phi_1^*, \ldots, \phi_n^*, \psi_1^*, \ldots, \psi_n^*)^T \,,$$

are skew-orthogonal. Suppose that the vector μ is skew-orthogonal to all eigenvectors of the matrix K. Since these vectors form a basis in $\mathbb{R}^{2n}$, μ is skew-orthogonal to all vectors in $\mathbb{R}^n$. Hence, $\mu = 0$. Thus, there always exists a vector μ^* such that the sum (3.19) is non-zero. This completes the proof.

Suppose that the degrees of the Hamiltonian and the additional integral are integer. Then among the Kovalevskaya exponents there is an additional pair of integers. One of them is the degree of the new integral, and the other is equal but opposite in sign to the degree of the Hamiltonian symmetry field generated by this integral.

3.5 In order to clarify the meaning of Theorems 1 and 2, we pass to the new independent variable τ given by $t = \exp(i\tau)$. Then the system (3.1) has the particular solution

$$z_j = c_j \exp(-ig_j\tau) , \qquad 1 \le j \le n . \tag{3.20}$$

In general this solution is quasi-periodic; the weights $g_1, \dots, g_n$ play the role of constant frequencies. The variational Eqs. (3.7) can be rewritten as follows:

$$\xi_j' = i \sum_k \frac{\partial v_j}{\partial z_k}(c) e^{i(g_k - g_j)\tau} \xi_k , \tag{3.21}$$

where a prime denotes differentiation with respect to τ. This linear system can be reduced to a system with constant coefficients by a linear transformation with quasi-periodic coefficients. In the new variables $\eta_j = \xi_j \exp(ig_j\tau)$, the system (3.1) takes the following form:

$$\eta' = iK\eta , \tag{3.22}$$

where K is the Kovalevskaya matrix. The characteristic exponents of the linear system (3.22) are equal to $i\rho_1, \dots, i\rho_n$, where ρ_j are the Kovalevskaya exponents.

Consider the particular case when system (3.1) is homogeneous: $g_1 = \dots = g_n = g$. Then the solution (3.20) is periodic with period $p = 2\pi/g$. The multipliers of this solution are equal to $\exp[p(i\rho_j)]$. If system (3.1) has an integral with no critical points on the trajectory of the solution (3.21), then at least one multiplier equals one. This assertion is a consequence of the arguments of §8, Chap. IV. (Actually, in §8 only real systems were considered. However, all the results hold for systems with complex variables and real time.) Let s be the degree of a homogeneous integral of system (3.1). By (3.8), its weight degree m equals sg. By Yoshida's theorem, the number $m = sg$ is among Kovalevskaya exponents ρ_j. The corresponding multiplier $\exp(ipm)$ equals one. Therefore, Yoshida's theorem is similar to Poincaré's theorem on degeneracy of periodic solutions of differential equations with integrals having no critical points. However, Yoshida's theorem contains additional information on the degree of the quasi-homogeneous integral. This analogy extends to the case of incommensurable weights $g_1, \dots, g_n$ (compare with §9, Chap. II).

3.6 As an example, consider the homogeneous system of three differential equations

$$\dot z_1 = \varepsilon_1 z_2 z_3 , \quad \dot z_2 = \varepsilon_2 z_3 z_1 , \quad \dot z_3 = \varepsilon_3 z_1 z_2 , \tag{3.23}$$

where ε_k are non-zero real constants [237].

In this example, we have $g_1 = g_2 = g_3 = 1$. Hence, we are to look for particular solutions of the form

$$z_k = c_k / t , \qquad k = 1, 2, 3 .$$

The constants c_k satisfy the system of algebraic equations

$$\varepsilon_1 c_2 c_3 = -c_1 , \quad \varepsilon_2 c_3 c_1 = -c_2 , \quad \varepsilon_3 c_1 c_2 = -c_3 .$$

Obviously,

$$c_1 = \pm(\varepsilon_2\varepsilon_3)^{-1/2} , \quad c_2 = \pm(\varepsilon_3\varepsilon_1)^{-1/2} , \quad c_3 = \pm(\varepsilon_1\varepsilon_2)^{-1/2} . \tag{3.24}$$

It is easy to calculate the Kovalevskaya exponents: regardless of the choice of signs in (3.24), the characteristic equation $\det \|K - \rho E\| = 0$ has the trivial root $\rho = -1$ and a root $\rho = 2$ of multiplicity two.

The latter indicates the possibility of the existence of two independent quadratic integrals of the system (3.23). Apparently, the function

$$f = \sum a_j z_j^2 , \quad \sum a_j \varepsilon_j = 0 , \quad \sum a_j^2 \neq 0 \tag{3.25}$$

is an integral of the system (3.23) with no critical points of type (3.24). Since there are two linearly independent vectors (a_1, a_2, a_3) satisfying the first condition in (3.25), the system (3.23) admits two independent quadratic integrals.

Similar reasoning can be applied in the search for integrals and symmetry groups of more complicated quasi-homogeneous systems. However, this approach does not always accomplish the goal, since Theorems 1 and 2 provide only necessary existence conditions. Moreover, they provide information only on integrals and symmetry fields with additional properties at the point $z = c$.

4 Kovalevskaya Numbers for Generalized Toda Lattices.

4.1 In §9 of Chap. II we introduced the Kovalevskaya numbers: they are the numbers of distinct "complete" families of meromorphic solutions of analytic differential equations. This section deals with calculation of the Kovalevskaya numbers for a class of Hamiltonian systems which generalizes the Toda lattices. We are going to show that the systems with a maximal Kovalevskaya number are completely integrable. This is analogous to the classical Kovalevskaya result in rigid body dynamics.

Consider Hamilton system with Hamiltonian

$$H = \frac{1}{2}\sum_{k=1}^{n} y_k^2 + \sum_{l=1}^{N} v_l \exp(a_l, x) . \tag{4.1}$$

Here $v_l \in \mathbb{R}$, $a_1, \ldots, a_N$ are vectors of $\mathbb{R}^n$, $x = (x_1, \ldots, x_n)$ are canonical coordinates conjugate to $y = (y_1, \ldots, y_n)$, and $(\, , \,)$ is the standard Euclidean product in $\mathbb{R}^n$. Such systems often appear in applications (see [26], [228]).

The system with Hamiltonian (4.1) is called *a generalized Toda lattice* if the following conditions are satisfied:

1) any n vectors of the set $a_1, \ldots, a_{n+1}$ are linearly independent and for some $p_s > 0$ we have $\sum_1^{n+1} p_s a_s = 0$;

2) the vectors $a_1, \dots, a_N$ gather in families F_s, $(s = 1, \dots, n-1)$ such that each vector a_j from F_s is collinear with a_s and $|a_j| \le |a_s|$;
3) $v_s \neq 0$ for all $s = 1, \dots, n+1$.

The generalized Toda lattices include the usual closed Toda lattices and their integrable generalizations found in [2], [25].

We note that in all integrable cases the momenta $y_1, \dots, y_n$ as well as the exponents $\exp(a_1, x), \dots, \exp(a_N, x)$ turn out to be meromorphic functions of the complexified time t. Therefore, we consider the existence problem of k distinct families of formally meromorphic solutions in the system

$$\dot{x} = H'_y , \qquad \dot{y} = -H'_x$$

with the Hamiltonian function (4.1). The solutions are assumed to have the form of series

$$y = \sum_{s=-M}^{\infty} b_s t^s , \quad \exp(a_l, x) = \sum_{s=-M_l}^{\infty} A_s^l t^s , \tag{4.2}$$
$$b_s \in \mathbb{C}^n , \quad A_s^l \in \mathbb{C} , \quad b_{-M} \neq 0 , \quad M > 0 ,$$

whose coefficients depend on $2n-1$ "free" parameters. This problem was considered in [38] for some types of lattices on the plane ($n = 2$ and $k = 1$) and in [3] for arbitrary n and $k = n+1$ under the condition that each set F_s consists of a single vector a_s. It turned out that if the Hamiltonian equations possess sufficiently many distinct families of meromorphic solutions, then they admit n independent integrals polynomial in the momenta and, therefore, are integrable by the Liouville theorem. Note that the inverse statement is not valid. Indeed, for $n = 1$ any system is integrable by the same theorem, however, as we shall see later, the assumption $k \ge 1$ imposes strong restrictions on the structure of the set $a_1, \dots, a_N$.

Theorem 1 [157]. *The inequality $k \ge k_0$ is valid if and only if there exists a set $I \subset \{1, 2, \dots, n+1\}$, $\operatorname{card} I = k_0$, such that*
A) for any index $s \in I$ the set $F_s \setminus \{a_s\}$ is empty or contains only the vector $a_s/2$;
B) for all $s \in I$ and $1 \le r \le N$ such that $a_r \notin F_s$ we have

$$2(a_s, a_r)/(a_s, a_s) \in -\mathbb{Z}_+ , \qquad \mathbb{Z}_+ = \{0, 1, 2, \dots\} . \tag{4.3}$$

Corollary. $k \le n+1$.

Now consider in detail the generalized Toda lattices with the maximal Kovalevskaya number $k = n+1$. In this case conditions A) and B) of Theorem 1 take the form

A) for any index $1 \le s \le n+1$ the set $F_s \setminus \{a_s\}$ is empty or contains only the vector $a_s/2$;
B) relation (4.3) is valid for any two linearly independent vectors a_r and a_s, $1 \le s \le n+1$.

These two properties enables us to classify all generalized Toda lattices with the maximal Kovalevskaya number $k = n+1$. A generalized Toda lattice will be called *complete* if the exponential term $u \exp(b, x)$ can not be added to the Hamiltonian (4.1) without breaking conditions 1) – 3) of this section as well as conditions A) and B) of Theorem 1. It is obvious that any generalized lattice with $k = n + 1$ is obtained from a certain complete lattice by rejecting a part of the vectors $a_s/2$ $(1 \leq s \leq n+1)$.

Let $n = 1$. Then, according to Theorem 1, the set of vectors $\{a_j\}$ of a complete one-dimensional Toda lattice with the maximal Kovalevskaya number $k = 2$ coincides with one of the sets

$$1)\ -2\mu,\ -\mu,\ \mu,\ 2\mu;\quad 2)\ -\mu,\ \mu,\ 2\mu;\quad 3)\ -2\mu,\ \mu,\ \mu,$$

where μ is a positive number. We may not distinguish cases 2) and 3) since the canonical change $x \to -x$, $y \to -y$ transforms the corresponding Hamiltonians into each other.

4.2 Theorem 1 enables us to find all generalized Toda lattices with the maximal Kovalevskaya number. In fact, the problem is reduced to the classification of the sets of $n+1$ vectors in n-dimensional Euclidean space such that
a) the vectors of each proper subset are linearly independent and

$$\sum_{1}^{n+1} p_s a_s = 0\,, \qquad p_s > 0\,,$$

b) relation (4.3) is valid for all linearly independent vectors $a_r \neq a_s$.

This problem, in turn, is closely connected with the theory of root systems which play an important role in modern mathematics (in such fields as finite groups of reflections of Euclidean spaces, semi-simple Lie algebras, etc; see, for example, [41]). The unexpected connection between generalized Toda lattices and root systems, which was first noticed by Bogoyavlensky [25], may seem quite mysterious.

In this connection it is worth recalling the basic ideas of the theory of root systems. Let V be an n-dimensional linear space with the scalar product (,), and let α be a non-zero vector in V. The reflection with respect to the vector α is the orthogonal transformation s of the space V such that
1) $s(\alpha) = -\alpha$,
2) $s(x) = x$ if and only if $(x, \alpha) = 0$.

A finite set S of non-zero vectors from V is called a root system if
a) the linear span of S is V,
b) for any $\alpha \in S$ the reflection s_α with respect to α maps the set S into itself,
c) for any $\alpha, \beta \in S$

$$s_\alpha(\beta) - \beta = n\alpha\,, \qquad n \in \mathbb{Z}\,.$$

Elements of the set S are the "roots" of the space V. Since

$$s_\alpha(x) = x - 2\frac{(x, \alpha)}{(\alpha, \alpha)}\alpha , \qquad x \in V ,$$

condition c) is equivalent to

$$2\frac{(x, \alpha)}{(\alpha, \alpha)} \in \mathbb{Z} , \qquad \alpha, \beta \in S \tag{4.4}$$

(cf. (4.3)). The subset $B \subset S$ is called a basis of simple roots (or a fundamental system) if

1) B is a basis of the vector space V,
2) all the roots can be represented as linear combinations

$$\sum_{i=1}^{n} m_i \alpha_i, \qquad \alpha_i \in B$$

with integer coefficients m_i of the same sign (i.e., either all $m_i \geq 0$ or $m_i \leq 0$).

It turns out that for each root system the basis of simple roots always exists. It is useful also to introduce the maximal root

$$\alpha^* = \sum p_i \alpha_i, \qquad \alpha_i \in B \tag{4.5}$$

such that for each root $\sum q_i \alpha_i$ we have $p_1 \geq q_1, \ldots, p_n \geq q_n$.

According to (4.4) and (4.5), the system of $n + 1$ root vectors $B^* = \{\alpha_1, \ldots, \alpha_n, \alpha^*\}$ satisfies conditions a) and b) formulated above. The classification of the "complemented" systems of simple roots B^* uses the Coxeter graphs: each root vector is represented by a vertex, and each pair of vertices corresponding to the vectors α and β is joined by $k = 4\cos^2 \phi$ edges, ϕ being the angle between α and β. Since

$$2\frac{(\alpha, \beta)}{(\alpha, \alpha)} \quad \text{and} \quad 2\frac{(\beta, \alpha)}{(\beta, \beta)}$$

are integer numbers, their product (exactly $4\cos^2 \phi$) is integer as well, and does not exceed 4. The connected Coxeter graph of the complemented systems of simple roots is isomorphic to one of the following graphs

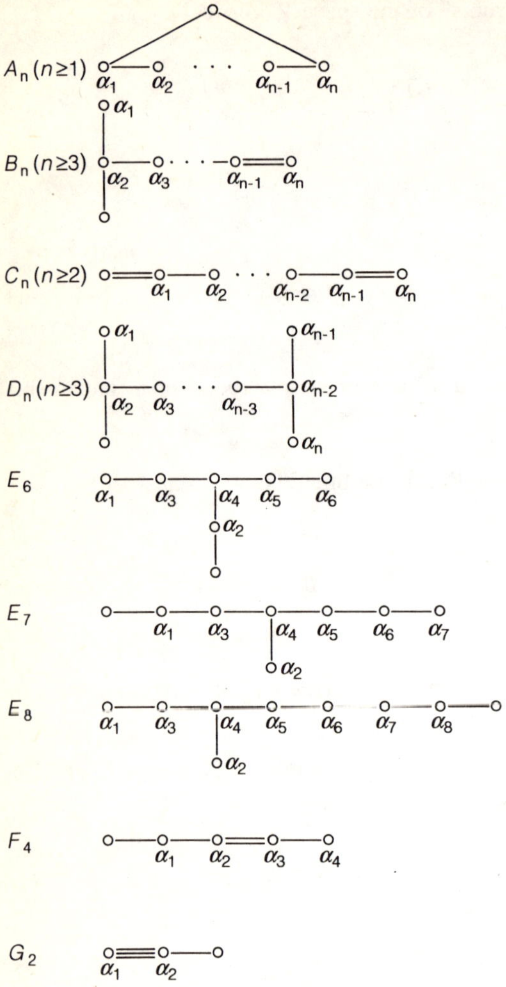

Coxeter graphs give no information on the ratios of lengths of the root vectors. For this reason, mathematicians usually consider the "extended" Coxeter graphs (known as Dynkin diagrams): each vertex is assigned to a coefficient proportional to the squared length of the corresponding vector in B^*.

When $a_1, \ldots, a_{n+1}$ belong to the complemented systems of simple roots, complete integrability of the generalized Toda lattices was proved in [25]. In [2] this result was extended to the case when the root system satisfies conditions a) and b). The classification of such root systems is a similar but more complicated problem (see, for example, [97]). Their Coxeter graphs are the same, but the Dynkin diagrams are more complex. For example, to the graph G_2 there correspond two possible diagrams

Theorem 2 [157]. *Consider a complete Toda lattice with the Kovalevskaya number $k = n + 1$. If $n \geq 2$ then the Dynkin diagram of the vector system $a_1, \ldots, a_N \in \mathbb{R}^n$ is one of the following ones:*

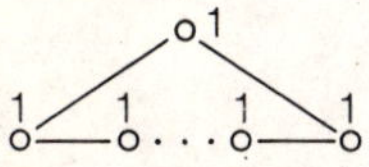

a

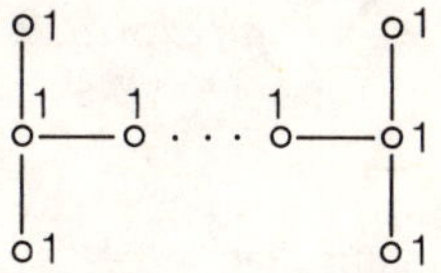

b

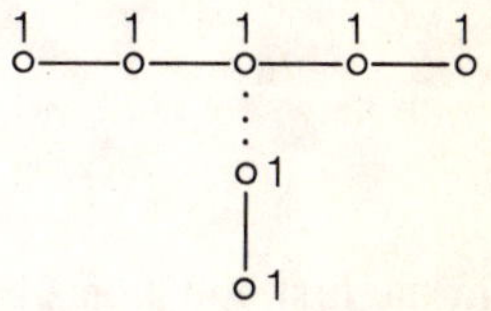

c

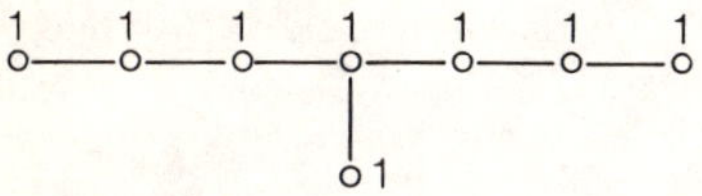

d

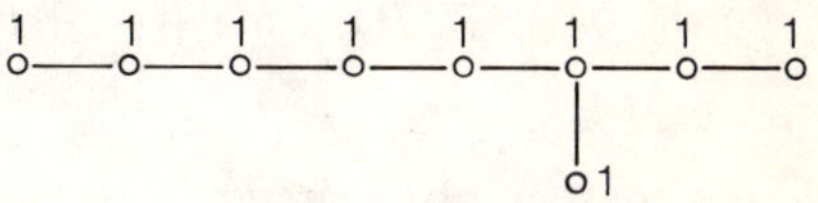

e

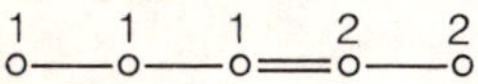

f

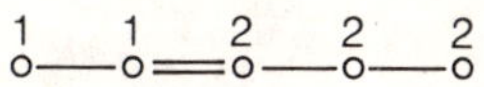

g

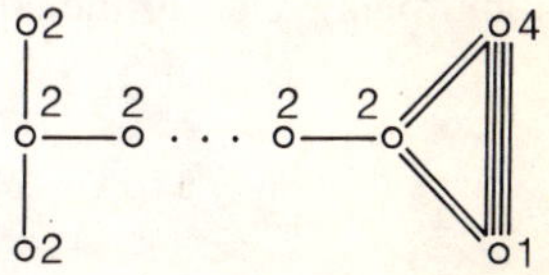

h

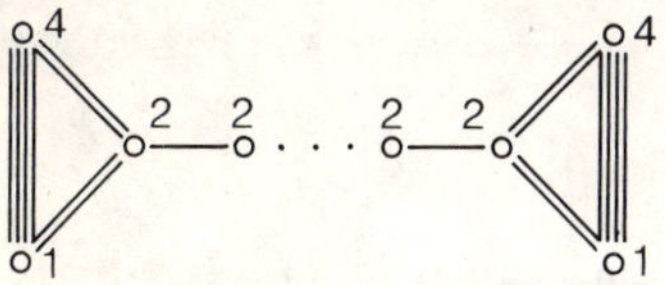

i

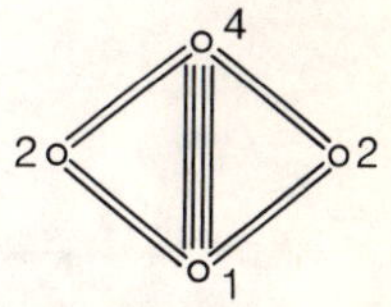

j

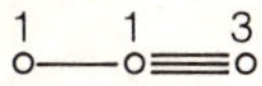

k

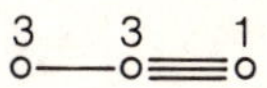

l

The proof is based on the results of [97, Appendix]. The Hamiltonian

$$H = \frac{1}{2} \sum\nolimits_1^n y_s^2 + \sum\nolimits_1^{n+1} v_s \exp(a_s, x)$$

corresponds to the "non-complete" lattice with a maximal Kovalevskaya number. We obtain the "complete" Dynkin diagrams from the diagrams given in [3]. Here we take into account that vectors of the form $a_s/2$ may be added to those diagrams.

4.3 Now turn to the proof of Theorem 1. First, we verify that its conditions are necessary. Let $y(t)$, $x(t)$ be a solution (4.2). The explicit form of the Hamiltonian equations is

$$\dot{x} = y \,, \quad \dot{y} = -\sum_{l=1}^{N} v_l a_l \exp(a_l, x) \,. \tag{4.6}$$

From the first equation (4.6) we obtain that

$$x = b_{-M} t^{1-M}/(1-M) + \cdots + b_{-1} \ln t + b' + b_0 t + \cdots \,,$$

where b' is a constant of integration.

Since the functions $\exp(a_l, x)$ must be formally meromorphic, we have $b_j = 0$ for all $j < -1$. The coefficient b_{-1} does not vanish (otherwise the solution is holomorphic). Due to the equality

$$\exp(a_l, x) = t^{(a_l, b_{-1})} \exp(a_l, b' + b_0 t + \cdots),$$

all the products (a_l, b_{-1}), $l = 1, \ldots, N$ are integer. The second equation (4.6) takes the form

$$\begin{aligned} &-b_{-1} t^{-2} + b_1 + 2b_2 t + \cdots \\ &\quad = -\sum_{l=1}^{N} a_l v_l t^{(a_l, b_{-1})} \exp(a_l, b' + b_0 t + \cdots) \,. \end{aligned} \tag{4.7}$$

Lemma 1. *Let* $m = \min_l (a_l, b_{-1})$. *Then*
1) *if* $(a_j, b_{-1}) = m$, *then* $j \leq n+1$;
2) $(a_s, b_{-1}) > 0$ *for some* $s \leq n+1$;
3) $m = -2$.

Proof.
1) If $(a_j, b_{-1}) = m$ for $j > n+1$, then due to condition 2) there exists a vector a_s $(s \le n+1)$ such that $a_s = ua_j$, $u > 1$. Consequently, $(a_s, b_{-1}) < m$.
2) Let we have $(a_l, b_{-1}) < 0$ for all $1 \le l \le n+1$. We take the scalar product of the equality $\sum p_j a_j = 0$ and the vector b_{-1}, and obtain $\sum p_j(a_j, b_{-1}) = 0$. This equality contradicts the positiveness of the values p_j.
3) Since $b_{-1} \neq 0$, due to the equality (4.7) the magnitude m is not greater than -2. Let the vectors $a_{j_1}, \dots, a_{j_\mu}$ be such that

$$(a_{j_s}, b_{-1}) = m \,, \quad s = 1, \dots, \mu \,; \quad (a_l, b_{-1}) > m \,, \quad l \notin \{j_1, \dots, j_\mu\} \,.$$

Using 1) and 2) we obtain: $j_s \le n+1$, $\mu < n+1$. Thus, we may put $j_s = s$, $s = 1, \dots, \mu$. Let $m < -2$. Then, according to (4.7),

$$\sum_{j=1}^{\mu} a_j v_j \exp(b', a_j) = 0 \,.$$

Due to property 1), this equality can hold only in the case $\mu = n+1$, consequently $m \ge -2$. Thus, $m = -2$ and Lemma 1 is proved completely.

We may assume that $a_1, \dots, a_\mu$ $(\mu < n+1)$ are vectors such that

$$(a_1, b_{-1}) = \dots = (a_\mu, b_{-1}) = -2 \,; \quad (a_l, b_{-1}) > -2 \,, \qquad l > \mu \,.$$

Let the set $W = \{c_1, \dots, c_\nu\} \subset \{a_1, \dots, a_N\}$ be such that $(c_1, b_{-1}) = \dots = (c_\nu, b_{-1}) = -1$ and for all $a_j \notin W$ we have $(a_j, b_{-1}) \neq -1$.

Lemma 2. *$\nu \le \mu$ and $c_j = a_{l(j)}/2$, where $j = 1, \dots, \nu$ and $l(j) \le \mu$.*

Proof. According to (4.7), the following equality holds

$$0 = \sum_{j=1}^{\mu} a_j v_j (a_j, b_0) \exp(b', a_j) + \sum_{j=1}^{\nu} c_j v'_j \exp(b', c_j) \,,$$

where v'_j are coefficients of $\exp(c_j, x)$ in (4.1). Since any vector c_s is parallel to some vector a_l, $l \in \{1, \dots, n+1\}$, the last equality has the form $\sum_1^{n+1} a_s w_s = 0$, where w_s are some coefficients. This sum can not contain all vectors of the set $\{a_1, \dots, a_{n+1}\}$ with non-zero coefficients (indeed, according to assertion 2) of Lemma 1, for at least one of these vectors its scalar product with b_{-1} is positive). Consequently, all the values w_s vanish. This implies

$$\begin{aligned} & c_j = a_{l(j)}/2 \,, \qquad j = 1, \dots, \nu \,; \quad l(j) \le \mu \,, \\ & v'_j = -2 v_{l(j)} (a_{l(j)}, b_0) \exp[(b', a_{l(j)})/2] \,, \\ & (a_s, b_0) = 0 \quad \text{if} \quad s \le \mu \quad \text{and} \quad s \neq l(q) \,, \quad q \in \{1, \dots, \nu\} \,. \end{aligned} \tag{4.8}$$

Lemma 2 is proved.

Lemma 3. $\mu = 1$.

Proof. We show that in the case $\mu > 1$ solutions of equations (4.6) depend on fewer than $2n-1$ arbitrary constants. Indeed, the vector b_{-1} is fixed; according to the equality

$$b_{-1} = \sum_{j=1}^{\mu} a_j v_j \exp(b', a_j) \tag{4.9}$$

(see (4.7)) μ conditions are imposed on the vector b'. Hence, b' contains not more than $n-\mu$ free parameters. Equalities (4.8) yield μ conditions on the components of the vector b_0. Therefore, b_0 contains not more than $n-\mu$ free parameters.

We equate the coefficients of t^j, $j = 1, 2, \ldots$ in equation (4.7) and obtain the following relations:

$$b_j + Gb_j/[j(j+1)] = Q_j\ , \tag{4.10}$$

where the linear operator G has the form

$$Gb = \sum_{s=1}^{\mu} a_s v_s(a_s, b) \exp(b', a_s)\ , \tag{4.11}$$

and the functions Q_j do not depend on b_s, $s \geq j$.

The operator G vanishes on the linear space that is orthogonal to the linear hull of the vectors $a_1, \ldots, a_\mu$. Consequently, the equation $\det(I - \lambda G) = 0$ has not more than μ real roots. Thus, equations (4.10) give us not more than μ free parameters. We emphasize that now we are not interested in the solvability of these equations.

Thus, we have not more than $n - \mu + n - \mu + \mu = 2n - \mu$ free parameters, hence $\mu \leq 1$. Equality (4.9) implies $\mu \geq 1$. The lemma is proved.

We showed that the following conditions are necessary for the existence of $(2n-1)$-parameter formally meromorphic solutions of the system (4.6):

a) $(b_{-1}, a_s) = -2$ for some $s \in \{1, 2, \ldots, n+1\}$,

b) $(b_{-1}, a_s) \in \mathbb{Z}_+$ for all $a_l \notin F_s$,

c) the family F_s may contain apart from a_s only one vector $a_s/2$.

Using equality (4.9) and condition a), we see that

$$(a_s, a_s) v_s \exp(b', a_j) = -2\ , \tag{4.12}$$

$$b_{-1} = -2a_s/(a_s, a_s)\ . \tag{4.13}$$

Thus, condition b) takes the form

$$\frac{2(a_s, a_l)}{(a_s, a_s)} \in -\mathbb{Z}_+ \quad \text{for} \quad a_l \notin F_s\ .$$

Hence an index s, satisfying conditions 1) and 2) of Theorem 1, corresponds to each $(2n-1)$-parameter formally meromorphic solution of Eqs. (4.6). The necessity of conditions 1) and 2) is proved.

Now we verify the sufficiency of these conditions. For each index $s \in I$ we consider the corresponding $(2n-1)$-parameter formally meromorphic solution. Let b_{-1} satisfy equality (4.13) and b', b_0 satisfy relations (4.8) and (4.9) with $j = l(j) = \mu = 1$. It is necessary to prove that Eqs. (4.10) are solvable and the solutions depend on one free parameter.

Let $s = 1$. We assume that $a_1/2 \in \{a_1, \dots, a_N\}$ (since otherwise we may add to the Hamiltonian (4.1) the term $v_1' \exp(a_1/2, x)$, where $v_1' = 0$). Using relations (4.12) and (4.11), we obtain:

$$Gb = -2a_1(a_1, b)/(a_1, a_1) .$$

The rank of the operator G equals 1, and the non-zero eigenvalue of G equals -2: $Ga_1 = -2a_1$. Thus, the operators $E + G/[j(j+1)]$ are non-degenerate for $j = 2, 3, \dots$, and equations (4.10) have a unique solution for $j \neq 1$.

We write down Eq. (4.10) with $j = 1$ in the more detailed form:

$$\begin{aligned} b_1 - a_1(a_1, b_1)/(a_1, a_1) &= -a_1 v_1 (a_1, b_0)^2 \exp(b', a_1)/2 \\ &\quad - \frac{a_1}{2} v_1' \left(\frac{a_1}{2}, b_0\right) \exp\left(b', \frac{a_1}{2}\right) - \sum f_l v_l'' \exp(f_l, b_{-1}) , \end{aligned} \tag{4.14}$$

where f_l are vectors of the family $\{a_1, \dots, a_N\}$ such that $(f_l, a_1) = 0$, and v_l'' are the corresponding coefficients in the Hamiltonian (4.7). According to equality (4.8), the first two terms in the right-hand side of (4.14) cancel one another. Consequently, the vector

$$b_1 = -\sum f_l v_l'' \exp(f_l, b_{-1}) + \xi a_1$$

is a solution of equation (4.14), where ξ is the desired free parameter.

Theorem 1 is proved.

4.4 It turns out that there exist simple necessary conditions for the general solution of systems with exponential interaction to be single-valued. (The generalized Toda lattices from §4.1 are particular cases of such systems.) These conditions are connected with the existence of additional polynomial integrals.

So consider the system with Hamiltonian

$$H = \frac{1}{2} \sum_1^n y_i^2 + \sum_1^N \exp(a_i, x) , \qquad c_i \neq 0 .$$

We introduce "excessive" coordinates

$$v_j = \exp(a_j, x) , \quad u_j = (a_j, y) , \qquad 1 \leq j \leq N ,$$

and write down the Hamiltonian differential equations

$$\dot{x}_i = \partial H/\partial y_i , \quad \dot{y}_i = -\partial H/\partial x_i , \qquad 1 \leq i \leq n$$

in the form of a system with polynomial right-hand sides:

$$\dot{u}_i = \sum_1^N M_{ij} v_j , \quad \dot{v}_i = v_i u_i , \qquad M_{ij} = -c_j(a_i, a_j) . \tag{4.15}$$

In addition to the energy integral these equations have a number of trivial integrals. Indeed, let $\sum_1^N \alpha_k a_k = 0$, $\alpha_k \in \mathbb{R}$. Then the functions

$$F = \sum_1^N \alpha_k u_k \,, \quad \Phi = \prod_1^N v_k^{\alpha_k} \tag{4.16}$$

are obviously integrals of (4.15). For real motions we have $F = 0$ and $\Phi = 1$. Equations of the form (4.15) were used in [3]. We already considered them in the case of the Toda lattice in §9 of Chap. II.

The system (4.15) is quasi-homogeneous; its weight in the variables u equals 1, and the weight in the variables v equals 2. In order to establish whether its general solution is single-valued, we can use the arguments of §9 of Chap. II. Equations (4.15) have particular meromorphic solutions

$$\begin{aligned} &u_i = U_i/t \,, \quad v_i = V_i/t^2 \,, \qquad 1 \le i \le N \,; \\ &U_i = -2M_{ik}/M_{kk} \,; \qquad V_i = 0 \,, \quad i \neq k \,, \quad V_k = 2/M_{kk} \,. \end{aligned} \tag{4.17}$$

Write down the variational equations for the system (4.15) in a neighborhood of the solutions (4.17):

$$\dot{\xi}_i = \sum_1^N M_{ij}\eta_j \,, \quad \dot{\eta}_i = U_i\eta_i/t + V_i\xi_i/t^2 \,.$$

We search for their solutions in the form

$$\xi_i = \phi_i t^{\rho-1} \,, \quad \eta_i = \psi_i t^{\rho-2} \,.$$

We obtain the following linear homogeneous system with the spectral parameter ρ for ϕ_i and ψ_i:

$$(\rho-2)\psi_i = \psi_i U_i + \phi_i V_i \,, \quad (\rho-1)\phi_i = \sum_j M_{ij}\psi_j \,, \qquad 1 \le i, j \le N \,.$$

Using relations (4.17), we find all roots of the corresponding characteristic Kovalevskaya equation: $\rho = 1$ (of multiplicity $N-1$), $\rho = -1$, $\rho = 2$, and also $\rho = 2 - 2(a_i, a_k)/(a_k, a_k)$, $i \neq k$.

If the general solution of the system (4.15) is single-valued for complex values of t, then all roots ρ must be integer. This implies that for all i, j

$$2(a_i, a_j)/(a_j, a_j) \in \mathbb{Z} \,. \tag{4.18}$$

Condition (4.18) is valid if the vectors $a_1, \ldots, a_n$ form a root system. Probably, in this case the general solution of (4.15) is actually a single-valued (but not necessarily meromorphic) function of t (cf. Theorem 1 of §4.1).

Now we discuss also the question of the existence in the system (4.17) of additional first integrals which are polynomials in the variables u, v. It is not difficult to understand that every such integral is a finite sum of quasi-homogeneous polynomial integrals whose weights in the variables u and v equal 1 and 2 respectively. So let $F(u, v)$ be a quasi-homogeneous integral of degree m of system (4.15). According to Theorem 1 of §3, if the point $u_i = U_i$, $v_i = V_i$, where U_i and V_i are defined by (4.17), is not a critical point of the function F, then the number m

coincides with one of the characteristic roots ρ. It is worth remarking that not all integrals satisfy this condition. The trivial integrals Φ from the series (4.16) are an exception. If there are k quasi-homogeneous integrals of the same degree m, which are independent at the point $(u, v) = (U, V)$, then the root $\rho = m$ has multiplicity not less than k.

Suppose that there exists a "good" quasi-homogeneous integral of degree $m \geq 2$ which is independent of the energy integral at the points $u_i = -2M_{ik}/M_{kk}$; $v_i = 0$, $i \neq k$, $v_k = 2/M_{kk}$. Then for each $i = 1, \dots, N$ there is a number $j \neq i$ such that

$$2(a_i, a_j)/(a_j, a_j) \in -\mathbb{Z}_+ ,$$

and all these numbers coincide. It is not clear whether this condition is sufficient for a "good" quasi-homogeneous integral to exist. All integrals of degree $m = 1$ have the form (4.16).

It is interesting to compare the conditions for the existence of k "good" polynomial integrals of degree $m \geq 2$ with those of k "complete" families of meromorphic solutions. We perform such a comparison for the generalized Toda lattices with $N = n + 1$. Consider the square $(n + 1) \times (n + 1)$ matrix L with elements

$$L_{ij} = 2(a_i, a_j)/(a_j, a_j) , \quad i \neq j , \quad L_{ii} = 0 .$$

Suppose that there exist k independent quasi-homogeneous integrals of degree $m \geq 2$ which are additional to the energy integral. Then, according to the results of §9 of Chap. II there are at least k non-positive integers in each row of the matrix L. If the Kovalevskaya number of this system is not less than k, then due to Theorem 1, in the matrix L there are at least k rows whose elements are non-positive integers. These conditions coincide only for $k = n + 1$.

5 Monodromy Groups of Hamiltonian Systems with Single-Valued Integrals

5.1 Besides Poincaré's method of a small parameter (see §1), there is another effective method for investigating branching of solutions of analytic differential equations. This approach, proposed by Lyapunov [165] in 1894, is based on studying the variational equations for known particular solutions.

Let

$$\dot{z} = v(z) \tag{5.1}$$

be a system of differential equations in $\mathbb{C}^n$ with holomorphic right-hand side. Let $z_0(t)$ be a particular solution. Put $z = z_0 + \xi$. Then (5.1) transforms into

$$\dot{\xi} = A(t)\xi + \cdots , \qquad A = \frac{\partial v}{\partial z}(z_0(t)) . \tag{5.2}$$

Dots denote terms of order ≥ 2. The linearized system

$$\dot{\xi} = A(t)\xi \tag{5.3}$$

is a system of variational equations for the solution $z_0(\cdot)$. It turns out that if the system (5.3) has a multivalued solution, then the same holds for the original system (5.1). Indeed, let us replace ξ by $\varepsilon\xi$ and regard ε as a small parameter. Then we get the following system

$$\dot{\xi} = A(t)\xi + O(\varepsilon) .$$

Its solutions can be represented in the form of a series $\xi_0 + \varepsilon\xi_1 + \cdots$, where ξ_0 is a solution of the variational Eqs. (5.3). Hence, $z = z_0 + \varepsilon\xi_0 + o(\varepsilon)$. In particular, if the function $\xi_0(\cdot)$ changes its value after tracing a closed curve in the complex plane, then for small $\varepsilon \neq 0$ the same holds for the solution $z(\cdot)$ (see Poincaré's theorem in §1).

In fact, we have already used Lyapunov's method when analyzing quasi-homogeneous systems in §9 of Chap. II.

5.2 Therefore, it is useful to consider the linear system (5.3), where elements of the matrix A are holomorphic functions defined on a connected Riemann surface X. For example, if the elements of $A(\cdot)$ are meromorphic functions on $\mathbb{C}$, then X is the complex plane with a certain number of points removed (poles).

Locally, for given initial condition $\xi(t_0) = \xi_0$ ($t_0 \in X$) there always exists a unique holomorphic solution of the system (5.3). It can be continued along any curve in X, but its continuation is, in general, a multivalued function. Let γ be a closed path with end points at $t_0 \in X$. Since the system (5.3) is linear, every solution $\xi(t)$ (first defined only in a small neighborhood of t_0) can be continued analytically along γ. Therefore, in the same neighborhood of t_0 we obtain a function $\xi_*(t)$ which also satisfies (5.3). From linearity of (5.3) it follows that there exists a complex $(n \times n)$ matrix T_γ such that $\xi_*(t) \equiv T_\gamma \xi(t)$. If T_γ is not a unit matrix, then system (5.3) has branching solutions.

It turns out that the set of matrices $G = \{T_\gamma\}$ corresponding to all closed curves in X is a group. This group is called the *monodromy group* of the linear system (5.3).

Note that the matrices $T \in G$ depend on the point $t_0 \in X$. Thus, the correct notation is $G(t_0)$. However, for all $t_0 \in X$ the corresponding groups $G(t_0)$ are isomorphic.

In order to understand the group structure of the set G, consider the fundamental group $\pi_1(X)$ of the Riemann surface X. Its elements are classes of homotopic paths with end points at t_0.

For two closed paths σ and τ, their composition γ is defined by the following rule: the initial point of γ is the initial point of σ, the end point of σ is attached to the initial point of τ, and, finally, the end point of τ is the end point of γ. The path γ is called the product of the paths σ and τ and denoted by $\sigma\tau$. If the paths σ and τ are homotopic to the paths σ' and τ', then the products $\sigma\tau$ and $\sigma'\tau'$ are homotopic. The set $\pi_1(X, t_0)$ of classes of equivalent curves with the given product is a group. Its unity corresponds to contractible loops, and the curve σ^{-1} is obtained from σ by inversion of direction. It turns out that for all $t_0 \in X$ the groups $\pi_1(X, t_0)$ are isomorphic. Indeed, let $t_1 \in X$ and let λ be a curve joining t_0 and t_1. Obviously, any closed path with end points in t_1 is homotopic to some loop of type $\lambda^{-1}\gamma\lambda$, where γ is a closed path with end points in t_0. The correspondence $\gamma \to \lambda^{-1}\gamma\lambda$ is an isomorphism of the groups $\pi_1(X, t_0)$ and $\pi_1(X, t_1)$. Therefore, we can speak about the fundamental group of the surface X independent of the choice of the point t_0.

It is possible to show that:

1) analytic continuation of a solution along homotopic curves yields the same result,
2) if $\tau, \sigma \in \pi_1(X)$, then $T_{\tau\sigma} = T_\tau T_\sigma$.

Therefore, the correspondence $\gamma \to T_\gamma$ defines a group homomorphism $\pi_1(X) \to G$.

Let $\Xi(t)$ be a solution of the matrix equation

$$\dot{\Xi} = A(t)\Xi$$

with the initial condition $\Xi(t_0) = I$. We continue the function $\Xi(t)$ analytically onto a neighborhood of the point t_1 along a curve λ connecting t_0 with t_1. Put $\Lambda = \Xi(t_1)$. Let $\gamma' = \lambda^{-1}\gamma\lambda$ be a curve with end points at t_1, and let $T_{\gamma'}$ be the corresponding matrix from the group $G(t_1)$. It is easy to verify that

$$T_{\gamma'} = \Lambda T_\gamma \Lambda^{-1} ,$$

where $T_\gamma \in G(t_0)$ (see §8, Chap. II). This relation establishes an isomorphism of the groups $G(t_0)$ and $G(t_1)$. In particular, the spectrum of matrices from the monodromy group $G(t)$ is independent of $t \in X$.

For a detailed presentation of these results see, for example, [88].

In §8 of Chap. II we studied real systems such that X is a circle. Thus, $\pi_1(X)$ is the infinite cyclic group, and the monodromy group is formed by iterates of the monodromy matrix of the corresponding periodic solution.

5.3 Let $f(z)$ be a holomorphic integral of the system (5.1). We expand the function $f(z_0 + \xi)$ in Taylor series:

$$\sum_{m \geq 0} F_m(\xi, t) . \tag{5.4}$$

Here F_m is a homogeneous form in ξ which is single-valued on the Riemann surface X of the particular solution $z_0(t)$, and $F_0(t) = f(z_0) = \text{const}$. The series (5.4) is an integral of the system (5.2). Since the function F_m is constant on all solutions of (5.3), for every $t_0 \in X$ the homogeneous form $F_m(\xi, t_0)$ is invariant under the action of the monodromy group:

$$F_m(T\xi, t_0) = F_m(\xi, t_0) , \qquad T \in G .$$

This property imposes strong restrictions on the form of first integrals: if the group G is sufficiently "large", then all invariant functions (integrals) are constants.

Lyapunov's method makes it possible to reduce the integrability problem of system (5.1) to the following problem of invariant theory: find all homogeneous polynomials invariant under linear transformations from the given group.

5.4 Consider the vector-function $v_0(t) = v(z_0(t))$. It is holomorphic on the Riemann surface X and satisfies the variational equations (5.3). Hence, $Tv_0(t) = v_0(t)$, where T is any matrix in the monodromy group $G(t)$. Therefore, if $v(z_0) \not\equiv 0$, then at least one eigenvalue of the matrix T equals one.

We generalize this remark. Let $u(z)$ be a symmetry field of the system (5.1) with holomorphic components. Then the vector-function $u_0(t) = u(z_0(t))$ also satisfies (5.3). Since the vector fields u and v commute, we have

$$\frac{\partial u}{\partial z} v = \frac{\partial v}{\partial z} u . \tag{5.5}$$

Hence,

$$\dot{u}_0 = \frac{\partial u}{\partial z}(z_0)\dot{z}_0 = \frac{\partial u}{\partial z} v \Big|_{z_0} = \frac{\partial v}{\partial z} u \Big|_{z_0} = \frac{\partial v}{\partial z}(z_0) u_0 .$$

Thus, $Tu_0(t) = u_0(t)$ for all $T \in G(t)$. In particular, if there are m symmetry fields $u_1, \ldots, u_m$ with holomorphic components, and the vectors $u_1, \ldots, u_m$ are linearly independent at some point of the complex curve $z = z_0(t)$, $t \in X$, then at least $m + 1$ eigenvalues of the matrix T equal one.

Similarly, if the system (5.1) admits k holomorphic integrals $f_1, \ldots, f_k$ such that their differentials are linearly independent at a point $z = z_0(t)$, $t \in X$, then k eigenvalues of the matrix T equal one. Moreover, under the condition

$$\left(\frac{\partial f_i}{\partial z}, u_j \right) = df_i(u_j) = 0 ,$$

the spectrum of the matrix T contains at least $m + k + 1$ ones (compare with §8, Chap. IV).

5.5 The existence of a unit eigenvalue for all matrices from the monodromy group is responsible for considerable difficulties in dealing with problems on integrals and symmetry groups. Therefore, it is useful to reduce the number of independent variables in Eqs. (5.3). We introduce coordinates $\xi_1, \ldots, \xi_{n-1}, \xi_n$ in a neighborhood of the complex curve $\Gamma = \{z = z_0(t) : t \in X\}$ in such a way that the coordinate axes of the variables $\xi_1, \ldots, \xi_{n-1}$ are transverse to Γ, and, locally, Γ is given by the equations $\xi_1 = \cdots = \xi_{n-1}$. Linearizing the differential equations (5.1) in $\xi_1, \ldots, \xi_{n-1}$, we obtain a closed system of linear differential equations with holomorphic coefficients on X. We call this system the reduced system of variational equations, and its monodromy group the *reduced monodromy group* of the particular solution $z_0(\cdot)$. In §8 of Chap. IV we considered only reduced variational equations of real periodic solutions. In general, all eigenvalues of matrices from the reduced monodromy group are not equal to one.

Let F be an integral of the system (5.1) holomorphic in a neighborhood of the complex curve Γ. We expand F in a series in $\xi_1, \ldots, \xi_{n-1}$. Its coefficients are holomorphic functions of $t \in X$. Clearly, the first non-zero homogeneous form of the series is an integral of the reduced linear variational equations. Hence, there is a homogeneous form of $n-1$ variables which is invariant under the action of the reduced monodromy group.

Proposition 1. *If the system* (5.1) *admits a non-constant holomorphic integral, then the eigenvalues* $\lambda_1, \ldots, \lambda_{n-1}$ *of any matrix from the reduced monodromy group satisfy a relation of the type*

$$\lambda_1^{m_1} \cdots \lambda_{n-1}^{m_{n-1}} = 1 , \tag{5.6}$$

where m_k *are nonnegative integer numbers with sum* ≥ 1.

For simplicity, we assume that all the eigenvalues $\lambda_1, \ldots, \lambda_{n-1}$ are distinct. Then in appropriate variables $\xi_1, \ldots, \xi_{n-1}$ the monodromy transformation $\xi \to T\xi$ takes the following form:

$$\xi_j \to \lambda_j \xi_j , \qquad 1 \leq j \leq n-1 . \tag{5.7}$$

Let

$$F(\xi) = \sum a_m \xi_1^{m_1} \cdots \xi_{n-1}^{m_{n-1}} \tag{5.8}$$

be an invariant homogeneous form of degree $m = \sum m_j \geq 1$. After the transformation (5.7) this form becomes

$$F(T\xi) = \sum a_m \lambda_1^{m_1} \cdots \lambda_{n-1}^{m_{n-1}} \xi_1^{m_1} \cdots \xi_{n-1}^{m_{n-1}} . \tag{5.9}$$

Since the forms (5.8) and (5.9) coincide, at least one of the products $\lambda_1^{m_1} \cdots \lambda_{n-1}^{m_{n-1}}$ equals one. This completes the proof.

If some of the eigenvalues $\lambda_1, \ldots, \lambda_n$ coincide, then to prove Proposition 1 we need to reduce the matrix T to Jordan's normal form.

Corollary. *If the system* (5.1) *has an integral with no critical points on* Γ, *then* $\lambda = 1$ *is an eigenvalue of every monodromy matrix.*

This follows from the existence of an integral of the variational equations which is linear in $\xi_1, \ldots, \xi_{n-1}$.

Now we turn to the conditions for the existence of holomorphic symmetry fields of the system (5.1) that are independent of the field v. The components u_j of a symmetry field satisfy the relations

$$\sum_j \frac{\partial u_i}{\partial \xi_j} v_j = \sum_j \frac{\partial v_i}{\partial \xi_j} u_j , \qquad 1 \leq i \leq n . \tag{5.10}$$

Put $i = n$ and $\xi_1 = \ldots = \xi_{n-1} = 0$. Then

$$\frac{\partial u_n^0}{\partial \xi_n} v_n^0 = \frac{\partial v_n^0}{\partial \xi_n} u_n^0 , \tag{5.11}$$

where u_n^0 and v_n^0 are the values of the functions u_n and v_n on the curve $\Gamma \subset \mathbb{C}^n$. Since $v_n^0 \neq 0$, from (5.1) it follows that $u_n^0 = c v_n^0$, $c = \text{const}$.

The field $u^* = u - cv$ is also a symmetry field, and u_n^* vanishes on the curve Γ. We expand the components u_i^* in power series in $\xi_1, \ldots, \xi_{n-1}$. The coefficients of the series are single-valued holomorphic functions on X. Let $m \geq 0$ be the order of the first nontrivial homogeneous form of the functions $u_1^*, \ldots, u_n^*$:

$$u_j^* = u_j^{(m)} + u_j^{(m+1)} + \cdots .$$

In what follows we consider only symmetry fields such that at least one form $u_j^{(m)}$ ($j < n$) does not vanish completely. In this case the fields v and u^* are, obviously, independent.

Relations (5.10) can be represented in the following form:

$$\dot{u}_i = \sum_{j<n} \frac{\partial v_i}{\partial \xi_j} u_j + \frac{\partial v_i}{\partial \xi_n} u_n , \qquad i < n .$$

Since $v_1 = \cdots = v_{n-1}$ for $\xi_1 = \cdots = \xi_{n-1} = 0$ and the expansion of u_n in a series in $\xi_1, \ldots, \xi_{n-1}$ starts with terms of order m, the vector field

$$u^{(m)} = \left(u_1^{(m)}, \ldots, u_{n-1}^{(m)}\right)^T$$

satisfies the reduced system of variational equations.

Let T be a matrix from the reduced monodromy group having $n-1$ different eigendirections. In appropriate coordinates the monodromy transformation $\xi \to T\xi$ takes the form (5.7). Put

$$u_j^{(m)} = \sum U_{\alpha.j}(t) \xi_1^{\alpha_1} \cdots \xi_{n-1}^{\alpha_{n-1}} , \qquad \sum \alpha_k = m .$$

After going around the corresponding closed curve in X, the j-th component of the field $u^{(m)}$ becomes

$$\sum U_{\alpha.j}(t) \lambda_1^{\alpha_1} \cdots \lambda_{n-1}^{\alpha_{n-1}} \xi_1^{\alpha_1} \cdots \xi_{n-1}^{\alpha_{n-1}} . \tag{5.12}$$

On the other hand, since $u^{(m)}$ satisfies the reduced variational equations, the components $u_j^{(m)}$ are multiplied by λ_j:

$$u_j^{(m)} \to \sum U_{\alpha.j}(t) \lambda_j \xi_1^{\alpha_1} \cdots \xi_{n-1}^{\alpha_{n-1}} . \tag{5.13}$$

Comparing (5.12) and (5.13), we find that for at least one set of non-negative integers $\alpha_1, \ldots, \alpha_{n-1}$ there is a relation

$$\lambda_j = \lambda_1^{\alpha_1} \cdots \lambda_{n-1}^{\alpha_{n-1}} , \qquad \sum \alpha_k = m \geq 0 . \tag{5.14}$$

Thus, we obtain

Proposition 2. *If the system* (5.1) *admits a holomorphic symmetry field independent of v, then the eigenvalues of any matrix from the reduced monodromy group satisfy an equality of type* (5.14).

Corollary. *If there exists a symmetry field which is linearly independent of the field v at the points of the curve Γ, then $\lambda = 1$ is an eigenvalue of every monodromy matrix* (*in this case* $m = 0$).

Conditions of type (5.14) are well known in the problem of equivalence of a system of differential equations with its linear part.

5.6 Now consider a Hamiltonian system

$$\dot{z} = JH'_z , \qquad z \in \mathbb{C}^n , \tag{5.15}$$

with a holomorphic Hamiltonian $H(z)$. Here J is the unit symplectic matrix

$$\begin{Vmatrix} 0 & I_n \\ -I_n & 0 \end{Vmatrix} .$$

Suppose that the system (5.15) has a particular solution $z_0(t)$ which is single-valued on its Riemann surface X. Let $u = z - z_0(t)$. Then we can rewrite equation (5.15) in the form

$$\dot{u} = JH''(t)u + \cdots .$$

The linear non-autonomous system

$$\dot{u} = JH''(t)u \tag{5.16}$$

is the variational equation for the solution $z_0(t)$. Of course, it is Hamiltonian: the Hamilton function is the quadratic form $(u, H''(t)u)/2$. To the integral $H(z)$ of the autonomous system (5.15) there corresponds the linear integral $(H'(z_0(t)), u)$ of the variational equations.

Since the system (5.16) is Hamiltonian, the transformations from the monodromy group are symplectic. Indeed, consider two solutions $u_1(t)$ and $u_2(t)$ of the system (5.16). It is easy to verify that the function (u_1, Ju_2) is constant. Let T_γ be the matrix from the monodromy group corresponding to some closed curve γ in the Riemann surface X. After tracing the loop γ the functions u_1, u_2 turn into Tu_1, Tu_2 respectively. Hence,

$$(u_1, Ju_2) = (Tu_1, JTu_2) = (u_1, T^T JTu_2)$$

and, therefore, $J = T^T JT$. Thus, the linear transformation $u \to Tu$ is symplectic. By the arguments of §5.4, two eigenvalues of any matrix $T \in G$ are always equal to 1: one because the system is autonomous, and the other because of the presence of the energy integral.

Fix the value of the energy integral corresponding to the particular solution $z_0(t)$ and restrict the Hamiltonian system (5.15) to the $(2n - 1)$-dimensional energy surface $H(z) = H(z_0(\,\cdot\,)) = \text{const}$. Then we obtain an autonomous system of differential equations with the same particular solution. To this solution there correspond the reduced variational equations (of order $2n-2$) and the reduced monodromy group. From Whittaker's theorem on reduction of Hamiltonian equations by means of the energy integrals it follows that the reduced variational equations are Hamiltonian. Consequently, the matrices of the reduced monodromy group are symplectic.

5.7 The problem of polynomial invariants for groups of symplectic transformations was studied by Ziglin in [245]. We quote his main results.

According to the Poincaré–Lyapunov theorem, the eigenvalues of any symplectic transformation $g : \mathbb{C}^{2n-2} \to \mathbb{C}^{2n-2}$ split into $n-1$ pairs $\lambda_1 = \lambda_n^{-1}, \dots, \lambda_{n-1} = \lambda_{2n-2}^{-1}$. Thus, in the Hamiltonian case there are $n-1$ different relations of type (5.5). We call a transformation $g \in G$ non-resonant if from any equality $\lambda_1^{m_1} \cdots \lambda_{n-1}^{m_{n-1}} = 1$ with integer $m_1, \dots, m_{n-1}$ it follows that all $m_s = 0$. For $n = 1$, this condition means that λ is not a root of unity. Since none of the eigenvalues of T equal 1, the equation $Tz = z$ has only the trivial solution $z = 0$.

It is convenient to pass to the symplectic basis of the map g: if $z = (x, y)$, $x = (x_1, \dots, x_n)$, $y = (y_1, \dots, y_n)$ are coordinates in this basis, then the map g is given by $(x, y) \to (\lambda x, \lambda^{-1} y)$. Such a basis exists if all $\lambda_s \neq 1$ $(1 \leq s \leq n-1)$. This fact is proved, for example, in [217].

Let $F(z) = \sum_{s \geq 1} F_s(z)$ be an integral of the map g. Then all homogeneous forms F_s are also integrals. Let $F_s(x, y) = \sum_{k+l=s} f_{kl} x^k y^l$. Then, obviously,

$$\sum f_{kl} x^k y^l = \sum \lambda^{k-l} f_{kl} x^k y^l \,.$$

If g is non-resonant, then s is even and $f_{kl} = 0$ for $k \neq l$.

Theorem 1 [245]. *Suppose that the reduced monodromy group of the solution* $z_0(\cdot)$ *of a Hamiltonian system contains a non-resonant transformation* g. *If the system* (5.15) *has* n *independent holomorphic integrals in a connected neighborhood of the curve* $\Gamma = \{z = z_0(t) : t \in X\}$, *then every transformation* g' *from the monodromy group has the same fixed point as* g *and takes the eigendirections of* g *into eigendirections. If, in addition, no set of eigenvalues forms a regular polygon with* ≥ 2 *vertices centered at zero in the complex plane, then* g' *conserves the eigendirections of the transformation* g (g' *commutes with* g).

The last condition is evidently satisfied if g' is also non-resonant.

We shall prove Theorem 1 in the simple case $n = 1$ which is important for applications. Suppose that eigenvalues of g are not roots of unity and let x, y be the coordinates in the symplectic basis for g. The eigendirections for g are the coordinate axes $x = 0$ and $y = 0$. We have already shown that every homogeneous integral of g has the form $c(xy)^s$, $s \in \mathbb{N}$. Let g' be another element of the group G. Since the function $(xy)^s$ is g'-invariant, the set $xy = 0$ is invariant under the map g'. Besides, since g' is a non-degenerate linear map, the origin $x = y = 0$ is a fixed point of g', and g' either preserves the eigendirections of g, or permutes them. In the first case g', obviously, commutes with g, while in the second case it has the form

$$x \to \alpha y\,, \quad y \to \beta x\,.$$

Since g' is symplectic, its matrix

$$T' = \begin{Vmatrix} 0 & \alpha \\ \beta & 0 \end{Vmatrix}$$

satisfies the condition

$$(T')^T J T' = J\,.$$

Hence, $\alpha\beta = -1$. On the other hand, in this case the eigenvalues of the matrix T' are $\pm i$. The points $\pm i$ form the exceptional regular polygon introduced in the statement of the theorem. This completes the proof.

Ziglin [245] proved that the conditions of the theorem are also sufficient for the existence of n independent integrals in a neighborhood of the complex curve Γ. It would be interesting to find sufficient conditions for the existence of n independent symmetry vector fields with holomorphic components.

5.8 Now consider the case when the elements of the matrix $A(t)$ are single-valued doubly periodic meromorphic functions of time $t \in \mathbb{C}$ having only one double pole inside the parallelogram of periods. We may regard $A(t)$ as a meromorphic function on the complex torus X obtained by factorization of the complex plane by the lattice of periods. Consider the symplectic maps g and g' over the periods of the matrix $A(t)$. We assume that their eigenvalues satisfy the conditions of Theorem 1. Therefore, if the system (5.16) has n independent analytic integrals, then g and g' commute. Hence, to a circuit around the singular point (i.e., to the element $gg'g^{-1}g'^{-1}$) there corresponds the identity map of $\mathbb{C}^{2n-2}$.

We apply this argument to the linear differential equation

$$\ddot{z} + (\omega^2 + \varepsilon f(t))z = 0\,, \tag{5.17}$$

where ω and ε are real constants. Suppose that f is an elliptic function with periods 2π and $2\pi i$ having a unique pole of order 2 in the rectangle of periods. We can assume that for real t the function f takes real values. An example of such a function is the Weierstrass $\wp$-function.

Equation (5.17) can be interpreted as the linearized equation of oscillations of a pendulum with a vibrating suspension point.

Let us find the eigenvalues of the map g from the monodromy group generated by a circuit around the pole of f. For simplicity, let the double pole be at $t = 0$. The Laurent series of the function $f(t)$ in a neighborhood of the point $t = 0$ has the form

$$\frac{\alpha}{t^2} + \sum_{n \ge 0} f_n t^n\,, \qquad \alpha \neq 0\,.$$

We are looking for linearly independent solutions of Eq. (5.17) in the form of a series

$$z(t) = t^\rho \sum_{n\geq 0} c_n t^n\,, \qquad \rho \in \mathbb{C}\,, \quad c_0 \neq 0\,.$$

Since

$$\ddot{z}(t) = t^\rho \sum_{n\geq 0} (\rho+n)(\rho+n-1)c_n t^{n-2}\,,$$

we get

$$\sum_{n\geq 0} (\rho+n)(\rho+n-1)c_n t^{n-2} + (\omega^2 + \varepsilon\alpha t^{-2} + \varepsilon \sum_{s\geq 0} f_s t^s) \sum_{n\geq 0} c_n t^n = 0\,.$$

Equating the coefficient of t^{-2} to zero, we obtain the equation

$$[\rho(\rho-1) + \varepsilon\alpha]c_0 = 0\,.$$

This equation gives us two values ρ_1 and ρ_2 to which there correspond two linearly independent solutions of the linear Eq. (5.17). After circling around the pole these solutions are multiplied by $e^{2\pi i\rho_1}$ and $e^{2\pi i\rho_2}$ respectively. The corresponding monodromy matrix is the unit matrix provided ρ_1 and ρ_2 are integers. In particular, $\varepsilon\alpha$ must be an integer.

For $\varepsilon = 0$, the eigenvalues of the monodromy matrices of (5.17) corresponding to the periods 2π and $2\pi i$ are, respectively, $\lambda_{1,2} = e^{\pm 2\pi\omega i}$ and $\mu_{1,2} = e^{\pm 2\pi\omega}$. Obviously, $|\mu_{1,2}| \neq 1$ for $\omega \neq 0$, and $\lambda_{1,2} \neq \pm i$ for $\omega \neq 1/4 + k\pi$, $k \in \mathbb{Z}$. By continuity, if $\omega \neq 1/4 + k\pi$, then for sufficiently small $\varepsilon \neq 0$ the eigenvalues $\mu_{1,2}$ are not roots of unity, and $\lambda_{1,2} \neq \pm i$ (apparently, this property holds for almost all ω and ε). Hence, by Theorem 1, in these cases Eq. (5.17) is not integrable in the complex domain. Note that in the real domain this equation is completely integrable: it has an analytic integral F which is 2π-periodic in t. Indeed, by a linear canonical transformation that is 2π-periodic in t, Eq. (5.17) can be reduced to a linear autonomous Hamiltonian system with one degree of freedom. For F we can take the Hamiltonian function of the autonomous system.

Now consider the nonlinear equation of oscillations of a mathematical pendulum

$$\ddot{z} + (\omega^2 + \varepsilon f(t)) \sin z = 0\,.$$

We claim that it can have an integral $F(\dot{z}, z, t)$, analytic and doubly periodic in $t \in \mathbb{C}$, only for those values of the parameters ω and ε for which the linear equation (5.17) is integrable. To prove this, assume that (5.17) is not integrable and expand F in a convergent power series

$$\sum_{s\geq m} \sum_{k+l=s} f_{kl} \dot{z}^k z^l\,, \tag{5.18}$$

where the coefficients f_{kl} are elliptic functions with periods 2π and $2\pi i$. The first homogeneous form in the expansion (5.18) (with $s = m$) is, obviously, a single-valued integral of the linear equation (5.17). By the assumption, this form is constant. Then the next form (with $s = m$) is an integral of (5.17) and therefore also constant, and so on.

5.9 Ziglin's theorem makes it possible to prove nonintegrability of many Hamiltonian systems which are important in applications. Yoshida applied this method to reversible Hamiltonian systems with homogeneous potentials. The Hamiltonian function is as follows:

$$H = |y|^2/2 + V_k(x)\,, \qquad x \in \mathbb{R}^n\,, \quad y \in \mathbb{R}^n\,, \tag{5.19}$$

where V_k is a homogeneous form of degree $k \neq 0, \pm 2$. The Hamilton equations with Hamiltonian (5.19) are quasi-homogeneous: the quasi-homogeneous degrees f, g in the variables y, x are equal to $k/(k-2)$, $2/(k-2)$ respectively (see §3.4). This makes it possible to calculate the Kovalevskaya exponents $\rho_1, \dots, \rho_{2n}$ of the particular solution

$$x(t) = \alpha t^{-g}, \quad y(t) = -g\alpha t^{-f}, \qquad \alpha \in \mathbb{R}^n\,. \tag{5.20}$$

The Kovalevskaya exponents break into pairs ρ_i, ρ_{i+n} with the sum $f+g = 2g+1$. Let $\Delta\rho_i = \rho_{i+n} - \rho_i$. We can assume that $\rho_n = -1$, $\rho_{2n} = f + g + 1$. Hence, $\Delta\rho_n = (3k-2)/(k-2)$ is a rational number.

Theorem 2 [239]. *If the numbers $\Delta\rho_1, \dots, \Delta\rho_n$ are rationally independent, then the Hamiltonian system with Hamiltonian* (5.19) *does not have a complete set of n independent holomorphic integrals in $\mathbb{C}^{2n} = \{x, y\}$.*

The procedure for calculating Kovalevskaya exponents of the solution (5.20) is as follows. First we solve the system of algebraic equations

$$\frac{\partial V}{\partial x_j}(c) = c_j\,, \qquad c = (c_1, \dots, c_n)^T, \quad 1 \leq i \leq n\,. \tag{5.21}$$

The vector c is expressed in terms of the vector α introduced in (5.20):

$$\alpha = -[-g(g+1)]^{g/2} c\,.$$

Denote

$$\Gamma = \left\| \frac{\partial^2 V}{\partial x^2}(c) \right\|.$$

Let $\lambda_1, \dots, \lambda_n$ be the eigenvalues of the matrix Γ. Then the numbers ρ_i and ρ_{i+n} are the roots of the quadratic equations

$$\rho^2 - (2g+1)\rho + g(g+1)(1-\lambda_i) = 0\,.$$

Thus,

$$\Delta\rho_i = [1 + 8k\lambda_i/(k-2)^2]^{1/2}\,.$$

The proof of Theorem 2 is based on Ziglin's theorem in §5.7. The Newton equation

$$\ddot{x} = -\partial V/\partial x$$

admits the particular solution

$$x = c\Phi(t)\,, \tag{5.22}$$

where the vector $c \in \mathbb{R}^n$ is determined by (5.21), and the scalar function Φ satisfies the second order equation

$$\ddot{\Phi} + \Phi^{k-1} = 0\,.$$

It is sufficient to consider solutions of this equation with the energy constant equal to $1/k$:

$$\dot{\Phi}^2 = (2/k)(1 - \Phi^k)\,. \tag{5.23}$$

Every solution (5.22) is single-valued on its Riemann surface

$$\zeta^2 = (2/k)(1 - w^k)\,. \tag{5.24}$$

The variational equations of the solution (5.22) take the form:

$$\ddot{\xi} = -\Phi(t)^{k-2}\Gamma\xi\,.$$

In the eigenbasis of the matrix Γ, we obtain n linear second-order differential equations:

$$\ddot{\xi}_i + \lambda_i\Phi^{k-2}\xi_i = 0\,, \qquad 1 \le i \le n\,. \tag{5.25}$$

One of the eigenvalues of Γ is $\lambda_n = k - 1$. The corresponding equation from system (5.25) has a particular solution $\xi_n = \dot{\Phi}$ which is single-valued on the Riemann surface (5.24). Hence, the first $n - 1$ Eqs. (5.25) form a reduced system of variational equations. The matrices of the reduced monodromy group have the form

$$T = \operatorname{diag}\left[T(\lambda_1), \ldots, T(\lambda_{n-1})\right]\,,$$

where $T(\lambda_i)$ are unimodular (2×2) matrices. Equations (5.25) are equivalent to the Gauss hypergeometric equation. This makes it possible to calculate the matrices $T(\lambda_i)$.

It turns out that there exist two closed curves on the Riemann surface (5.24) such that the corresponding monodromy matrices T_1 and T_2 are non-resonant and do not commute. By Ziglin's theorem, these properties yield nonintegrability of the Hamiltonian system with Hamiltonian (5.19). The connection between the non-resonance condition and the properties of the Kovalevskaya exponents follows from analysis of the hypergeometric equation. The details can be found in [239]. In fact [239] contains the proof of a stronger result: under the assumptions of Theorem 2 there are no additional holomorphic integrals independent of the energy integral.

We apply Theorem 2 to the system of Yang–Mills equations of a homogeneous two-component field (§8, Chap. I). The Hamiltonian function has the form (5.19), where $V = x_1^2 x_2^2$. Equations (5.21) admit a solution $c = (1/\sqrt{2}, 1/\sqrt{2})^T$, and the eigenvalues of the Hessian are equal to -1 and 3. Hence, $\Delta\rho_1 = \sqrt{-7}$ and $\Delta\rho_2 = 5$. Since these numbers are rationally independent, by Theorem 2 the Yang–Mills equations have no new holomorphic integrals. This result was first obtained by Ziglin in [245].

A similar result holds for the three-component Yang–Mills model, where $V = x_1^2x_2^2 + x_2^2x_3^2 + x_3^2x_1^2$. In this case $c = (1/\sqrt{2}, 1/\sqrt{2}, 0)^T$, and the numbers $\Delta\rho_i$ are equal to $\sqrt{17}$, $\sqrt{7}$, and 5. Since they are rationally independent, the Hamiltonian system is nonintegrable.

Using Theorem 2, Ziglin proved nonintegrability of the Hénon–Heiles system (see §8 of Chap. II). The Hamiltonian function is as follows

$$H = (y_1^2 + y_2^2 + x_1^2 + x_2^2)/2 + x_1^3/3 - x_1 x_2^2 .$$

The corresponding Hamiltonian equations have a one-parameter family of elliptic solutions

$$\begin{aligned} &x_2 = y_2 = 0 , \quad x_1(t,k) = a_1 - (a_1 - a_2)\,\mathrm{sn}^2(\tau, k) , \\ &y_1(t,k) = -(a_1 - a_2)\sqrt{2(a_1 - a_3)/3}\,\mathrm{sn}(\tau,k)\,\mathrm{cn}(\tau,k)\,\mathrm{dn}(\tau,k) , \end{aligned} \tag{5.26}$$

where

$$\tau = t\sqrt{(a_1 - a_3)/6} , \quad k = k(H) = \sqrt{(a_1 - a_2)/(a_1 - a_3)} ,$$

and a_1, a_2, a_3 are the roots of the equation $z^3 + 3z^2/2 - 3H = 0$.

Therefore, the Riemann surface of the solution (5.26) is a two-dimensional torus with a point removed. Apparently, for small energy H the monodromy matrices corresponding to the basis cycles of this torus satisfy the conditions of Theorem 1.

In [238] Yoshida proposed a general approach to investigating the integrability of reversible systems with non-homogeneous potentials. It is based on studying the monodromy group as $h \to 0$ or $h \to \infty$, where h is the constant of the energy integral. As an application, he considered the existence of an additional integral for the system with Hamiltonian

$$\begin{aligned} &H_N = (y_1^2 + y_2^2)/2 + V_N(x_1, x_2) , \\ &V_N = \sum_{k=1}^{N} \left[(\sqrt{3}x_1 + x_2)^k + (-\sqrt{3}x_1 + x_2)^k + (-2x_2)^k\right]/k! . \end{aligned} \tag{5.27}$$

The function V_N is a truncated Maclaurin series of the Toda lattice potential:

$$\exp(\sqrt{3}x_1 + x_2) + \exp(-\sqrt{3}x_1 + x_2) + \exp(-2x_2) .$$

Thus, the system with Hamiltonian (5.27) can be called the "truncated" Toda lattice. For $N = 2$, we obtain the harmonic oscillator, and for $N = 3$ the Hénon–Heiles system. In [238] the nonexistence of new holomorphic integrals of the truncated Toda lattice was established for $N \geq 3$.

Another example of a reversible system with non-homogeneous potential is given by rigid body rotation with a non-holonomic constraint: the projection of the angular velocity on a certain axis l fixed in the body vanishes (see §8 of Chap. II). If the center of mass lies on the axis l, then rotation of the body with total energy h is governed by the Hamiltonian equations with Hamiltonian

$$H = \frac{1}{2}(p_1^2 + p_2^2) = \frac{1}{2}\left[h - \frac{1}{2}\left(\frac{q_1^2}{I_1} + \frac{q_2^2}{I_2}\right)\right]^2 .$$

(see Eqs. (7.7) of Chap. I). Here I_1, I_2 are the moments of inertia of the body. For $I_1 = I_2$, we obtain the integrable Lagrange case (with zero projection of the angular velocity on l).

VIII Polynomial Integrals of Hamiltonian Systems

In this chapter we present specific methods of searching for Hamiltonian systems which admit first integrals polynomial in momenta. This problem is actual because in Hamiltonian mechanics the majority of known integrals are either polynomials in momenta or functions of these polynomials (see §1 of Chap. II). The problem of the existence of linear and quadratic integrals is quite elementary. Usually, it can be easily solved. For integrals whose degree is not fixed in advance, however it is much more difficult. At present, it is solved only for some classes of Hamiltonian systems.

1 The Birkhoff Method

1.1 Consider a non-reversible system with two degrees of freedom. In local isothermic coordinates x, y the Lagrange function takes the form

$$L = \mu(\dot{x}^2 + \dot{y}^2)/2 + \alpha\dot{x} + \beta\dot{y} - V \, . \tag{1.1}$$

Here $\mu \geq 0$, α, β, V are functions of x, y. The Lagrange equations with Lagrangian (1.1) have the energy integral

$$\mu(\dot{x}^2 + \dot{y}^2) + 2V = 2h \, . \tag{1.2}$$

Following Birkhoff ([23, Chap. 2]), we perform the change of time $t \to \tau$ by the formula

$$d\tau = \mu^{-1} dt \, . \tag{1.3}$$

Denote by a prime differentiation with respect to τ and write down the Hamiltonian action:

$$\int_{t_1}^{t_2} L\, dt = \int_{\tau_1}^{\tau_2} \left[\frac{1}{2}(x'^2 + y'^2) + \alpha x' + \beta y' - \gamma\right] d\tau \, ,$$

where $\gamma = \mu V$. The functions $x(\tau), y(\tau)$ $(\tau_1 \leq \tau \leq \tau_2)$ are extremals of the following conditional variational problem:

$$\delta \int_{\tau_1}^{\tau_2} L^*\, d\tau = 0, \qquad \int_{\tau_1}^{\tau_2} \mu\, d\tau = \text{const} \, ,$$

where

$$L^* = (x'^2 + y'^2)/2 + \alpha x' + \beta y' - \gamma \, . \tag{1.4}$$

According to well-known results of variational calculus (see, for example, [24]), for each curve $x(\tau)$, $y(\tau)$ there exists a constant κ such that this curve is an extremal of the "unconditional" variational problem with the Lagrangian $L^* + \kappa\mu$. Taking into account relation (1.4), we write down the corresponding Lagrange equations:

$$x'' + \lambda y' = -\frac{\partial \gamma}{\partial x} + \kappa\frac{\partial \mu}{\partial x}\,, \quad y'' - \lambda x' = -\frac{\partial \gamma}{\partial y} + \kappa\frac{\partial \mu}{\partial y}\,,$$
$$\lambda = \frac{\partial \alpha}{\partial y} - \frac{\partial \beta}{\partial x}\,. \tag{1.5}$$

To understand a mechanical meaning of the multiplier κ, we write down the energy integral using the change of time (1.3):

$$(x'^2 + y'^2)/2 + \gamma - h\mu = 0\,. \tag{1.6}$$

On the other hand, equations (1.5) admit the integral

$$(x'^2 + y'^2)/2 + \gamma - \kappa\mu = \text{const}\,. \tag{1.7}$$

Relations (1.6) and (1.7) imply the equality $(h - \kappa)\mu = \text{const}$. Consequently, in the general case, when $\mu \neq \text{const}$, the Lagrange multiplier κ coincides with the total energy h.

1.2 Suppose that we are interested only in motions $x(t)$, $y(t)$ with a fixed value of energy h. For the sake of simplicity we assume $h = 0$ (otherwise we replace V by $V - h$). Trajectories of such motions parameterized by the new time τ can bc defined by the Birkhoff system

$$x'' + \lambda y' = -\partial\gamma/\partial x\,, \quad y'' - \lambda x' = -\partial\gamma/\partial x\,, \tag{1.8}$$

complemented by the energy relation

$$x'^2 + y'^2 + 2\gamma = 0\,. \tag{1.9}$$

This result was obtained by Birkhoff in [23]. He used relations (1.8) and (1.9) in order to solve the problem of the existence of conditional integrals (see §§7–9 of Chap. II) in the form of polynomials in the velocities of a degree which does not exceed two. It turns out that the existence of a conditionally linear integral is related to hidden cyclic coordinates, and the existence of a conditionally quadratic integral makes it possible to separate canonical variables. Below we present global versions of these assertions when the configuration space of the system is a two-dimensional torus. Note that in this case the isothermic coordinates can be introduced globally.

1.3 Consider a non-reversible system with the configuration space a torus. We concretize the classical Birkhoff result ([23], Chap. II) and give a criterion for the existence of a "multivalued" linear integral. We define a multivalued integral as a closed 1-form in the phase space, whose derivative along the vector field vanishes. There are two reasons for considering multivalued integrals:

1) integrals, which are polynomial in the velocity, with multi-valued coefficients, can exist in the simplest non-reversible systems,

2) the Liouville theorem on completely integrable systems can be easily extended to the case when closed 1-forms are considered instead of usual integrals.

Proposition 1. *Suppose that the system has a conditional linear integral (possibly multivalued) on the energy surface $H = h$, where $h > \max V$. Then on the configuration space one can choose angular coordinates $x_1, x_2 \bmod 2\pi$ and make a change of time $dt = \xi(x_1, x_2)\, d\tau$ in such a way that the trajectories corresponding to the total energy h are described by a Hamiltonian system with the following properties:*
1) *the kinetic energy is a quadratic form in x_1', x_2' with constant coefficients,*
2) *the 2-form of gyroscopic forces is $f = \lambda(x_1)\, dx_1 \wedge dx_2$,*
3) *the potential does not depend on the variable x_2.*

In the new variables x_1, x_2, τ the Lagrangian takes the form

$$L = \frac{1}{2} \sum a_{ij} x_i' x_j' + \mu(x_1) x_2' - \gamma(x_1)\,, \qquad \mu = \int \lambda(x)\, dx\,.$$

This function is single-valued only if

$$\int_{\mathbb{T}^2} f = 0\,.$$

The variable x_2 is cyclic: the Lagrangian does not depend on it. We have cyclic integral, linear in the velocities

$$\partial L / \partial x_2' = \sum a_{2j} x_i' + \mu(x_1)\,.$$

Proposition 1 does not assert that this integral coincides with the initial linear first integral. Here is a simple counterexample: the reversible system with kinetic energy $T = (\dot{x}_1^2 + \dot{x}_2^2)/2$ and zero potential has a linear integral $\dot{x}_1 + \sqrt{2}\dot{x}_2$ which cannot be made cyclic by any choice of angular coordinates.

To prove the proposition we use Birkhoff's equations (1.8). Let

$$lx' + my' + n \tag{1.10}$$

be a conditional linear integral. The differentials dl, dm and dn are single-valued on $\mathbb{T}^2 = \{(x, y) \bmod 2\pi\}$. We calculate the derivative of the integral (1.10) along solutions of the system (1.8):

$$\begin{aligned} &\frac{\partial l}{\partial x} x'^2 + \left(\frac{\partial l}{\partial y} + \frac{\partial m}{\partial x}\right) x' y' + \frac{\partial m}{\partial y} y'^2 \\ &\quad + \left(m\lambda + \frac{\partial n}{\partial x}\right) x' + \left(-l\lambda + \frac{\partial n}{\partial y}\right) y' - l \frac{\partial \gamma}{\partial x} - m \frac{\partial \gamma}{\partial y}\,. \end{aligned} \tag{1.11}$$

Since the function (1.9) is a first integral of the system (1.7) on the energy surface

$$x'^2 + y'^2 + 2\gamma = 0\,,$$

the major form of the polynomial (1.10) must be divisible by the Hamiltonian function. We get

$$\frac{\partial l}{\partial x} - \frac{\partial m}{\partial y} = 0, \qquad \frac{\partial l}{\partial y} + \frac{\partial m}{\partial x} = 0.$$

Hence, the forms dl and dm are harmonic on $\mathbb{T}^2$ and thus,

$$m = ax + by + m_0, \quad l = bx - ay + l_0; \qquad a, b, m_0, l_0 = \text{const}.$$

Now we put the coefficients of x' and y' in relation (1.11) equal to zero and obtain:

$$\frac{\partial n}{\partial x} = -m\lambda, \quad \frac{\partial n}{\partial y} = l\lambda. \tag{1.12}$$

Since the form dn is single-valued, we have $a = b = 0$ provided $\lambda \not\equiv 0$. Hence $m = m_0$, $l = l_0$. This is the case considered above. We equate (1.11) to zero and obtain one more relation for the potential:

$$l_0 \frac{\partial \gamma}{\partial x} + m_0 \frac{\partial \gamma}{\partial y} = 0. \tag{1.13}$$

Using (1.12) we find an analogous relation for the function λ:

$$l_0 \frac{\partial \lambda}{\partial x} + m_0 \frac{\partial \lambda}{\partial y} = 0. \tag{1.14}$$

If the numbers l_0 and m_0 are rationally incommensurable, then (1.13) and (1.14) imply that the functions γ and λ are constant, and the proposition is proved. Let $m_0/l_0 = p/q$, where the integer numbers p, q are relatively prime. With the help of (1.13) and (1.14) one can easily obtain that

$$\gamma = \phi(px - qy), \quad \lambda = \psi(px - qy),$$

where the functions ϕ and ψ are 2π-periodic. We perform the linear transformation

$$x_1 = px - qy, \quad x_2 = ux + vy$$

with integer numbers u, v which satisfy the relation $pv + qu = 1$. Such numbers exist, since p and q are relatively prime. The variables $x_1, x_2 \bmod 2\pi$ are the required ones.

1.4 Now consider a reversible system with two degrees of freedom, whose configuration space is again a two-dimensional torus.

Proposition 2. *Suppose that the system has a conditional quadratic integral with single-valued coefficients on the surface $H = h$, where $h > \max V$. Then on the configuration space one can choose angular coordinates $x_1, x_2 \bmod 2\pi$ and perform the change of time $dt = \xi(x_1, x_2)\, d\tau$ in such a way that the trajectories corresponding to the energy h will be described by the Lagrange system with Lagrangian*

$$(x_1'^2 + x_2'^2)/2 + \eta(px_1 + qx_2) + \zeta(qx_1 - px_2), \tag{1.15}$$

where $\eta(\cdot)$ and $\zeta(\cdot)$ are 2π-periodic functions, p and q are integer numbers.

Make the linear change of variables

$$x = px_1 + qx_2\,, \quad y = qx_1 - px_2\,.$$

Then Lagrangian (1.15) takes the form

$$\kappa(x'^2 + y'^2)/2 + \eta(x) + \zeta(y)\,, \qquad \kappa^{-1} = p^2 + q^2\,.$$

The variables x and y separate: the functions

$$\kappa x'^2/2 + \eta(x)\,, \quad \kappa y'^2/2 + \zeta(y)$$

are independent quadratic integrals.

Proposition 2 is a global version of the well-known Birkhoff result on conditionally quadratic integrals ([23, Chap. 2]).

Our proof uses again equations (1.8) where we put $\lambda \equiv 0$. Suppose that the system (1.8) has the quadratic integral

$$(ax'^2 + 2bx'y' + cy'^2)/2 + dx' + ey' + f \tag{1.16}$$

on the surface

$$x'^2 + y'^2 + 2\gamma = 0\,. \tag{1.17}$$

Differentiate the integral (1.16) with respect to the new time τ and write down terms of degree three in the velocities:

$$\frac{1}{2}\frac{\partial a}{\partial x}x'^3 + \left(\frac{\partial b}{\partial x} + \frac{1}{2}\frac{\partial a}{\partial y}\right)x'^2y' + \left(\frac{\partial b}{\partial y} + \frac{1}{2}\frac{\partial c}{\partial x}\right)x'y'^2 + \frac{1}{2}\frac{\partial c}{\partial y}y'^3\,.$$

Since (1.16) is a conditional integral, this polynomial must be divisible by $x'^2+y'^2$. Consequently,

$$\frac{\partial}{\partial x}(a-c) - \frac{2\partial b}{\partial y} = 0\,, \quad \frac{\partial}{\partial y}(a-c) + \frac{2\partial b}{\partial x} = 0\,,$$

and thus, the functions $a - c$ and b are harmonic on the two-dimensional torus. Since, by the assumption, these functions are single-valued, $a - c = \text{const}$ and $b = \text{const}$.

Using this fact and the energy integral (1.17), we can transform the quadratic integral (1.16) to a form with constant coefficients a, b, c. Differentiate the integral (1.16) along the solutions of the system and put the coefficients of x' and y' equal to zero. As a result we obtain the relations

$$a\frac{\partial \gamma}{\partial x} + b\frac{\partial \gamma}{\partial y} = \frac{\partial f}{\partial x}\,, \quad b\frac{\partial \gamma}{\partial x} + c\frac{\partial \gamma}{\partial y} = \frac{\partial f}{\partial y}\,.$$

Consequently,

$$(a-c)\frac{\partial^2 \gamma}{\partial x \partial y} + b\left(\frac{\partial^2 \gamma}{\partial y^2} - \frac{\partial^2 \gamma}{\partial x^2}\right) = 0\,. \tag{1.18}$$

Now we expand the potential γ in the Fourier series

$$\sum v_{mn} \exp[i(mx + ny)]$$

and obtain from (1.18) the sequence of equalities

$$[(a - c)mn + b(n^2 - m^2)]v_{mn} = 0 .$$

Suppose that $v_{m_1 n_1} \neq 0$ and $v_{m_2 n_2} \neq 0$. Then

$$(a - c)m_1 n_1 + b(n_1^2 - m_1^2) = (a - c)m_2 n_2 + b(n_2^2 - m_2^2) = 0 . \tag{1.19}$$

It is clear that among the numbers $a - c$ and b there is at least one distinct from zero (otherwise integral (1.16) is reduced to a linear one). From (1.19) we obtain

$$\frac{m_1 n_1}{n_1^2 - m_1^2} = \frac{m_2 n_2}{n_2^2 - m_2^2} .$$

Thus, either $m_1/n_1 = m_2/n_2$ or $(m_1/n_1)(m_2/n_2) = -1$. Consequently, the integer vectors (m_1, n_1) and (m_2, n_2) are either parallel or orthogonal. The proposition is proved.

1.5 The Birkhoff equations (1.8) contain two arbitrary functions λ and γ. The initial equations of motion can be simplified in a different way by reducing the 2-form of gyroscopic forces (instcad of the kinetic energy) to the simplest form.

Let $f = F\,dx \wedge dy$ be a 2-form on a two-dimensional torus, where $f \neq 0$. We put

$$F_0 = \frac{1}{4\pi^2} \int_{\mathbb{T}^2} f \neq 0 .$$

It turns out that there exists a diffeomorphism of the torus $(x, y) \to (u, v)$ such that in the new variables

$$f = F_0\, du \wedge dv . \tag{1.20}$$

Indeed, let α and β be "strongly incommensurable" real numbers and $g : \mathbb{T}^2 \to \mathbb{R}$ a solution of the differential equation

$$\alpha \frac{\partial g}{\partial x} + \beta \frac{\partial g}{\partial y} = F - F_0 .$$

We put

$$u = x + \frac{\alpha}{F_0} g , \qquad v = y + \frac{\beta}{F_0} g .$$

Since

$$\frac{\partial(u, v)}{\partial(x, y)} = \frac{F}{F_0} \neq 0 ,$$

we obtain

$$f = F\frac{\partial(x, y)}{\partial(u, v)}\, du \wedge dv = F_0\, du \wedge dv\,, \quad \text{Q.E.D.}$$

This result is a particular case of the Moser theorem which asserts that any two volume forms on a compact manifold M are reduced to each other if their integrals over M coincide [180].

If the form of gyroscopic forces is reduced to (1.20), then the transformed Birkhoff equations (1.8) again contain two "arbitrary" functions: g and λ.

2 Influence of Gyroscopic Forces on the Existence of Polynomial Integrals

Consider equations on the n-dimensional torus $\mathbb{T}^n = \{x \bmod 2\pi\}$ with gyroscopic forces

$$\ddot{x} = \Lambda(x)\dot{x} - \frac{\partial V}{\partial x}\,. \tag{2.1}$$

Here Λ is a skew-symmetric matrix, 2π-periodic in x and V is the potential of the force field. Birkhoff's equations (1.8) have evidently the form (2.1). We study the existence problem of integrals polynomial in the velocities with single-valued coefficients. We assume the integrals to be independent of the energy integral

$$H = (\dot{x}, \dot{x})/2 + V(x)\,. \tag{2.2}$$

We put

$$\Lambda_0 = \frac{1}{(2\pi)^n}\int_{\mathbb{T}^n} \Lambda(x)\, d^n x\,.$$

Like any skew-symmetric matrix, the matrix Λ_0 can be reduced to the form $\mathrm{diag}[a_1, \ldots, a_m, 0, \ldots, 0]$, where

$$a_j = \begin{Vmatrix} 0 & \lambda_j \\ -\lambda_j & 0 \end{Vmatrix}\,, \quad \lambda_j \in \mathbb{R}\,.$$

In particular, the spectrum of Λ_0 is as follows: $\pm i\lambda_1, \ldots, \pm i\lambda_m, 0, \ldots, 0$. If $\det \Lambda_0 \neq 0$, then $n = 2m$.

Introduce the additional group, formed by the numbers $\alpha_1\lambda_1 + \cdots + \alpha_m\lambda_m$, $\alpha_j \in \mathbb{Z}$. Denote its rank by k. We can say that among the numbers $\lambda_1, \ldots, \lambda_m$ there are k numbers independent over the field $\mathbb{Q}$.

Theorem 1 [147]. *Suppose that the system* (2.1) *has l independent polynomial integrals. Then $l + k \leq n$.*

In order to prove Theorem 1, we use the Poincaré method and the result of the paper [247]. Introduce the new time $t \to t/\varepsilon$, where ε is a parameter. This change transforms equations (2.1) into the equations

$$\ddot{x} = \varepsilon \Lambda \dot{x} - \varepsilon^2 \partial V / \partial x \tag{2.3}$$

which contain the small parameter ε. Since the velocities $\dot{x}$ pass to $\dot{x}/\varepsilon$, a polynomial integral of Eqs. (2.1) converts into an integral, analytic in ε, of the system (2.3):

$$F_0(\dot{x}, x) + \varepsilon F_1(\dot{x}, x) + \cdots .$$

The functions F_s are obviously 2π-periodic in x. The function F_0 is a first integral of the unperturbed system $\ddot{x} = 0$. Consequently,

$$\dot{F}_0 = \left(\frac{\partial F_0}{\partial \dot{x}}, \ddot{x}\right) + \left(\frac{\partial F_0}{\partial x}, \dot{x}\right) = \left(\frac{\partial F_0}{\partial x}, \dot{x}\right) \equiv 0 .$$

This implies that F_0 does not depend on the angular variables x. The function F_1 satisfies the equation

$$\left(\frac{\partial F_0}{\partial \dot{x}}, \Lambda \dot{x}\right) + \left(\frac{\partial F_1}{\partial x}, \dot{x}\right) = 0 .$$

Averaging this equation with respect to $x_1, \ldots, x_n$, we obtain

$$\left(\frac{\partial F_0}{\partial \dot{x}}, \Lambda_0 \dot{x}\right) = 0 .$$

Hence, the function $u \to F_0(u)$ is a first integral of the linear system with constant coefficients $\dot{u} = \Lambda_0 u$.

By a linear change this system can be transformed to the following form:

$$\dot{u}_1 = \lambda_1 u_2 , \quad \dot{u}_2 = -\lambda_1 u_1 , \ \ldots ,$$
$$\dot{u}_{2m-1} = \lambda_m u_{2m} , \quad \dot{u}_{2m} = -\lambda_m u_{2m-1} , \qquad \dot{u}_{2m+1} = \cdots = \dot{u}_n = 0 .$$

We put $u_1 = \rho_1 \sin \phi_1$, $u_2 = \rho_1 \cos \phi_1$, $\ldots$, $u_{2m-1} = \rho_m \sin \phi_m$, $u_{2m} = \rho_m \cos \phi_m$, and get:

$$\begin{aligned} &\dot{\phi}_1 = \lambda_1 , \ \ldots , \quad \dot{\phi}_m = \lambda_m , \\ &\dot{\rho}_1 = \cdots = \dot{\rho}_m = \dot{u}_{2m+1} = \cdots = \dot{u}_n = 0 . \end{aligned} \tag{2.4}$$

The function F_0 is an integral of these equations. It is 2π-periodic in $\phi_1, \ldots, \phi_m$. If the numbers $\lambda_1, \ldots, \lambda_m$ are rationally incommensurable, then F_0 is independent of $\phi_1, \ldots, \phi_m$. In this situation the system (4.4) may have at most $n - m$ independent integrals. In the general case the numbers $\lambda_1, \ldots, \lambda_m$ satisfy $m - k$ independent integer linear relations. Consequently, the system (2.4) has not more than $n - k$ independent integrals. Q.E.D.

For $n = 2$ Theorem 1 can be made more precise:

Theorem 2. *Let $n = 2$ and $\Lambda_0 \neq 0$. Then the system (2.1) has no polynomial integral independent of function (2.2) with coefficients, which are single-valued on $\mathbb{T}^2$.*

Indeed, suppose that the system (2.3) has an integral which is independent of the energy integral $(\dot{x}, \dot{x})/2 + \varepsilon^2 V$. Then (according to §1, Chap. IV) for $n = 2$ the system (2.3) has an integral in the form of a series in ε which is independent of the function $(\dot{x}, \dot{x})/2$ for $\varepsilon = 0$. Theorem 2 is a consequence of Theorem 3 of §4 in Chap. III.

3 Polynomial Integrals of Systems with One and a Half Degrees of Freedom

3.1 In this section we discuss integrability of the following second order non-autonomous equation

$$\ddot{x} = -\partial V/\partial x \tag{3.1}$$

with a potential $V(x, t)$ which is 2π-periodic in x. Equation (3.1) describes, in particular, oscillations of pendulum systems. An example is the Chaplygin equation

$$\ddot{x} = kt^2 \sin x\,, \qquad k = \text{const} \neq 0 \tag{3.2}$$

which describes the rotation of a heavy plate in an unbounded volume of an ideal fluid.

Following [146], we consider the existence problem of an integral of Eq. (3.1) in the form of a polynomial in the velocity

$$a_0(x, t) + a_1(x, t)\dot{x} + \cdots + a_n(x, t)\dot{x}^n \tag{3.3}$$

with coefficients a_k $(0 \le k \le n)$ which are 2π-periodic in x. The integral (3.3) is assumed to be a single-valued function on the extended phase space $(\dot{x}, x, t) \in \mathbb{R} \times \mathbb{T}^1 \times \mathbb{R}$.

Differentiating the function (3.3) along solutions of the system (3.1), we obtain the following system of first order partial differential equations for the potential energy V of the integrable system and the coefficients a_k:

$$(a_n)_x = 0\,, \tag{3.4.n+1}$$

$$(a_n)_t + (a_{n-1})_x = 0\,, \tag{3.4.n}$$

$$(a_{n-1})_t + (a_{n-2})_x = na_nV_x\,, \tag{3.4.n-1}$$

$$(a_{n-2})_t + (a_{n-3})_x = (n-1)a_{n-1}V_x\,, \tag{3.4.n-2}$$

$$\dots \quad \dots \quad \dots \quad \dots$$

$$(a_1)_t + (a_0)_x = 2a_2V_x\,, \tag{3.4.1}$$

$$(a_0)_t = a_1V_x\,. \tag{3.4.0}$$

This system consists of $n + 2$ equations and contains the same number of unknown functions $V, a_0, \dots, a_n$.

Note that equation (3.1) preserves its form after the change

$$x \to x + \alpha t\,, \quad V(x, t) \to V(x + \alpha t, t) + f(t)\,,$$

where $\alpha = \text{const}$, and f is an arbitrary function of time. We shall use this trivial gauge transformation below.

Lemma 1. $a_n = a_n^0 = \text{const} \neq 0$, $a_{n-1} = a_{n-1}^0 = \text{const}$, $a_{n-2} = na_n^0 V$ *(after a proper gauge transformation of* V*).*

Proof. Equation (3.4.$n+1$) implies that the function a_n does not depend on x. Consequently, $a_n = \langle a_n \rangle$, where

$$\langle g \rangle = \frac{1}{2\pi} \int_0^{2\pi} g(x,t)\,dx\,.$$

Taking the average of both parts of Eq. (3.4.n) with respect to x, we obtain $(\langle a_n \rangle)_t = 0$. Consequently, $a_n = \text{const}$. Thus, (3.4.n) implies that a_{n-1} does not depend on x, and hence $a_{n-1} = \langle a_{n-1} \rangle$. We take the average of (3.4.$n-1$) in x and obtain the equality $a_{n-1} = \langle a_{n-1} \rangle = \text{const}$. Consequently, (3.4.$n-1$) takes the form $(a_{n-2})_x = na_n^0 V_x$, and the difference $a_{n-2} - na_n^0 V_x$ depends only on time. Q.E.D.

We divide the integral (3.3) by the non-zero constant a_n^0 and arrive at the equality $a_n \equiv 1$. Performing the substitution $x \to x - (a_{n-1}^0/n)t$, we obtain an equation of the form (3.1) with the integral (3.3) in which $a_{n-1} \equiv 0$. Thus, we can assume that $a_n = 1$ and $a_{n-1} = 0$.

Using Lemma 1, it is easy to obtain conditions for the existence of linear and quadratic integrals. For $n = 1$ with the help of the relation $a_{n-2} = nV$ we find that $V \equiv 0$. Consequently, any linear integral can be reduced to the angular momentum integral $\dot{x}$. Using equation (3.4.0), for $n = 2$ we obtain the relation $V_t = 0$. Thus, $V = f(x)$, where $f(\cdot)$ is an arbitrary smooth 2π-periodic function. In this case the integral (3.3) converts to the usual energy integral of an autonomous system with one degree of freedom.

For $n = 3$ Eqs. (3.4.0) and (3.4.1) generate the system

$$(a_0)_x = -3V_t\,, \quad (a_0)_t = -3VV_x$$

which allows us to find the second order equation for the function V:

$$V_{tt} + (VV_x)_x = 0\,. \tag{3.5}$$

It coincides with the integrable stationary Khokhlov-Zabolotski equation. A method for its precise integration is presented e.g. in [115] (where one can find references to analogous results). We are interested only in solutions of (3.5) which are 2π-periodic in the variable x. Applying the well-known Cauchy–Kovalevskaya theorem, we obtain that for arbitrary analytic 2π-periodic functions f and g there exists an analytic solution $V(x,t)$, 2π-periodic in x, of (3.5) with the initial conditions $V(x,0) = f(x)$, $V_t(x,0) = g(x)$. We note that the Cauchy–Kovalevskaya theorem actually says nothing about the periodicity of V in the variable x. However, it is easy to prove that if the initial data are periodic in x, then the corresponding solution of (3.5) has the same property. Thus, there exists a family of potentials such that the corresponding Eq. (3.1) has a nontrivial integral of degree three in the velocity. This family depends on two arbitrary functions.

3.2 These observations can be generalized to the case of polynomial integrals of arbitrary degree. It turns out that for each $n \geq 1$ there exists a family of analytic 2π-periodic in x potentials $V(x, t)$ such that the corresponding Eq. (3.1) has a polynomial integral of degree n with single-valued analytic coefficients. This family depends on $n-1$ arbitrary analytic 2π-periodic functions. The proof is based on applications of the Cauchy–Kovalevskaya theorem. However, this theorem cannot be applied directly to the system (3.4). Transform (3.4) with the help of Lemma 1 and the agreement that $a_n = 1$:

$$nV_t = -(a_{n-3})_x + (n-1)a^0_{n-1}V_x\,, \tag{3.6.$n-2$}$$
$$(a_{n-3})_t = -(a_{n-4})_x + (n-2)nVV_x\,, \tag{3.6.$n-3$}$$
$$(a_{n-4})_t = -(a_{n-5})_x + (n-3)a_{n-3}V_x\,, \tag{3.6.$n-4$}$$
$$\dots \quad \dots \quad \dots \quad \dots \quad \dots$$
$$(a_1)_t = -(a_0)_x + 2a_2V_x\,, \tag{3.6.1}$$
$$(a_0)_t = a_1V_x\,. \tag{3.6.0}$$

This system consists of $n-1$ equations and contains $n-1$ unknown functions $a_0, \dots, a_{n-3}$, and V. The Cauchy–Kovalevskaya theorem is applicable to the system (3.6): it is necessary to define for $t = 0$ arbitrary analytic 2π-periodic in x functions $a_0, \dots, a_{n-3}, V$. Then for small t system (3.6) has analytic 2π-periodic in x solutions.

From the results of [114–115] it can be derived that the nonlinear evolutionary system of partial differential equations (3.6) is integrable. However, one cannot write down its solutions in an explicit form. Thus, it is unclear if solutions of system (3.6) can be continued to the whole time axis.

3.3 Solutions of system (3.6) can be sought in the class of potentials which are trigonometric polynomials in the variable x.

Now we do not assume that $a^0_{n-1} = 0$. Hence, one must add to the right-hand side of (3.6.$n-2$) the term $(n-1)a^0_{n-1}V_x$.

Theorem 1. *Suppose that system* (3.1) *with the potential*

$$V = \sum_{k=-m}^{m} v_k(t)\exp(ikx) \tag{3.7}$$

has a polynomial integral of degree $n \geq 1$. *The following assertions hold:*
1) *if* n *is odd, then* V *does not depend on* x,
2) *if* n *is even, then* $v_m = c\exp(i\beta t)$, *where* $c \in \mathbb{C}$ *and* $\beta = a^0_{n-1}/n$.

Corollary 1. *Suppose that* $|v_m| \neq$ const. *Then equation* (3.1) *has no non-trivial polynomial integral with single-valued* 2π*-periodic in* x *coefficients.*

Corollary 2. *Chaplygin's equation* (3.2) *has no single-valued polynomial integral.*

In [53] Chaplygin wrote about the importance of studying equation (3.2), "... I have not however performed its integration".

Corollary 3. *Equation* (3.1) *with the potential* $V = \lambda(t)\sin mx + \mu(t)\cos mx$ ($m \in \mathbb{N}$) *has a nontrivial single-valued polynomial integral if and only if it has a polynomial integral of degree* ≤ 2.

Indeed, if Eq. (3.1) admits an integral of odd degree, then there is a linear integral (assertion 1) of Theorem 2). If the degree of the polynomial integral is even, then from assertion 2) one can easily conclude that V is a 2π-periodic function of the variable $x + \alpha t$, $\alpha = \text{const}$. In this case equation (3.1) admits a generalized energy integral of degree 2 (see §3.1).

It is unclear if Corollary 3 holds for polynomial potentials (3.7) of general form (cf. §5 of Chap. IV).

3.4 Now turn to the proof of Theorem 1. If V is a trigonometric polynomial in the variable x, then using relations (3.4), it is easy to show by induction that the functions $a_{n-2}, \dots, a_1, a_0$ are also trigonometric polynomials, and

$$\begin{aligned}
a_{n-2} &= \overline{A}_{n-2}\exp(-imx) + \cdots + A_{n-2}\exp(imx)\,,\\
a_{n-3} &= \overline{A}_{n-3}\exp(-imx) + \cdots + A_{n-3}\exp(imx)\,,\\
a_{n-4} &= \overline{A}_{n-4}\exp(-i2mx) + \cdots + A_{n-4}\exp(i2mx)\,,\\
a_{n-5} &= \overline{A}_{n-5}\exp(-i2mx) + \cdots + A_{n-5}\exp(i2mx)\,,\\
&\dots \quad \dots \quad \dots \quad \dots \quad \dots \quad \dots
\end{aligned}$$

Expanding the left- and the right-hand sides of relations (3.4) in Fourier series and equating coefficients at identical harmonics, we obtain the following chain of equations for $A_{n-2}, A_{n-3}, \dots$:

$$A_{n-2} = nv_m\,, \tag{3.8.$n-2$}$$

$$\dot{A}_{n-2} + imA_{n-3} = (n-1)ima^0_{n-1}v_m\,, \tag{3.8.$n-3$}$$

$$2A_{n-4} = (n-2)A_{n-2}v_m\,, \tag{3.8.$n-4$}$$

$$\dot{A}_{n-4} + i2mA_{n-5} = (n-3)imA_{n-3}v_m\,, \tag{3.8.$n-5$}$$

$$\dots \quad \dots \quad \dots \quad \dots \quad \dots$$

Here the dots denote differentiation with respect to t. When deriving Eqs. (3.8), we supposed that $m \neq 0$.

First, consider the case when n is odd. Then the subsystem of Eqs. (3.8.$n-2$), (3.8.$n-4$), ..., (3.8.1) is closed:

$$\begin{aligned} A_{n-2} &= n v_m\,, \\ 2A_{n-4} &= (n-2)A_{n-2}v_m\,, \\ 3A_{n-6} &= (n-4)A_{n-4}v_m\,, \\ \cdots & \quad \cdots \quad \cdots \\ \frac{n-1}{2}A_1 &= 3A_3 v_m\,. \end{aligned}$$

Hence,

$$\left(\frac{n-1}{2}\right)!A_1 = n!!(v_m)^{(n-1)/2}\,. \tag{3.9}$$

The last equation of system (3.4)

$$(a_0)_t = a_1 V_x \tag{3.10}$$

yields $ima_1 v_m = 0$. Since, by assumption, $m \neq 0$, according to (3.9), we arrive at the equality $v_m = 0$. Applying the countdown induction procedure with respect to m, we see that V does not depend on x.

Now consider the case of even $n \geq 2$. First, using (3.8.$n-2$), (3.8.$n-4$), ..., (3.8.0), we find:

$$\begin{aligned} A_{n-2} &= n v_m\,, \\ 2A_{n-4} &= (n-2)A_{n-2}v_m\,, \\ \cdots & \quad \cdots \quad \cdots \quad \cdots \\ lA_{n-2l} &= (n-l+2)A_{n-l+2}v_m\,, \\ \cdots & \quad \cdots \quad \cdots \quad \cdots \\ \frac{n}{2}A_0 &= 2A_2 v_m\,. \end{aligned}$$

Thus,

$$A_{n-2l} = \frac{n(n-2)\cdots(n-l+2)}{l!}(v_m)^l\,, \qquad 1 \leq l \leq n/2\,. \tag{3.11}$$

In order to find A_{n-3}, A_{n-5}, ..., we use equations (3.8.$n-3$), (3.8.$n-5$), ... :

$$\begin{aligned} A_{n-3} &= (n-1)a_{n-1}^0 v_m - \dot{A}_{n-2}/im\,, \\ 2A_{n-5} &= (n-3)A_{n-3}v_m - \dot{A}_{n-4}/im\,, \\ \cdots & \quad \cdots \quad \cdots \quad \cdots \\ \frac{n-2}{2}A_1 &= 3A_3 v_m - \dot{A}_2/im\,. \end{aligned}$$

Taking into account relations (3.11) we find

$$\left(\frac{n-2}{2}\right)!A_1 = (n-1)!!\, a^0_{n-1}(v_m)^{(n-2)/2} + \frac{iF(n)}{m} v_m^{n/2-2}\dot{v}_m , \tag{3.12}$$

where F is a positive number (depending on n only).

For even n, equation (3.10) implies the relation

$$\dot{A}_0 = imA_1 v_m . \tag{3.13}$$

We put $l = n/2$ in (3.12) and obtain

$$A_0 = \frac{n!!}{(n/2)!!} v_m^{n/2} . \tag{3.14}$$

Using relations (3.12), (3.13) and (3.14), we find the following relation for v_m:

$$G(n)\dot{v}_m = ima^0_{n-1} v_m , \tag{3.15}$$

where, like F, $G(\cdot)$ is real and positive. Linear Eq. (3.15) implies the desired formula

$$v_m = c\exp(i\beta t) , \qquad \beta = ma^0_{n-1}/G(n) .$$

Simultaneously we established that β depends linearly on the parameter a^0_{n-1}. It remains to prove that $G(n) = n$. To do this, perform the substitution $x \to x-(a^0_{n-1}/n)t$. As a result we obtain an equation of the form (3.1) with the potential

$$V^*(x,t) = V(x - ta^0_{n-1}/n, t) \tag{3.16}$$

which admits integral (3.3) with $a_{n-1} = 0$. The function V^* is a trigonometric polynomial in x. According to (3.15) the coefficient of the leading harmonics of V^* equals some complex number c. Then relation (3.16) implies that the leading Fourier coefficient of the polynomial V equals $c\exp(ima^0_{n-1}/n)t$. Consequently, $G(n) = n$.

Theorem 1 is proved.

4 Polynomial Integrals of Hamiltonian Systems with Exponential Interaction

4.1 Let W and W^* be dual n-dimensional linear spaces over the reals. We shall denote their elements by x and y respectively. Let (y, x) be the value of the covector y on the vector x. Consider the function $V : W \to \mathbb{R}$ defined by the relation

$$V(x) = \sum_{k=1}^{m} v_k \exp(a_k, x) , \tag{4.1}$$

where v_k are non-zero real numbers and $a_1, \dots, a_m$ are nonzero vectors of W^*. The function V plays the role of the potential energy of an exponential interaction. We denote by Δ the collection of vectors $a_1, \dots, a_m$ – the "spectrum" of the sum (4.1). Let $\langle\ ,\ \rangle$ be an inner product in the space W^*. The metric $\langle\ ,\ \rangle$ allows one to identify the dual spaces W and W^*. More precisely, there exists a linear isomorphism $A : W^* \to W$ such that $(y, x) = \langle y, A^{-1}x\rangle$ for every $y \in W^*$. Knowing the metric $\langle\ ,\ \rangle$ and the potential V, we can write down the equations of motion of a system with the exponential interaction $(W, \langle\ ,\ \rangle, V)$:

$$\dot{x} = Ay\,, \quad \dot{y} = -\sum_k [v_k \exp(a_k, x)]a_k\,. \tag{4.2}$$

Let $e_1, \ldots, e_n$ and $e_1^*, \ldots, e_n^*$ be dual bases in W and W^*. We put $s = \sum x_i e_i$ and $y = \sum y_i e_i^*$. In the coordinates $x_1, \ldots, x_n$ and $y_1, \ldots, y_n$ on the phase space $W \times W^*$ equations (4.2) may be written as the Hamiltonian canonical equations

$$\dot{x}_i = \partial H/\partial y_i\,, \quad \dot{y}_i = -\partial H/\partial x_i\,; \qquad 1 \le i \le n \tag{4.3}$$

with Hamiltonian $H = T + V$, where $T = \langle y, y\rangle/2$ is the kinetic energy of the system.

Let $D : W \to W$ be a non-degenerate linear operator, $D^* : W^* \to W^*$ the adjoint operator. The map $W \times W^* \to W \times W^*$ given by the relations $x' = Dx$ and $y' = (D^*)^{-1}y$ is canonical. In particular, in the new variables $x_1', \ldots, x_n', y_1', \ldots, y_n'$ the Hamiltonian Eqs. (4.3) have again the canonical form with the same Hamiltonian. By a suitable choice of D we can reduce the kinetic energy to sum of squares:

$$T = \frac{1}{2}(y_1^2 + \cdots + y_n^2)\,.$$

Hamiltonian systems of the form (4.2) are frequently encountered in applications. For instance, the dynamics of the finite periodic Toda lattice is described by the system (4.3) with Hamiltonian function

$$H = \frac{1}{2}\sum_{s=1}^{n} y_s^2 + \sum_{s=1}^{n} \exp(x_s - x_{s+1})\,, \qquad x_{n+1} = x_1\,. \tag{4.4}$$

Equations (4.2) are also encountered in some homogeneous cosmological models of general relativity theory [26].

Many papers are concerned with the search for cases of integrability among the Hamiltonian systems (4.2). Hénon [98], Flaschka [78] and Manakov [168] established the complete integrability of the Toda lattice: the Hamilton equations with the Hamiltonian (4.4) have n independent first integrals polynomial in the momenta and pairwise in involution. In [25, 195, 121] this result was generalized to the case when the spectrum of Δ is a system of simple roots of a simple Lie algebra. From this point of view, the algebra of type A_n corresponds to the Hamiltonian (4.4). Sklyanin [218] indicated another integrable generalization of the Toda lattice:

$$\begin{aligned} H = \sum_{s=1}^{n} \frac{y_s^2}{2} + \sum_{s=1}^{n-1} \exp(x_{s+1} - x_s) + \alpha_1 \exp x_1 + \frac{\beta_1}{2}\exp(2x_1) \\ + \alpha_n \exp(-x_n) + \frac{\beta_n}{2}\exp(-2x_n)\,, \end{aligned} \tag{4.5}$$

where $\alpha_1, \beta_1, \alpha_n$, and β_n are arbitrary real constants. The method of [25] is based on the representation of the Hamiltonian equations (4.2) in the form of a Lax L–A pair. The elements of the matrices L and A are linear functions of the momenta $y_1, \ldots, y_n$ with coefficients which are finite sums of real exponents:

$$\sum f_\lambda \exp(c_\lambda, x) , \qquad f_\lambda \in \mathbb{R} , \quad c_\lambda \in \mathbb{R}^n . \tag{4.6}$$

Consequently, the traces of powers of the matrix L (integrals of Hamiltonian equations) are polynomials in the momenta with coefficients of the form (4.6).

Little was known about the integrability of systems (4.2) in the general case before paper [3]. In [3] the case is considered when Δ consists of $n+1$ vectors $a_1, \ldots, a_{n+1}$, and any n of them are assumed to be independent. It is proved that under these assumptions a criterion for algebraic integrability of (4.2) is

$$\frac{2\langle a_i, a_j\rangle}{\langle a_i, a_i\rangle} \in -\mathbb{Z}_+ , \qquad \mathbb{Z}_+ = \{0, 1, 2, \ldots\}$$

for every $i \neq j$. Algebraic integrability means, in particular, that the variables y_s and $\exp x_s$ $(1 \leq s \leq n)$ are meromorphic on the plane of complex time for almost all initial data. The search for necessary algebraic integrability conditions is based on the classical method developed and used by Kovalevskaya in rigid body dynamics. A generalization of this result is presented in §4 of Chap. VII. A simpler problem on polynomial integrals of the system (4.3), with some additional properties, is also considered there (see §4.4).

We observe that far from every completely integrable system of the form (4.2) is algebraically integrable in the sense of the definition given in [3]. Here is a simple example of a system with one degree of freedom described by the Hamiltonian

$$H = [y^2 + \exp(-2x) f_m(\exp x)]/2 ,$$

where $f_m(\cdot)$ is a polynomial of degree m with simple roots. This system is not algebraically integrable for $m \geq 5$. Indeed, the functions

$$\int (2hq^2 - f_m(q))^{-1/2}\, dq = t , \qquad q = \exp x , \quad y = \dot{q}/q$$

are solutions with total energy h. Clearly, for $m \geq 5$ and for almost each h the function $t \to q(t)$ is multivalued on the complex plane.

Following [156], we shall study the integrability of equations (4.2) for real values of the variables.

4.2 A Hamiltonian system (4.2) is called Birkhoff integrable if it has n integrals polynomial in the momenta with coefficients of form (4.6). The integrals are assumed to be independent almost everywhere in $W \times W^*$.

We call a vector in Δ maximal if it has the greatest length among all the vectors of Δ having the same direction.

The construction of integrable systems is defined by

Theorem 1. *Suppose that the Hamiltonian system* (4.2) *is Birkhoff integrable. Let a_i be a maximal vector in Δ and assume that the vector $a_j \in \Delta$ is linearly independent of a_i. Then*

$$\frac{2\langle a_i, a_j\rangle}{\langle a_i, a_i\rangle} \in -\mathbb{Z}_+ . \tag{4.7}$$

Corollary 1. *If the system* (4.2) *is Birkhoff integrable, then any two linearly independent maximal vectors $a_i, a_j \in \Delta$ satisfy condition* (4.7).

It is useful to compare this assertion with the result of [3] which deals with the case when Δ consists of $n+1$ vectors $a_1, \ldots, a_{n+1}$ such that any subsystem of n vectors is linearly independent. In [3] it was shown that a criterion for algebraic integrability of (4.2) is precisely condition (4.7). Corollary 1 asserts that a criterion for Birkhoff integrability is condition (4.7) as well. This situation is analogous to that in the classical problem on the rotation of a heavy rigid body with a fixed point: the equations of motion are algebraically integrable if and only if they have a complete set of independent polynomial integrals.

Corollary 2. *If the system is Birkhoff integrable, then the following assertions are true.*

1) *The angle between any two vectors of Δ has one of the following values: $0, \pi/2, 2\pi/3, 3\pi/4, 5\pi/6$ or π.*
2) *Let $a, a' \in \Delta$ and assume that the vector a is maximal. If the angle between a and a' is equal to $2\pi/3$, then $|a| = |a'|$; if it is $3\pi/4$, then $|a| = \sqrt{2}|a'|$ or $|a| = |a'|/\sqrt{2}$; and if it is $5\pi/6$, then either $|a| = \sqrt{3}|a'|$ or $|a| = 2|a'|/\sqrt{3}$, or $|a| = |a'|/\sqrt{3}$.*

Indeed, the angle between vectors of Δ coincides with the angle between two corresponding maximal vectors. Let a_i, a_j be maximal vectors, and ϕ the value of the angle between them. According to Theorem 1,

$$\frac{2\langle a_j, a_i\rangle}{\langle a_j, a_j\rangle} \in -\mathbb{Z}_+ . \tag{4.8}$$

Using (4.7) and (4.8), we obtain:

$$4\cos^2\phi = \frac{4\langle a_i, a_j\rangle^2}{\langle a_i, a_i\rangle\langle a_j, a_j\rangle} = l ,$$

where l is an integer number. Since $l \leq 4$, $\cos\phi$ may take one of the following values: $0, \pm 1/2, \pm\sqrt{2}/2, \pm\sqrt{3}/2, \pm 1$. Assertion 1) is proved. The proof of assertion 2) is presented in §4.5.

Below we give a classification of Birkhoff integrable systems which is based on Theorem 1.

4.3 Assume that W^* is the direct sum of orthogonal subspaces $W_1^*, \ldots, W_m^*$, and the spectrum of Δ lies in their union $\bigcup W_i^*$. Denote by $W_1, \ldots, W_m$ the images of $W_1^*, \dots, W_m^*$ under the map $A : W^* \to W$. It is easy to see that the system (4.2) splits into m closed subsystems with phase spaces $W_i \times W_i^* \subset W \times W^*$. Let H_i be the restriction of the Hamiltonian function H to $W_i \times W_i^*$. Then clearly $H = \sum H_i$. If the basis vectors $e_1, \dots, e_n$ belong to the union $W_1 \cup \dots \cup W_m$, then in the corresponding canonical variables $x_1, \dots, x_n, y_1, \dots, y_n$ equations (4.3) split into m closed Hamiltonian systems with Hamiltonian functions H_i (i.e. we have a partial separation of variables). We say that in this situation the original Hamiltonian system is a direct sum of its subsystems. If such a splitting is impossible (certainly, under the assumption that each subspace W_i is nontrivial), then we say that the Hamiltonian system is irreducible. As a result we come to the following obvious assertion.

Proposition 1. *Each Hamiltonian system* (4.2) *is a direct sum of its irreducible Hamiltonian subsystems.*

Let α and β be vectors of Δ and let ϕ be the angle between them. If the Hamiltonian system is Birkhoff integrable, then, according to Corollary 2 of Theorem 1, the value $4\cos^2\phi$ may be one of the following integers: 0,1,2,3 or 4. This fact hints that it is convenient to introduce the Coxeter graph whose vertices denote vectors of Δ, and where any two vertices α and β are joined by $4\cos^2\phi$ edges. Clearly, if an integrable system is irreducible, then its Coxeter graph is connected and non-empty.

If $\dim W_i^* = 1$, then $\Delta \cap W_i^*$ may be any finite set of vectors. This follows from the complete integrability of Hamiltonian systems with one degree of freedom. Leaving aside these trivial cases, we assume below that $\dim W_i^* > 1$.

Thus, consider the structure of an integrable irreducible Hamiltonian system with $n > 1$ degrees of freedom.

Proposition 2. *Any two linearly dependent vectors of Δ have the same direction.*

Proof. We put $\Pi_\alpha = \{b \in W^* : \langle \alpha, \beta \rangle \leq 0\}$. Let α and β be linearly dependent vectors of Δ. If $\gamma \in \Delta$ and $\gamma \neq k\alpha$, $k \in \mathbb{R}$, then by Theorem 1 we have $\gamma \in \Pi_\alpha$ and $\gamma \in \Pi_\beta$. If α and β have opposite directions, then the intersection $\Pi_\alpha \cap \Pi_\beta$ is a hyperplane in W^* orthogonal to both α and β. Therefore, in this case γ is orthogonal to the vectors α and β. But this contradicts the irreducibility assumption. The proposition is proved.

We say that an integrable system with the spectrum Δ is complete, if there is no non-zero vector $a \in W^*$ such that the set $\Delta \cup \{a\}$ satisfies the assumptions of Theorem 1. The spectrum of each Birkhoff integrable Hamiltonian system is obtained from a complete spectrum by dropping some elements. Certainly, in this reduction of Δ we must preserve the connectedness of the Coxeter graph, and the number of its vertices cannot be less than $\dim W^* = n$.

It is easy to understand that the Coxeter graph determines only the angles between pairs of vectors of Δ. In order to be able to reconstruct the ratios of the lengths of the vectors, we assign to every vertex of the Coxeter graph a coefficient proportional to the square of the length $\langle a, a \rangle$ of the corresponding vector $a \in \Delta$. This extended Coxeter graph is called a Dynkin diagram (as is customary in the theory of root systems). We agree to identify Dynkin diagrams which differ only by a positive proportionality coefficient. This convention can be given a clear dynamical sense.

Consider two dynamical systems such that the vectors of their spectra differ by a positive factor $k^2 > 0$. It is not difficult to check that the change $x \to kx$, $t \to kt$ transforms the Eqs. (4.2) of one system into those of another one. Using Corollary 2 of Theorem 1 and Proposition 2, one can prove that the Dynkin diagram uniquely (up to an isomorphism) defines the spectrum of an integrable irreducible Hamiltonian system.

Theorem 2. *The Dynkin diagram of a complete irreducible Birkhoff integrable Hamiltonian system is isomorphic to one of the following diagrams*:

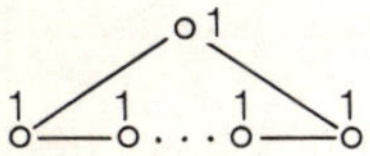

a

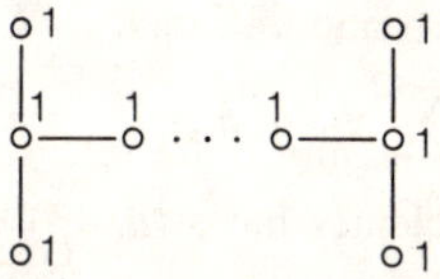

b

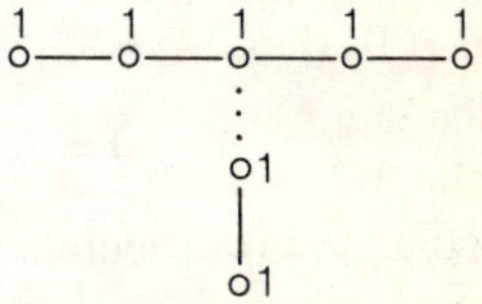

c

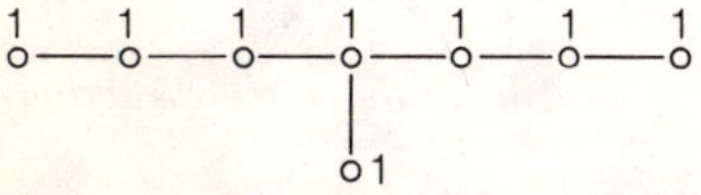

d

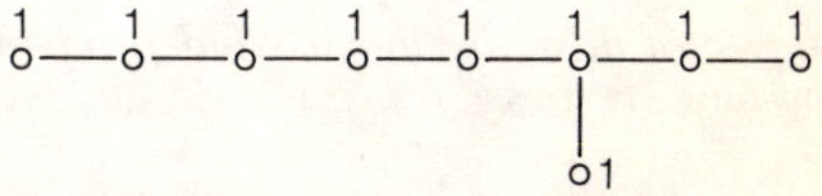

e

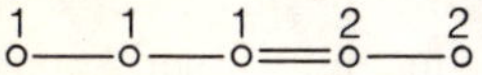

f

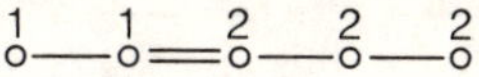

g

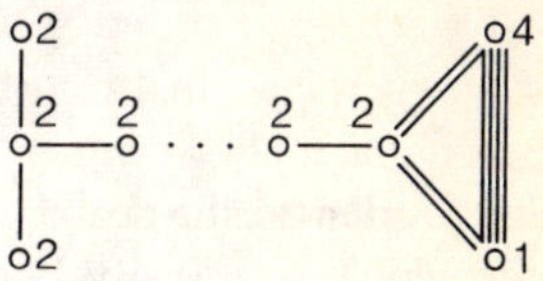

h

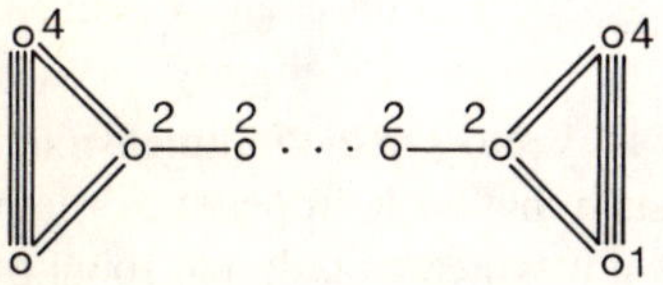

i

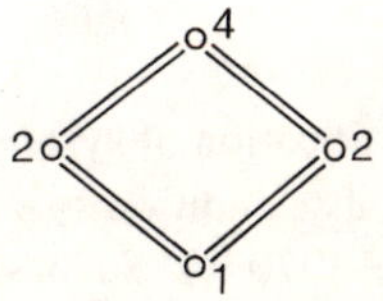

j

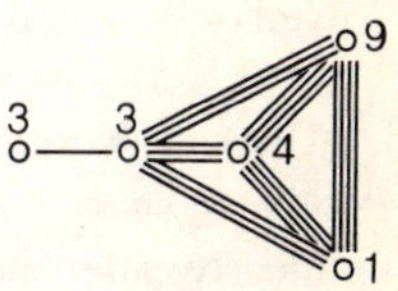

k

Corollary. *The spectrum of an irreducible Birkhoff integrable Hamiltonian system with $n \geq 2$ degrees of freedom contains at most $n+3$ distinct vectors.*

Note that the estimate $\operatorname{card}\Delta \leq n+3$ cannot be improved. This is shown by the example of a system with Hamiltonian function (4.5).

In order to prove Theorem 2 we need the following fact.

Lemma 1. *Assume that the vectors $a_\lambda \in \Delta$ are pairwise linearly independent and $\sum x_\lambda a_\lambda = 0$ for some real $x_\lambda \neq 0$. If a is a vector of Δ, linearly independent of each vector a_λ, then $\langle a, a_\lambda \rangle = 0$.*

Proof. The equality $\sum x_\lambda a_\lambda = 0$, $x_\lambda \neq 0$, may be rewritten as

$$\sum y_\mu a_\mu = \sum z_\nu a_\nu \,, \qquad y_\mu, z_\nu > 0 \,.$$

We put $b = \sum y_\mu a_\mu$ and compute

$$\langle b, b \rangle = \sum y_\mu z_\nu \langle a_\mu, a_\nu \rangle \,.$$

Since the vectors a_λ are pairwise linearly independent, by Theorem 1 we have $\langle b, b \rangle \leq 0$. Hence $b = 0$. Since

$$0 = \langle a, b \rangle = \sum y_\mu \langle a, a_\mu \rangle \,, \qquad y_\mu > 0$$

and $\langle a, a_\mu \rangle \leq 0$ (Theorem 1), we clearly have $\langle a, a_\mu \rangle = 0$. Q.E.D.

Proposition 3. *The set Δ contains not more than $n+1$ linearly independent vectors, and any n of them are linearly independent.*

Proof. Let Δ have $n+1$ pairwise linearly independent vectors $a_1, \ldots, a_{n+1}$. Since $\dim W^* = n$, we see that

$$\sum_{i=1}^{n+1} x_i a_i = 0 \,, \qquad \sum x_i^2 \neq 0 \,.$$

Let Δ' denote the set of vectors of Δ which are dependent on those $a_1, \ldots, a_{n+1}$ for which $x_i \neq 0$. By Lemma 1, the sets Δ' and $\Delta \setminus \Delta'$ are orthogonal. Since, by assumption, the system is irreducible, $\Delta = \Delta'$ and all x_i are non-zero. In particular, any n vectors of the set $a_1, \ldots, a_{n+1}$ are linearly independent. The proposition is proved.

There is a complete classification of systems of $n+1$ vectors in the n-dimensional Euclidean space which satisfy condition (4.7) and such that each proper subsystem is linearly independent (see [97, 3]). Such systems are systems of simple roots of graduated Kac–Moody algebras. The complete Dynkin diagrams a)–k) presented in Theorem 2 were obtained from the well-known root systems of Kac–Moody algebras, taking into account the fact that the spectrum of an integrable system may contain vectors with the same direction. We omit the simple arguments. Assume now that Δ contains n linearly independent maximal vectors satisfying condition (4.7). It turns out that such a system is not complete in the sense of our definition: to these n vectors we can add one vector so that condition (4.7) would be preserved and any subsystem of n vectors would be linearly independent. This follows, for instance, from the fact that a Dynkin diagram of a system of simple roots is obtained from diagram of a root system of a certain Kac–Moody algebra by removing one vertex [97].

4.4 Now consider the Birkhoff integrability of the Hamiltonian systems given in Theorem 2. In the cases a)–g) complete integrability was established in [25] and [2]. The Hamiltonian of a system with Dynkin diagram i) can always be reduced to the form (4.5). This system has been integrated in [218].

The graph j) corresponds to the Hamiltonian system with two degrees of freedom and the Hamiltonian function

$$H = (y_1^2 + y_2^2)/2 + v_1 e_1 + v_2 e_2 + v_3 e_3 + v_4 e_4 ,$$

where

$$e_1 = \exp x_1 , \quad e_2 = \exp x_2 ,$$
$$e_3 = \exp(-x_1 - x_2) , \quad e_4 = \exp\left(- \frac{x_1 + x_2}{2} \right) .$$

For all values of the coefficients $v_1, \ldots, v_4$ it has an additional integral of fourth degree in the momenta:

$$\begin{aligned} F = {} & y_1^2 y_2^2 + 2v_2 y_1^2 e_2 + 2v_3 y_1 y_2 e_3 + 2v_1 y_2^2 e_1 + 2v_2 v_3 e_2 e_3 \\ & + 2v_1 v_3 e_1 e_3 + 4v_1 v_2 e_1 e_2 + v_3^2 e_3^2 + 2v_3 v_4 e_3 e_4 + v_4^2 e_4^2 . \end{aligned}$$

This is a new integrable lattice. For $v_3 = 0$ or $v_4 = 0$ we obtain the integrable lattices found in [25] and [2]. Clearly, the leading homogeneous forms of the functions H and F are independent.

A Hamiltonian function with Dynkin diagrams h) has the following form in suitable canonical variables:

$$\begin{aligned} H = {} & \frac{1}{2} \sum_{s=1}^{n} y_s^2 + \sum_{s=1}^{n-1} \exp(x_s - x_{s+1}) + \exp(-x_1 - x_2) \\ & + \alpha_n \exp x_n + \beta_n \exp(2x_n) ; \qquad \alpha_n, \beta_n \in \mathbb{R} , \quad n \geq 4 . \end{aligned} \tag{4.9}$$

If either $\alpha_n = 0$ or $\beta = 0$, then complete integrability of this system has been established in [25] and [2]: for $\beta_n = 0$ this is a system of type B_n (according to the classification given in [25]), and for $\alpha_n = 0$ we obtain the system of type $b^{(n)}$ (according to the classification in [2]). The question of integrability of the Hamiltonian equations with Hamiltonian (4.9) in the general case, when $\alpha_n \neq 0$ and $\beta_n \neq 0$, remains unclear, but the answer is apparently positive.

The integrability problem of the Hamiltonian system with graph k) turns out to be more complicated. In [25] and [2] Hamiltonian systems with two degrees of freedom are integrated whose Dynkin diagrams have the following form:

They are obtained from graph k) by eliminating two vertices. We have studied the Birkhoff integrability of the system with Hamiltonian

$$\begin{aligned} H = (y_1^2 + y_2^2)/2 + v_1 \exp x_1 + v_2 \exp(-x_1/2 - \sqrt{3}x_2/2) \\ + v_3 \exp(\sqrt{3}x_2/3) + v_4 \exp(\sqrt{3}x_2) . \qquad (4.10) \end{aligned}$$

Its Coxeter graph is obtained from graph k) by dropping one vertex. This case differs from the general one in that all linearly independent vectors of Δ satisfy (4.7). It turned out that if all the coefficients v_i are non-zero then the Hamiltonian system with Hamiltonian (4.10) does not have an additional integral whose degree does not exceed 6. The number 6 is not accidental: this is the rank of the Coxeter group generated by reflections with respect to vectors of the spectrum Δ. We note that in the remaining integrable systems with two degrees of freedom the degree of the additional polynomial integral is precisely equal to the rank of the corresponding Coxeter group. Apparently the system with Hamiltonian (4.10) (and, moreover, the system corresponding to graph k)) does not have an additional analytic integral. This assumption is confirmed by numerical calculations performed by Borisov for the system with Hamiltonian (4.10) with the additional term $v_5 \exp(2x_2/\sqrt{3})$ and $v_1 = \cdots = v_5 = 1$. Note that the integrability of the closed three-particle Toda lattice was first observed as a result of numerical computations (see [81]).

It is worth emphasizing that, according to the results of §4, Chap. VII, the Hamiltonian system with graph k) is algebraically integrable only under the conditions $v_3 = v_5 = 0$, or $v_4 = v_5 = 0$.

4.5 Now turn to the proof of Theorem 1. Let $e_1 \dots, e_n$ and $e_1^* \dots, e_n^*$ be dual bases in W and W^*. We introduce an order relation $\prec$ in W^*. Let $\sigma = \sum \sigma_i e_i^*$ and $\delta = \sum \delta_i e_i^*$. We say that $\sigma \prec \delta$ if for the least index k such that $\sigma_k \neq \delta_k$ we have $\sigma_k < \delta_k$. Clearly, $\prec$ is the usual lexicographic order relation in W^* with respect to the basis $e_1^*, \dots, e_n^*$. This relation induces in a natural way an order relation in W. We say that $\sigma \preceq \delta$ if either $\sigma \prec \delta$ or $\sigma = \delta$.

The proof of Theorem 1 uses the following assertion which is interesting in itself.

Theorem 3. *Let α be a maximal element in Δ with respect to the order relation $\prec$. Assume that there exists a vector $\beta \in \Delta$ satisfying the following conditions:*
1) *$\beta \succ 0$ and β is independent of α,*
2) *the equality $s\alpha + \beta = \tau_1 + \cdots + \tau_{s+1}$, where $\tau_i \in \Delta$, implies $\tau_k = \beta$ for some k and $\tau_j = \alpha$ for each $j \neq k$,*

3) $m\langle\alpha, \alpha\rangle + 2\langle\alpha, \beta\rangle \neq 0$ *for all integers* $m \geq 0$.

Then the Hamiltonian system (4.2) *is not Birkhoff integrable.*

We make two remarks. 1) For condition 2) to hold it is sufficient to require that $\beta \in \Delta$ is a maximal vector linearly independent of α.

2) Theorem 3 is also valid when the metric $\langle\ ,\ \rangle$ is pseudo-Euclidean.

Theorem 3 will be proved in the next section. Now we deduce Theorem 1 from it.

Introduce a natural affine structure on the vector space W^*; therefore Δ may be regarded as a finite collection of points of an affine space. We denote by $\mathscr{E}(\Delta)$ the convex hull of Δ. Clearly, $\mathscr{E}(\Delta)$ is a convex polyhedron in the n-dimensional affine space. The notions of the theory of convex figures that we use here may be found, for example, in [21].

Lemma 2. *Let* α *and* β *be two neighboring vertices of* $\mathscr{E}(\Delta)$ *(joined by an edge), linearly independent as vectors in* W^*. *If the Hamiltonian system is integrable, then for some integer* $m \geq 0$ *we have* $m\langle\alpha, \alpha\rangle + 2\langle\alpha, \beta\rangle = 0$.

Proof. Let γ be a segment joining α and β. Clearly, γ is an edge of the convex polyhedron $\mathscr{E}(\Delta)$. Since α and β are linearly independent as vectors in W^*, and $\mathscr{E}(\Delta)$ is a convex polyhedron, there exists a hyperplane Γ not containing $0 \in W^*$ such that $\gamma \cap \mathscr{E}(\Delta)$ coincides with the edge γ. There are two possibilities: either Γ separates the polyhedron $\mathscr{E}(\Delta)$ and the point $0 \in W^*$, or it does not separate them. In the first case we choose the following basis in W^*: $e_1^* = \alpha$, e_2^* is the vector starting at β and ending at α, and $e_3^*, \ldots, e_n^*$ are vectors in Γ, linearly independent of e_2^*. In the second case we choose the vectors $(-e_1^*), e_2^*, \ldots, e_n^*$ as a basis.

Clearly, in this basis the vector α is the maximal element of the set Δ. Let α' be the greatest element in $\Delta \setminus \{\alpha\}$ with respect to the relation $\prec$. It is obvious that α', as a point of an affine space, is the nearest point to α in the set $(\Delta \setminus \{\alpha\}) \cap \gamma$. We claim that the angle between α and α' is not less than $\pi/2$. Indeed, otherwise both α and α' satisfy conditions 1)–3) of Theorem 3, and, therefore, the Hamiltonian system under consideration is not integrable. Let β' be the nearest point to β in $(\Delta \setminus \{\beta\}) \cap \gamma$. Analogously we prove that the angle between β and β' is not less than $\pi/2$. Since the vectors α and β are independent by the assumption, the angle between them is strictly less than π. Consequently, the point α' coincides with the vertex β, and β' coincides with α. Hence, it follows that the vectors α and β satisfy conditions 1) and 2) of Theorem 3. If 3) also holds, the system is nonintegrable, which contradicts the assumption. The lemma is proved.

Lemma 3. *Assume that the original Hamiltonian system is Birkhoff integrable. Then the following assertions are true:*

1) *all the points of* Δ *lie on rays starting at* $0 \in W^*$ *and passing through the vertices of the convex polyhedron* $\mathscr{E}(\Delta)$,
2) *the angles between these rays are not less than* $\pi/2$.

Proof. Let δ be a point of Δ. Then there exists a vertex σ of the polyhedron $\mathscr{E}(\Delta)$ such that the angle between σ and δ is acute. For otherwise all vertices lie in the half-space $\Pi_\delta = \{y \in W^* : \langle y, \delta\rangle \leq 0\}$. Hence $\mathscr{E}(\Delta) \subset \Pi_\delta$, and therefore $\delta \notin \mathscr{E}(\Delta)$. We obtain a contradiction. We shall assume that δ does not lie on the ray given by the vector σ. Then the angle between σ and δ is non-zero. We may also assume that σ has maximal length among all the vectors of Δ with the same direction as σ.

Let Γ be the hyperplane passing through σ (as a point of affine space) and orthogonal to the vector σ. Denote by Π the closed half-space with boundary Γ which does not contain 0. We show that $\mathscr{E}(\Delta)$ has only the point σ in common with Π. Indeed, otherwise there is another such point $\tau \neq \sigma$. But then there is a point μ, a vertex of $\mathscr{E}(\Delta)$ which is connected to σ by an edge and lies in Π. Now we prove this assertion by contradiction. It is clear that $\mathscr{E}(\Delta) = M_1 \cup M_2$, where M_1 is the convex hull of the vertices of $\mathscr{E}(\Delta)$ except σ, and M_2 is the convex hull of the edges issuing from σ.

Consider the segment γ joining σ and τ. Assume that there is no such point μ. Then $\gamma \cap M_2 = \Pi \cap M_2 = \sigma$, and the set $\gamma \cap M_1$ is closed and does not contain σ. But this contradicts the relation $\gamma \subset (M_1 \cup M_2)$. Hence there is indeed such a point μ. The angle between σ and μ is acute and non-zero. Consequently, by Lemma 2, the system is nonintegrable.

Thus, $\mathscr{E}(\Delta) \cap \Pi = \{\sigma\}$. We consider the following basis $\{e_i^*\}$ in W^*: $e_1^* = \sigma$, and $e_2^*, \dots, e_n^*$ are independent vectors in Π. Let $\prec$ be the corresponding lexicographic order relation. It is obvious that σ is the maximal element in Δ and $\delta \succ 0$. Let δ' be the maximal element of Δ which is linearly independent of σ. Since $\delta' \succeq \delta$, we have $\langle \sigma, \delta'\rangle \geq 0$ and therefore the angle between σ and δ' is also acute. By Remark 1 which follows Theorem 3 the system is nonintegrable. The lemma is proved.

Now it is not hard to prove Theorem 1. By Lemma 3, for any two linearly independent vectors σ and τ of Δ we have $\langle \sigma, \tau\rangle \leq 0$. Let α and β be linearly independent vectors of Δ, and assume that α is maximal. We choose the following basis $\{e_i^*\}$ in W^*: $e_1^* = \alpha$, $e_2^* = \beta$, and the vectors $e_3^*, \dots, e_n^*$ are orthogonal to α and β. Let $\tau = \sum \tau_i e_i^* \in \Delta$. We prove that $\tau_1 \leq 0$ if τ and α are linearly independent. Indeed,

$$\langle \tau, \alpha\rangle = \tau_1 \langle e_1^*, e_1^*\rangle + \tau_2 \langle e_2^*, e_1^*\rangle \leq 0\,,$$
$$\langle \tau, \beta\rangle = \tau_1 \langle e_1^*, e_2^*\rangle + \tau_2 \langle e_2^*, e_2^*\rangle \leq 0\,.$$

If $\tau_1 > 0$, then the first inequality implies $\tau_2 > 0$. Therefore

$$\tau_1 \langle \tau, e_1^*\rangle + \tau_2 \langle \tau, e_2^*\rangle \leq 0\,.$$

On the other hand, the last inequality may be written as

$$\tau_1^2\langle e_1^*, e_1^*\rangle + 2\tau_1\tau_2\langle e_1^*, e_2^*\rangle + \tau_2^2\langle e_2^*, e_2^*\rangle \leq 0\,.$$

However, this relation contradicts the positive definiteness of the Gram matrix $\|\langle e_i^*, e_j^*\rangle\|$, $i, j = 1, 2$.

Thus, the order relation $\prec$ corresponding to the basis $\{e_i^*\}$ is such that α is the greatest element of Δ and $\beta \succ 0$. Moreover, if $\tau \succeq \beta$ then τ has the same direction as either α or β. These properties of the vectors α and β imply conditions 1) and 2) of Theorem 3. Thus by Theorem 3, if the Hamiltonian system is integrable, then for some integer $m \geq 0$ we have

$$m\langle\alpha, \alpha\rangle + 2\langle\alpha, \beta\rangle = 0\,. \tag{4.11}$$

Now let $\beta \in \Delta$ be the vector of maximal length having the same direction as α. For some integer $k \geq 0$ we obtain the analogous relation

$$k\langle\beta, \beta\rangle + 2\langle\beta, \alpha\rangle = 0\,. \tag{4.12}$$

From (4.11) and (4.12) we conclude that the angle ϕ between α and β has one of the following values: $2\pi/3$, $3\pi/4$ or $5\pi/6$, and if $\phi = 2\pi/3$, then $|\alpha| = |\beta|$; if $\phi = 3\pi/4$, then either $|\alpha| = \sqrt{2}|\beta|$ or $|\beta| = \sqrt{2}\alpha$; finally, if $\phi = 5\pi/6$, then either $|\alpha| = \sqrt{3}|\beta|$ or $|\beta| = \sqrt{3}\alpha$ (cf. Corollary 2 of Theorem 1).

Now let us consider the case when there is a vector $\beta' \in \Delta$ having the same direction as β, with $|\beta'| < |\beta|$. As we have just shown, we must have

$$l\langle\alpha, \alpha\rangle + 2\langle\alpha, \beta'\rangle = 0 \tag{4.13}$$

for some integer $l \geq 0$.

Let $\phi = 2\pi/3$. Then we obtain from (4.13) the relation $|\beta'| = l|\alpha|$. We arrive at a contradiction, since in this case $|\alpha| = |\beta| > |\beta'|$ and $l \geq 1$. Thus, if $\phi = 2\pi/3$, there is no such β'.

Assume now that $\phi = 3\pi/4$. Then we obtain from (4.13) the equality $|\beta'| = l|\alpha|/\sqrt{2}$. If $|\alpha| = \sqrt{2}|\beta|$, then $|\beta'| = l|\beta|$. This equality contradicts the initial assumption $|\beta'| < |\beta|$, since $l \geq 1$. If $|\beta| = \sqrt{2}|\alpha|$, then $|\beta'| = l|\beta|/2$. Since $|\beta'| < |\beta|$, $l = 1$. Therefore, when $\phi = 3\pi/4$, we have $\beta' = \beta/2$.

Consider the remaining case $\phi = 5\pi/6$. Equality (4.13) yields $|\beta'| = l|\alpha|/\sqrt{3}$. If $|\alpha| = \sqrt{3}|\beta|$, then $|\beta'| = l|\beta|$. This is impossible since $l \geq 1$. Assume now that $|\beta| = \sqrt{3}|\alpha|$. Then $|\beta'| = l|\beta|/3$. Since $|\beta'| < |\beta|$, either $l = 1$ or $l = 2$. Thus, if $\phi = 5\pi/6$, then β' may be either $\beta/3$ or $2\beta/3$. Theorem 1 is completely proved.

The reader may have already noticed an analogy between the conditions of integrability of reversible systems whose potentials are trigonometric polynomials and finite sums of real exponents (see relations (5.3) of Chap. IV and (4.7)). Certainly, this analogy is not accidental. We discuss its origin in the next section.

5 Perturbations of Hamiltonian Systems with Non-Compact Invariant Surfaces

5.1 The geometric version of the Liouville theorem on complete integrability (Theorem 1 of §4 of Chap. II) asserts that non-critical common level surfaces of n commuting integrals of a Hamiltonian system with n degrees of freedom are diffeomorphic to $\mathbb{T}^k \times \mathbb{R}^{n-k}$ $(0 \leq k \leq n)$. Moreover, in proper variables $x_1 \bmod 2\pi, \dots, x_k \bmod 2\pi, x_{k+1}, \dots, x_n$ the Hamiltonian equations have a very simple form: $\dot{x}_s = \omega_s =$ const. In the compact case (when $k = n$) there is a sufficiently detailed theory of behavior of Hamiltonian systems that are close to integrable ones. Below we discuss, following [142], some analytic aspects of this theory for the non-compact case, and its connection with the existence problem of a complete set of independent integrals.

Let $M^n = \mathbb{T}^k \times \mathbb{R}^{n-k}$ be the configuration space of the reversible Hamiltonian system with Hamiltonian $H_0 + \varepsilon H_1$, where

$$H_0 = \frac{1}{2} \sum a_{ij} y_i y_j$$

is a positive definite quadratic form with constant coefficients, H_1 is a single-valued function on $M^n = \{x\}$, and ε is a small parameter. Introduce in $\mathbb{R}^n$ two scalar products:

$$\langle \xi, \eta \rangle = \sum a_{ij} \xi_i \eta_j \,, \quad (\xi, \eta) = \sum \xi_\iota \eta_\iota \,.$$

We see that $H_0 = \langle y, y \rangle / 2$.

Following the well-known method of classical perturbation theory, we try to find an ε-dependent canonical transformation $x, y \to u, v$ of the form

$$y = \partial S / \partial x \,, \quad u = \partial S / \partial v \,; \qquad S = S_0(v, x) + \varepsilon S_1(v, x) + \cdots$$

which transforms the Hamiltonian $H_0 + \varepsilon H_1$ to the function $K_0(v) + \varepsilon K_1(v) + \cdots$. We put $S_0 = (v, x)$; then for $\varepsilon = 0$ we have the identity mapping. The generating function S satisfies the Hamilton partial differential equation

$$H_0(\partial S / \partial x) + \varepsilon H_1(x) = K_0(v) + \varepsilon K_1(v) + \cdots \,.$$

We obtain an infinite chain of equations for the functions $S_1, S_2, \dots$ and $K_1, K_2, \dots$:

$$\begin{gathered} \left\langle v, \frac{\partial S_1}{\partial x} \right\rangle + H_1(x) = K_1(v) \,, \\ \cdots \quad \cdots \quad \cdots \quad \cdots \\ \left\langle v, \frac{\partial S_m}{\partial x} \right\rangle + \frac{1}{2} \sum_{p+q=m} \left\langle \frac{\partial S_p}{\partial x}, \frac{\partial S_q}{\partial x} \right\rangle = K_m(v) \,, \\ \cdots \quad \cdots \quad \cdots \quad \cdots \quad \cdots \end{gathered} \tag{5.1}$$

In the compact case $(k = n)$ it is usually supposed that the average values of the functions S_m $(m \geq 1)$ over $\mathbb{T}^n$ vanish. Thus, one can (on a formal level) uniquely find $S_1, S_2, \cdots$ and $K_1, K_2, \ldots$ from system (5.1) (see §10 of Chap. II). As a rule, the resulting formal series do not converge due to the presence of small denominators.

Each Eq. (5.1) can be written in the following form:

$$(w, \partial f/\partial x) = g\,, \tag{5.2}$$

where w is a constant vector field on M^n (its components are linear combinations of v_i), g is a known function, and f an unknown one. There is a simple

Proposition 1. *If integral curves of the vector field w are unbounded on M^n, then equation* (5.2) *is globally solvable.*

Indeed, the function f can be found by integration of g along trajectories of the field w. Moreover, f is defined uniquely by its values in the $(n-1)$-dimensional section $x_j = \text{const}$, where x_j is any linear coordinate in M^n that is a monotone function in time. Each trajectory of the field w intersects this section only at one point.

Now we consider a solution of equation (5.2) as a function of x and v. In general it has a singularity at the point $v = 0$. It is worth emphasizing that if $M \neq \mathbb{R}^n$, then equation (5.2) does not have a solution which is analytic in v in the domain $\mathbb{R}^n \setminus \{0\}$. The reason is that the fields v have closed trajectories on some invariant surfaces.

5.2 It turns out that the chain of Eqs. (5.1) can be easily solved explicitly for Hamiltonian systems with exponential interaction. We put

$$H_1 = \sum h^\alpha \exp(\alpha, x)\,.$$

Here the summation is performed over a finite set of vectors $\alpha \in \Delta$. We assume that $\alpha \neq 0$: the constant terms in this relation can be included in K_1. We look for a solution in the form of a sum of exponents

$$S_1 = \sum S_1^\alpha \exp(\alpha, x)\,.$$

Then, obviously

$$S_1^\alpha = -h^\alpha/\langle v, \alpha\rangle\,. \tag{5.3}$$

The coefficients of the sum S_1 are not defined on the "resonance" hyperplanes $\langle v, \alpha\rangle$, $\alpha \in \Delta$. We denote the union of all these hyperplanes by $\mathbf{B}_1$, and call it a resonance set of the first order.

The equation for S_2 has the same form as that for S_1. The function

$$\frac{1}{2}\langle \frac{\partial S_1}{\partial x}, \frac{\partial S_1}{\partial x}\rangle \tag{5.4}$$

is a finite sum of exponents, but its coefficients already depend on the new momenta. We look for S_2 in the form of a sum

$$\sum S_2^\tau \exp(\tau, x) .$$

We include in the function K_2 the terms of (5.4) that do not depend on x. From the second equation of the system (5.1) and relations (5.3) we find that

$$S_2^\tau = -\frac{1}{2\langle v, \tau\rangle} \sum_{\sigma+\delta=\tau} \frac{\langle \sigma, \delta\rangle h^\sigma h^\delta}{\langle v, \sigma\rangle \langle v, \delta\rangle} .$$

We call the set of all $v \in \mathbb{R}^n$ such that $\langle v, \tau\rangle = 0$ ($\tau \neq 0$) and $\langle v, \tau\rangle S_2^\tau \not\equiv 0$ a second order resonance set $\mathbf{B}_2$.

Equations for $S_3, S_4, \ldots$ are solved in the same way. We set

$$S_m = \sum S_m^\tau(v) \exp(\tau, x) , \qquad S_m^\tau \not\equiv 0 . \tag{5.5}$$

The coefficients in this sum are found using the following recurrent relation

$$S_m^\tau = -\frac{1}{2\langle v, \tau\rangle} \sum_{p+q=m, \sigma+\delta=\tau} \langle \sigma, \delta\rangle S_p^\sigma S_q^\delta .$$

This formula is a consequence of the systems (5.1) and (5.5).

Introduce the resonance set of m-th order $\mathbf{B_m}$. It consists of all $v \in \mathbb{R}^n$ such that

1) $\langle v, \tau\rangle = 0 , \quad \tau \neq 0 ,$
2) $\langle v, \tau\rangle S_m^\tau(v) \not\equiv 0 .$

We put

$$\mathbf{B} = \bigcup_{k=1}^{\infty} \mathbf{B}_k$$

and call the set $\mathbf{B}$ the resonance set of the Hamiltonian system. The resonance set is defined in $\mathbb{R}^n = \{v\}$. We identify the Euclidean spaces $\mathbb{R}^n = \{v\}$ and $\mathbb{R}^n = \{y\}$, and obtain the set of points in the space $\{y\}$. This set is also denoted henceforth by $\mathbf{B}$.

From the analytic point of view the cause of the appearance of "small denominators" for potentials with real exponents is the same as in the case of a compact surface M^n. The only difference is as follows. The analytic assumption that solutions of the system (5.1) have the form of multiple Fourier series has a geometric meaning in the compact case: these solutions must be single-valued on M^n.

The resonance set $\mathbf{B}$ plays the same role as the Poincaré set $\mathbf{P}$ does in the classical method of perturbation theory for invariant tori (cf. §4 of Chap. IV). The expediency of introducing and studying the resonance set follows from

Proposition 2. *Suppose that the Hamiltonian system* (4.3) *has n integrals polynomial in the momenta with coefficients of the form* (4.6). *Then their leading*

homogeneous forms do not depend on the coordinates x and are dependent at all points of the set **B**.

The proof of Proposition 2 repeats the corresponding arguments from §1 and §5 of Chap. IV (the detailed proof is contained in [156]).

The construction of the resonance set of Hamiltonian systems with exponential interaction is described by

Lemma 1 (Basic Lemma). *Suppose that all conditions of Theorem* 3, §4 *hold. Then the set* $\mathbf{B}_k$ *contains the hyperplane* $\langle k\alpha+\beta, y\rangle = 0$. *In particular, the resonance set* **B** *consists of an infinite number of distinct hyperplanes, and its closure contains the hyperplane* $\langle y, \alpha\rangle = 0$.

Since the leading homogeneous forms of polynomial integrals are analytic functions in $\mathbb{R}^n = \{y\}$ (see Proposition 2), then Proposition 2 and the Basic Lemma imply Theorem 3 of §4. Indeed, the Jacobian J of the leading homogeneous forms is an analytic function in $\mathbb{R}^n = \{y\}$. The function J vanishes on an infinite set of hyperplanes which pass through the origin. Consequently, J vanishes identically, and the leading homogeneous forms of n polynomial integrals are dependent everywhere.

The proof of Lemma 1 repeats with inessential complications the proof of the Basic Lemma from §5, Chap. IV (there are some changes only in Lemma 5; the detailed proof is contained in [156]).

5.3 One should not expect that each solution of the system (5.1) has singularities on resonance hyperplanes. As a simple example consider the equation

$$v_1 \frac{\partial S}{\partial x_1} + v_2 \frac{\partial S}{\partial x_2} = e^{x_1} .$$

It has a solution e^{x_1}/v_1; the straight line $v_1 = 0$ is resonance. However, we can indicate a solution that has no singularities in the domain $\mathbb{R}^2 \setminus \{0\}$:

$$S = \begin{cases} \dfrac{1}{v_1}\left(\exp x_1 - \exp \dfrac{v_2(v_2 x_1 - v_1 x_2)}{v_1^2 + v_2^2} \right) ; & v_1 \neq 0 , \\ \dfrac{x_2}{v_2} \exp x_1 ; \quad v_1 = 0 , \quad v_2 \neq 0 . \end{cases}$$

Certainly it is not a polynomial of exponents for all $v_1^2 + v_2^2 \neq 0$.

This remark shows that one can avoid the problem of "small denominators". We see that apparently Hamiltonian systems with exponential interaction may have nontrivial integrals of a different analytic nature. Indeed, the existence of analytic integrals of some Hamiltonian systems of the form (4.2) is established in paper [90]. These systems are not described by the classification of Theorem 2, §4. The corresponding integrals are certainly not polynomials in momenta, whose coefficients are series of real exponents. These integrals are not written down explicitly; they exist due to special properties of solutions of the system (4.2) as $t \to \pm\infty$.

5.4 The method of solving the system (5.1), developed in §5.2, is applicable in the more general case when $M^n = \mathbb{T}^k \times \mathbb{R}^{n-k}$. The perturbing function H_1 is 2π-periodic in the first k coordinates x. Consequently, we must modify the representation of H_1 in the form of a sum of exponents: the first k components of each vector a must be numbers from $i\mathbb{Z}$ (i is the imaginary unit). Since H_1 is a real function, for each a there exists a vector a' such that

1) the first k components of the vector $a + a'$ vanish,
2) the last $n - k$ components of the vector $a - a'$ vanish,
3) $\overline{h}_a = h_{a'}$.

In this more general case one may try to solve the system (5.1) by the method of §5.2. Any equation $\langle a, v\rangle = 0$ in this case defines two hyperplanes $\langle a', v\rangle = \langle a'', v\rangle = 0$. Here $a' + a'' = a$, and the components of the vectors a' and a'' are real and purely imaginary respectively. Thus, in the general case the codimension of resonance planes is greater than one.

References

1. D.L. Abrarov: Topological obstructions to the existence of conditionally-linear integrals. Vestnik Moskov. Univ. Ser. I Mat. Mekh. **6** , 72–75 (1984) (Russian)
2. M. Adler, P. van Moerbeke: Completely integrable systems, Euclidean Lie algebras and curves. Adv. Math. **38**, 267–317 (1980)
3. M. Adler, P. van Moerbeke: Kowalewski's asymptotic method, Kac-Moody Lie algebras and regularization. Comm. Math. Phys. **83**, 83–106 (1982)
4. M. Adler, P. van Moerbeke: A systematic approach towards solving integrable systems. Perspectives in Mathematics. Academic Press, New York 1987
5. V.M. Alekseev: Quasi-random dynamical systems I, II, III. Mat. Sb. **76**, 72–134 (1968); **77**, 545–601 (1968); **78**, 3–50 (1969) (Russian). [English transl.: Math. USSR-Sb. **5**, 73–128 (1968); **6**, (1968); **7**, 1–43 (1969)]
6. V.M. Alekseev: Perron sets and topological Markov chains. Uspekhi Mat. Nauk **24**, 227–228 (1969) (Russian)
7. V.M. Alekseev: Symbolic Dynamics. Proc. XI Mathematical Summer School, Kiev 1976 (Russian)
8. V.M. Alekseev: Final motions in the three-body problem and symbolic dynamics. Uspekhi Mat. Nauk **36**, No. 4 161–176 (1981) (Russian). [English transl.: Russian Math. Surveys **36**, 181–200 (1981)]
9. D.V. Anosov: Geodesic flows on closed Riemann manifolds with negative curvature. Trudy Mat. Inst. Steklov **90**, 1967 (Russian). [English transl.: Amer. Math. Soc., Providence, R.I., 1969]
10. D.V. Anosov: On the typical properties of closed geodesics. Izv. Akad. Nauk SSSR Ser. Mat. **46**, 675–709 (1982) (Russian). [English transl.: Math. USSR-Izv. **46**, 555–574 (1982)]
11. V.I. Arnol'd: Instability of dynamical systems with several degrees of freedom. Dokl. Akad. Nauk SSSR **156**, 9–12 (1964) (Russian). [English transl.: Sov. Math. Dokl. **5**, 581–585 (1964)]
12. V.I. Arnol'd: Mathematical methods in classical mechanics. Nauka, Moscow 1974 (Russian). [English transl.: Springer-Verlag, Berlin, New York 1978]
13. V.I. Arnol'd: Geometrical methods in the theory of ordinary differential equations. Nauka, Moscow 1974 (Russian). [English transl.: Springer-Verlag, New-York, Heidelberg, Berlin 1982]
14. V.I. Arnol'd, V.V. Kozlov, A.I. Neishtadt: Mathematical aspects of classical and celestial mechanics. Itogi Nauki i Tekhniki. Sovremennye Problemy Matematiki. Fundamental'nye Napravleniya, Vol. **3** VINITI, Moscow 1985 (Russian). [English transl.: Encyclopedia of Math. Sciences, Vol. **3**, Springer–Verlag 1989]
15. V.I. Arnol'd, A.B. Givental': Symplectic geometry. Itogi Nauki i Tekhniki. Sovremennye Problemy Matematiki. Fundamental'nye Napravleniya, Vol. **4**, 5–137 VINITI, Moscow 1985 (Russian). [English transl.: Encyclopedia of Math. Sciences, Vol. **4**, Springer–Verlag 1989]
16. Yu.A. Arkhangelsky: Analytic dynamics of a rigid body. Nauka, Moscow 1977 (Russian)
17. I.S. Arzhanuch: Field of momenta. Nauka, Tashkent 1965 (Russian)
18. Yu.V. Barkin, A.V. Borisov: Nonintegrability of Kirchhoff's equations and similar problems of dynamics. VINITI SSSR **5037–1389** (1989) (Russian)
19. C. Bechlivanidis, P. van Moerbeke: The Goryachev–Chaplygin top and the Toda lattice. Comm. Math. Phys. **110**, 317–324 (1987)

20. V.V. Beletsky: Motion of an artificial satellite around the center of mass. Nauka, Moscow 1965 (Russian)
21. M. Berger: Géométrie. CEDIC, Paris 1977, 1978
22. M.L. Bialy: On polynomial in momenta first integrals of a mechanical system on the two-dimensional torus. Funktsional. Anal. i Prilozhen. **21**, 64–65 (1982) (Russian)
23. G.D. Birkhoff: Dynamical Systems. Amer. Math. Soc. Colloquium Publications **9**, New York 1927
24. G.A. Bliss: Lectures on the calculus of variations. Univ. of Chicago Press, Chicago 1947
25. O.I. Bogoyavlenskiĭ: On perturbation of the periodic Toda lattice. Comm. Math. Phys. **51**, 201–209 (1976)
26. O.I. Bogoyavlenskiĭ: Methods in the qualitative theory of dynamical systems in astrophysics and gas dynamics. Nauka, Moscow 1980 (Russian). [English transl.: Springer-Verlag, Berlin, New York 1985]
27. O.I. Bogoyavlenskiĭ: New integrable problem of classical mechanics. Comm. Math. Phys. **94**, 255–269 (1984)
28. O.I. Bogoyavlenskiĭ: Integrable cases of rigid-body dynamics and integrable systems on spheres S^n. Izv. Akad. Nauk SSSR Ser. Mat. **49**, 899–915 (1985) (Russian). [English transl.: Math. USSR–Izv. **27**, 203–218 (1986)]
29. O.I. Bogoyavlenskiĭ: Integrable cases of the rotation of a rigid body in axisymmetric force fields. Dokl. Akad. Nauk SSSR **288**, no 3, 593–596 (1986) (Russian). [English transl.: Sov. Phys. Dokl. **31** 339–401 (1986)]
30. S.V. Bolotin: The existence of homoclinic motions. Vestnik Moskov. Univ. Ser. I Mat. Mekh. **6**, 98–103 (1983)
31. S.V. Bolotin: Nonintegrability of the n-center problem for $n > 2$. Vestnik Moskov. Univ. Ser. I, Mat. Mekh. **3**, 65–68 (1984) (Russian)
32. S.V. Bolotin: The effect of singularities of the potential energy on the integrability of mechanical systems. Prikl. Mat. Mekh. **48**, 356–362 (1984) (Russian). [English transl.: J. Appl. Math. Mech. **48** No. 3 (1984), 255–260 (1985)]
33. S.V. Bolotin: On first integrals of systems with gyroscopic forces. Vestnik Moskov. Univ. Ser. I Mat. Mekh. **6**, 75–82 (1984) (Russian)
34. S.V. Bolotin: Condition for the Liouville nonintegrability of Hamiltonian systems. Vestnik Moskov. Univ. Ser. I Mat. Mekh. **3**, 58–64 (1986) (Russian)
35. S.V. Bolotin: Doubly asymptotic trajectories and conditions for nonintegrability of Hamiltonian systems. Vestnik Moskov. Univ. Ser. I Mat. Mekh. **1**, 55–63 (1990) (Russian)
36. S.V. Bolotin: Doubly asymptotic trajectories of minimal geodesics. Vestnik Moskov. Univ. Ser. I Mat. Mekh. **1**, 92–98 (1992) (Russian)
37. S.V. Bolotin, V.V. Kozlov: On asymptotic solutions of the equations of dynamics. Vestnik Moskov. Univ. Ser. I Mat. Mekh. **4**, 84–89 (1980) (Russian)
38. T. Bountis, H. Segur, F. Vivaldi: Integrable Hamiltonian systems and the Painlevé property. Phys. Rev. A. **25**, 1257–1264 (1982)
39. H. Bourbaki: Élements de Mathématique. Algébre. Hermann, Paris 1970
40. H. Bourbaki: Élements de Mathématique. Integration. Hermann, Paris 1952–1963
41. H. Bourbaki: Élements de Mathématique. Groupes et algébres de Lie. Hermann, Paris 1971–1972
42. F. Brun: Rotation kring fix punkt. II, III. Ark. Mat., Astron., Fis. **4**, 1–4 (1907); **6**, 1–10 (1909)
43. H. Bruns: Über die Integrale des Vielkörperproblems. Acta Math. **11**, 25–96 (1887)
44. A.D. Bryuno: Local method of nonlinear analysis of differential equations. Nauka, Moskva 1979 (Russian)
45. R.M. Bulatovich [Bulatović]: Existence of solutions of the Hamilton-Jacobi equation in the neighborhood of a nondegenerate equilibrium position. Prikl. Mat. Mekh. **47**, 330–333 (1983) (Russian). [English transl.: J. Appl. Math. Mech. **47** (1983), 277–280 (1984)]

46. R.M. Bulatovich: Analytic solutions of the Hamilton–Jacobi equations of an irreversible system in a neighborhood of a nondegenerate maximum of the potential energy. Prikl. Mat. Mekh. **53**, 739–742 (1989) (Russian). [English transl.: J. Appl. Math. Mech. **53** (1989), 578–581 (1990)]
47. A.A. Burov: Nonintegrability of the equation of planar oscillations of a satellite on an elliptic orbit. Vestnik Moskov. Univ. Ser. I Mat. Mekh. **1**, 71–73 (1984)
48. A.A. Burov, A.V. Karapetyan: On the nonexistence of the additional integral in the problem of the motion of a heavy rigid ellipsoid along a smooth plane. Prikl. Mat. Mekh. **49**, 501–503 (1985) (Russian). [English transl.: J. Appl. Math. Mech. **49** (1985), 387–389 (1986)]
49. A.A. Burov, G.I. Subkhankulov: On the motion of a rigid body in a magnetic field. Prikl. Mat. Mekh. **50**, 960–966 (1986) (Russian)
50. F. Calogero: Exactly solvable one-dimensional many body problems. Lett. Nuovo Cimento. **13**, 411–416 (1975)
51. E. Cartan: Leçons sur les invariants integraux. Hermann, Paris 1922
52. J.W.S. Cassels: An introduction to Diophantine approximation. Cambridge Univ. Press 1957
53. S.A. Chaplygin: On motion of heavy bodies in an incompressible fluid. Complete Collection of Works, Vol. 1, Izd. Acad. Nauk SSSR, Leningrad 133–150 (1933) (Russian)
54. S.A. Chaplygin: On a paraboloid pendulum. Complete Collection of Works, Vol. 1, Izd. Akad. Nauk SSSR, Leningrad 194–199 (1933) (Russian)
55. S.A. Chaplygin: On the theory of motion of non-holonomic systems. Complete Collection of Works, Vol. 1, Izd. Akad. Nauk SSSR, Leningrad 207–215 (1933) (Russian)
56. S.A. Chaplygin: On rolling of a ball on a horizontal plane. Complete Collection of Works, Vol. 1, Izd. Akad. Nauk SSSR, Leningrad 216–234 (1933) (Russian)
57. C.L. Charlier: Die Mechanik des Himmels. Vorlesungen Bd. 1–2. Leipzig Veit. 1902–1907.
58. N. Chetayev: Sur les équations de Poincaré. C. R. Acad. Sci. Paris. **185**, 1577–1578 (1927)
59. B.V. Chirikov: A universal instability of many-dimensional oscillator systems. Phys. Rep. **52**, 263–389 (1979)
60. G. Contopoulos: On the existence of a third integral of motion. Astron. J. **68**, 1–14 (1963)
61. R. Cushman: Examples of nonintegrable analytic Hamiltonian vector fields with no small divisors. Trans. Amer. Math. Soc. **238**, 45–55 (1978)
62. J.M.A. Danby: Two notes on the Copenhagen problem. Celestial. Mech. **33**, 251–260 (1984)
63. A. Delshams, T.M. Seara: An asymptotic expressions for the splitting of separatrices of the rapidly forced pendulum. Comm. Math. Phys. **150**, 433–463 (1992)
64. B.P. Demidovich: Lectures on mathematical stability theory. Nauka, Moscow 1967 (Russian)
65. R.L. Devaney: Homoclinic orbits in Hamiltonian systems. J. Differ. Equat. **21**, 431–438 (1976)
66. R.L. Devaney: Transversal homoclinic orbits in an integrable systems. Amer. J. Math. **100**, 631–642 (1978)
67. P.A. Dirac: On generalized Hamiltonian dynamics. Canad. J. Math. **2**, 129–148 (1950)
68. S.A. Dovbysh: Intersection of asymptotic surfaces of the perturbed Euler–Poinsot problem. Prikl. Mat. Mekh. **51**, 363–370 (1987) (Russian). [English transl.: J. Appl. Math. Mech. **51**, 283–288 (1988)]
69. S.A. Dovbysh: Splitting of separatrices and birth of isolated periodic solutions in Hamiltonian systems with one-and-a-half degrees of freedom. Uspekhi Mat. Nauk **44**, 229–230 (1989) (Russian). [English transl.: Russian Math. Surveys **44**, 281–282 (1989)]
70. S.A. Dovbysh: Splitting of separatrices of unstable permanent rotations and nonintegrability of the perturbed Lagrange problem. Vestnik Moskov. Univ. Ser. I Mat. Mekh. **3**, 70–77 (1990) (Russian)
71. B.A. Dubrovin: Theta-functions and nonlinear equations. Uspekhi Mat. Nauk **36**, No 2 11–80 (1981) (Russian). [English transl.: Russian Math. Surveys **36** (1981), No 2 11–92 (1982)]
72. B.A. Dubrovin, I.M. Krichever, S.P. Novikov: Integrable systems. I. Itogi Nauki i Tekhniki. Sovremennye Problemy Matematiki. Fund. Naprav. Vol. **4** VINITI, Moscow 1985 (Russian). [English transl.: Encyclopedia of Math. Sciences, Vol. **4**, Springer–Verlag 1989]

73. B.A. Dubrovin, I.M. Krichever, S.P. Novikov: Integrable systems. II. Itogi Nauki i Tekhniki. Sovremennye Problemy Matematiki. Fund. Naprav. Vol. **15** VINITI, Moscow 1989 (Russian). [English transl.: Encyclopedia of Math. Sciences, Vol. **15**, Springer–Verlag 1991]
74. B.A. Dubrovin, S.P. Novikov, A.T. Fomenko: Modern geometry. Methods and applications. "Nauka", Moscow 1979 (Russian). [English transl.: Parts I,II, Springer–Verlag, Berlin, New York 1984, 1985]
75. B.A. Dubrovin, S.P. Novikov, A.T. Fomenko: Modern geometry. Methods of homology theory. "Nauka", Moscow 1984 (Russian)
76. L.H. Eliasson: Perturbations of stable invariant tori. Ann. Scuola Norm. Sup. Pisa. Cl. Sci., **IV** Ser. **15**, 115–147 (1988)
77. E. Fermi, J. Pasta, S. Ulam: Investigation on nonlinear problems. Collected Papers of Enrico Fermi. Univ. of Chicago Press, Chicago **2**, 978 (1965)
78. H. Flaschka: The Toda lattice. I. Existence of integrals. Phys. Rev. **9**, 1924–1925 (1974)
79. A.T. Fomenko: A Morse theory for integrable Hamiltonian systems. Dokl. Akad. Nauk SSSR **287**, 1071–1075 (1986) (Russian). [English transl.: Soviet Math. Dokl. **33**, 502–506 (1986)]
80. A.T. Fomenko: Topological invariants of Hamiltonian systems that are integrable in the sense of Liouville. Funktsional. Anal. i Prilozhen. **22**, 38–51 (1988) (Russian). [English transl.: Functional Anal. Appl. **22**, 286–296 (1988)]
81. J. Ford, S.D. Stoddard, J.S. Turner: On the integrability of the Toda lattice. Prog. Theor. Phys. **50**, 1547–1560 (1973)
82. O. Forster: Riemannsche Flächen. Springer-Verlag, Berlin, Heidelberg, New York 1977
83. E.V. Gaıdukov: Asymptotic geodesics on a Riemann manifold nonhomeomorphic to a sphere. Dokl. Akad. Nauk SSSR **169**, 999–1001 (1966) (Russian). [English transl.: Soviet Math. Dokl. **7**, 1033–1035 (1966)]
84. L. Galgani, A. Giorgilli, J.M. Strelcyn: Chaotic motions and transition to stochasticity in the classical problem of the heavy rigid body with a fixed point. Nuovo Cimento. **61B**, 1–20 (1981)
85. V.G. Gelfreich: Splitting of separatrices of the rapidly forced pendulum. Preprint, St. Petersburg 1990
86. V.G. Gelfreich, V.F. Lazutkin, N.V. Svanidze: Refined formula to separatrix splitting for the standard map. Preprint Warwick **59** (1992)
87. C. Godbillon: Géométrie differentielle et mechanique analytique. Hermann, Paris 1969
88. V.V. Golubev: Lectures on analytic theory of differential equations. Gostekhizdat, Moscow 1950 (Russian)
89. V.V. Golubev: Lectures on integration of the equations of motion of a rigid body about a fixed point. Gostekhizdat, Moscow 1953 (Russian)
90. G. Gorni, G. Zampieri: Complete integrability for Hamiltonian systems with a cone potential. J. Differential Equations (1989)
91. D.I. Goryachev: New integrable cases of integrability of the Euler dynamical equations. Warsaw Univ. Izv., Warsaw 1910 (Russian)
92. S.M. Graff: On the conservation of hyperbolic invariant tori for Hamiltonian systems. J. Differential Equations **15**, 1–69 (1974)
93. P.A. Griffiths: Exterior differential systems and the calculus of variations. Progress in math./25 Birkhäuser, Boston, Basel, Stuttgart 1983
94. F. Gustavson: On constructing formal integrals of a Hamiltonian system near an equilibrium point. Astron. J. **71**, 670–686 (1966)
95. L. Hall: A theory of exact and approximate configurational invariants. Phys. D. **8**, 90–116 (1983)
96. L. Heine, E. Khorozov: A Lax pair for Kowalewski's top. Phys. D. **29**, 173–180 (1987)
97. S. Helgason: Differential geometry, Lie groups and symmetric spaces. Academic Press, New York 1978
98. M. Hénon: Integrals of the Toda lattice. Phys. Rev. **B 9**, 1921–1923 (1974)

99. M. Hénon, C. Heiles: The applicability of the third integral of motion; some numerical experiments. Astron. J. **69**, 73–79 (1964)

100. P.J. Holmes, J.E. Marsden: Horseshoes and Arnold diffusion for Hamiltonian systems on Lie groups. Indiana Univ. Math. J. **32**, 273–309 (1983)

101. P. Holmes, J. Marsden, J. Scheurle: Exponentially small splittings of separatrices with applications to KAM theory and degenerate bifurcations. Contemp. Math. **81**, 213–244 (1988)

102. A. Hurwitz, R. Courant: Allgemeine Funktionentheorie und Elliptische Funktionen. Geometrische Funktionentheorie. Springer-Verlag, 1964

103. H. Ito: Convergence of Birkhoff normal forms for integrable systems. Comment. Math. Helv. **64**, 412–461 (1989)

104. E.A. Ivin: On the problem of integrability of the problem of inertial motion of a bunch of rigid bodies. Vestnik Moskov. Univ. Ser. I Mat. Mekh. **2**, 63–66 (1986) (Russian)

105. C.G.J. Jacobi: Vorlesungen über Dynamik. Druck und Verlag von Reimer. Berlin 1884

106. N. Jacobson: Lie Algebras. Wiley, New York, London 1962

107. I. Kaplansky: An introduction to differential algebra. Hermann, Paris 1957

108. A. Katok: Entropy and closed geodesics. Ergodic Theory and Dynamical Systems **2**, 339–367 (1982)

109. E.I. Kharlamova-Zabelina: Rapid rotation of a rigid body about a fixed point in the presence of a nonholonomic constraint. Vestnik Moskov. Univ. Ser. I Mat. Mekh. Astron. Fiz. Khim. **12**, No. 6, 25–34 (1958) (Russian)

110. S.V. Kirgetova: Some problems of perturbation theory of Hamiltonian systems with Hamiltonian a trigonometric polynomial. Vestnik Moskov. Univ. Ser. I Mat. Mekh. **1**, 62–68 (1991) (Russian)

111. F. Klein, A. Sommerfeld: Über die Theorie des Kreisels. Teubner, Leipzig 1910

112. S. Kobayashi: Fixed points of isometries. Nagoya Math. J. **13**, 63–68 (1958)

113. P.Ya. Kochina: On univalent solutions and algebraic integrals of the problem on rotation of a heavy rigid body about a fixed point. In: Rigid body rotation around a fixed point. Collect. in honor of S.V. Kovalevskaya, 157–186. AN SSSR, Moscow 1940 (Russian)

114. Y. Kodama: Exact solutions of hydrodynamic type equations having infinitely many conserved densities. Phys. Lett. A. **135**, 171–174 (1989)

115. Y. Kodama, J. Gibbons: A method for solving the dispersionless KP hierarchy and its exact solutions II. Phys. Lett. A. **135**, 167–170 (1989)

116. A.N. Kolmogorov: On dynamical systems with integral invariant on the torus. Dokl. Akad. Nauk SSSR **93**, No. 5, 763–766 (1953) (Russian)

117. A.N. Kolmogorov: On conservation of conditionally periodic motions under small perturbations of the Hamiltonian. Dokl. Akad. Nauk SSSR **98**, 527–530 (1954) (Russian). [English transl.: G. Casati and G. Ford (Editors), Lecture Notes in Phys. **93**, Springer-Verlag, Berlin 1979]

118. A.N. Kolmogorov: New metric invariant of transitive dynamical systems and automorphisms of Lebesgue spaces. Dokl. Akad. Nauk SSSR **119**, 861–864 (1958) (Russian)

119. A.N. Kolmogorov: General theory of dynamical systems and classical mechanics. In Proc. 1954 Intern. Congr. Math., North Holland, Amsterdam 1957 (Russian). [English transl.: in R. Abraham, J.E. Marsden: Foundations of mechanics. Benjamin, Reading 1978]

120. V.N. Kolokol'tsov: Geodesic flows on two-dimensional manifolds with an additional first integral that is polynomial in the velocities. Izv. Akad. Nauk SSSR, Ser. Mat. **46**, 994–1010 (1982) (Russian). [English transl. in Math. USSR Izv. **21**, No. 2 (1983)]

121. B. Kostant: The solution of a generalized Toda lattice and representation theory. Adv. in Math. **34** 195–338 (1979)

122. S. Kovalevskaya [Kowalewski]: Sur le problème de la rotation d'un corps solide autour d'un point fixe. Acta Math. **12**, 177–232 (1889)

123. A.N. Kozhevnikov, A.M. Shulgin: On normalization of Hamiltonian systems depending on a parameter. Dokl. Akad. Nauk UZSSR **7**, 19–21 (1985) (Russian)

124. V.V. Kozlov: On the nonexistence of analytic integrals of canonical systems close to integrable. Vestnik Moskov. Univ. Ser. I Mat. Mekh. **2**, 77–82 (1974) (Russian)

125. V.V. Kozlov: The nonexistence of an additional analytic integral in the problem of the motion of an asymmetric rigid body around a fixed point. Vestnik Moskov. Univ. Ser. I Mat. Mekh. **1**, 105–110 (1975) (Russian)

126. V.V. Kozlov: The nonexistence of analytic integrals near equilibrium positions of Hamiltonian systems. Vestnik Moskov. Univ. Ser. I Mat. Mekh. **1**, 110–115 (1976) (Russian)

127. V.V. Kozlov: Splitting of the separatrices in the perturbed Euler–Poinsot problem. Vestnik Moskov. Univ. Ser. I Mat. Mekh. **6**, 99–104 (1976) (Russian)

128. V.V. Kozlov: Nonexistence of univalued integrals and branching of solutions in rigid body dynamics. Prikl. Mat. Mekh. **42**, No. 3, 400–406 (1978) (Russian). [English transl.: J. Appl. Math. Mech. **42**, No. 3 420–426 (1979)]

129. V.V. Kozlov: Integrable and nonintegrable cases in the problem on rotation of a heavy rigid body with a fixed point. Uspekhi Mat. Nauk **34**, 241 (1979) (Russian)

130. V.V. Kozlov: Topological obstructions to the integrability of natural mechanical systems. Dokl. Akad. Nauk SSSR **249**, 1299–1302 (1979) (Russian). [English transl.: Soviet Math. Dokl. **20**, No. 6 (1979)]

131. V.V. Kozlov: Methods of qualitative analysis in the dynamics of a rigid body. Moskov. Gos. Univ., Moscow 1980

132. V.V. Kozlov: Two integrable problems of classical dynamics. Vestnik Moskov. Univ. Ser. I Mat. Mekh. **4**, 80–84 (1981) (Russian)

133. V.V. Kozlov: Dynamics of systems with nonintegrable constraints. I–V. Vestnik Moskov. Univ. Ser. I Mat. Mekh. **3**, 92–100 (1982); **4**, 70–76 (1982); **3**, 102–111 (1983); **5**, 76–83 (1987); **6**, 51–54 (1988) (Russian)

134. V.V. Kozlov: Nonintegrability of Kirchhoff's equations. Short Commun. Int. Cong. Math. Warszawa. Section 13: Math. Phys. and Mech. Warsaw 1982

135. V.V. Kozlov: Integrability and nonintegrability in Hamiltonian mechanics. Uspekhi Mat. Nauk **38**, No. 1, 3–67 (1983) (Russian). [English transl.: Russian Math. Surveys **38**, No. 1, 1–76 (1983)]

136. V.V. Kozlov: Hydrodynamics of Hamiltonian systems. Vestnik Moskov. Univ. Ser. I Mat. Mekh. **6**, 10–22 (1983) (Russian)

137. V.V. Kozlov: On integrability theory of the equations of non-holonomic mechanics. Adv. in Mech. **8**, No. 3, 85–107 (1985) (Russian)

138. V.V. Kozlov: On the problem of rotation of a rigid body in a magnetic field. Izv. Akad. Nauk SSSR Ser. Mekh. Tverd. Tela. **6**, 28–33 (1985) (Russian)

139. V.V. Kozlov: Splitting of separatrices and generation of isolated periodic solutions in Hamiltonian systems with one and a half degrees of freedom. Uspekhi Mat. Nauk **41**, 177–178 (1986) (Russian). [English transl.: Russian Math. Surveys **41**, 149–150 (1986)]

140. V.V. Kozlov: On the existence of an integral invariant of a smooth dynamic system. Prikl. Mat. Mekh. **51**, 538–545 (1987) (Russian). [English transl.: J. Appl. Math. Mech. **51**, (1987) 420–426 (1988)]

141. V.V. Kozlov: Invariant measures of the Euler–Poincaré equations on Lie algebras. Funktsional. Anal. i Prilozhen. **22**, 69–70 (1988) (Russian). [English transl.: Functional Anal. Appl. **22**, 58–59 (1988)]

142. V.V. Kozlov: On perturbation theory for Hamiltonian systems with non-compact invariant surfaces. Vestnik Moskov. Univ. Ser. I Mat. Mekh. **2**, 55–61 (1988) (Russian)

143. V.V. Kozlov: On symmetry groups of dynamical systems. Prikl. Mat. Mekh. **52**, 531–541 (1988) (Russian). [English transl.: J. Appl. Math. Mech. **52**, 413–420 (1988)]

144. V.V. Kozlov: Bifurcation of solutions and polynomial integrals in an invertible system on a torus. Mat. Zametki **44**, 100–104 (1988) (Russian). [English transl.: Math. Notes **44**, 543–545 (1988)]

145. V.V. Kozlov: Polynomial integrals of a system of interacting particles. Dokl. Akad. Nauk SSSR **301**, 785–788 (1988) (Russian). [English transl.: Soviet Math. Dokl. **38**, 143–146 (1989)]

146. V.V. Kozlov: Polynomial integrals of dynamical systems with one-and-a-half degrees of freedom. Mat. Zametki **45**, 46–52 (1989) (Russian). [English transl.: Math. Notes **45**, 296–300 (1989)]

147. V.V. Kozlov: Integrable and nonintegrable Hamiltonian systems. Soviet Sci. Rev. Sect. C. Math. Phys. Rev. **8**, 1–81 (1989)

148. V.V. Kozlov: Curl theory of a top. Vestnik Moskov. Univ. Ser. I Mat. Mekh. **4**, 56–62 (1990) (Russian)

149. V.V. Kozlov: Symmetry groups of geodesic flows on closed surfaces. Mat. Zametki **48**, 62–67 (1990) (Russian). [English transl.: Math. Notes **48** (1990), 1119–1122 (1991)]

150. V.V. Kozlov: On stochastization of plane flows of an ideal fluid. Vestnik Moskov. Univ. Ser. I Mat. Mekh. **1**, 72–76 (1991) (Russian)

151. V.V. Kozlov, N.N. Kolesnikov: On integrability of Hamiltonian systems. Vestnik Moskov. Univ. Ser. I Mat. Mekh. **6**, 88–91 (1979) (Russian)

152. V.V. Kozlov, E.M. Nikishin: The relativistic version of the Hamiltonian formalism and wave functions of hydrogen atoms. Vestnik Moskov. Univ. Ser. I Mat. Mekh. **5**, 11–20 (1986) (Russian)

153. V.V. Kozlov, D.A. Onishchenko: Nonintegrability of Kirchhoff's equations. Dokl. Akad. Nauk SSSR **266**, 1298–1300 (1982) (Russian). [English transl.: Soviet Math. Dokl. **26**, No. 2, 495–498 (1983)]

154. V.V. Kozlov, D.V. Treshchëv: Nonintegrability of the general problem of rotation of a dynamically symmetric heavy rigid body with a fixed point I, II. Vestnik Moskov. Univ. Ser. I Mat. Mekh. **6**, 73–81 (1985); **1**, 39–44 (1986) (Russian)

155. V.V. Kozlov, D.V. Treshchëv: The integrability of Hamiltonian systems with configuration space a torus. Mat. Sb. **135**, No. 1. 119–138 (1988) (Russian). [English transl.: Math. USSR-Sb. **63**, No. 1, 121–139 (1989)]

156. V.V. Kozlov, D.V. Treshchëv: Polynomial integrals of Hamiltonian systems with exponential interaction. Izv. Akad. Nauk SSSR Ser. Mat. **53**, No. 3, 537–556 (1989) (Russian). [English transl.: Math. USSR-Izv. **34**, No. 3, 555–574 (1990)]

157. V.V. Kozlov and D.V. Treshchëv: Kovalevskaya numbers of generalized Toda chains. Mat. Zametki **46**, No. 5, 17–28 (1989) (Russian). [English transl.: Math. Notes **46**, No. 5–6, 840–848 (1990).]

158. G. Lamb: Hydrodynamics. 6-th ed. Dover, New York 1945

159. P.D. Lax: Integrals of nonlinear equations of evolution and solitary waves. Comm. Pure Appl. Math. **21**, 467–490 (1968)

160. V.F. Lazutkin: Splitting of separatrices for standard Chirikov's mapping. VINITI No. 6372-84, 24 Sept. 1984 (Russian)

161. V.F. Lazutkin, I.G. Schachmannsky, M.B. Tabanov: Splitting of separatrices for standard and semistandard mappings. Phys. D **40**, 235–248 (1989)

162. T. Levi-Civita, E. Amaldi: Lezioni di Meccanica Razionale. Volume secondo, Parte seconda. Zanichelli, Bologna 1952

163. A.J. Lichtenberg, M.A. Lieberman: Regular and stochastic motion. Springer, New York 1983

164. J. Llibre, C. Simo: Oscillatory solutions in the planar restricted three-body problem. Math. Ann. **248**, 153–184 (1980)

165. A.M. Lyapunov: On one property of the differential equations in the problem of heavy rigid body motion with a fixed point. Collected works of A.M. Lyapunov. Gostekhizdat, Moscow 1954 (Russian)

166. A.M. Lyapunov: New integrable case of the equations of motion of a rigid body in a fluid. Fortsh. der Mathem. **Bd. 25**, 1501 (1897)

167. A.M. Lyapunov: The general problem of stability of motion. Khark. Mat. Obshch., Kharkiv 1892 (Russian). [French transl.: Ann. Math. Studies. No 17, Princeton Univ. Press, Princeton, New York 1947]
168. S.V. Manakov: Complete integrability and stochastization of discrete dynamical systems. Zh. Eksper. Teoret. Fiz. **67**, 543–555 (1974) (Russian). [English transl.: Soviet Physics JETP 40 (1974), No. 2, 269–274 (1975)]
169. A.P. Markeev: Libration points in celestial mechanics and cosmodynamics. Nauka, Moscow 1978 (Russian)
170. A.P. Markeev: On the integrability of the problem of rolling of a ball with a non simply connected cavity filled by a fluid. Izv. Akad. Nauk SSSR Mekh. Tverd. Tela. **1**, 64–65 (1986) (Russian)
171. A.P. Markeev, N.I. Churkina: On the Poincaré periodic solutions of the canonical system with one degree of freedom. Pism. Astr. Zhurn. **11**, 634–639 (1985) (Russian)
172. A.I. Markushevich: Introduction to the classical theory of Abelian functions. Nauka, Moscow 1979
173. V.K. Melnikov: On the stability of the center for time-periodic perturbations. Trudy Moskov Mat. Obshch. **12**, 3–52 (1963) (Russian). [English transl.: Trans. Moscow Math. Soc. 1963, 1–56 (1965)
174. V.K. Melnikov: On some cases of conservation of quasi-periodic motions under small perturbations of the Hamilton function. Dokl. Akad. Nauk SSSR **165**, 1245–1248 (1965) (Russian)
175. A.S. Mishchenko, A.T. Fomenko: A generalized Liouville method for the integration of Hamiltonian systems. Funktsional. Anal. i Prilozhen. **12**, No. 2, 46–56 (1978) (Russian). [English transl.: Functional Anal. Appl. **12**, No. 2, 113–121 (1978)]
176. A.D. Morozov, L.P. Shilnikov. On the mathematical theory of synchronisation of auto-oscillations. Dokl. Akad. Nauk SSSR **223**, 1340–1343 (1975) (Russian). [English transl.: Sov. Phys. Dokl. **20**, No. 8, 551–553 (1975)]
177. M. Morse, H. Feshbach: Methods of Theoretical Physics. McGraw–Hill Book Company, New York, Toronto, London 1953
178. J. Moser: The analytical invariants of an area-preserving mapping near a hyperbolic fixed point. Comm. Pure Appl. Math. **9**, 673–692 (1956)
179. J. Moser: Nonexistence of integrals for canonical systems of differential equations. Comm. Pure Appl. Math. **8**, 409–436 (1955)
180. J. Moser: On the volume elements of a manifold. Trans. Amer. Math. Soc. **120**, 286–294 (1965)
181. J. Moser: Convergent series expansions for quasi-periodic motions. Math. Ann. **169**, 136–176 (1967)
182. J. Moser: Lectures on Hamiltonian systems. Memoires of Amer. Math. Soc. No. 81, 1–60 (1968)
183. J. Moser: Regularization of the Kepler's problem and the averaging method on a manifold. Comm. Pure Appl. Math. **23**, 609–636 (1970)
184. J. Moser: Stable and random motions in dynamical systems. Ann. Math. Studies No. 77. Princeton Univ. Press, Princeton, New York 1973
185. J. Moser: Three integrable Hamiltonian systems connected with isospectral deformations. Adv. in Math. **16** 197–220 (1975)
186. J.Moser: Various aspects of integrable Hamiltonian systems. Progr. Math. **8**, 223–289. Birkhäuser, Boston 1980
187. N.G. Moshchevitin: On the existence and smoothness of an integral for a Hamiltonian system of a particular form. Mat. Zametki **49**, 80–85 (1991) (Russian)
188. M.A. Neimark: Linear representations of the Lorentz group. Gostekhizdat, Moscow 1958
189. A.I. Neishtadt: On accuracy of the perturbation theory for systems with one fast variable. Prikl. Mat. Mekh. **45**, 80–87 (1981) (Russian). [English transl.: J. Appl. Math. Mech. **45**, No. 1, (1981), 58–63 (1982)]

190. N.N. Nekhoroshev: Action–angle variables and their generalization. Trudy Moskov. Mat. Obshch. **26**, 181–198 (1972) (Russian). [English transl.: Trans. Moscow Math. Soc. **26** 180–198 (1972)]
191. E.S. Nikolayevsky, L.N. Shur: Nonintegrability of classical Yang–Mills fields. Pism. Zh. Eksper. Teoret. Fiz. **36**, 176–179 (1982) (Russian)
192. Z. Nitecki: Differentiable Dynamics. The MIT Press, Cambridge, Massachusetts, and London, England 1971
193. S.P. Novikov, I. Shmeltser: Periodic solutions of Kirchhoff's equations of free motion of a rigid body in a fluid and the extended Lusternik–Shnirelman–Morse theory I. Funktsional Anal. Prilozhen. **15**, 54–66 (1982) (Russian)
194. M. Olshanetsky, A. Perelomov: Completely integrable Hamiltonian systems connected with semisimple Lie algebras. Invent. Math. **37**, 93–108 (1976)
195. M. Olshanetsky, A. Perelomov: Explicit solutions of classical generalized Toda models. Invent. Math. **54**, 261–269 (1979)
196. P.J. Olver: Applications of Lie groups to differential equations. Springer–Verlag, New York, Berlin, Heidelberg, Tokyo 1986
197. D.A. Onishchenko: Normalization of canonical systems depending on a parameter. Vestnik Moskov. Univ. Ser. I Mat. Mekh. **3**, 78–81 (1982) (Russian)
198. V.I. Orekhov: Topological analysis of natural systems with quadratic integrals. Prikl. Mat. Mekh. **49**, 10–15 (1985) (Russian). [English transl.: J. Appl. Math. Mech. **49** (1985), 221–224 (1985)]
199. G.P. Paternain: On the topology of manifolds with completely integrable geodesic flows. Ergodic Theory Dynamical Systems **12**, 109–121 (1992)
200. A.M. Perelomov: Integrable systems of classical mechanics and Lie algebras. Nauka, Moscow 1990 (Russian). [English transl.: Birkhäuser Verlag, Basel Boston Berlin 1990]
201. S.I. Pidkuiko, A.M. Stepin: Polynomial integrals of Hamiltonian systems. Dokl. Akad. Nauk SSSR **239**, 50–51 (1978) (Russian). [English transl.: Sov. Math. Dokl. **19** 282–286 (1978)]
202. H. Poincaré: Sur le problème des trois corps et les équations de la dynamique. Acta Math. **13**, 1–270 (1890)
203. H. Poincaré: Les méthodes nouvelles de la mécanique celeste. Vol. 1–3. Gauthier–Villars, Paris 1892, 1893, 1899
204. H. Poincaré: Sur la méthode de Bruns. C. R. Acad. Sci. Paris. **123**, 1224–1228 (1896)
205. H. Poincaré: Sur une forme nouvelle des équations de la mécanique. C.R. Acad. Sci. Paris. **132**, 369–371 (1901)
206. H. Poincaré: Sur les lignes géodésiques des surfaces convexes. Trans. Amer. Math. Soc. **6**, 237–274 (1905)
207. J. Pöschel: On elliptic lower dimensional tori in Hamiltonian systems. Math. Z. **202**, 559–608 (1989)
208. H. Rüssmann: Über das verhalten analytischer Hamiltonscher differentialgleichungen in der nähe einer gleichgewichtslösung. Math. Ann. **154**, 284–300 (1964)
209. S.T. Sadetov: Conditions for integrability of the Kirkhoff equations. Vestnik Moskov. Univ. Ser. I Mat. Mekh. **3**, 56–62 (1990) (Russian)
210. T.V. Salnikova: Nonintegrability of the perturbed Lagrange problem. Vestnik Moskov. Univ. Ser. I Mat. Mekh. **4**, 62–66 (1984) (Russian)
211. T.V. Salnikova: On nonintegrability of Kirkhoff's equations in the non-symmetric case. Vestnik Moskov. Univ. Ser. I Mat. Mekh. **2**, 68–71 (1985) (Russian)
212. V.A. Samsonov: On rotation of a rigid body in a magnetic field. Izv. Akad. Nauk SSSR Ser. Mekh. Tverd. Tela. **4**, 32–34 (1984) (Russian)
213. H. Seifert, W. Threlfall: A textbook of topology. Academic Press, New York 1980
214. C.L. Siegel: Über die algebraischen Integrale des restringierten Dreikörperproblems. Trans. Amer. Math. Soc. **39**, 225–233 (1936)
215. C.L. Siegel: On the integrals of canonical systems. Ann. of Math. **42**, 806–822 (1941)

216. C.L. Siegel: Über die normalform analytischer differentialgleichungen in der nähe einer gleichgewichtslösung. Nachr. Acad. Wiss. Gottingen, Math.–Phys. Kl. IIa, Jahrg. 21–30 (1952)
217. C.L. Siegel, J.K. Moser: Lectures on celestial mechanics. Springer, Berlin, Heidelberg, New York 1971
218. E.K. Sklyanin: Boundary conditions for integrable quantum systems. Preprint LOMI. Leningrad, 1986.
219. S. Smale: Diffeomorphisms with many periodic points. Differential and Combinatorial Topology, Princeton Univ. Press, 63–80 (1965)
220. J. Součec, V. Součec: Morse–Sard theorem for real-analytic functions. Comment. Math. Univ. Carolin. **13**, 45–51 (1972)
221. V.A. Steklov: On motion of a rigid body in a fluid. Kharkov 1893 (Russian)
222. G.K. Suslov: Theoretical mechanics. Gostekhizdat, Moscow 1944 (Russian)
223. J.L. Synge: Classical dynamics. Handbuch der Physic, vol. III/1, Principles of classical mechanics and field theory. Springer, Berlin 1960
224. Ya.V. Tatarinov: Lectures on classical dynamics. Moscow Univ. Publ., Moscow 1984 (Russian)
225. Ya.V. Tatarinov: Separable variables and new topological phenomena in holonomic and non-holonomic systems. Trudy Sem. Vektor. Tenzor. Anal. **23**, 160–174 (1988) (Russian)
226. I.A. Taımanov: Topological obstructions to the integrability of geodesic flows on nonsimply connected manifolds. Izv. Akad. Nauk SSSR Ser. Mat. **51**, 429–435 (1987) (Russian). [English transl.: Math. USSR–Izv. **30**, 403–409 (1988)]
227. I.A. Taımanov: Topological properties of integrable geodesic flows. Mat. Zametki **44**, 283–284 (1988) (Russian)
228. M. Toda: Theory of Nonlinear Lattices. Springer–Verlag, Berlin, Heidelberg, New York 1981
229. D.V. Treshchëv: On the existence of an infinite number of nondegenerate periodic solutions of a near-integrable Hamiltonian system. Geometry, Differential equations and mechanics. Moskov. Gos. Univ., Mekh.-Mat. Fak., Moscow 121–127 (1985) (Russian)
230. D.V. Treshchëv: A mechanism for the destruction of resonance tori in Hamiltonian systems. Mat. Sb. **180**, No. 10, 1325–1346 (1989) (Russian). [English transl.: Math. USSR-Sb. **68**, No. 1, 181–203 (1991)
231. J. Vey: Sur certain systemès dynamiques séparables. Amer. J. Math. **22**, 201–358 (1899)
232. V. Volterra: Sur la théorie des variations des latitudes. Acta Math. **22**, 201–358 (1899)
233. E.T. Whittaker: A treatise on analytical dynamics. 3-d ed. Cambridge Univ. Press, Cambridge 1927
234. A. Wintner: The Analytical Foundations of Celestial Mechanics. Princeton Univ. Press, Princeton 1941
235. S. Wiggins: Global bifurcations and chaos. Applied Mathematical Sciences **73**, Springer–Verlag, New York, Heidelberg, Berlin 1991
236. H.M. Yehia: New integrable cases of the problem of motion of a gyrostat. Vestnik Moskov. Univ. Ser. I Mat. Mekh. **4**, 88–90 (1987) (Russian)
237. H. Yoshida: Necessary condition for the existence of algebraic first integrals. Celestial Mech. **31**, 363–399 (1983)
238. H. Yoshida: Nonintegrability of the truncated Toda lattice Hamiltonian at any order. Comm. Math. Phys. **116**, 529–538 (1988)
239. H. Yoshida: A criterion for the nonexistence of an additional analytic integral in Hamiltonian systems with n degrees of freedom. Phys. Lett. A. **141**, 108–112 (1989)
240. V.E. Zakharov, M.F. Ivanov, P.N. Shur: On anomalously slow stochastisation in some two-dimensional models of the field theory. Soviet Phys. Lett. **30**, 39–44 (1979)
241. G.M. Zaslavsky, R.Z. Sagdeev: Introduction to nonlinear physics. From the pendulum to turbulence and chaos. Nauka, Moscow 1988 (Russian)